PRINCIPLES OF GENE MANIPULATION

Principles of
Gene Manipulation

Sandy Primrose

*High Wycombe
Buckinghamshire, UK*

Richard Twyman

*John Innes Centre
Norwich Research Park
Norwich, UK*

Bob Old

*Department of Biological Sciences
University of Warwick
Coventry, UK*

SIXTH EDITION

b

**Blackwell
Science**

© 2001 by
Blackwell Science Ltd
Editorial Offices:
Osney Mead, Oxford OX2 0EL
25 John Street, London WC1N 2BS
23 Ainslie Place, Edinburgh EH3 6AJ
350 Main Street, Malden
  MA 02148-5018, USA
54 University Street, Carlton
  Victoria 3053, Australia
10, rue Casimir Delavigne
  75006 Paris, France

Other Editorial Offices:
Blackwell Wissenschafts-Verlag
  GmbH
Kurfürstendamm 57
10707 Berlin, Germany

Blackwell Science KK
MG Kodenmacho Building
7–10 Kodenmacho Nihombashi
Chuo-ku, Tokyo 104, Japan

Iowa State University Press
A Blackwell Science Company
2121 S. State Avenue
Ames, Iowa 50014-8300, USA

First published in 1980
Second edition 1981
Reprinted 1982, 1983 (twice)
Third edition 1985
Reprinted 1986 (twice), 1987, 1988
Fourth edition 1989
Reprinted 1990, 1991, 1992, 1993
  (twice)
Fifth edition 1994
Reprinted 1995 (twice), 1996, 1998,
  2000, 2001

ISBN 0–632–05954 0

Library of Congress
Cataloging-in-Publication Data
has been applied for

Set by Graphicraft Limited, Hong Kong
Printed and bound in Italy
by Rotolito Lombarda

DISTRIBUTORS

Marston Book Services Ltd
PO Box 269
Abingdon, Oxon OX14 4YN
(*Orders*: Tel: 01235 465500
           Fax: 01235 465555)

The Americas
Blackwell Publishing
c/o AIDC
PO Box 20
50 Winter Sport Lane
Williston, VT 05495-0020
(*Orders*: Tel: 800 216 2522
           Fax: 802 864 7626)

Australia
Blackwell Science Pty Ltd
54 University Street
Carlton, Victoria 3053
(*Orders*: Tel: 3 9347 0300
           Fax: 3 9347 5001)

A catalogue record for this title is
available from the British Library

For further information on
Blackwell Science, visit our website:
www.blackwell-science.com

Contents

Preface vii
The genetic code and single-letter amino acid
designations viii
Conversion scale viii

**1 Gene manipulation: an all-embracing
technique, 1**
Introduction, 1
Sequence analysis, 1
In vivo biochemistry, 2
The new medicine, 3
Biotechnology: the new industry, 4
The central role of *E. coli*, 5
Outline of the rest of the book, 6

2 Basic techniques, 8
Introduction, 8
The basic problems, 8
The solutions: basic techniques, 8
Agarose gel electrophoresis, 9
Nucleic acid blotting, 11
Transformation of *E. coli*, 17
The polymerase chain reaction (PCR), 19

**3 Cutting and joining DNA
molecules, 26**
Cutting DNA molecules, 26
Joining DNA molecules, 36

**4 Basic biology of plasmid and
phage vectors, 43**
Plasmid biology and simple plasmid
vectors, 43
The purification of plasmid DNA, 48

Desirable properties of plasmid cloning
vehicles, 49
Bacteriophage λ, 53
DNA cloning with single-stranded DNA
vectors, 60

**5 Cosmids, phasmids and other advanced
vectors, 64**
Introduction, 64
Vectors for cloning large fragments
of DNA, 64
Specialist-purpose vectors, 70
Putting it all together: vectors with
combinations of features, 84

6 Cloning strategies, 85
Introduction, 85
Cloning genomic DNA, 86
cDNA cloning, 92
Screening strategies, 101
Difference cloning, 114

7 Sequencing and mutagenesis, 120
Introduction, 120
Basic DNA sequencing, 120
Whole-genome sequencing, 126
Analysing sequence data, 126
Changing genes: site-directed
mutagenesis, 132

**8 Cloning in bacteria other than
Escherichia coli, 139**
Introduction, 139
Introducing DNA into bacterial cells, 139

Cloning in Gram-negative bacteria
other than *E. coli*, 144
Cloning in Gram-positive bacteria, 148
Cloning in streptomycetes, 153
Homoeologous recombination, 155

**9   Cloning in *Saccharomyces cerevisiae* and
other fungi, 156**
Introduction, 156
Introducing DNA into fungi, 156
The fate of DNA introduced into
fungi, 156
Plasmid vectors for use in fungi, 158
Retrovirus-like vectors, 159
Expression of cloned genes, 163
Overexpression of proteins in fungi, 165
Specialist vectors, 166
Yeast surface display, 168
Identifying genes encoding particular
cellular activities, 171
Determining functions associated
with particular genes, 171

10   Gene transfer to animal cells, 174
Introduction, 174
DNA-mediated transformation, 174
Gene transfer by viral transduction, 187
Summary of expression systems for
animal cells, 199

11   Genetic manipulation of animals, 202
Introduction, 202
Genetic manipulation of mammals, 203
DNA transfer to other vertebrates, 215
DNA transfer to invertebrates, 218

12   Gene transfer to plants, 221
Introduction, 221
Agrobacterium-mediated transformation, 224
Direct DNA transfer to plants, 237
In planta transformation, 239
Chloroplast transformation, 240
Plant viruses as vectors, 241

13   Advances in transgenic technology, 247
Introduction, 247
Inducible expression systems, 247
Applications of site-specific
recombination, 253
Further transgenic strategies for gene
inhibition, 260
Transgenic technology for functional
genomics, 266

**14   Applications of recombinant DNA
technology, 274**
Introduction, 274
Theme 1: Nucleic acid sequences as
diagnostic tools, 274
Theme 2: New drugs and new therapies for
genetic diseases, 283
Theme 3: Combating infectious disease, 293
Theme 4: Protein engineering, 299
Theme 5: Metabolic engineering, 303
Theme 6: Plant breeding in the
twenty-first century, 311
Epilogue: from genes to genomes, 319

References, 320

Index, 377

Preface

It is 22 years since the first edition of *Principles of Gene Manipulation* was published. In writing the first edition, our aim was to explain a new and rapidly growing technology. Our basic philosophy was to present the principles of gene manipulation, and its associated techniques, in sufficient detail to enable the non-specialist reader to understand them. In this new edition we have stuck to our original philosophy which we feel has been justified with the passage of time. Although originally intended as an advanced undergraduate textbook we know that many experienced researchers also used the previous editions as an informative reference source. In preparing this edition we have been conscious of the need to serve these two types of reader but herein lies a problem: the needs of experienced researchers are now vastly different from those of the undergraduate or early postgraduate student whereas this was not true a generation ago. Deliberately, we have not changed the level at which the book is written nor the general style. Rather, our solution has been to present the basic methodologies in the first half of the book, which is devoted to cloning in *Escherichia coli*, and to present the advanced techniques in the second half. The reader who has read and understood the material in the first half, or already knows it, should have no difficulty in understanding any of the material in the second half of the book. In the final chapter we present an overview of some important applications of recombinant DNA technology in a format that should appeal to all readers.

In writing the first half of the book much thought was given as to the amount of the early work on gene manipulation to include. Although older readers may feel that some of the material selected is 'dated' we have included it because an understanding of it is assumed in many publications today and is required if older papers are to be comprehended. In the second half of the book the emphasis is on cloning in eukaryotes, particularly in higher eukaryotes, and reflects current scientific focus. What is not included is methodology for analysing genomes since this is outwith the scope of 'gene manipulation'. However, this topic is covered in the third edition of our companion book *Principles of Genome Analysis and Genomics*.

We would like to thank our friends and colleagues for reading and commenting on various chapters prepared for this new edition. In particular we would like to thank Dylan Sweetman, Jane Pritchard, James Drummond, Roz Drummond, Phil Gardner, Gavin Craig, Ajay Kohli, Eva Stoger, Mark Leech, Victoria James, Paul Christou, Bruce Whitelaw, and Sue Goddard and her staff at CAMR for providing excellent library assistance.

Richard Twyman would like to dedicate this book to his parents, Peter and Irene, and his children, Emily and Lucy.

S.B. Primrose
R.M. Twyman
R.W. Old

The genetic code and single-letter amino acid designations

5'-OH terminal base	Middle base				3'-OH terminal base
	U	C	A	G	
U	Phe	Ser	Tyr	Cys	U
	Phe	Ser	Tyr	Cys	C
	Leu	Ser	STOP	STOP	A
	Leu	Ser	STOP	Trp	G
C	Leu	Pro	His	Arg	U
	Leu	Pro	His	Arg	C
	Leu	Pro	Gln	Arg	A
	Leu	Pro	Gln	Arg	G
A	Ile	Thr	Asn	Ser	U
	Ile	Thr	Asn	Ser	C
	Ile	Thr	Lys	Arg	A
	Met	Thr	Lys	Arg	G
G	Val	Ala	Asp	Gly	U
	Val	Ala	Asp	Gly	C
	Val	Ala	Glu	Gly	A
	Val*	Ala	Glu	Gly	G

*Codes for Met if in the initiator position.

Alanine	A	Leucine	L	
Arginine	R	Lysine	K	
Asparagine	N	Methionine	M	
Aspartic acid	D	Phenylalanine	F	
Cysteine	C	Proline	P	
Glycine	G	Serine	S	
Glutamic acid	E	Threonine	T	
Glutamine	Q	Tryptophan	W	
Histidine	H	Tyrosine	Y	
Isoleucine	I	Valine	V	

Scale for conversion between kilobase pairs of duplex DNA and molecular weight

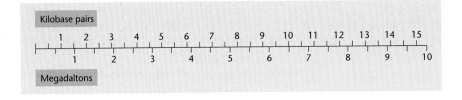

CHAPTER 1

Gene manipulation: an all-embracing technique

Introduction

Occasionally technical developments in science occur that enable leaps forward in our knowledge and increase the potential for innovation. Molecular biology and biomedical research experienced such a revolutionary change in the mid-70s with the development of gene manipulation. Although the initial experiments generated much excitement, it is unlikely that any of the early workers in the field could have predicted the breadth of applications to which the technique has been put. Nor could they have envisaged that the methods they developed would spawn an entire industry comprising several hundred companies, of varying sizes, in the USA alone.

The term gene manipulation can be applied to a variety of sophisticated *in vivo* genetics as well as to *in vitro* techniques. In fact, in most Western countries there is a precise *legal* definition of gene manipulation as a result of government legislation to control it. In the UK, gene manipulation is defined as

> the formation of new combinations of heritable material by the insertion of nucleic acid molecules, produced by whatever means outside the cell, into any virus, bacterial plasmid or other vector system so as to allow their incorporation into a host organism in which they do not naturally occur but in which they are capable of continued propagation.

The definitions adopted by other countries are similar and all adequately describe the subject-matter of this book. Simply put, gene manipulation permits stretches of DNA to be isolated from their host organism and propagated in the same or a different host, a technique known as *cloning*. The ability to clone DNA has far-reaching consequences, as will be shown below.

Sequence analysis

Cloning permits the isolation of discrete pieces of a genome and their amplification. This in turn enables the DNA to be sequenced. Analysis of the sequences of some genetically well-characterized genes led to the identification of the sequences and structures which characterize the principal control elements of gene expression, e.g. promoters, ribosome binding sites, etc. As this information built up it became possible to scan new DNA sequences and identify potential new genes, or *open reading frames*, because they were bounded by characteristic motifs. Initially this sequence analysis was done manually but to the eye long runs of nucleotides have little meaning and patterns evade recognition. Fortunately such analyses have been facilitated by rapid increases in the power of computers and improvements in software which have taken place contemporaneously with advances in gene cloning. Now sequences can be scanned quickly for a whole series of structural features, e.g. restriction enzyme recognition sites, start and stop signals for transcription, inverted palindromes, sequence repeats, Z-DNA, etc., using programs available on the Internet.

From the nucleotide sequence of a gene it is easy to deduce the protein sequence which it encodes. Unfortunately, we are unable to formulate a set of general rules that allows us to predict a protein's three-dimensional structure from the amino acid sequence of its polypeptide chain. However, based on crystallographic data from over 300 proteins, certain structural motifs can be predicted. Nor does an amino acid sequence on its own give any clue to function. The solution is to compare the amino acid sequence with that of other better-characterized proteins: a high degree of homology suggests similarity in function. Again, computers are of great value since algorithms exist for comparing two sequences or for comparing one sequence with a group of other sequences simultaneously. The Internet has made such comparisons easy because researchers can access all the protein sequence data that are stored in central databases, which are updated daily.

In vivo biochemistry

Any living cell, regardless of its origin, carries out a plethora of biochemical reactions. To analyse these different reactions, biochemists break open cells, isolate the key components of interest and measure their levels. They purify these components and try to determine their performance characteristics. For example, in the case of an enzyme, they might determine its substrate specificity and kinetic parameters, such as K_m and V_{max}, and identify inhibitors and their mode of action. From these data they try to build up a picture of what happens inside the cell. However, the properties of a purified enzyme in a test-tube may bear little resemblance to its behaviour when it shares the cell cytoplasm or a cell compartment with thousands of other enzymes and chemical compounds. Understanding what happens inside cells has been facilitated by the use of mutants. These permit the determination of the consequences of altered regulation or loss of a particular component or activity. Mutants have also been useful in elucidating macromolecule structure and function. However, the use of mutants is limited by the fact that with classical technologies one usually has little control over the type of mutant isolated and/or location of the mutation.

Gene cloning provides elegant solutions to the above problems. Once isolated, entire genes or groups of genes can be introduced back into the cell type whence they came or into different cell types or completely new organisms, e.g. bacterial genes in plants or animals. The levels of gene expression can be measured directly or through the use of reporter molecules and can be modulated up or down at the whim of the experimenter. Also, specific mutations, ranging from a single base-pair to large deletions or additions, can be built into the gene at any position to permit all kinds of structural and functional analyses. Function in different cell types can also be analysed, e.g. do those structural features of a protein which result in its secretion from a yeast cell enable it to be exported from bacteria or higher eukaryotes? Experiments like these permit comparative studies of macromolecular processes and, in some cases, gene cloning and sequencing provides the only way to begin to understand such events as mitosis, cell division, telomere structure, intron splicing, etc. Again, the Internet has made such comparisons easy because researchers can access all the protein sequence data that are stored in central databases, which are updated daily.

The original goal of sequencing was to determine the precise order of nucleotides in a gene. Then the goal became the sequence of a small genome. First it was that of a small virus (φX174, 5386 nucleotides). Then the goal was larger plasmid and viral genomes, then chromosomes and microbial genomes until ultimately the complete genomes of higher eukaryotes (humans, *Arabidopsis*) were sequenced (Table 1.1).

Table 1.1 Increases in sizes of genomes sequenced.

Genome sequenced	Year	Genome size	Comment
Bacteriophage φX174	1977	5.38 kb	First genome sequenced
Plasmid pBR322	1979	4.3 kb	First plasmid sequenced
Bacteriophage λ	1982	48.5 kb	
Epstein–Barr virus	1984	172 kb	
Yeast chromosome III	1992	315 kb	First chromosome sequenced
Haemophilus influenzae	1995	1.8 Mb	First genome of cellular organism to be sequenced
Saccharomyces cerevisiae	1996	12 Mb	First eukaryotic genome to be sequenced
Ceanorhabditis elegans	1998	97 Mb	First genome of multicellular organism to be sequenced
Drosophila melanogaster	2000	165 Mb	
Homo sapiens	2000	3000 Mb	First mammalian genome to be sequenced
Arabidopsis thaliana	2000	125 Mb	First plant genome to be sequenced

Now the sequencing of large genomes has become routine, albeit in specialist laboratories. Having the complete genome sequence of an organism provides us with fascinating insights into certain aspects of its biology. For example, we can determine the metabolic capabilities of a new microbe without knowing anything about its physiology. However, there are many aspects of cellular biology that cannot be ascertained from sequence data alone. For example, what RNA species are made when in the cell or organism life cycle and how fast do they turn over? What proteins are made when and how do the different proteins in a cell interact? How does environment affect gene expression? The answers to these questions are being provided by the new disciplines of genomics, proteomics and environomics which rely heavily on the *techniques* of gene manipulation, which are discussed in later chapters. A detailed presentation of whole-genome sequencing, genomics and proteomics can be found in Primrose and Twyman (2002).

The new medicine

The developments in gene manipulation that have taken place in the last 25 years have revolutionized the study of biology. There is no subject area within biology where recombinant DNA is not being used and as a result the old divisions between subject areas such as botany, genetics, zoology, biochemistry, etc. are fast breaking down. Nowhere has the impact of recombinant DNA technology been greater than on the practice of medicine.

The first medical benefit to arise from recombinant DNA technology was the availability of significant quantities of therapeutic proteins, such as human growth hormone (HGH). This protein is used to treat adolescents suffering from pituitary dwarfism to enable them to achieve a normal height. Originally HGH was purified from pituitary glands removed from cadavers. However, a very large number of pituitary glands are required to produce sufficient HGH to treat just one child. Furthermore, some children treated with pituitary-derived HGH have developed Creutzfeld–Jakob syndrome. Following the cloning and expression of the HGH gene in *Escherichia coli*, it is possible to produce enough HGH in a 10 litre fermenter to treat hundreds of children. Since then, many different therapeutic proteins have become available for the first time. Many of these proteins are also manufactured in *E. coli* but others are made in yeast or animal cells and some in plants or the milk of animals. The only common factor is that the relevant gene has been cloned and overexpressed using the techniques of gene manipulation.

Medicine has benefited from recombinant DNA technology in other ways (Fig. 1.1). New routes to vaccines have been developed. The current hepatitis B vaccine is based on the expression of a viral antigen on the surface of yeast cells and a recombinant vaccine has been used to eliminate rabies from foxes in a large part of Europe. Gene manipulation can

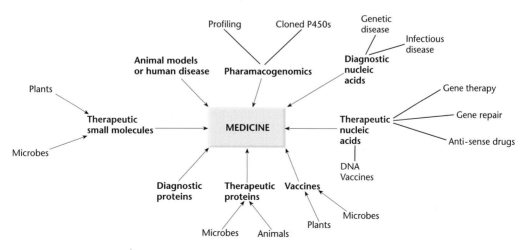

Fig. 1.1 The impact of gene manipulation on the practice of medicine.

also be used to increase the levels of small molecules within microbial cells. This can be done by cloning all the genes for a particular biosynthetic pathway and overexpressing them. Alternatively, it is possible to shut down particular metabolic pathways and thus redirect particular intermediates towards the desired end-product. This approach has been used to facilitate production of chiral intermediates and antibiotics. Novel antibiotics can also be created by mixing and matching genes from organisms producing different but related molecules in a technique known as combinatorial biosynthesis.

Gene cloning enables nucleic acid probes to be produced readily and such probes have many uses in medicine. For example, they can be used to determine or confirm the identity of a microbial pathogen or to diagnose pre- or perinatally an inherited genetic disease. Increasingly, probes are being used to determine the likelihood of adverse reactions to drugs or to select the best class of drug to treat a particular illness (pharmacogenomics). A variant of this technique is to use cloned cytochrome P450s to determine how a new drug will be metabolized and if any potentially toxic by-products will result.

Nucleic acids are also being used as therapeutic entities in their own right. For example, antisense nucleic acids are being used to down-regulate gene expression in certain diseases. In other cases, nucleic acids are being administered to correct or repair inherited gene defects (gene therapy/gene repair) or as vaccines. In the reverse of gene repair, animals are being generated that have mutations identical to those found in human disease. Note that the use of antisense nucleic acids and gene therapy/repair depends on the availability of information on the exact *cause* of a disease. For most medical conditions such information is lacking and currently available drugs are used to treat *symptoms*. This situation will change significantly in the next decade.

Biotechnology: the new industry

The early successes in overproducing mammalian proteins in *E. coli* suggested to a few entrepreneurial individuals that a new company should be formed to exploit the potential of recombinant DNA technology. Thus was Genentech born (Box 1.1). Since then thousands of biotechnology companies have been formed worldwide. As soon as major new developments in the science of gene manipulation are reported, a rash of new companies are formed to commercialize the new technology. For example, many recently formed companies are hoping the data from the Human Genome Sequencing Project will result in the identification of a large number of new proteins with potential for human therapy. Others are using gene manipulation to understand the regulation of transcription of particular genes, arguing that it would make better therapeutic sense to modulate the process with low-molecular-weight, orally active drugs.

Although there are thousands of biotechnology companies, fewer than 100 have sales of their products and even fewer are profitable. Already many biotechnology companies have failed, but the technology advances at such a rate that there is no shortage of new company start-ups to take their place. One group of biotechnology companies that has prospered is those supplying specialist reagents to laboratory workers engaged in gene manipulation. In the very beginning, researchers had to make their own restriction enzymes and this restricted the technology to those with protein chemistry skills. Soon a number of companies were formed which catered to the needs of researchers by supplying high-quality enzymes for DNA manipulation. Despite the availability of these enzymes, many people had great difficulty in cloning DNA. The reason for this was the need for careful quality control of all the components used in the preparation of reagents, something researchers are not good at! The supply companies responded by making easy-to-use cloning kits in addition to enzymes. Today, these supply companies can provide almost everything that is needed to clone, express and analyse DNA and have thereby accelerated the use of recombinant DNA technology in all biological disciplines. In the early days of recombinant DNA technology, the development of methodology was an end in itself for many academic researchers. This is no longer true. The researchers have gone back to using the tools to further our

Box 1.1 The birth of an industry

Biotechnology is not new. Cheese, bread and yoghurt are products of biotechnology and have been known for centuries. However, the stock-market excitement about biotechnology stems from the potential of gene manipulation, which is the subject of this book. The birth of this modern version of biotechnology can be traced to the founding of the company Genentech.

In 1976, a 27-year-old venture capitalist called Robert Swanson had a discussion over a few beers with a University of California professor, Herb Boyer. The discussion centred on the commercial potential of gene manipulation. Swanson's enthusiasm for the technology and his faith in it was contagious. By the close of the meeting the decision was taken to found Genentech (Genetic Engineering Technology). Though Swanson and Boyer faced scepticism from both the academic and business communities they forged ahead with their idea. Successes came thick and fast (see Table B1.1) and within a few years they had proved their detractors wrong. Over 1000 biotechnology companies have been set up in the USA alone since the founding of Genentech but very, very few have been as successful.

Table B1.1 Key events at Genentech.

1976	Genentech founded
1977	Genentech produced first human protein (somatostatin) in a microorganism
1978	Human insulin cloned by Genentech scientists
1979	Human growth hormone cloned by Genentech scientists
1980	Genentech went public, raising $35 million
1982	First recombinant DNA drug (human insulin) marketed (Genentech product licensed to Eli Lilly & Co.)
1984	First laboratory production of factor VIII for therapy of haemophilia. Licence granted to Cutter Biological
1985	Genentech launched its first product, Protropin (human growth hormone), for growth hormone deficiency in children
1987	Genentech launched Activase (tissue plasminogen activator) for dissolving blood clots in heart-attack patients
1990	Genentech launched Actimmune (interferon-$\gamma_{1\beta}$) for treatment of chronic granulomatous disease
1990	Genentech and the Swiss pharmaceutical company Roche complete a $2.1 billion merger

knowledge of biology, and the development of new methodologies has largely fallen to the supply companies.

The central role of *E. coli*

E. coli has always been a popular model system for molecular geneticists. Prior to the development of recombinant DNA technology, there existed a large number of well-characterized mutants, gene regulation was understood and there was a ready availability of a wide selection of plasmids. Compared with other microbial systems it was matchless. It is not surprising, therefore, that the first cloning experiments were undertaken in *E. coli*. Subsequently, cloning techniques were extended to a range of other microorganisms, such as *Bacillus subtilis*, *Pseudomonas* sp., yeasts and filamentous fungi, and then to higher eukaryotes. Curiously, cloning in *E. coli* is technically easier than in any other organism. As a result, it is rare for researchers to clone DNA directly in other organisms. Rather, DNA from the organism of choice is first manipulated in *E. coli* and subsequently transferred back to the original host. Without the ability to clone and manipulate DNA in *E. coli*, the application of recombinant DNA technology to other organisms would be greatly hindered.

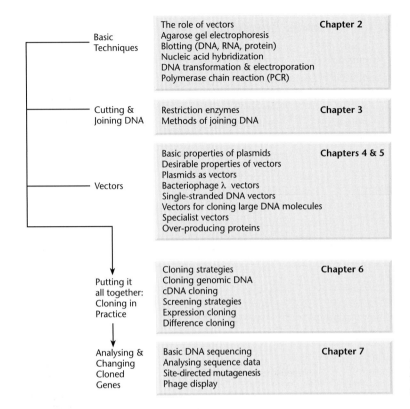

Fig. 1.2 'Roadmap' outlining the basic techniques in gene manipulation and their relationships.

Outline of the rest of the book

As noted above, *E. coli* has an essential role in recombinant DNA technology. Therefore, the first half of the book is devoted to the methodology for manipulating genes in this organism (Fig. 1.2). Chapter 2 covers many of the techniques that are common to all cloning experiments and are fundamental to the success of the technology. Chapter 3 is devoted to methods for selectively cutting DNA molecules into fragments that can be readily joined together again. Without the ability to do this, there would be no recombinant DNA technology. If fragments of DNA are inserted into cells, they fail to replicate except in those rare cases where they integrate into the chromosome. To enable such fragments to be propagated, they are inserted into DNA molecules (vectors) that are capable of extrachromosomal replication. These vectors are derived from plasmids and bacteriophages and their basic properties are described in Chapter 4. Originally, the purpose of vectors was the propagation of cloned DNA but today vectors fulfil many other roles, such as facilitating DNA sequencing, promoting expression of cloned genes, facilitating purification of cloned gene products, etc. The specialist vectors for these tasks are described in Chapter 5. With this background in place it is possible to describe in detail how to clone the particular DNA sequences that one wants. There are two basic strategies. Either one clones all the DNA from an organism and then selects the very small number of clones of interest or one amplifies the DNA sequences of interest and then clones these. Both these strategies are described in Chapter 6. Once the DNA of interest has been cloned, it can be sequenced and this will yield information on the proteins that are encoded and any regulatory signals that are present. There might also be a wish to modify the DNA and/or protein sequence and determine the biological effects of such changes. The techniques for sequencing and changing cloned genes are described in Chapter 7.

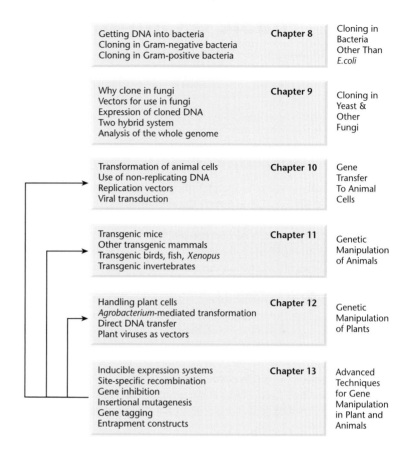

Getting DNA into bacteria **Chapter 8** Cloning in Gram-negative bacteria Cloning in Gram-positive bacteria	Cloning in Bacteria Other Than *E.coli*
Why clone in fungi **Chapter 9** Vectors for use in fungi Expression of cloned DNA Two hybrid system Analysis of the whole genome	Cloning in Yeast & Other Fungi
Transformation of animal cells **Chapter 10** Use of non-replicating DNA Replication vectors Viral transduction	Gene Transfer To Animal Cells
Transgenic mice **Chapter 11** Other transgenic mammals Transgenic birds, fish, *Xenopus* Transgenic invertebrates	Genetic Manipulation of Animals
Handling plant cells **Chapter 12** *Agrobacterium*-mediated transformation Direct DNA transfer Plant viruses as vectors	Genetic Manipulation of Plants
Inducible expression systems **Chapter 13** Site-specific recombination Gene inhibition Insertional mutagenesis Gene tagging Entrapment constructs	Advanced Techniques for Gene Manipulation in Plant and Animals

Fig. 1.3 'Roadmap' of the advanced techniques in gene manipulation and their application to organisms other than *E. coli*.

In the second half of the book the specialist techniques for cloning in organisms other than *E. coli* are described (Fig. 1.3). Each of these chapters can be read in isolation from the other chapters in this section, provided that there is a thorough understanding of the material from the first half of the book. Chapter 8 details the methods for cloning in other bacteria. Originally it was thought that some of these bacteria, e.g. *B. subtilis*, would usurp the position of *E. coli*. This has not happened and gene manipulation techniques are used simply to better understand the biology of these bacteria. Chapter 9 focuses on cloning in fungi, although the emphasis is on the yeast *Saccharomyces cerevisiae*. Fungi are eukaryotes and are useful model systems for investigating topics such as meiosis and mitosis, control of cell division, etc. Animal cells can be cultured like microorganisms and the techniques for cloning in them are described in Chapter 10. Chapters 11 and 12 are devoted to the intricacies of cloning in animal and plant representatives of higher eukaryotes and Chapter 13 covers some cutting-edge techniques for these same systems.

The concluding chapter is a survey of the different applications of recombinant DNA technology that are being exploited by the biotechnology industry. Rather than going through application after application, we have opted to show the interplay of different technologies by focusing on six themes:

• Nucleic acid sequences as diagnostic tools.
• New drugs and new therapies for genetic diseases.
• Combating infectious disease.
• Protein engineering.
• Metabolic engineering.
• Plant breeding in the twenty-first century.

By treating the topic in this way we have been able to show the interplay between some of the basic techniques and the sophisticated analysis now possible with genome sequence information.

CHAPTER 2

Basic techniques

Introduction

The initial impetus for gene manipulation *in vitro* came about in the early 1970s with the simultaneous development of techniques for:
- genetic transformation of *Escherichia coli*;
- cutting and joining DNA molecules;
- monitoring the cutting and joining reactions.

In order to explain the significance of these developments we must first consider the essential requirements of a successful gene-manipulation procedure.

The basic problems

Before the advent of modern gene-manipulation methods there had been many early attempts at transforming pro- and eukaryotic cells with foreign DNA. But, in general, little progress could be made. The reasons for this are as follows. Let us assume that the exogenous DNA is taken up by the recipient cells. There are then two basic difficulties. First, where detection of uptake is dependent on gene expression, failure could be due to lack of accurate transcription or translation. Secondly, and more importantly, the exogenous DNA may not be maintained in the transformed cells. If the exogenous DNA is integrated into the host genome, there is no problem. The exact mechanism whereby this integration occurs is not clear and it is usually a rare event. However this occurs, the result is that the foreign DNA sequence becomes incorporated into the host cell's genetic material and will subsequently be propagated as part of that genome. If, however, the exogenous DNA fails to be integrated, it will probably be lost during subsequent multiplication of the host cells. The reason for this is simple. In order to be replicated, DNA molecules must contain an *origin of replication*, and in bacteria and viruses there is usually only one per genome. Such molecules are called *replicons*. Fragments of DNA are not replicons

and in the absence of replication will be diluted out of their host cells. It should be noted that, even if a DNA molecule contains an origin of replication, this may not function in a foreign host cell.

There is an additional, subsequent problem. If the early experiments were to proceed, a method was required for assessing the fate of the donor DNA. In particular, in circumstances where the foreign DNA was maintained because it had become integrated in the host DNA, a method was required for mapping the foreign DNA and the surrounding host sequences.

The solutions: basic techniques

If fragments of DNA are not replicated, the obvious solution is to attach them to a suitable replicon. Such replicons are known as *vectors* or *cloning vehicles*. Small plasmids and bacteriophages are the most suitable vectors for they are replicons in their own right, their maintenance does not necessarily require integration into the host genome and their DNA can be readily isolated in an intact form. The different plasmids and phages which are used as vectors are described in detail in Chapters 4 and 5. Suffice it to say at this point that initially plasmids and phages suitable as vectors were only found in *E. coli*. An important consequence follows from the use of a vector to carry the foreign DNA: simple methods become available for purifying the vector molecule, complete with its foreign DNA insert, from transformed host cells. Thus not only does the vector provide the replicon function, but it also permits the easy bulk preparation of the foreign DNA sequence, free from host-cell DNA.

Composite molecules in which foreign DNA has been inserted into a vector molecule are sometimes called DNA *chimeras* because of their analogy with the Chimaera of mythology – a creature with the head of a lion, body of a goat and tail of a serpent. The construction of such composite or *artificial recombinant* molecules has also been termed *genetic engineering* or *gene manipulation* because of the potential for creating novel genetic combinations by biochemical means. The process has also been termed *molecular cloning* or *gene cloning* because a line of genetically identical organisms, all of which contain the composite molecule, can be propagated and grown in bulk, hence *amplifying* the composite molecule and *any gene product whose synthesis it directs*.

Although conceptually very simple, cloning of a fragment of foreign, or *passenger*, or *target* DNA in a vector demands that the following can be accomplished.

• The vector DNA must be purified and cut open.

• The passenger DNA must be inserted into the vector molecule to create the artificial recombinant. DNA joining reactions must therefore be performed. Methods for cutting and joining DNA molecules are now so sophisticated that they warrant a chapter of their own (Chapter 3).

• The cutting and joining reactions must be readily monitored. This is achieved by the use of gel electrophoresis.

• Finally, the artificial recombinant must be transformed into *E. coli* or another host cell. Further details on the use of gel electrophoresis and transformation of *E. coli* are given in the next section. As we have noted, the necessary techniques became available at about the same time and quickly led to many cloning experiments, the first of which were reported in 1972 (Jackson *et al.* 1972, Lobban & Kaiser 1973).

Agarose gel electrophoresis

The progress of the first experiments on cutting and joining of DNA molecules was monitored by velocity sedimentation in sucrose gradients. However, this has been entirely superseded by gel electrophoresis. Gel electrophoresis is not only used as an analytical method, it is routinely used preparatively for the purification of specific DNA fragments. The gel is composed of polyacrylamide or agarose. Agarose is convenient for separating DNA fragments ranging in size from a few hundred base pairs to about 20 kb (Fig. 2.1). Polyacrylamide is preferred for smaller DNA fragments.

The mechanism responsible for the separation of DNA molecules by molecular weight during gel

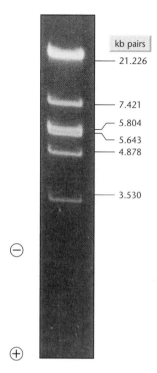

Fig. 2.1 Electrophoresis of DNA in agarose gels. The direction of migration is indicated by the arrow. DNA bands have been visualized by soaking the gel in a solution of ethidium bromide (see Fig. 2.3), which complexes with DNA by intercalating between stacked base-pairs, and photographing the orange fluorescence which results upon ultraviolet irradiation.

electrophoresis is not well understood (Holmes & Stellwagen 1990). The migration of the DNA molecules through the pores of the matrix must play an important role in molecular-weight separations since the electrophoretic mobility of DNA in free solution is independent of molecular weight. An agarose gel is a complex network of polymeric molecules whose average pore size depends on the buffer composition and the type and concentration of agarose used. DNA movement through the gel was originally thought to resemble the motion of a snake (reptation). However, real-time fluorescence microscopy of stained molecules undergoing electrophoresis has revealed more subtle dynamics (Schwartz & Koval 1989, Smith *et al.* 1989). DNA molecules display elastic behaviour by stretching in the direction of the applied field and then contracting into dense balls. The larger the pore size of the gel, the greater the ball of DNA which can pass through and hence the larger the molecules which can be separated. Once the globular volume of the DNA molecule exceeds the pore size, the DNA molecule can only pass through by reptation. This occurs with molecules about 20 kb in size and it is difficult to separate molecules larger than this without recourse to pulsed electrical fields.

In pulsed-field gel electrophoresis (PFGE) (Schwartz & Cantor 1984) molecules as large as 10 Mb can be separated in agarose gels. This is achieved by causing the DNA to periodically alter its direction of migration by regular changes in the orientation of the electric field with respect to the gel. With each change in the electric-field orientation, the DNA must realign its axis prior to migrating in the new direction. Electric-field parameters, such as the direction, intensity and duration of the electric field, are set independently for each of the different fields and are chosen so that the net migration of the DNA is down the gel. The difference between the direction of migration induced by each of the electric fields is the *reorientation angle* and corresponds to the angle that the DNA must turn as it changes its direction of migration each time the fields are switched.

A major disadvantage of PFGE, as originally described, is that the samples do not run in straight lines. This makes subsequent analysis difficult. This problem has been overcome by the development of improved methods for alternating the electrical field. The most popular of these is contour-clamped homogeneous electrical-field electrophoresis (CHEF) (Chu *et al.* 1986). In early CHEF-type systems (Fig. 2.2) the reorientation angle was fixed at 120°. However, in newer systems, the reorientation angle can be varied and it has been found that for whole-yeast chromosomes the migration rate is much faster with an angle of 106° (Birren *et al.* 1988). Fragments of DNA as large as 200–300 kb are routinely handled in genomics work and these can be separated in a matter of hours using CHEF systems with a reorientation angle of 90° or less (Birren & Lai 1994).

Aaij and Borst (1972) showed that the migration rates of the DNA molecules were inversely proportional to the logarithms of the molecular weights. Subsequently, Southern (1979a,b) showed that plotting fragment length or molecular weight against the reciprocal of mobility gives a straight

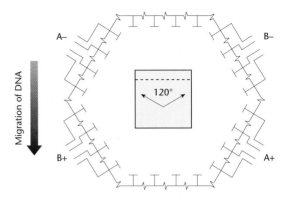

Fig. 2.2 Schematic representation of CHEF (contour-clamped homogeneous electrical field) pulsed-field gel electrophoresis.

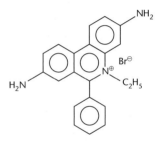

Fig. 2.3 Ethidium bromide.

line over a wider range than the semilogarithmic plot. In any event, gel electrophoresis is frequently performed with marker DNA fragments of known size, which allow accurate size determination of an unknown DNA molecule by interpolation. A particular advantage of gel electrophoresis is that the DNA bands can be readily detected at high sensitivity. The bands of DNA in the gel are stained with the intercalating dye ethidium bromide (Fig. 2.3), and as little as 0.05 μg of DNA in one band can be detected as visible fluorescence when the gel is illuminated with ultraviolet light.

In addition to resolving DNA fragments of different lengths, gel electrophoresis can be used to separate different molecular configurations of a DNA molecule. Examples of this are given in Chapter 4 (see p. 44). Gel electrophoresis can also be used for investigating protein–nucleic acid interactions in the so-called *gel retardation* or *band shift* assay. It is based on the observation that binding of a protein to DNA fragments usually leads to a reduction in electrophoretic mobility. The assay typically involves the addition of protein to linear double-stranded DNA fragments, separation of complex and naked DNA by gel electrophoresis and visualization. A review of the physical basis of electrophoretic mobility shifts and their application is provided by Lane *et al.* (1992).

Nucleic acid blotting

Nucleic acid labelling and hybridization on membranes have formed the basis for a range of experimental techniques central to recent advances in our understanding of the organization and expression of the genetic material. These techniques may be applied in the isolation and quantification of specific nucleic acid sequences and in the study of their organization, intracellular localization, expression and regulation. A variety of specific applications includes the diagnosis of infectious and inherited disease. Each of these topics is covered in depth in subsequent chapters.

An overview of the steps involved in nucleic acid blotting and membrane hybridization procedures is shown in Fig. 2.4. *Blotting* describes the immobilization of sample nucleic acids on to a solid support, generally nylon or nitrocellulose membranes. The blotted nucleic acids are then used as 'targets' in subsequent hybridization experiments. The main blotting procedures are:

- blotting of nucleic acids from gels;
- dot and slot blotting;
- colony and plaque blotting.

Colony and plaque blotting are described in detail on pp. 104–105 and dot and slot blotting in Chapter 14.

Southern blotting

The original method of blotting was developed by Southern (1975, 1979b) for detecting fragments in an agarose gel that are complementary to a given RNA or DNA sequence. In this procedure, referred to as Southern blotting, the agarose gel is mounted on a filter-paper wick which dips into a reservoir containing transfer buffer (Fig. 2.5). The hybridization membrane is sandwiched between the gel and a stack of paper towels (or other absorbent material), which serves to draw the transfer buffer through the gel by capillary action. The DNA molecules are carried out of the gel by the buffer flow and immobilized on the membrane. Initially, the membrane material used was nitrocellulose. The main drawback with this membrane is its fragile nature. Supported nylon membranes have since been developed which have greater binding capacity for nucleic acids in addition to high tensile strength.

For efficient Southern blotting, gel pretreatment is important. Large DNA fragments (> 10 kb) require a longer transfer time than short fragments. To allow

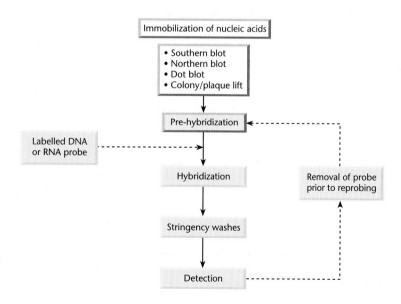

Fig. 2.4 Overview of nucleic acid blotting and hybridization (reproduced courtesy of Amersham Pharmacia Biotech).

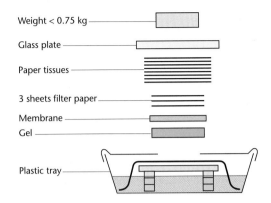

Weight < 0.75 kg

Glass plate

Paper tissues

3 sheets filter paper

Membrane

Gel

Plastic tray

Fig. 2.5 A typical capillary blotting apparatus.

uniform transfer of a wide range of DNA fragment sizes, the electrophoresed DNA is exposed to a short depurination treatment (0.25 mol/l HCl) followed by alkali. This shortens the DNA fragments by alkaline hydrolysis at depurinated sites. It also denatures the fragments prior to transfer, ensuring that they are in the single-stranded state and accessible for probing. Finally, the gel is equilibrated in neutralizing solution prior to blotting. An alternative method uses positively charged nylon membranes, which remove the need for extended gel pretreatment. With them the DNA is transferred in native (non-denatured) form and then alkali-denatured *in situ* on the membrane.

After transfer, the nucleic acid needs to be fixed to the membrane and a number of methods are available. Oven baking at 80°C is the recommended method for nitrocellulose membranes and this can also be used with nylon membranes. Due to the flammable nature of nitrocellulose, it is important that it is baked in a vacuum oven. An alternative fixation method utilizes ultraviolet cross-linking. It is based on the formation of cross-links between a small fraction of the thymine residues in the DNA and positively charged amino groups on the surface of nylon membranes. A calibration experiment must be performed to determine the optimal fixation period.

Following the fixation step, the membrane is placed in a solution of labelled (radioactive or non-radioactive) RNA, single-stranded DNA or oligodeoxynucleotide which is complementary in sequence to the blot-transferred DNA band or bands to be detected. Conditions are chosen so that the labelled nucleic acid hybridizes with the DNA on the membrane. Since this labelled nucleic acid is used to detect and locate the complementary sequence, it is called the *probe*. Conditions are chosen which maximize the rate of hybridization, compatible with a low background of non-specific binding on the membrane (see Box 2.1). After the hybridization reaction has been carried out, the membrane is washed to remove unbound radioactivity and regions of hybridization

---

Box 2.1 Hybridization of nucleic acids on membranes

The hybridization of nucleic acids on membranes is a widely used technique in gene manipulation and analysis. Unlike solution hybridizations, membrane hybridizations tend not to proceed to completion. One reason for this is that some of the bound nucleic acid is embedded in the membrane and is inaccessible

to the probe. Prolonged incubations may not generate any significant increase in detection sensitivity.

The composition of the hybridization buffer can greatly affect the speed of the reaction and the sensitivity of detection. The key components of these buffers are shown below:

Rate enhancers	Dextran sulphate and other polymers act as volume excluders to increase both the rate and the extent of hybridization
Detergents and blocking agents	Dried milk, heparin and detergents such as sodium dodecyl sulphate (SDS) have been used to depress non-specific binding of the probe to the membrane. Denhardt's solution (Denhardt 1966) uses Ficoll, polyvinylpyrrolidone and bovine serum albumin
Denaturants	Urea or formamide can be used to depress the melting temperature of the hybrid so that reduced temperatures of hybridization can be used
Heterologous DNA	This can reduce non-specific binding of probes to non-homologous DNA on the blot

continued

Box 2.1 *continued*

Stringency control

Stringency can be regarded as the specificity with which a particular target sequence is detected by hybridization to a probe. Thus, at high stringency, only completely complementary sequences will be bound, whereas low-stringency conditions will allow hybridization to partially matched sequences. Stringency is most commonly controlled by the temperature and salt concentration in the post-hybridization washes, although these parameters can also be utilized in the hybridization step. In practice, the stringency washes are performed under successively more stringent conditions (lower salt or higher temperature) until the desired result is obtained.

The melting temperature (T_m) of a probe–target hybrid can be calculated to provide a starting-point for the determination of correct stringency. The T_m is the temperature at which the probe and target are 50% dissociated. For probes longer than 100 base pairs:

$$T_m = 81.5\,°C + 16.6 \log M + 0.41\,(\% \text{ G} + \text{C})$$

where M = ionic strength of buffer in moles/litre. With long probes, the hybridization is usually carried out at $T_m - 25\,°C$. When the probe is used to detect partially matched sequences, the hybridization temperature is reduced by 1°C for every 1% sequence divergence between probe and target.

Oligonucleotides can give a more rapid hybridization rate than long probes as they can be used at a higher molarity. Also, in situations where target is in excess to the probe, for example dot blots, the hybridization rate is diffusion-limited and longer probes diffuse more slowly than oligonucleotides. It is standard practice to use oligonucleotides to analyse putative mutants following a site-directed mutagenesis experiment where the difference between parental and mutant progeny is often only a single base-pair change (see p. 132 *et seq.*).

The availability of the exact sequence of oligonucleotides allows conditions for hybridization and stringency washing to be tightly controlled so that the probe will only remain hybridized when it is 100% homologous to the target. Stringency is commonly controlled by adjusting the temperature of the wash buffer. The 'Wallace rule' (Lay Thein & Wallace 1986) is used to determine the appropriate stringency wash temperature:

$$T_m = 4 \times (\text{number of GC base pairs}) + 2 \times (\text{number of AT base pairs})$$

In filter hybridizations with oligonucleotide probes, the hybridization step is usually performed at 5°C below T_m for perfectly matched sequences. For every mismatched base pair, a further 5°C reduction is necessary to maintain hybrid stability.

The design of oligonucleotides for hybridization experiments is critical to maximize hybridization specificity. Consideration should be given to:

- probe length – the longer the oligonucleotide, the less chance there is of it binding to sequences other than the desired target sequence under conditions of high stringency;
- oligonucleotide composition – the GC content will influence the stability of the resultant hybrid and hence the determination of the appropriate stringency washing conditions. Also the presence of any non-complementary bases will have an effect on the hybridization conditions.

are detected autoradiographically by placing the membrane in contact with X-ray film (see Box 2.2). A common approach is to carry out the hybridization under conditions of relatively low stringency which permit a high rate of hybridization, followed by a series of post-hybridization washes of increasing stringency (i.e. higher temperature or, more commonly, lower ionic strength). Autoradiography following each washing stage will reveal any DNA bands that are related to, but not perfectly complementary with, the probe and will also permit an estimate of the degree of mismatching to be made.

Box 2.2 The principles of autoradiography

The localization and recording of a radiolabel within a solid specimen is known as autoradiography and involves the production of an image in a photographic emulsion. Such emulsions consist of silver halide crystals suspended in a clear phase composed mainly of gelatin. When a β-particle or γ-ray from a radionuclide passes through the emulsion, the silver ions are converted to silver atoms. This results in a latent image being produced, which is converted to a visible image when the image is developed. Development is a system of amplification in which the silver atoms cause the entire silver halide crystal to be reduced to metallic silver. Unexposed crystals are removed by dissolution in fixer, giving an autoradiographic image which represents the distribution of radiolabel in the original sample.

In direct autoradiography, the sample is placed in intimate contact with the film and the radioactive emissions produce black areas on the developed autoradiograph. It is best suited to detection of weak- to medium-strength β-emitting radionuclides (^{3}H, ^{14}C, ^{35}S). Direct autoradiography is not suited to the detection of highly energetic β-particles, such as those from ^{32}P, or for γ-rays emitted from isotopes like ^{125}I. These emissions pass through and beyond the film, with the majority of the energy being wasted. Both ^{32}P and ^{125}I are best detected by indirect autoradiography.

Indirect autoradiography describes the technique by which emitted energy is converted to light by means of a scintillator, using fluorography or intensifying screens. In fluorography the sample is impregnated with a liquid scintillator. The radioactive emissions transfer their energy to the scintillator molecules, which then emit photons which expose the photographic emulsion. Fluorography is mostly used to improve the detection of weak β-emitters (Fig. B2.1). Intensifying screens are

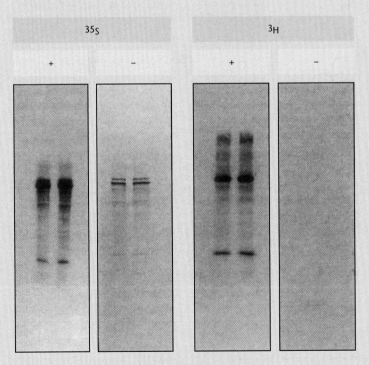

Fig. B2.1 Autoradiographs showing the detection of ^{35}S- and ^{3}H-labelled proteins in acrylamide gels with (+) and without (−) fluorography. (Photo courtesy of Amersham Pharmacia Biotech.)

continued

Box 2.2 *continued*

sheets of a solid inorganic scintillator which are placed behind the film. Any emissions passing through the photographic emulsion are absorbed by the screen and converted to light, effectively superimposing a photographic image upon the direct autoradiographic image.

The gain in sensitivity which is achieved by use of indirect autoradiography is offset by non-linearity of film response. A single hit by a β-particle or γ-ray can produce hundreds of silver atoms, but a single hit by a photon of light produces only a single silver atom. Although two or more silver atoms in a silver halide crystal are stable, a single silver atom is unstable and reverts to a silver ion very rapidly.

This means that the probability of a second photon being captured before the first silver atom has reverted is greater for large amounts of radioactivity than for small amounts. Hence small amounts of radioactivity are under-represented with the use of fluorography and intensifying screens. This problem can be overcome by a combination of pre-exposing a film to an instantaneous flash of light (pre-flashing) and exposing the autoradiograph at −70°C. Pre-flashing provides many of the silver halide crystals of the film with a stable pair of silver atoms. Lowering the temperature to −70°C increases the stability of a single silver atom, increasing the time available to capture a second photon (Fig. B2.2).

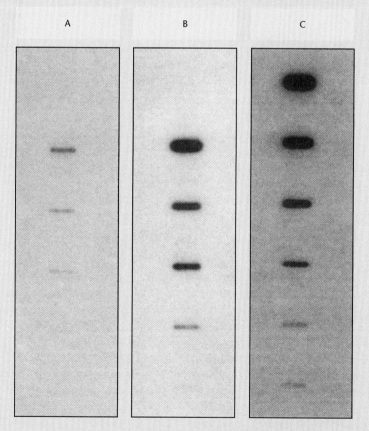

A B C

Fig. B2.2 The improvement in sensitivity of detection of ^{125}I-labelled IgG by autoradiography obtained by using an intensifying screen and pre-flashed film. A, no screen and no pre-flashing; B, screen present but film not pre-flashed; C, use of screen and pre-flashed film. (Photo courtesy of Amersham Pharmacia Biotech.)

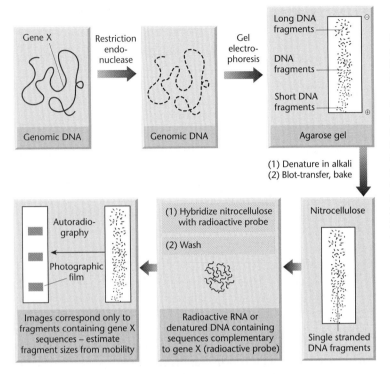

Fig. 2.6 Mapping restriction sites around a hypothetical gene sequence in total genomic DNA by the Southern blot method.

Genomic DNA is cleaved with a restriction endonuclease into hundreds of thousands of fragments of various sizes. The fragments are separated according to size by gel electrophoresis and blot-transferred on to nitrocellulose paper. Highly radioactive RNA or denatured DNA complementary in sequence to gene *X* is applied to the nitrocellulose paper bearing the blotted DNA. The radiolabelled RNA or DNA will hybridize with gene *X* sequences and can be detected subsequently by autoradiography, so enabling the sizes of restriction fragments containing gene *X* sequences to be estimated from their electrophoretic mobility. By using several restriction endonucleases singly and in combination, a map of restriction sites in and around gene *X* can be built up.

The Southern blotting methodology can be extremely sensitive. It can be applied to mapping restriction sites around a single-copy gene sequence in a complex genome such as that of humans (Fig. 2.6), and when a 'mini-satellite' probe is used it can be applied forensically to minute amounts of DNA (see Chapter 14).

Northern blotting

Southern's technique has been of enormous value, but it was thought that it could not be applied directly to the blot-transfer of RNAs separated by gel electrophoresis, since RNA was found not to bind to nitrocellulose. Alwine *et al.* (1979) therefore devised a procedure in which RNA bands are blot-transferred from the gel on to chemically reactive paper, where they are bound covalently. The reactive paper is prepared by diazotization of aminobenzyloxymethyl paper (creating diazobenzyloxymethyl (DBM) paper), which itself can be prepared from Whatman 540 paper by a series of uncomplicated reactions. Once covalently bound, the RNA is available for hybrid-

ization with radiolabelled DNA probes. As before, hybridizing bands are located by autoradiography. Alwine *et al.*'s method thus extends that of Southern and for this reason it has acquired the jargon term *northern* blotting.

Subsequently it was found that RNA bands can indeed be blotted on to nitrocellulose membranes under appropriate conditions (Thomas 1980) and suitable nylon membranes have been developed. Because of the convenience of these more recent methods, which do not require freshly activated paper, the use of DBM paper has been superseded.

Western blotting

The term 'western' blotting (Burnette 1981) refers to a procedure which does not directly involve nucleic acids, but which is of importance in gene manipulation. It involves the transfer of electrophoresed protein bands from a polyacrylamide gel on to a membrane of nitrocellulose or nylon, to which they bind strongly (Gershoni & Palade 1982, Renart & Sandoval 1984). The bound proteins are then available for analysis by a variety of specific protein–ligand interactions. Most commonly, antibodies are used to detect specific antigens. Lectins have been used to identify glycoproteins. In these cases the probe may itself be labelled with radioactivity, or some other 'tag' may be employed. Often, however, the probe is unlabelled and is itself detected in a 'sandwich' reaction, using a second molecule which is labelled, for instance a species-specific second antibody, or protein A of *Staphylococcus aureus* (which binds to certain subclasses of IgG antibodies), or strept-avidin (which binds to antibody probes that have been biotinylated). These second molecules may be labelled in a variety of ways with radioactive, enzyme or fluorescent tags. An advantage of the sandwich approach is that a single preparation of labelled second molecule can be employed as a general detector for different probes. For example, an antiserum may be raised in rabbits which reacts with a range of mouse immunoglobins. Such a rabbit anti-mouse (RAM) antiserum may be radio-labelled and used in a number of different applica-tions to identify polypeptide bands probed with different, specific, monoclonal antibodies, each mono-clonal antibody being of mouse origin. The sand-wich method may also give a substantial increase in sensitivity, owing to the multivalent binding of antibody molecules.

Alternative blotting techniques

The original blotting technique employed capillary blotting but nowadays the blotting is usually accom-plished by electrophoretic transfer of polypeptides from an SDS-polyacrylamide gel on to the membrane (Towbin *et al.* 1979). Electrophoretic transfer is also the method of choice for transferring DNA or RNA from low-pore-size polyacrylamide gels. It can also be used with agarose gels. However, in this case, the rapid electrophoretic transfer process requires high currents, which can lead to extensive heating effects, resulting in distortion of agarose gels. The use of an external cooling system is necessary to prevent this.

Another alternative to capillary blotting is vacuum-driven blotting (Olszewska & Jones 1988), for which several devices are commercially available. Vacuum blotting has several advantages over capillary or

electrophoretic transfer methods: transfer is very rapid and gel treatment can be performed *in situ* on the vacuum apparatus. This ensures minimal gel handling and, together with the rapid transfer, pre-vents significant DNA diffusion.

Transformation of *E. coli*

Early attempts to achieve transformation of *E. coli* were unsuccessful and it was generally believed that *E. coli* was refractory to transformation. However, Mandel and Higa (1970) found that treatment with $CaCl_2$ allowed *E. coli* cells to take up DNA from bac-teriophage λ. A few years later Cohen *et al.* (1972) showed that $CaCl_2$-treated *E. coli* cells are also effect-ive recipients for plasmid DNA. Almost any strain of *E. coli* can be transformed with plasmid DNA, albeit with varying efficiency, whereas it was thought that only *rec*BC$^-$ mutants could be transformed with lin-ear bacterial DNA (Cosloy & Oishi 1973). Later, Hoekstra *et al.* (1980) showed that *rec*BC$^+$ cells can be transformed with linear DNA, but the efficiency is only 10% of that in otherwise isogenic *rec*BC$^-$ cells. Transformation of *rec*BC$^-$ cells with linear DNA is only possible if the cells are rendered recombination-proficient by the addition of a *sbc*A or *sbc*B muta-tion. The fact that the *rec*BC gene product is an exonuclease explains the difference in transforma-tion efficiency of circular and linear DNA in *rec*BC$^+$ cells.

As will be seen from the next chapter, many bac-teria contain restriction systems which can influence the efficiency of transformation. Although the com-plete function of these restriction systems is not yet known, one role they do play is the recognition and degradation of foreign DNA. For this reason it is usual to use a restriction-deficient strain of *E. coli* as a transformable host.

Since transformation of *E. coli* is an essential step in many cloning experiments, it is desirable that it be as efficient as possible. Several groups of workers have examined the factors affecting the efficiency of transformation. It has been found that *E. coli* cells and plasmid DNA interact productively in an en-vironment of calcium ions and low temperature (0–5°C), and that a subsequent heat shock (37–45°C) is important, but not strictly required. Several other factors, especially the inclusion of metal ions in addition to calcium, have been shown to stimulate the process.

A very simple, moderately efficient transformation procedure for use with *E. coli* involves resuspending log-phase cells in ice-cold 50 mmol/l calcium chloride at about 10^{10} cells/ml and keeping them on ice for about 30 min. Plasmid DNA (0. 1 µg) is then added to a small aliquot (0.2 ml) of these now *competent* (i.e. competent for transformation) cells, and the incubation on ice continued for a further 30 min, followed by a heat shock of 2 min at 42°C. The cells are then usually transferred to nutrient medium and incubated for some time (30 min to 1 h) to allow phenotypic properties conferred by the plasmid to be expressed, e.g. antibiotic resistance commonly used as a selectable marker for plasmid-containing cells. (This so-called *phenotypic lag* may not need to be taken into consideration with high-level ampicillin resistance. With this marker, significant resistance builds up very rapidly, and ampicillin exerts its effect on cell-wall biosynthesis only in cells which have progressed into active growth.) Finally the cells are plated out on selective medium. Just why such a transformation procedure is effective is not fully understood (Huang & Reusch 1995). The calcium chloride affects the cell wall and may also be responsible for binding DNA to the cell surface. The actual uptake of DNA is stimulated by the brief heat shock.

Hanahan (1983) has re-examined factors that affect the efficiency of transformation, and has devised a set of conditions for optimal efficiency (expressed as transformants per µg plasmid DNA) applicable to most *E. coli* K12 strains. Typically, efficiencies of 10^7 to 10^9 transformants/µg can be achieved depending on the strain of *E. coli* and the method used (Liu & Rashidbaigi 1990). Ideally, one wishes to make a large batch of competent cells and store them frozen for future use. Unfortunately, competent cells made by the Hanahan procedure rapidly lose their competence on storage. Inoue *et al.* (1990) have optimized the conditions for the preparation of competent cells. Not only could they store cells for up to 40 days at −70°C while retaining efficiencies of $1-5 \times 10^9$ cfu/µg, but competence was affected only minimally by salts in the DNA preparation.

There are many enzymic activities in *E. coli* which can destroy incoming DNA from non-homologous sources (see Chapter 3) and reduce the transformation efficiency. Large DNAs transform less efficiently, on a molar basis, than small DNAs. Even with such improved transformation procedures, certain potential gene-cloning experiments requiring large numbers of clones are not reliable. One approach which can be used to circumvent the problem of low transformation efficiencies is to package recombinant DNA into virus particles *in vitro*. A particular form of this approach, the use of cosmids, is described in detail in Chapter 5. Another approach is electroporation, which is described below.

Electroporation

A rapid and simple technique for introducing cloned genes into a wide variety of microbial, plant and animal cells, including *E. coli*, is electroporation. This technique depends on the original observation by Zimmerman & Vienken (1983) that high-voltage electric pulses can induce cell plasma membranes to fuse. Subsequently it was found that, when subjected to electric shock, the cells take up exogenous DNA from the suspending solution. A proportion of these cells become stably transformed and can be selected if a suitable marker gene is carried on the transforming DNA. Many different factors affect the efficiency of electroporation, including temperature, various electric-field parameters (voltage, resistance and capacitance), topological form of the DNA, and various host-cell factors (genetic background, growth conditions and post-pulse treatment). Some of these factors have been reviewed by Hanahan *et al.* (1991).

With *E. coli*, electroporation has been found to give plasmid transformation efficiencies (10^9 cfu/µg DNA) comparable with the best CaCl$_2$ methods (Dower *et al.* 1988). More recently, Zhu and Dean (1999) have reported 10-fold higher transformation efficiencies with plasmids (9×10^9 transformants/µg) by co-precipitating the DNA with transfer RNA (tRNA) prior to electroporation. With conventional CaCl$_2$-mediated transformation, the efficiency falls off rapidly as the size of the DNA molecule increases and is almost negligible when the size exceeds 50 kb. While size also affects the efficiency of electroporation (Sheng *et al.* 1995), it is possible to get transformation efficiencies of 10^6 cfu/µg DNA with molecules as big as 240 kb. Molecules three to four times this size also can be electroporated successfully. This is important because much of the work on mapping and sequencing of genomes demands the ability to handle large fragments of DNA (see p. 64 and p. 126).

Transformation of other organisms

Although *E. coli* often remains the host organism of choice for cloning experiments, many other hosts are now used, and with them transformation may still be a critical step. In the case of Gram-positive bacteria, the two most important groups of organisms are *Bacillus* spp. and actinomycetes. That *B. subtilis* is naturally competent for transformation has been known for a long time and hence the genetics of this organism are fairly advanced. For this reason *B. subtilis* is a particularly attractive alternative prokaryotic cloning host. The significant features of transformation with this organism are detailed in Chapter 8. Of particular relevance here is that it is possible to transform protoplasts of *B. subtilis*, a technique which leads to improved transformation frequencies. A similar technique is used to transform actinomycetes, and recently it has been shown that the frequency can be increased considerably by first entrapping the DNA in liposomes, which then fuse with the host-cell membrane.

In later chapters we discuss ways, including electroporation, in which cloned DNA can be introduced into eukaryotic cells. With animal cells there is no great problem as only the membrane has to be crossed. In the case of yeast, protoplasts are required (Hinnen *et al.* 1978). With higher plants one strategy that has been adopted is either to package the DNA in a plant virus or to use a bacterial plant pathogen as the donor. It has also been shown that protoplasts prepared from plant cells are competent for transformation. A further remarkable approach that has been demonstrated with plants and animals (Klein & Fitzpatrick-McElligott 1993) is the use of microprojectiles shot from a gun (p. 238).

Animal cells, and protoplasts of yeast, plant and bacterial cells are susceptible to transformation by liposomes (Deshayes *et al.* 1985). A simple transformation system has been developed which makes use of liposomes prepared from a cationic lipid (Felgner *et al.* 1987). Small unilamellar (single-bilayer) vesicles are produced. DNA in solution spontaneously and efficiently complexes with these liposomes (in contrast to previously employed liposome encapsidation procedures involving non-ionic lipids). The positively charged liposomes not only complex with DNA, but also bind to cultured animal cells and are efficient in transforming them, probably by fusion with the plasma membrane. The use of liposomes as a transformation or transfection system is called *lipofection*.

The polymerase chain reaction (PCR)

The impact of the PCR upon molecular biology has been profound. The reaction is easily performed, and leads to the amplification of specific DNA sequences by an enormous factor. From a simple basic principle, many variations have been developed with applications throughout gene technology (Erlich 1989, Innis *et al.* 1990). Very importantly, the PCR has revolutionized prenatal diagnosis by allowing tests to be performed using small samples of fetal tissue. In forensic science, the enormous sensitivity of PCR-based procedures is exploited in DNA profiling; following the publicity surrounding *Jurassic Park*, virtually everyone is aware of potential applications in palaeontology and archaeology. Many other processes have been described which should produce equivalent results to a PCR (for review, see Landegran 1996) but as yet none has found widespread use.

In many applications of the PCR to gene manipulation, the enormous amplification is secondary to the aim of altering the amplified sequence. This often involves incorporating extra sequences at the ends of the amplified DNA. In this section we shall consider only the amplification process. The applications of the PCR will be described in appropriate places.

Basic reaction

First we need to consider the basic PCR. The principle is illustrated in Fig. 2.7. The PCR involves two oligonucleotide primers, 17–30 nucleotides in length, which flank the DNA sequence that is to be amplified. The primers hybridize to opposite strands of the DNA after it has been denatured, and are

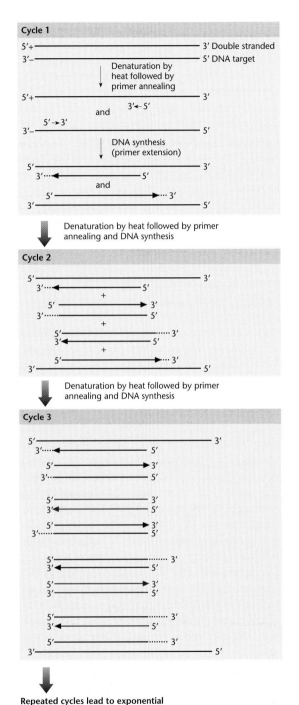

Cycle 1

5'+ ———————————————————— 3' Double stranded
3'– ———————————————————— 5' DNA target

    Denaturation by
    heat followed by
    primer annealing

5'+ ———————————————————— 3'
       and   3'←5'
3'– ———————————————————— 5'
   5'→3'

    DNA synthesis
    (primer extension)

5' ———————————————————— 3'
3'····◄——————— 5'
      and
    5' ———————►··· 3'
3' ———————————————————— 5'

Denaturation by heat followed by primer
annealing and DNA synthesis

Cycle 2

5' ———————————————————— 3'
3'····◄——————— 5'
    +
  5' ———————————► 3'
3' ———————————— 5'
    +
    5'——————————····3'
    3'◄——————— 5'
    +
    5' ———————————►··· 3'
3' ———————————————————— 5'

Denaturation by heat followed by primer
annealing and DNA synthesis

Cycle 3

5' ———————————————————— 3'
3'·····◄——————— 5'

5' ———————————► 3'
3'···———————————— 5'

5' ———————————— 3'
3'◄——————— 5'

5' ———————————— 3'
3'·····——————————— 5'

5' ——————————········ 3'
3'◄——————— 5'

5' ———————————► 3'
3' ———————————— 5'

5' ——————————········ 3'
3'◄——————— 5'

5' ———————————►········ 3'
3' ———————————————————— 5'

Repeated cycles lead to exponential
doubling of the target sequence

orientated so that DNA synthesis by the polymerase proceeds through the region between the two primers. The extension reactions create two double-stranded target regions, each of which can again be denatured ready for a second cycle of hybridization and extension. The third cycle produces two double-stranded molecules that comprise precisely the target region in double-stranded form. By repeated cycles of heat denaturation, primer hybridization and extension, there follows a rapid exponential accumulation of the specific target fragment of DNA. After 22 cycles, an amplification of about 10^6-fold is expected (Fig. 2.8), and amplifications of this order are actually attained in practice.

In the original description of the PCR method (Mullis & Faloona 1987, Saiki *et al.* 1988, Mullis 1990), Klenow DNA polymerase was used and, because of the heat-denaturation step, fresh enzyme had to be added during each cycle. A breakthrough came with the introduction of *Taq* DNA polymerase (Lawyer *et al.* 1989) from the thermophilic bacterium *Thermus aquaticus*. The *Taq* DNA polymerase is resistant to high temperatures and so does not need to be replenished during the PCR (Erlich *et al.* 1988, Sakai *et al.* 1988). Furthermore, by enabling the extension reaction to be performed at higher temperatures, the specificity of the primer annealing is not compromised. As a consequence of employing the heat-resistant enzyme, the PCR could be automated very simply by placing the assembled reaction in a heating block with a suitable thermal cycling programme (see Box 2.3).

Fig. 2.7 (*left*) The polymerase chain reaction. In cycle 1 two primers anneal to denatured DNA at opposite sides of the target region, and are extended by DNA polymerase to give new strands of variable length. In cycle 2, the original strands and the new strands from cycle 1 are separated, yielding a total of four primer sites with which primers anneal. The primers that are hybridized to the new strands from cycle 1 are extended by polymerase as far as the end of the template, leading to a precise copy of the target region. In cycle 3, double-stranded DNA molecules are produced (highlighted in colour) that are precisely identical to the target region. Further cycles lead to exponential doubling of the target region. The original DNA strands and the variably extended strands become negligible after the exponential increase of target fragments.

Cycle number	Number of double-stranded target molecules
1	0
2	0
3	2
4	4
5	8
6	16
7	32
8	64
9	128
10	256
11	512
12	1024
13	2048
14	4096
15	8192
16	16,384
17	32,768
18	65,536
19	131,072
20	262,144
21	524,288
22	1,048,576
23	2,097,152
24	4,194,304
25	8,388,608
26	16,777,216
27	33,554,432
28	67,108,864
29	134,217,728
30	268,435,456

Fig. 2.8 Theoretical PCR amplification of a target fragment with increasing number of cycles.

Recent developments have sought to minimize amplification times. Such systems have used small reaction volumes in glass capillaries to give large surface area-to-volume ratios. This results in almost instantaneous temperature equilibration and minimal annealing and denaturation times. This, accompanied by temperature ramp rates of 10–20°C/s, made possible by the use of turbulent forced hot-air systems to heat the sample, results in an amplification reaction completed in tens of minutes.

While the PCR is simple in concept, practically there are a large number of variables which can influence the outcome of the reaction. This is especially important when the method is being used with rare samples of starting material or if the end result has diagnostic or forensic implications. For a detailed analysis of the factors affecting the PCR, the reader should consult McDowell (1999). There are many substances present in natural samples (e.g. blood, faeces, environmental materials) which can inter-

fere with the PCR, and ways of eliminating them have been reviewed by Bickley and Hopkins (1999).

RT-PCR

The thermostable polymerase used in the basic PCR requires a DNA template and hence is limited to the amplification of DNA samples. There are numerous instances in which the amplification of RNA would be preferred. For example, in analyses involving the differential expression of genes in tissues during development or the cloning of DNA derived from an mRNA (complementary DNA or *cDNA*), particularly a rare mRNA. In order to apply PCR methodology to the study of RNA, the RNA sample must first be reverse-transcribed to cDNA to provide the necessary DNA template for the thermostable polymerase. This process is called reverse transcription (RT), hence the name RT-PCR.

Avian myeloblastosis virus (AMV) or Moloney murine leukaemia virus (MuLV) reverse transcriptases are generally used to produce a DNA copy of the RNA template. Various strategies can be adopted for first-strand cDNA synthesis (Fig. 2.9).

Long accurate PCR (LA-PCR)

Amplification of long DNA fragments is desirable for numerous applications of gene manipulation. The basic PCR works well when small fragments are amplified. The efficiency of amplification and therefore the yield of amplified fragments decrease significantly as the size of the amplicon increases over 5 kb. This decrease in yield of longer amplified fragments is attributable to partial synthesis across the desired sequence, which is not a suitable substrate for the subsequent cycles. This is demonstrated by the presence of smeared, as opposed to discrete, bands on a gel.

Barnes (1994) and Cheng *et al.* (1994) examined the factors affecting the thermostable polymerization across larger regions of DNA and identified key variables affecting the yield of longer PCR fragments. Most significant of these was the absence of a 3′–5′ exonuclease (proofreading) activity in *Taq* polymerase. Presumably, when the *Taq* polymerase misincorporates a dNTP, subsequent extension of the strand either proceeds very slowly or stops completely. To overcome this problem, a second

Box 2.3 The polymerase chain reaction achieves enormous amplifications, of specific target sequence, very simply

The reaction is assembled in a single tube, and then placed in a thermal cycler (a programmable heating/cooling block), as described below.

A typical PCR for amplifying a human genomic DNA sequence has the following composition. The reaction volume is 100 μl.

Input genomic DNA, 0.1–1 μg
Primer 1, 20 pmol
Primer 2, 20 pmol
20 mmol/l Tris-HCl, pH 8.3 (at 20°C)
1.5 mmol/l magnesium chloride
25 mmol/l potassium chloride
50 mmol/l each deoxynucleoside triphosphate
(dATP, dCTP, dGTP, dTTP)
2 units *Taq* DNA polymerase

A layer of mineral oil is placed over the reaction mix to prevent evaporation.

The reaction is cycled 25–35 times, with the following temperature programme:

Denaturation   94°C, 0.5 min
Primer annealing   55°C, 1.5 min
Extension   72°C, 1 min

Typically, the reaction takes some 2–3 h overall.

Notes:
• The optimal temperature for the annealing step will depend upon the primers used.
• The pH of the Tris-HCl buffer decreases markedly with increasing temperature. The actual pH varies between about 6.8 and 7.8 during the thermal cycle.
• The time taken for each cycle is considerably longer than 3 min (0.5 + 1.5 + 1 min), depending upon the rates of heating and cooling between steps, but can be reduced considerably by using turbo systems (p. 21).
• The standard PCR does not efficiently amplify sequences much longer than about 3 kb.

Random primer

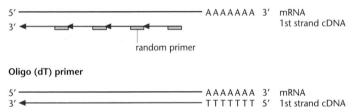

Oligo (dT) primer

Sequence-specific primer

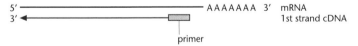

Fig. 2.9 Three strategies for synthesis of first-strand cDNA. (a) Random primer; (b) oligo (dT) primer; (c) sequence-specific primer.

thermostable polymerase with proofreading capability is added. Thermostable DNA polymerases with proofreading capabilities are listed in Table 2.1.

Key factors affecting the PCR

The specificity of the PCR depends crucially upon the primers. The following factors are important in choosing effective primers.

• Primers should be 17 to 30 nucleotides in length.
• A GC content of about 50% is ideal. For primers with a low GC content, it is desirable to choose a long primer so as to avoid a low melting temperature.
• Sequences with long runs (i.e. more than three or four) of a single nucleotide should be avoided.
• Primers with significant secondary structure are undesirable.

Table 2.1 Sources of thermostable DNA polymerases with proofreading (3′–5′ exonuclease) activity.

DNA polymerase	Source
Tma	*Thermotoga maritima*
Deep Vent™	*Pyrococcus* sp.
Tli	*Thermococcus litoralis*
Pfu	*Pyrococcus furiosus*
Pwo	*Pyrococcus woesi*

• There should be no complementarity between the two primers. The great majority of primers which conform with these guidelines can be made to work, although not all comparable primer sets are equally effective even under optimized conditions.

In carrying out a PCR it is usual to employ a *hot-start* protocol. This entails adding the DNA polymerase after the heat-denaturation step of the first cycle, the addition taking place at a temperature at or above the annealing temperature and just prior to the annealing step of the first cycle. The hot start overcomes the problem that would arise if the DNA polymerase were added to complete the assembly of the PCR reaction mixture at a relatively low temperature. At low temperature, below the desired hybridization temperature for the primer (typically in the region 45–60°C), mismatched primers will form and may be extended somewhat by the polymerase. Once extended, the mismatched primer is stabilized at the unintended position. Having been incorporated into the extended DNA during the first cycle, the primer will hybridize efficiently in subsequent cycles and hence may cause the amplification of a spurious product.

Alternatives to the hot-start protocol include the use of *Taq* polymerase antibodies, which are inactivated as the temperature rises (Taylor & Logan 1995), and AmpliTaq Gold™, a modified *Taq* polymerase that is inactive until heated to 95°C (Birch 1996). Yet another means of inactivating *Taq* DNA polymerase at ambient temperatures is the SELEX method (systematic evolution of ligands by exponential enrichment). Here the polymerase is reversibly inactivated by the binding of nanomolar amounts of a 70-mer, which is itself a poor polymerase substrate and should not interfere with the amplification primers (Dang & Jayasena 1996).

In order to minimize further the amplification of spurious products, the strategy of *nested primers* may be deployed. Here the products of an initial PCR amplification are used to seed a second PCR amplification, in which one or both primers are located internally with respect to the primers of the first PCR. Since there is little chance of the spurious products containing sequences capable of hybridizing with the second primer set, the PCR with these nested primers selectively amplifies the sought-after DNA.

As noted above, the *Taq* DNA polymerase lacks a 3′–5′ proofreading exonuclease. This lack appears to contribute to errors during PCR amplification due to misincorporation of nucleotides (Eckert & Kunkel 1990). Partly to overcome this problem, other thermostable DNA polymerases with improved fidelity have been sought, although the *Taq* DNA polymerase remains the most widely used enzyme for PCR. In certain applications, especially where amplified DNA is cloned, it is important to check the nucleotide sequence of the cloned product to reveal any mutations that may have occurred during the PCR. The fidelity of the amplification reaction can be assessed by cloning, sequencing and comparing several independently amplified molecules.

Real-time quantitative PCR

There are many applications of the PCR where it would be advantageous to be able to quantify the amount of starting material. Theoretically, there is a quantitative relationship between the amount of starting material (target sequence) and the amount of PCR product at any given cycle. In practice, replicate reactions yield different amounts of product, making quantitation unreliable. Higuchi *et al.* (1992, 1993) pioneered the use of ethidium bromide to quantify PCR products as they accumulate. Amplification produces increasing amounts of double-stranded DNA, which binds ethidium bromide, resulting in an increase in fluorescence. By plotting the increase in fluorescence versus cycle number it is possible to analyse the PCR kinetics in real time. This is much more satisfactory than analysing product accumulation after a fixed number of cycles.

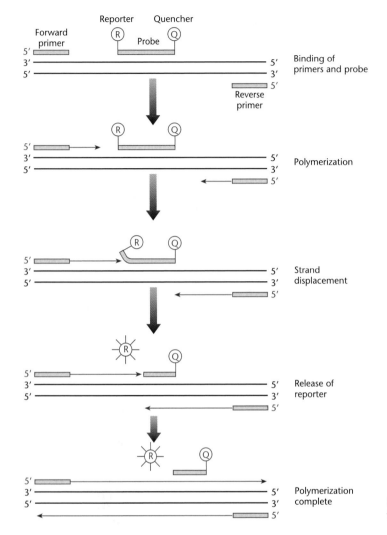

Fig. 2.10 Real-time quantitative PCR. See text for details.

The principal drawback to the use of ethidium bromide is that both specific and non-specific products generate a signal. This can be overcome by the use of probe-based methods for assaying product accumulation (Livak *et al.* 1995). The probes used are oligonucleotides with a reporter fluorescent dye attached at the 5′ end and a quencher dye at the 3′ end. While the probe is intact, the proximity of the quencher reduces the fluorescence emitted by the reporter dye. If the target sequence is present, the probe anneals downstream from one of the primer sites. As the primer is extended, the probe is cleaved by the 5′ nuclease activity of the *Taq* polymerase

(Fig. 2.10). This cleavage of the probe separates the reporter dye from the quencher dye, thereby increasing the reporter-dye signal. Cleavage removes the probe from the target strand, allowing primer extension to continue to the end of the template strand. Additional reporter-dye molecules are cleaved from their respective probes with each cycle, effecting an increase in fluorescence intensity proportional to the amount of amplicon produced.

Instrumentation has been developed which combines thermal cycling with measurement of fluorescence, thereby enabling the progress of the PCR to be monitored in real time. This revolutionizes the way one approaches PCR-based quantitation of DNA. Reactions are characterized by the point in time during cycling when amplification of a product is first detected, rather than by the amount of PCR product accumulated after a fixed number of cycles. The higher the starting copy number of the target, the sooner a significant increase in fluorescence is noted. Quantitation of the amount of target in unknown samples is achieved by preparing a standard curve, using different starting copy numbers of the target sequence.

CHAPTER 3
Cutting and joining DNA molecules

Cutting DNA molecules

Before 1970 there was no method of cleaving DNA at discrete points. All the available methods for fragmenting DNA were non-specific. The available endonucleases had little site specificity and chemical methods produced very small fragments of DNA. The only method where any degree of control could be exercised was the use of mechanical shearing. The long, thin threads which constitute duplex DNA molecules are sufficiently rigid to be very easily broken by shear forces in solution. Intense sonication with ultrasound can reduce the length to about 300 nucleotide pairs. More controlled shearing can be achieved by high-speed stirring in a blender. Typically, high-molecular-weight DNA is sheared to a population of molecules with a mean size of about 8 kb by stirring at 1500 rev/min for 30 min (Wensink *et al.* 1974). Breakage occurs essentially at random with respect to DNA sequence. The termini consist of short, single-stranded regions which may have to be taken into account in subsequent joining procedures.

During the 1960s, phage biologists elucidated the biochemical basis of the phenomenon of host restriction and modification. The culmination of this work was the purification of the restriction endonuclease of *Escherichia coli* K12 by Meselson and Yuan (1968). Since this endonuclease cuts unmodified DNA into large discrete fragments, it was reasoned that it must recognize a target sequence. This in turn raised the prospect of controlled manipulation of DNA. Unfortunately, the K12 endonuclease turned out to be perverse in its properties. While the enzyme does bind to a defined recognition sequence, cleavage occurs at a 'random' site several kilobases away (Yuan *et al.* 1980). The much sought-after breakthrough finally came in 1970 with the discovery in *Haemophilus influenzae* (Kelly & Smith 1970, Smith & Wilcox 1970) of an enzyme that behaves more simply. That is, the enzyme recognizes a particular target sequence in a duplex DNA molecule and breaks the polynucleotide chain within that sequence to give rise to discrete fragments of defined length and sequence.

The presence of restriction and modification systems is a double-edged sword. On the one hand, they provide a rich source of useful enzymes for DNA manipulation. On the other, these systems can significantly affect the recovery of recombinant DNA in cloning hosts. For this reason, some knowledge of restriction and modification is essential.

Host-controlled restriction and modification

Restriction systems allow bacteria to monitor the origin of incoming DNA and to destroy it if it is recognized as foreign. Restriction endonucleases recognize specific sequences in the incoming DNA and cleave the DNA into fragments, either at specific sites or more randomly. When the incoming DNA is a bacteriophage genome, the effect is to reduce the efficiency of plating, i.e. to reduce the number of plaques formed in plating tests. The phenomena of restriction and modification were well illustrated and studied by the behaviour of phage λ on two *E. coli* host strains.

If a stock preparation of phage λ, for example, is made by growth upon *E. coli* strain C and this stock is then titred upon *E. coli* C and *E. coli* K, the titres observed on these two strains will differ by several orders of magnitude, the titre on *E. coli* K being the lower. The phage are said to be *restricted* by the second host strain (*E. coli* K). When those phage that do result from the infection of *E. coli* K are now replated on *E. coli* K they are no longer restricted; but if they are first cycled through *E. coli* C they are once again restricted when plated upon *E. coli* K (Fig. 3.1). Thus the efficiency with which phage λ plates upon a

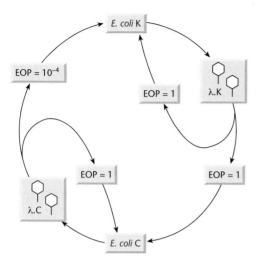

Fig. 3.1 Host-controlled restriction and modification of phage λ in *E. coli* strain K, analysed by efficiency of plating (EOP). Phage propagated by growth on strains K or C (i.e. λ.K or λ.C) have EOPs on the two strains, as indicated by arrows.

responsible for this degradation is called a *restriction endonuclease* or restriction enzyme (Lederberg & Meselson 1964). The restrictive host must, of course, protect its own DNA from the potentially lethal effects of the restriction endonuclease and so its DNA must be appropriately modified. Modification involves methylation of certain bases at a very limited number of sequences within DNA, which constitute the recognition sequences for the restriction endonuclease. This explains why phage that survive one cycle of growth upon the restrictive host can subsequently reinfect that host efficiently; their DNA has been replicated in the presence of the modifying methylase and so it, like the host DNA, becomes methylated and protected from the restriction system. Although phage infection has been chosen as our example to illustrate restriction and modification, these processes can occur whenever DNA is transferred from one bacterial strain to another.

particular host strain depends upon the strain on which it was last propagated. This non-heritable change conferred upon the phage by the second host strain (*E. coli* K) that allows it to be replated on that strain without further restriction is called *modification*.

The restricted phages adsorb to restrictive hosts and inject their DNA normally. When the phage are labelled with ^{32}P, it is apparent that their DNA is degraded soon after injection (Dussoix & Arber 1962) and the endonuclease that is primarily

Types of restriction and modification (R-M) system

At least four different kinds of R-M system are known: type I, type II, type III and type IIs. The essential differences between them are summarized in Table 3.1.

The type I systems were the first to be characterized and a typical example is that from *E. coli* K12. The active enzyme consists of two restriction subunits, two modification (methylation) subunits and one recognition subunit. These subunits are the

Table 3.1 Characteristics of the different types of endonucleases.

System	Key features
Type I	One enzyme with different subunits for recognition, cleavage and methylation. Recognizes and methylates a single sequence but cleaves DNA up to 1000 bp away
Type II	Two different enzymes which both recognize the same target sequence, which is symmetrical. The two enzymes either cleave or modify the recognition sequence
Type III	One enzyme with two different subunits, one for recognition and modification and one for cleavage. Recognizes and methylates same sequence but cleaves 24–26 bp away
Type IIs	Two different enzymes but recognition sequence is asymmetric. Cleavage occurs on one side of recognition sequence up to 20 bp away

Box 3.1 Restriction: from a phenomenon in bacterial genetics to a biological revolution

In the two related phenomena of host-controlled restriction and modification, a single cycle of phage growth, in a particular host bacterium, alters the host range of the progeny virus. The virus may fail to plate efficiently on a second host; its host range is restricted. This modification of the virus differs fundamentally from mutation because it is imposed by the host cell on which the virus has been grown but it is not inherited; when the phage is grown in some other host, the modification may be lost. In the 1950s, restriction and modification were recognized as common phenomena, affecting many virulent and temperate (i.e. capable of forming lysogens) phages and involving various bacterial species (Luria 1953, Lederberg 1957).

Arber and Dussoix clarified the molecular basis of the phenomena. They showed that restriction of phage λ is associated with rapid breakdown of the phage DNA in the host bacterium. They also showed that modification results from an alteration of the phage DNA which renders the DNA insensitive to restriction. They deduced that a single modified strand in the DNA duplex is sufficient to prevent restriction (Arber & Dussoix 1962, Dussoix & Arber 1962). Subsequent experiments implicated methylation of the DNA in the modification process (Arber 1965).

Detailed genetic analysis, in the 1960s, of the bacterial genes (in *E. coli* K and *E. coli* B) responsible for restriction and modification supported the duality of the two phenomena. Mutants of the bacteria could be isolated that were both restriction-deficient and modification-deficient (R$^-$M$^-$), or that were R$^-$M$^+$. The failure to recover R$^+$M$^-$ mutants was correctly ascribed to the suicidal failure to confer protective modification upon the host's own DNA.

The biochemistry of restriction advanced with the isolation of the restriction endonuclease from *E. coli* K (Meselson & Yuan 1968). It was evident that the restriction endonucleases from *E. coli* K

and *E. coli* B were important examples of proteins that recognize specific structures in DNA, but the properties of these type I enzymes as they are now known, were complex. Although the recognition sites in the phage could be mapped genetically (Murray *et al.* 1973a), determined efforts to define the DNA sequences cleaved were unsuccessful (Eskin & Linn 1972, Murray *et al.* 1973b).

The breakthrough came with Hamilton Smith's discovery of a restriction endonuclease from *Haemophilus influenzae* strain Rd (Smith & Wilcox 1970) and the elucidation of the nucleotide sequence at its cleavage sites in phage T7 DNA (Kelly & Smith 1970). This enzyme is now known as *Hind*II. The choice of T7 DNA as the substrate for cleavage was a good one, because the bacterium also contains another type II restriction enzyme, *Hind*III, in abundance. Fortunately, *Hind*III does not cleave T7 DNA, and so any contaminating *Hind*III in the *Hind*II preparation could not be problematical (Old *et al.* 1975). Shortly after the discovery of *Hind*II, several other type II restriction endonucleases were isolated and characterized. *Eco*RI was foremost among these (Hedgepeth *et al.* 1972). They were rapidly exploited in the first recombinant DNA experiments.

By the mid-1960s, restriction and modification had been recognized as important and interesting phenomena within the field of bacterial genetics (see, for example, Hayes 1968), but who could have foreseen the astonishing impact of restriction enzymes upon biology?

products of the *hsd*R, *hsd*M and *hsd*S genes. The methylation and cutting reactions both require ATP and *S*-adenosylmethionine as cofactors. The recognition sequences are quite long with no recognizable features such as symmetry. The enzyme also cuts unmodified DNA at some distance from the recognition sequence. However, because the methylation reaction is performed by the same enzyme which mediates cleavage, the target DNA may be modified before it is cut. These features mean that type I systems are of little value for gene manipulation (see also Box 3.1). However, their presence in *E. coli* strains can affect recovery of recombinants (see p. 33). Type III enzymes have symmetrical recognition sequences but otherwise resemble type I systems and are of little value.

Most of the useful R-M systems are of type II. They have a number of advantages over type I and III systems. First, restriction and modification are mediated by separate enzymes so it is possible to cleave DNA in the absence of modification. Secondly, the restriction activities do not require cofactors such as ATP or *S*-adenosylmethionine, making them easier to use. Most important of all, type II enzymes recognize a defined, usually symmetrical, sequence *and cut within it*. Many of them also make a staggered break in the DNA and the usefulness of this will become apparent. Although type IIs systems have similar cofactors and macromolecular structure to those of type II systems, the fact that restriction occurs at a distance from the recognition site limits their usefulness.

The classification of R-M systems into types I to III is convenient but may require modification as new discoveries are made. For example, the *Eco*571 system comprises a single polypeptide which has both restriction and modification activities (Petrusyte *et al.* 1988). Other restriction systems are known which fall outside the above classification. Examples include the *mcr* and *mrr* systems (see p. 34) and homing endonucleases. The latter are double-stranded DNases derived from introns or inteins (Belfort & Roberts 1997). They have large, asymmetric recognition sequences and, unlike standard restriction endonucleases, tolerate some sequence degeneracy within their recognition sequence.

Nomenclature

The discovery of a large number of restriction and modification systems called for a uniform system of nomenclature. A suitable system was proposed by Smith and Nathans (1973) and a simplified version of this is in use today. The key features are:

• The species name of the host organism is identified by the first letter of the genus name and the first two letters of the specific epithet to generate a three-letter abbreviation. This abbreviation is always written in italics.

• Where a particular strain has been the source then this is identified.

• When a particular host strain has several different R-M systems, these are identified by roman numerals. Some examples are given in Table 3.2.

Homing endonucleases are named in a similar fashion except that intron-encoded endonucleases are given the prefix 'I-' (e.g. I-*Ceu*I) and intein endonucleases have the prefix 'PI-' (e.g. Pl-*Psp*I). Where it is necessary to distinguish between the restriction and methylating activities, they are given the prefixes 'R' and 'M', respectively, e.g. R.*Sma*I and M.*Sma*I.

Table 3.2 Examples of restriction endonuclease nomenclature.

Enzyme	Enzyme source	Recognition sequence
*Sma*I	*Serratia marcescens*, 1st enzyme	CCCGGG
*Hae*III	*Haemophilus aegyptius*, 3rd enzyme	GGCC
*Hind*II	*Haemophilus influenzae*, strain d, 2nd enzyme	GTPyPuAC
*Hind*III	*Haemophilus influenzae*, strain d, 3rd enzyme	AAGCTT
*Bam*HI	*Bacillus amyloliquefaciens*, strain H, 1st enzyme	GGATCC

Recognition sequences

Most, but not all, type II restriction endonucleases recognize and cleave DNA within particular sequences of four to eight nucleotides which have a twofold axis of *rotational symmetry*. Such sequences are often referred to as *palindromes* because of their similarity to words that read the same backwards as forwards. For example, the restriction and modification enzymes R.*Eco*RI and M.*Eco*RI recognize the sequence:

5′-GAA┊TTC-3′
3′-CTT┊AAG-5′

Axis of symmetry

The position at which the restricting enzyme cuts is usually shown by the symbol '/' and the nucleotides methylated by the modification enzyme are usually marked with an asterisk. For *Eco*RI these would be represented thus:

5′-G/AA*T T C-3′
3′-C TT A*A/G-5′

For convenience it is usual practice to simplify the description of recognition sequences by showing only one strand of DNA, that which runs in the 5′ to 3′ direction. Thus the *Eco*RI recognition sequence would be shown as G/AATTC.

From the information shown above we can see that *Eco*RI makes single-stranded breaks four bases apart in the opposite strands of its target sequence so generating fragments with protruding 5′ termini:

5′-G   5′-AATTC-3′
3′-CTTAA-5′   G-5′

These DNA fragments can associate by hydrogen bonding between overlapping 5′ termini, or the fragments can circularize by intramolecular reaction (Fig. 3.2). For this reason the fragments are said to have *sticky* or *cohesive* ends. In principle, DNA fragments from diverse sources can be joined by means of the cohesive ends and, as we shall see later, the nicks in the molecules can be sealed to form an intact *artificially recombinant* DNA molecule.

Not all type II enzymes cleave their target sites like *Eco*RI. Some, such as *Pst*I (CTGCA/G), produce fragments bearing 3′ overhangs, while others, such as *Sma*I (CCC/GGG), produce *blunt* or *flush* ends.

To date, over 10 000 microbes from around the world have been screened for restriction enzymes. From these, over 3000 enzymes have been found representing approximately 200 different sequence specificities. Some representative examples are shown in Table 3.3. For a comprehensive database of information on restriction endonucleases and their associated methylases, including cleavage sites, commercial availability and literature references, the reader should consult the website maintained by New England Biolabs (www.rebase.neb.com).

Occasionally enzymes with novel DNA sequence specificities are found but most prove to have the same specificity as enzymes already known. Restriction enzymes with the same sequence specificity and cut site are known as *isoschizomers*. Enzymes that recognize the same sequence but cleave at different points, for example *Sma*I (CCC/GGG) and *Xma*I C/CCGGG), are sometimes known as *neoschizomers*.

Under extreme conditions, such as elevated pH or low ionic strength, restriction endonucleases are

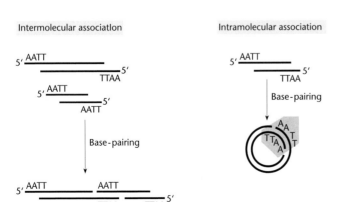

Fig. 3.2 Cohesive ends of DNA fragments produced by digestion with *Eco*RI.

Table 3.3 Some restriction endonucleases and their recognition sites.

Enzyme	Recognition sequence
4-Base cutters	
MboI, DpnI, Sau3AI	/GATC
MspI, HpaII	C/CGG
AluI	AG/CT
HaeIII	GG/CC
TaiI	ACGT/
6-Base cutters	
BglII	A/GATCT
ClaI	AT/CGAT
PvuII	CAG/CTG
PvuI	CGAT/CG
KpnI	GGTAC/C
8-Base cutters	
NotI	GC/GGCCGC
SbfI	CCTGCA/GG

capable of cleaving sequences which are similar but not identical to their defined recognition sequence. This altered specificity is known as *star* activity. The most common types of altered activity are acceptance of base substitutions and truncation of the number of bases in the recognition sequence. For example, *Eco*RI* (*Eco*RI star activity) cleaves the sequence N/AATTN, where N is any base, whereas *Eco*RI cleaves the sequence GAATTC.

Number and size of restriction fragments

The number and size of the fragments generated by a restriction enzyme depend on the frequency of occurrence of the target site in the DNA to be cut. Assuming a DNA molecule with a 50% G+C content and a random distribution of the four bases, a four-base recognition site occurs every 4^4 (256) bp. Similarly, a six-base recognition site occurs every 4^6 (4096) bp and an eight-base recognition sequence every 4^8 (65 536) bp. In practice, there is not a random distribution of the four bases and many organisms can be AT- or GC-rich, e.g. the nuclear genome of mammals is 40% G+C and the dinucleotide CG is fivefold less common than statistically expected. Similarly, CCG and CGG are the rarest trinucleotides in most A+T-rich bacterial genomes and CTAG is the rarest tetranucleotide in G+C-rich bacterial genomes. Thus different restriction endonucleases with six-base recognition sites can produce average fragment sizes significantly different from the expected 4096 bp (Table 3.4).

Certain restriction endonucleases show preferential cleavage of some sites in the same DNA molecule. For example, phage λ DNA has five sites for *Eco*RI but the different sites are cleaved non-randomly (Thomas & Davis 1975). The site nearest the right terminus is cleaved 10 times faster than the sites in the middle of the molecule. There are four sites for *Sac*II in λ DNA but the three sites in the centre of the molecule are cleaved 50 times faster than the remaining site. There is a group of three restriction enzymes which show an even more dramatic site preference. These are *Nar*I, *Nae*I and *Sac*II and they require simultaneous interaction with two copies of their recognition sequence before they will cleave DNA (Kruger 1988, Conrad & Topal, 1989). Thus *Nar*I will rapidly cleave two of the four recognition sites on plasmid pBR322 DNA but will seldom cleave the remaining two sites.

Table 3.4 Average fragment size (bp) produced by different enzymes with DNA from different sources.

Enzyme	Target	*Arabidopsis*	Nematode	*Drosophila*	*E. coli*	Human
*Apa*I	GGGCCC	25 000	40 000	6 000	15 000	2 000
*Avr*II	CCTAGG	15 000	20 000	20 000	150 000	8 000
*Bam*HI	GGATCC	6 000	9 000	4 000	5 000	5 000
*Dra*I	TTTAAA	2 000	1 000	1 000	2 000	2 000
*Spe*I	ACTAGT	8 000	8 000	9 000	60 000	10 000

Variations on cutting and joining DNA molecules

In order to join two fragments of DNA together, it is not essential that they are produced by the same restriction endonuclease. Many different restriction endonucleases produce compatible cohesive ends. For example, *Age*I (A/CCGGT) and *Ava*I (C/CCGGG) produce molecules with identical 5′ overhangs and so can be ligated together. There are many other examples of compatible cohesive ends. What is more, if the cohesive ends were produced by six-base cutters, the ligation products are often recleavable by four-base cutters. Thus, in the example cited above, the hybrid site ACCGGG can be cleaved by *Hpa*II (C/CGG), *Nci*I (CC/GGG) or *Scr*FI (CC/NGG).

New restriction sites can be generated by filling in the overhangs generated by restriction endonucleases and ligating the products together. Figure 3.3 shows that after filling in the cohesive ends produced by *Eco*RI, ligation produces restriction sites recognized by four other enzymes. Many other examples of creating new target sites by filling and ligation are known.

There are also many examples of combinations of blunt-end restriction endonucleases that produce recleavable ligation products. For example, when molecules generated by cleavage with *Alu*I (AG/CT) are joined to ones produced by *Eco*RV (GAT/ATC), some of the ligation sites will have the sequence GATCT and others will have the sequence AGATC. Both can be cleaved by *Mbo*I (GATC).

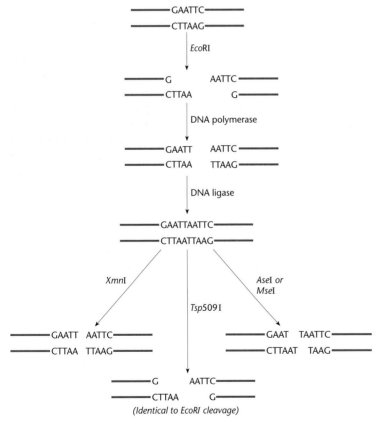

Fig. 3.3 The generation of three new restriction sites after filling in the overhangs produced by endonuclease *Eco*RI and ligating the products together. Note that there are two target sites, 4 bp apart, in the reconstituted molecule for endonuclease *Tsp*509I.

A methyltransferase, M.*Sss*I, that methylates the dinucleotide CpG (Nur *et al.* 1985) has been isolated from *Spiroplasma*. This enzyme can be used to modify *in vitro* restriction endonuclease target sites which contain the CG sequence. Some of the target sequences modified in this way will be resistant to endonuclease cleavage, while others will remain sensitive. For example, if the sequence CCGG is modified with *Sss*I, it will be resistant to *Hpa*II but sensitive to *Msp*I. Since 90% of the methyl groups in the genomic DNA of many animals, including vertebrates and echinoderms, occur as 5-methylcytosine in the sequence CG, M.*Sss* can be used to imprint DNA from other sources with a vertebrate pattern.

The Dam and Dcm methylases of *E. coli*

Most laboratory strains of *E. coli* contain three site-specific DNA methylases. The methylase encoded by the *dam* gene transfers a methyl group from *S*-adenosylmethionine to the N^6 position of the adenine residue in the sequence GATC. The methylase encoded by the *dcm* gene (the Dcm methylase, previously called the Mec methylase) modifies the internal cytosine residues in the sequences CCAGG and CCTGG at the C^5 position (Marinus *et al.* 1983). In DNA in which the GC content is 50%, the sites for these two methylases occur, on average, every 256–512 bp. The third methylase is the enzyme M.*Eco*KI but the sites for this enzyme are much rarer and occur about once every 8 kb.

These enzymes are of interest for two reasons. First, some or all of the sites for a restriction endonuclease may be resistant to cleavage when isolated from strains expressing the Dcm or Dam methylases. This occurs when a particular base in the recognition site of a restriction endonuclease is methylated. The relevant base may be methylated by one of the *E. coli* methylases if the methylase recognition site overlaps the endonuclease recognition site. For example, DNA isolated from Dam⁺ *E. coli* is completely resistant to cleavage by *Mbo*I, but not *Sau*3AI, both of which recognize the sequence GATC. Similarly, DNA from a Dcm⁺ strain will be cleaved by *Bst*NI but not by *Eco*RII, even though both recognize the sequence CCATGG. It is worth noting that most cloning strains of *E. coli* are Dam⁺ Dcm⁺ but double mutants are available (Marinus *et al.* 1983).

The second reason these methylases are of interest is that the modification state of plasmid DNA can affect the frequency of transformation in special situations. Transformation efficiency will be reduced when Dam-modified plasmid DNA is introduced into Dam⁻ *E. coli* or Dam- or Dcm-modified DNA is introduced into other species (Russell & Zinder 1987). When DNA is to be moved from *E. coli* to another species it is best to use a strain lacking the Dam and Dcm methylases.

As will be seen later, it is difficult to stably clone DNA that contains short, direct repetitive sequences. Deletion of the repeat units occurs quickly, even when the host strain is deficient in recombination. However, the deletion mechanism appears to involve Dam methylation, for it does not occur in *dam* mutants (Troester *et al.* 2000).

The importance of eliminating restriction systems in *E. coli* strains used as hosts for recombinant molecules

If foreign DNA is introduced into an *E. coli* host it may be attacked by restriction systems active in the host cell. An important feature of these systems is that the fate of the incoming DNA in the restrictive host depends not only on the sequence of the DNA but also upon its history: the DNA sequence may or may not be restricted, depending upon its source immediately prior to transforming the *E. coli* host strain. As we have seen, post-replication modifications of the DNA, usually in the form of methylation of particular adenine or cytosine residues in the target sequence, protect against cognate restriction systems but not, in general, against different restriction systems.

Because restriction provides a natural defence against invasion by foreign DNA, it is usual to employ a K restriction-deficient *E. coli* K12 strain as a host in transformation with newly created recombinant molecules. Thus where, for example, mammalian DNA has been ligated into a plasmid vector, transformation of the *Eco*K restriction-deficient host eliminates the possibility that the incoming sequence will be restricted, even if the mammalian sequence contains an unmodified *Eco*K target site. If the host happens to be *Eco*K restriction-deficient but *Eco*K modification-*proficient*, propagation on the host will confer modification methylation and hence allow subsequent propagation of the recombinant in *Eco*K restriction-proficient strains, if desired.

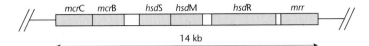

Fig. 3.4 The immigration control region of
E. coli strain K12.

Whereas the *Eco*KI restriction system, encoded by
the *hsd*RMS genes, cleaves DNA that is not protected
by methylation at the target site, the *Mcr*A, *Mcr*BC
and *Mrr* endonucleases cut DNA that is methylated
at specific positions. All three endonucleases restrict
DNA modified by CpC methylase (M.*Sss*I) and the
Mrr endonuclease will attack DNA with methylade-
nine in specific sequences. The significance of these
restriction enzymes is that DNA from many bacteria,
and from all plants and higher animals, is extensively
methylated and its recovery in cloning experiments
will be greatly reduced if the restriction activity is
not eliminated. There is no problem with DNA from
Saccharomyces cerevisiae or *Drosophila melanogaster*
since there is little methylation of their DNA.

All the restriction systems in *E. coli* are clustered
together in an 'immigration control region' about
14 kb in length (Fig. 3.4). Some strains carry mutations in one of the genes. For example, strains DH1
and DH5 have a mutation in the *hsd*R gene and so
are defective for the *Eco*KI endonuclease but still
mediate the *Eco*KI modification of DNA. Strain DP50
has a mutation in the *hsd*S gene and so lacks both the
*Eco*KI restriction and modification activities. Other
strains, such as *E. coli* C and the widely used cloning
strain HB101, have a deletion of the entire *mcrC–
mrr* region and hence lack all restriction activities.

The importance of enzyme quality

Restriction enzymes are available from many different commercial sources. In choosing a source of
enzyme, it is important to consider the quality of the
enzyme supplied. High-quality enzymes are purified
extensively to remove contaminating exonucleases
and endonucleases and tests for the absence of such
contaminants form part of routine quality control
(QC) on the finished product (see below). The absence
of exonucleases is particularly important. If they are
present, they can nibble away the overhangs of
cohesive ends, thereby eliminating or reducing the
production of subsequent recombinants. Contaminating phosphatases can remove the terminal phosphate residues, thereby preventing ligation. Even

where subsequent ligation is achieved, the resulting
product may contain small deletions. The message is
clear: cheap restriction enzymes are in reality poor
value for money!

A typical QC procedure is as follows. DNA fragments are produced by an excessive overdigestion of
substrate DNA with each restriction endonuclease.
These fragments are then ligated and recut with the
same restriction endonuclease. Ligation can occur
only if the 3′ and 5′ termini are left intact, and only
those molecules with a perfectly restored recognition site can be recleaved. A normal banding pattern
after cleavage indicates that both the 3′ and 5′ termini
are intact and the enzyme preparation is free of detectable exonucleases and phosphatases (Fig. 3.5).

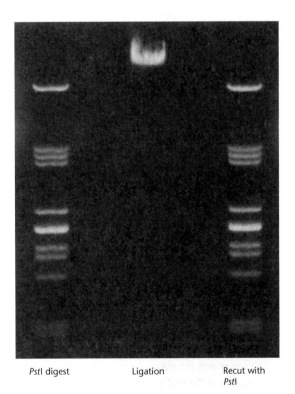

*Pst*I digest   Ligation   Recut with
                                            *Pst*I

Fig. 3.5 Quality control of the enzyme *Pst*I. DNA was
overdigested with the endonuclease and the fragments were
ligated together and then recut. Note that the two digests give
an identical banding pattern upon agarose gel electrophoresis.

An additional QC test is the *blue/white screening assay*. In this, the starting material is a plasmid carrying the E. *coli lacZ* gene in which there is a single recognition site for the enzyme under test. The plasmid is overdigested with the restriction enzyme, religated and transformed into a LacZ⁻ strain of E. *coli*. The transformants are plated on media containing the β-galactosidase substrate Xgal. If the *lacZ* gene remains intact after digestion and ligation, it will give rise to a blue colony. If any degradation of the cut ends occurred, then a white colony will be produced (Box 3.2).

Box 3.2 α-Complementation of β-galactosidase and the use of Xgal

The activity of the enzyme β-galactosidase is easily monitored by including in the growth medium the chromogenic substrate 5-bromo-4-chloro-3-indolyl-β-D-galactoside (Xgal). This compound is colourless but on cleavage releases a blue indolyl derivative. On solid medium, colonies that are expressing active β-galactosidase are blue in colour while those without the activity are white in colour. This is often referred to as blue/white screening. Since Xgal is not an inducer of β-galactosidase, the non-substrate (*gratuitous*) inducer isopropyl-β-D-thiogalactoside (IPTG) is also added to the medium.

The phenomenon of α-complementation of β-galactosidase is widely used in molecular genetics. The starting-point for α-complementation is the M15 mutant of E. *coli*. This has a deletion of residues 11–41 in the *lacZ* gene and shows no β-galactosidase activity. Enzyme activity can be restored to the mutant enzyme *in vitro* by adding a cyanogen bromide peptide derived from amino acid residues 3–92 (Langley *et al.* 1975, Langley & Zabin 1976). Complementation can also be shown *in vivo*. If a plasmid carrying the N-terminal fragment of the *lacZ* gene encompassing the missing region is introduced into the M15 mutant, then β-galactosidase is produced, as demonstrated by the production of a blue colour on medium containing Xgal. In practice, the plasmid usually carries the *lacI* gene and the first 146 codons of the *lacZ* gene, because in the early days of genetic engineering this was a convenient fragment of DNA to manipulate.

Since wild-type β-galactosidase has 1021 amino acids, it is encoded by a gene 3.1 kb in length. While a gene of this length is easily manipulated *in vitro*, there are practical disadvantages to using the whole gene. As will be seen later, it is preferable to keep cloning vectors and their inserts as small as possible. The phenomenon of α-complementation allows genetic engineers to take advantage of the *lac* system without having to have the entire Z gene on the vector.

Joining DNA molecules

Having described the methods available for cutting DNA molecules, we must consider the ways in which DNA fragments can be joined to create artificially recombinant molecules. There are currently three methods for joining DNA fragments *in vitro*. The first of these capitalizes on the ability of DNA ligase to join covalently the annealed cohesive ends produced by certain restriction enzymes. The second depends upon the ability of DNA ligase from phage T4-infected *E. coli* to catalyse the formation of phosphodiester bonds between blunt-ended fragments. The third utilizes the enzyme terminal deoxynucleotidyltransferase to synthesize homopolymeric 3′ single-stranded tails at the ends of fragments. We can now look at these three methods a little more deeply.

DNA ligase

E. coli and phage T4 encode an enzyme, DNA ligase, which seals single-stranded nicks between adjacent nucleotides in a duplex DNA chain (Olivera *et al.* 1968, Gumport & Lehman 1971). Although the reactions catalysed by the enzymes of *E. coli* and T4-infected *E. coli* are very similar, they differ in their cofactor requirements. The T4 enzyme requires ATP, while the *E. coli* enzyme requires NAD^+. In each case the cofactor is split and forms an enzyme–AMP complex. The complex binds to the nick, which must expose a 5′ phosphate and 3′ OH group, and makes a covalent bond in the phosphodiester chain, as shown in Fig. 3.6.

When termini created by a restriction endonuclease that creates cohesive ends associate, the joint has nicks a few base pairs apart in opposite strands. DNA ligase can then repair these nicks to form an intact duplex. This reaction, performed *in vitro* with purified DNA ligase, is fundamental to many gene-manipulation procedures, such as that shown in Fig. 3.7.

The optimum temperature for ligation of nicked DNA is 37°C, but at this temperature the hydrogen-bonded join between the sticky ends is unstable. *Eco*RI-generated termini associate through only

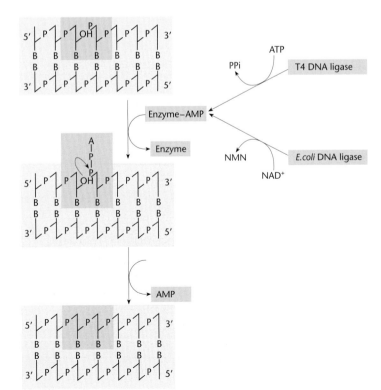

Fig. 3.6 Action of DNA ligase. An enzyme – AMP complex binds to a nick bearing 3′ OH and 5′ P groups. The AMP reacts with the phosphate group. Attack by the 3′ OH group on this moiety generates a new phosphodiester bond, which seals the nick.

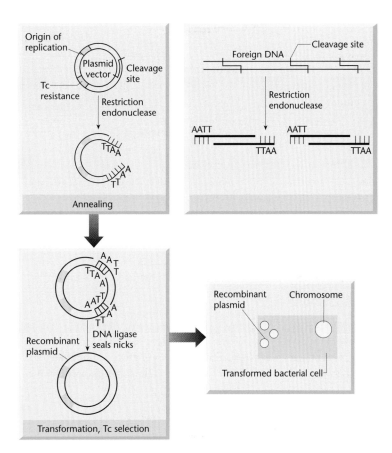

Fig. 3.7 Use of DNA ligase to create a covalent DNA recombinant joined through association of termini generated by *Eco*RI.

four AT base pairs and these are not sufficient to resist thermal disruption at such a high temperature. The optimum temperature for ligating the cohesive termini is therefore a compromise between the rate of enzyme action and association of the termini, and has been found experimentally to be in the range 4–15°C (Dugaiczyk *et al.* 1975, Ferretti & Sgaramella 1981).

The ligation reaction can be performed so as to favour the formation of recombinants. First, the population of recombinants can be increased by performing the reaction at a high DNA concentration; in dilute solutions *circularization* of linear fragments is relatively favoured because of the reduced frequency of intermolecular reactions. Secondly, by treating linearized plasmid vector DNA with alkaline phosphatase to remove 5′-terminal phosphate groups, both recircularization and plasmid dimer formation are prevented (Fig. 3.8). In this case,

circularization of the vector can occur only by insertion of non-phosphatase-treated foreign DNA which provides one 5′-terminal phosphate at each join. One nick at each join remains unligated, but, after transformation of host bacteria, cellular repair mechanisms reconstitute the intact duplex.

Joining DNA fragments with cohesive ends by DNA ligase is a relatively efficient process which has been used extensively to create artificial recombinants. A modification of this procedure depends upon the ability of T4 DNA ligase to join blunt-ended DNA molecules (Sgaramella 1972). The *E. coli* DNA ligase will not catalyse blunt ligation except under special reaction conditions of macromolecular crowding (Zimmerman & Pheiffer 1983). Blunt ligation is most usefully applied to joining blunt-ended fragments via *linker* molecules; in an early example of this, Scheller *et al.* (1977) synthesized self-complementary decameric oligonucleotides, which contain sites for

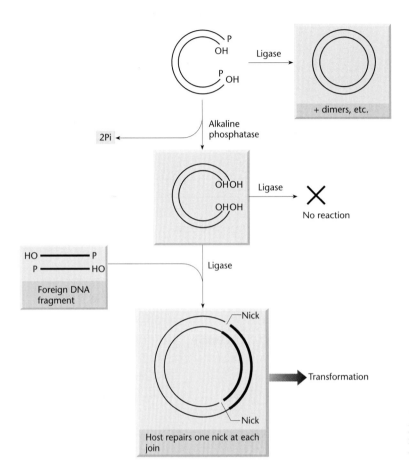

Fig. 3.8 Application of alkaline phosphatase treatment to prevent recircularization of vector plasmid without insertion of foreign DNA.

one or more restriction endonucleases. One such molecule is shown in Fig. 3.9. The molecule can be ligated to both ends of the foreign DNA to be cloned, and then treated with restriction endonuclease to produce a sticky-ended fragment, which can be incorporated into a vector molecule that has been cut with the same restriction endonuclease. Insertion by means of the linker creates restriction-enzyme target sites at each end of the foreign DNA and so enables the foreign DNA to be excised and recovered after cloning and amplification in the host bacterium.

Adaptors

It may be the case that the restriction enzyme used to generate the cohesive ends in the linker will also cut the foreign DNA at internal sites. In this situation,

the foreign DNA will be cloned as two or more sub-fragments. One solution to this problem is to choose another restriction enzyme, but there may not be a suitable choice if the foreign DNA is large and has sites for several restriction enzymes. Another solution is to methylate internal restriction sites with the appropriate modification methylase. An example of this is described in Fig. 6.2. Alternatively, a general solution to the problem is provided by chemically synthesized adaptor molecules which have a *pre-formed* cohesive end (Wu *et al.* 1978). Consider a blunt-ended foreign DNA containing an internal *Bam*HI site (Fig. 3.10), which is to be cloned in a *Bam*HI-cut vector. The *Bam* adaptor molecule has one blunt end bearing a 5′ phosphate group and a *Bam* cohesive end which is not phosphorylated. The adaptor can be ligated to the foreign DNA ends. The foreign DNA plus added adaptors is then phosphory-

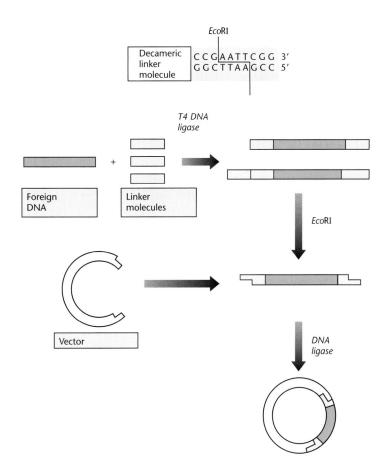

Fig. 3.9 A decameric linker molecule containing an *Eco*RI target site is joined by T4 DNA ligase to both ends of flush-ended foreign DNA. Cohesive ends are then generated by *Eco*RI. This DNA can then be incorporated into a vector that has been treated with the same restriction endonuclease.

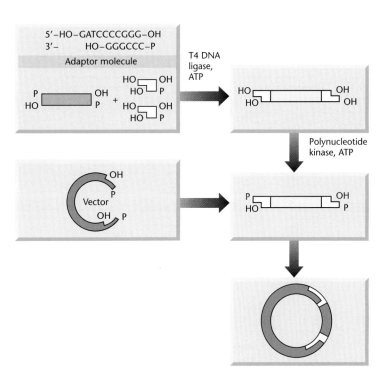

Fig. 3.10 Use of a *Bam*HI adaptor molecule. A synthetic adaptor molecule is ligated to the foreign DNA. The adaptor is used in the 5′-hydroxyl form to prevent self-polymerization. The foreign DNA plus ligated adaptors is phosphorylated at the 5′-termini and ligated into the vector previously cut with *Bam*HI.

lated at the 5′ termini and ligated into the *Bam*HI site of the vector. If the foreign DNA were to be recovered from the recombinant with *Bam*HI, it would be obtained in two fragments. However, the adaptor is designed to contain two other restriction sites (*Sma*I, *Hpa*II), which may enable the foreign DNA to be recovered intact.

Note that the only difference between an adaptor and a linker is that the former has cohesive ends and the latter has blunt ends. A wide range of adaptors are available commercially.

Homopolymer tailing

A general method for joining DNA molecules makes use of the annealing of complementary homopolymer sequences. Thus, by adding oligo(dA) sequences to the 3′ ends of one population of DNA molecules and oligo(dT) blocks to the 3′ ends of another population, the two types of molecule can anneal to form mixed dimeric circles (Fig. 3.11).

An enzyme purified from calf thymus, terminal deoxynucleotidyltransferase, provides the means

by which the homopolymeric extensions can be synthesized, for if presented with a single deoxynucleotide triphosphate it will repeatedly add nucleotides to the 3′ OH termini of a population of DNA molecules (Chang & Bollum 1971). DNA with exposed 3′ OH groups, such as arise from pretreatment with phage λ exonuclease or restriction with an enzyme such as *Pst*I, is a very good substrate for the transferase. However, conditions have been found in which the enzyme will extend even the shielded 3′ OH of 5′ cohesive termini generated by *Eco*RI (Roychoudhury *et al.* 1976, Humphries *et al.* 1978).

The terminal transferase reactions have been characterized in detail with regard to their use in gene manipulation (Deng & Wu 1981, Michelson & Orkin 1982). Typically, 10–40 homopolymeric residues are added to each end.

One of the earliest examples of the construction of recombinant molecules, the insertion of a piece of λ DNA into SV40 viral DNA, made use of homopolymer tailing (Jackson *et al.* 1972). In their experiments, the single-stranded gaps which remained

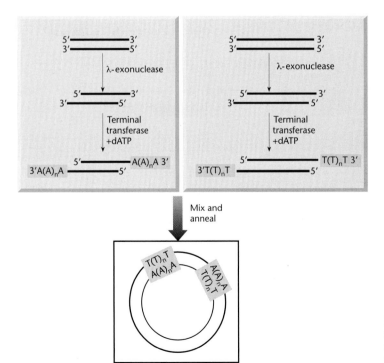

Fig. 3.11 Use of calf-thymus terminal deoxynucleotidyltransferase to add complementary homopolymer tails to two DNA molecules.

in the two strands at each join were repaired *in vitro* with DNA polymerase and DNA ligase so as to produce covalent-ly closed circular molecules. The recombinants were then transfected into susceptible mammalian cells (see Chapter 10). Subsequently, the homopolymer method, using either dA.dT or dG.dC homopolymers was used extensively to construct recombinant plasmids for cloning in *E. coli*. In recent years, homopolymer tailing has been largely replaced as a result of the availability of a much wider range of restriction endonucleases and other DNA-modifying enzymes. However, it is still important for cDNA cloning (see p. 95 *et seq.*).

Joining polymerase chain reaction (PCR) products

Many of the strategies for cloning DNA fragments do not work well with PCR products. The reason for this is that the polymerases used in the PCR have a terminal transferase activity. For example, the *Taq* polymerase adds a single 3′ A overhang to each end of the PCR product. Thus PCR products cannot be blunt-end-ligated unless the ends are first *polished* (blunted). A DNA polymerase like Klenow can be used to fill in the ends. Alternatively, *Pfu* DNA polymerase can be used to remove extended bases with its 3′ to 5′ exonuclease activity. However, even when the PCR fragments are polished, blunt-end-ligating them into a vector still may be very inefficient. One solution to this problem is to use T/A cloning (Mead *et al.* 1991). In this method, the PCR fragment is ligated to a vector DNA molecule with a single 3′ deoxythymidylate extension (Fig. 3.12).

Incorporation of extra sequence at the 5′ end of a primer into amplified DNA

A PCR primer may be designed which, in addition to the sequence required for hybridization with the input DNA, includes an extra sequence at its 5′ end. The extra sequence does not participate in the first hybridization step – only the 3′ portion of the primer hybridizes – but it subsequently becomes incorporated into the amplified DNA fragment (Fig. 3.13). Because the extra sequence can be chosen at the will of the experimenter, great flexibility is available here.

A common application of this principle is the incorporation of restriction sites at each end of the amplified product. Figure 3.13 illustrates the addition of a *Hind*III site and an *Eco*RI site to the ends of an amplified DNA fragment. In order to ensure that the restriction sites are good substrates for the restriction endonucleases, four nucleotides are placed between the hexanucleotide restriction sites and the extreme ends of the DNA. The incorporation of these restriction sites provides one method for cloning amplified DNA fragments (see below).

Joining DNA molecules without DNA ligase

In all the cutting and joining reactions described above, two separate protein components were required: a site-specific endonuclease and a DNA ligase. Shuman (1994) has described a novel approach to the synthesis of recombinant molecules in which a single enzyme, vaccinia DNA topoisomerase, both

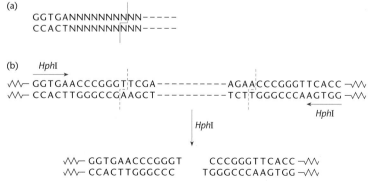

Fig. 3.12 Cleavage of a vector DNA molecule to generate single thymidylate overhangs. (a) The recognition sequence and cleavage point for the restriction endonuclease *Hph*I. (b) Sequences in the vector DNA which result in desired overhangs after cleavage with *Hph*I.

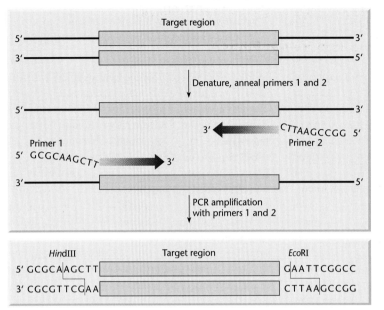

Fig. 3.13 Incorporation of extra sequence at the 5′ end of a primer. Two primers have sequences designed to hybridize at the ends of the target region. Primer 1 has an extra sequence near its 5′ end which forms a *Hind*III site (AAGCTT), and primer 2 has an extra sequence near its 5′ end which forms an *Eco*RI (GAATTC) site. Each primer has an additional 5′-terminal sequence of four nucleotides so that the hexanucleotide restriction sites are placed within the extreme ends of the amplified DNA, and so present good substrates for endonuclease cleavage.

cleaves and rejoins DNA molecules. Placement of the CCCTT cleavage motif for vaccinia topoisomerase near the end of a duplex DNA permits efficient generation of a stable, highly recombinogenic protein DNA adduct that can only religate to acceptor DNAs that contain complementary single-strand extensions. Linear DNAs containing CCCTT cleavage sites at both ends can be activated by topoisomerase and inserted into a plasmid vector.

Heyman *et al.* (1999) have used the properties of vaccinia topoisomerase to develop a ligase-free technology for the covalent joining of DNA fragments to plasmid vectors. Whereas joining molecules with DNA ligase requires an overnight incubation, topoisomerase-mediated ligation occurs in 5 min. The method is particularly suited to cloning PCR fragments. A linearized vector with single 3′ T extensions is activated with the topoisomerase. On addition of the PCR product with 3′ A overhangs, ligation is very rapid. In addition, the high substrate specificity of the enzyme means that there is a low rate of formation of vectors without inserts.

CHAPTER 4

Basic biology of plasmid and phage vectors

Plasmid biology and simple plasmid vectors

Plasmids are widely used as cloning vehicles but, before discussing their use in this context, it is appropriate to review some of their basic properties. Plasmids are replicons which are stably inherited in an extrachromosomal state. Most plasmids exist as double-stranded circular DNA molecules. If both strands of DNA are intact circles the molecules are described as covalently closed circles or CCC DNA (Fig. 4.1). If only one strand is intact, then the molecules are described as open circles or OC DNA. When isolated from cells, covalently closed circles often have a deficiency of turns in the double helix, such that they have a supercoiled configuration.

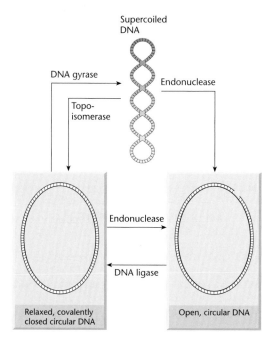

Fig. 4.1 The interconversion of supercoiled, relaxed covalently closed circular DNA and open circular DNA.

The enzymatic interconversion of supercoiled, relaxed CCC DNA* and OC DNA is shown in Fig. 4.1. Because of their different structural configurations, supercoiled and OC DNA separate upon electrophoresis in agarose gels (Fig. 4.2). Addition of an intercalating agent, such as ethidium bromide, to supercoiled DNA causes the plasmid to unwind. If excess ethidium bromide is added, the plasmid will rewind in the opposite direction (Fig. 4.3). Use of this fact is made in the isolation of plasmid DNA (see p. 48).

Not all plasmids exist as circular molecules. Linear plasmids have been found in a variety of bacteria, e.g. *Streptomyces* sp. and *Borrelia burgdorferi*. To prevent nuclease digestion, the ends of linear plasmids need to be protected and two general mechanisms have evolved. Either there are repeated sequences ending in a terminal DNA hairpin loop (*Borrelia*) or the ends are protected by covalent attachment of a protein (*Streptomyces*). For more details of linear plasmids the reader should consult Hinnebusch and Tilly (1993).

Plasmids are widely distributed throughout the prokaryotes, vary in size from less than 1×10^6 daltons to greater than 200×10^6, and are generally dispensable. Some of the phenotypes which these plasmids confer on their host cells are listed in Table 4.1. Plasmids to which phenotypic traits have not yet been ascribed are called *cryptic* plasmids.

Plasmids can be categorized into one of two major type – conjugative or non-conjugative – depending upon whether or not they carry a set of transfer genes, called the *tra* genes, which promote bacterial conjugation. Plasmids can also be categorized on the basis of their being maintained as multiple copies per

---

* The reader should not be confused by the terms *relaxed circle* and *relaxed plasmid*. Relaxed circles are CCC DNA that does not have a supercoiled configuration. Relaxed plasmids are plasmids with multiple copies per cell.

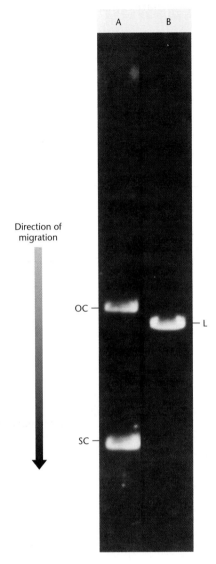

Direction of migration

OC —

SC —

A   B

— L

Fig. 4.2 Electrophoresis of DNA in agarose gels. The direction of migration is indicated by the arrow. DNA bands have been visualized by soaking the gel in a solution of ethidium bromide (which complexes with DNA by intercalating between stacked base pairs) and photographing the orange fluorescence which results upon ultraviolet irradiation. (A) Open circular (OC) and supercoiled (SC) forms of a plasmid of 6.4 kb pairs. Note that the compact supercoils migrate considerably faster than open circles (B). Linear plasmid (L) DNA is produced by treatment of the preparation shown in lane (A) with *Eco*RI, for which there is a single target site. Under the conditions of electrophoresis employed here, the linear form migrates just ahead of the open circular form.

cell (*relaxed* plasmids) or as a limited number of copies per cell (*stringent* plasmids). Generally, conjugative plasmids are of relatively high molecular weight and are present as one to three copies per chromosome, whereas non-conjugative plasmids are of low molecular weight and present as multiple copies per chromosome (Table 4.2). An exception is the conjugative plasmid R6K, which has a molecular weight of 25×10^6 daltons and is maintained as a relaxed plasmid.

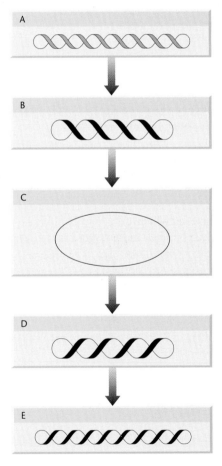

Fig. 4.3 Effect of intercalation of ethidium bromide on supercoiling of DNA. As the amount of intercalated ethidium bromide increases, the double helix untwists, with the result that the supercoiling decreases until the open form of the circular molecule is produced. Further intercalation introduces excess turns in the double helix, resulting in supercoiling in the opposite sense (note the direction of coiling at B and D). For clarity, only a single line represents the double helix.

Table 4.1 Some phenotypic traits exhibited by plasmid-carried genes.

Antibiotic resistance
Antibiotic production
Degradation of aromatic compounds
Haemolysin production
Sugar fermentation
Enterotoxin production
Heavy-metal resistance
Bacteriocin production
Induction of plant tumours
Hydrogen sulphide production
Host-controlled restriction and modification

Host range of plasmids

Plasmids encode only a few of the proteins required for their own replication and in many cases encode only one of them. All the other proteins required for replication, e.g. DNA polymerases, DNA ligase, helicases, etc., are provided by the host cell. Those replication proteins that are plasmid-encoded are located very close to the *ori* (origin of replication) sequences at which they act. Thus, only a small region surrounding the *ori* site is required for replication. Other parts of the plasmid can be deleted and foreign sequences can be added to the plasmid and replication will still occur. This feature of plasmids has greatly simplified the construction of versatile cloning vectors.

The host range of a plasmid is determined by its *ori* region. Plasmids whose *ori* region is derived from plasmid Col E1 have a restricted host range: they only replicate in enteric bacteria, such as *E. coli*, *Salmonella*, etc. Other *promiscuous* plasmids have a broad host range and these include RP4 and RSF1010. Plasmids of the RP4 type will replicate in most Gram-negative bacteria, to which they are readily transmitted by conjugation. Such promiscuous plasmids offer the potential of readily transferring cloned DNA molecules into a wide range of genetic backgrounds. Plasmids like RSF1010 are not conjugative but can be transformed into a wide range of Gram-negative and Gram-positive bacteria, where they are stably maintained. Many of the plasmids isolated from *Staphylococcus aureus* also have a broad host range and can replicate in many other Gram-positive bacteria. Plasmids with a broad host range encode most, if not all, of the proteins required for replication. They must also be able to express these genes and thus their promoters and ribosome binding sites must have evolved such that they can be recognized in a diversity of bacterial families.

Plasmid copy number

The copy number of a plasmid is determined by regulating the initiation of plasmid replication. Two major mechanisms of control of initiation have been recognized: regulation by antisense RNA and regulation by binding of essential proteins to iterons (for review, see Del Solar *et al.* 1998). Most of the cloning vectors in current use carry an *ori* region derived from plasmid Col E1 and copy-number control is mediated by antisense RNA. In this type of plasmid, the primer for DNA replication is a 555-base

Table 4.2 Properties of some conjugative and non-conjugative plasmids of Gram-negative organisms.

Plasmid	Size (MDa)	Conjugative	No. of plasmid copies/chromosome equivalent	Phenotype
Col E1	4.2	No	10–15	Col E1 production
RSF1030	5.6	No	20–40	Ampicillin resistance
clo DF13	6	No	10	Cloacin production
R6K	25	Yes	13–38	Ampicillin and streptomycin resistance
F	62	Yes	1–2	–
RI	62.5	Yes	3–6	Multiple drug resistance
Ent P 307	65	Yes	1–3	Enterotoxin production

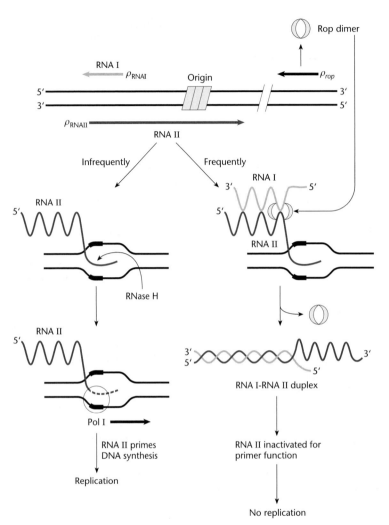

Fig. 4.4 Regulation of replication of Col E1-derived plasmids. RNA II must be processed by RNase H before it can prime replication. 'Origin' indicates the transition point between the RNA primer and DNA. Most of the time, RNA I binds to RNA II and inhibits the processing, thereby regulating the copy number. ρ_{RNAI} and ρ_{RNAII} are the promoters for RNA I and RNA II transcription, respectively. RNA I is coloured pale red and RNA II dark red. The Rop protein dimer enhances the initial pairing of RNA I and RNA II.

ribonucleotide molecule called RNA II, which forms an RNA–DNA hybrid at the replication origin (Tomizawa & Itoh 1982), RNA II can only act as a primer if it is cleaved by RNase H to leave a free 3′ hydroxyl group. Unless RNA II is processed in this way, it will not function as a primer and replication will not ensue. Replication control is mediated by another small (108-base) RNA molecule called RNA I (Tomizawa & Itoh 1981), which is encoded by the same region of DNA as RNA II but by the complementary strand. Thus RNA I and RNA II are complementary to each other and can hybridize to form a double-stranded RNA helix. The formation of this duplex interferes with the processing of RNA II by RNase H and hence replication does not ensue

(Fig. 4.4). Since RNA I is encoded by the plasmid, more of it will be synthesized when the copy number of the plasmid is high. As the host cell grows and divides, so the concentration of RNA I will fall and the plasmid will begin to replicate again (Cesarini *et al.* 1991, Eguchi *et al.* 1991).

In addition to RNA I, a plasmid-encoded protein called Rop helps maintain the copy number (Cesarini *et al.* 1982). This protein, which forms a dimer, enhances the pairing between RNA I and RNA II so that processing of the primer can be inhibited even at relatively low concentrations of RNA I. Deletion of the *ROP* gene (Twigg & Sherratt 1980) or mutations in RNA I (Muesing *et al.* 1981) result in increased copy numbers.

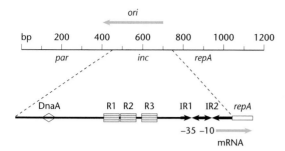

Fig. 4.5 The *ori* region of pSC101. R1, R2, and R3 are the three iteron sequences (CAAAGGTCTAGCAGCAGAATTTACAGA for R3) to which RepA binds to handcuff two plasmids. RepA autoregulates its own synthesis by binding to the inverted repeats IR1 and IR2. The location of the partitioning site *par* and the binding sites for the host protein DnaA are also shown.

In plasmid pSC101 and many of the broad-host-range plasmids, the *ori* region contains three to seven copies of an iteron sequence which is 17 to 22 bp long. Close to the *ori* region there is a gene, called *repA* in pSC101, which encodes the RepA protein. This protein, which is the only plasmid-encoded protein required for replication, binds to the iterons and initiates DNA synthesis (Fig. 4.5).

Copy-number control is exerted by two superimposed mechanisms. First, the RepA protein represses its own synthesis by binding to its own promoter region and blocking transcription of its own gene (Ingmer & Cohen 1993). If the copy number is high, synthesis of RepA will be repressed. After cell division, the copy number and concentration of RepA will drop and replication will be initiated. Mutations in the RepA protein can lead to increased copy number (Ingmer & Cohen 1993, Cereghino *et al.* 1994). Secondly, the RepA protein can link two plasmids together, by binding to their iteron sequences, thereby preventing them from initiating replication. By this mechanism, known as *handcuffing* (McEachern *et al.* 1989), the replication of iteron plasmids will depend both on the concentration of RepA protein and on the concentration of the plasmids themselves.

Partitioning and segregative stability of plasmids

The loss of plasmids due to defective partitioning is called segregative instability. Naturally occurring plasmids are stably maintained because they contain a partitioning function, *par*, which ensures that they are stably maintained at each cell division. Such *par* regions are essential for stability of low-copy-number plasmids (for review, see Bingle & Thomas 2001). The higher-copy-number plasmid Col E1 also contains a *par* region but this is deleted in many Col E1-derived cloning vectors, e.g. pBR322. Although the copy number of vectors such as pBR322 is usually high, plasmid-free cells arise under nutrient limitation or other stress conditions (Jones *et al.* 1980, Nugent *et al.* 1983). The *par* region from a plasmid such as pSC101 can be cloned into pBR322, thereby stabilizing the plasmid (Primrose *et al.* 1983).

DNA superhelicity is involved in the partitioning mechanism (Miller *et al.* 1990). pSC101 derivatives lacking the *par* locus show decreased overall superhelical density as compared with wild-type pSC101. Partition-defective mutants of pSC101 and similar mutants of unrelated plasmids are stabilized in *Eseherichia coli* by *top*A mutations, which increase negative DNA supercoiling. Conversely, DNA gyrase inhibitors and mutations in DNA gyrase increase the rate of loss of *par*-defective pSC101 derivatives.

Plasmid instability may also arise due to the formation of multimeric forms of a plasmid. The mechanism that controls the copy number of a plasmid ensures a fixed number of plasmid origins per bacterium. Cells containing multimeric plasmids have the same number of plasmid origins but fewer plasmid molecules, which leads to segregative instability if they lack a partitioning function. These multimeric forms are not seen with Col E1, which has a natural method of resolving multimers back to monomers. It contains a highly recombinogenic site (*cer*). If the *cer* sequence occurs more than once in a plasmid, as in a multimer, the host-cell Xer protein promotes recombination, thereby regenerating monomers (Summers & Sherratt 1984, Guhathakurta *et al.* 1996; for review, see Summers 1998).

Incompatibility of plasmids

Plasmid *incompatibility* is the inability of two different plasmids to coexist in the same cell in the absence of selection pressure. The term incompatibility can only be used when it is certain that entry of the second plasmid has taken place and that DNA restriction is not involved. Groups of plasmids which are mutually incompatible are considered to belong to the same incompatibility (Inc) group. Over 30 incompatibility groups have been defined in *E. coli* and 13 for plasmids of *S. aureus*. Plasmids will be incompatible if they have the same mechanism of replication control. Not surprisingly, by changing the sequence of the RNA I/RNA II region of plasmids with antisense control of copy number, it is possible to change their incompatibility group. Alternatively, they will be incompatible if they share the same *par* region (Austin & Nordstrom 1990, Firsheim & Kim 1997).

The purification of plasmid DNA

An obvious prerequisite for cloning in plasmids is the purification of the plasmid DNA. Although a wide range of plasmid DNAs are now routinely purified, the methods used are not without their problems. Undoubtedly the trickiest stage is the lysis of the host cells; both incomplete lysis and total dissolution of the cells result in greatly reduced recoveries of plasmid DNA. The ideal situation occurs when each cell is just sufficiently broken to permit the plasmid DNA to escape without too much contaminating chromosomal DNA. Provided the lysis is done gently, most of the chromosomal DNA released will be of high molecular weight and can be removed, along with cell debris, by high-speed centrifugation to yield a *cleared lysate*. The production of satisfactory cleared lysates from bacteria other than *E. coli*, particularly if large plasmids are to be isolated, is frequently a combination of skill, luck and patience.

Many methods are available for isolating pure plasmid DNA from cleared lysates but only two will be described here. The first of these is the 'classical' method and is due to Vinograd (Radloff *et al.* 1967). This method involves isopycnic centrifugation of cleared lysates in a solution of CsCl containing ethidium bromide (EtBr). EtBr binds by intercalating between the DNA base pairs, and in so doing causes the DNA to unwind. A CCC DNA molecule, such as a plasmid, has no free ends and can only unwind to a limited extent, thus limiting the amount of EtBr bound. A linear DNA molecule, such as fragmented chromosomal DNA, has no such topological con-

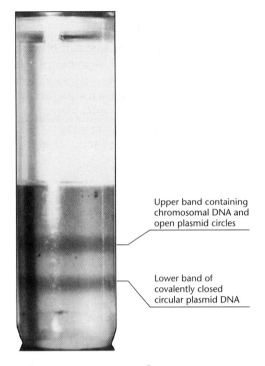

Upper band containing chromosomal DNA and open plasmid circles

Lower band of covalently closed circular plasmid DNA

Fig. 4.6 Purification of Col E1 *Kan*^R plasmid DNA by isopycnic centrifugation in a CsCl–EtBr gradient. (Photograph by courtesy of Dr G. Birnie.)

straints and can therefore bind more of the EtBr molecules. Because the density of the DNA–EtBr complex decreases as more EtBr is bound, and because more EtBr can be bound to a linear molecule than to a covalent circle, the covalent circle has a higher density at saturating concentrations of EtBr. Thus covalent circles (i.e. plasmids) can be separated from linear chromosomal DNA (Fig. 4.6).

Currently the most popular method of extracting and purifying plasmid DNA is that of Birnboim and Doly (1979). This method makes use of the observation that there is a narrow range of pH (12.0–12.5) within which denaturation of linear DNA, but not covalently closed circular DNA, occurs. Plasmid-containing cells are treated with lysozyme to weaken the cell wall and then lysed with sodium hydroxide and sodium dodecyl sulphate (SDS). Chromosomal DNA remains in a high-molecular-weight form but is denatured. Upon neutralization with acidic sodium acetate, the chromosomal DNA renatures and aggregates to form an insoluble network. Simultaneously, the high concentration of sodium acetate causes precipitation of protein–SDS complexes and of high-molecular-weight RNA. Provided the pH of the alkaline denaturation step has been carefully controlled, the CCC plasmid DNA molecules will remain in a native state and in solution, while the contaminating macromolecules co-precipitate. The precipitate can be removed by centrifugation and the plasmid concentrated by ethanol precipitation. If necessary, the plasmid DNA can be purified further by gel filtration.

Recently a number of commercial suppliers of convenience molecular-biology products have developed kits to improve the yield and purity of plasmid DNA. All of them take advantage of the benefits of alkaline lysis and have as their starting-point the cleared lysate. The plasmid DNA is selectively bound to an ion-exchange material, prepacked in columns or tubes, in the presence of a chaotropic agent (e.g. guanidinium hydrochloride). After washing away the contaminants, the purified plasmid is eluted in a small volume.

The yield of plasmid is affected by a number of factors. The first of these is the actual copy number inside the cells at the time of harvest. The copy-number control systems described earlier are not the only factors affecting yield. The copy number is also affected by the growth medium, the stage of growth and the genotype of the host cell (Nugent *et al.* 1983, Seelke *et al.* 1987, Duttweiler & Gross 1998). The second and most important factor is the care taken in making the cleared lysate. Unfortunately, the commercially available kits have not removed the vagaries of this procedure. Finally, the presence in the host cell of a wild-type *endA* gene can affect the recovery of plasmid. The product of the *endA* gene is endonuclease I, a periplasmic protein whose substrate is double-stranded DNA. The function of endonuclease I is not fully understood. Strains bearing *endA* mutations have no obvious phenotype other than improved stability and yield of plasmid obtained from them.

Although most cloning vehicles are of low molecular weight (see next section), it is sometimes necessary to use the much larger conjugative plasmids. Although these high-molecular-weight plasmids can be isolated by the methods just described, the yields are often very low. Either there is inefficient

release of the plasmids from the cells as a consequence of their size or there is physical destruction caused by shear forces during the various manipulative steps. A number of alternative procedures have been described (Gowland & Hardmann 1986), many of which are a variation on that of Eckhardt (1978). Bacteria are suspended in a mixture of Ficoll and lysozyme and this results in a weakening of the cell walls. The samples are then placed in the slots of an agarose gel, where the cells are lysed by the addition of detergent. The plasmids are subsequently extracted from the gel following electrophoresis. The use of agarose, which melts at low temperature, facilitates extraction of the plasmid from the gel.

Desirable properties of plasmid cloning vehicles

An ideal cloning vehicle would have the following three properties:
• low molecular weight;
• ability to confer readily selectable phenotypic traits on host cells;
• single sites for a large number of restriction endonucleases, preferably in genes with a readily scorable phenotype.

The advantages of a low molecular weight are several. First, the plasmid is much easier to handle, i.e. it is more resistant to damage by shearing, and is readily isolated from host cells. Secondly, low-molecular-weight plasmids are usually present as multiple copies (see Table 4.2), and this not only facilitates their isolation but leads to gene dosage effects for all cloned genes. Finally, with a low molecular weight there is less chance that the vector will have multiple substrate sites for any restriction endonuclease (see below).

After a piece of foreign DNA is inserted into a vector, the resulting chimeric molecules have to be transformed into a suitable recipient. Since the efficiency of transformation is so low, it is essential that the chimeras have some readily scorable phenotype. Usually this results from some gene, e.g. antibiotic resistance, carried on the vector, but could also be produced by a gene carried on the inserted DNA.

One of the first steps in cloning is to cut the vector DNA and the DNA to be inserted with either the same endonuclease or ones producing the same ends. If the vector has more than one site for the endonuclease, more than one fragment will be produced. When the two samples of cleaved DNA are subsequently mixed and ligated, the resulting chimeras will, in all probability, lack one of the vector fragments. It is advantageous if insertion of foreign DNA at endonuclease-sensitive sites inactivates a gene whose phenotype is readily scorable, for in this way it is possible to distinguish chimeras from cleaved plasmid molecules which have self-annealed. Of course, readily detectable insertional inactivation is not essential if the vector and insert are to be joined by the homopolymer tailing method (see p. 40) or if the insert confers a new phenotype on host cells.

pBR322, a purpose-built cloning vehicle

In early cloning experiments, the cloning vectors used were natural plasmids, such as Col E1 and pSC101. While these plasmids are small and have single sites for the common restriction endonucleases, they have limited genetic markers for selecting transformants. For this reason, considerable effort was expended on constructing, *in vitro*, superior cloning vectors. The best, and most widely used of these early purpose-built vectors is pBR322. Plasmid pBR322 contains the Ap^R and Tc^R genes of RSF2124 and pSC101, respectively, combined with replication elements of pMB1, a Col E1-like plasmid (Fig. 4.7a). The origins of pBR322 and its progenitor, pBR313, are shown in Fig. 4.7b, and details of its construction can be found in the papers of Bolivar *et al.* (1977a,b).

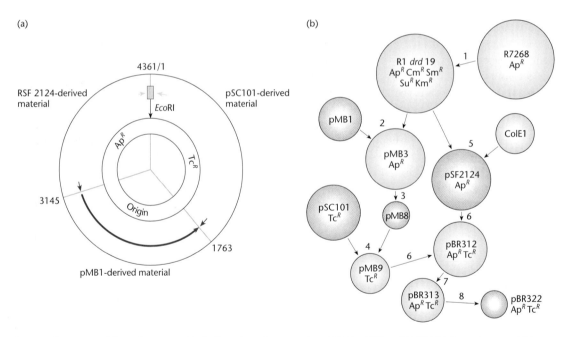

Fig. 4.7 The origins of plasmid pBR322. (a) The boundaries between the pSC101, pMB1 and RSF2124-derived material. The numbers indicate the positions of the junctions in base pairs from the unique *Eco*RI site. (b) The molecular origins of plasmid pBR322. R7268 was isolated in London in 1963 and later renamed R1. 1, A variant, R1*drd*19, which was derepressed for mating transfer, was isolated. 2, The Ap^R transposon, Tn3, from this plasmid was transposed on to pMB1 to form pMB3. 3, This plasmid was reduced in size by *Eco*RI* rearrangement to form a tiny plasmid, pMB8, which carries only colicin immunity. 4, *Eco*RI* fragments from pSC101 were combined with pMB8 opened at its unique *Eco*RI site and the resulting chimeric molecule rearranged by *Eco*RI* activity to generate pMB9. 5, In a separate event, the Tn3 of R1*drd*19 was transposed to Col E1 to form pSF2124. 6, The Tn3 element was then transposed to pMB9 to form pBR312. 7, *Eco*RI* rearrangement of pBR312 led to the formation of pBR313, from which (8) two separate fragments were isolated and ligated together to form pBR322. During this series of constructions, R1 and Col E1 served only as carries for Tn3. (Reproduced by courtesy of Dr G. Sutcliffe and Cold Spring Harbor Laboratory.)

Plasmid pBR322 has been completely sequenced. The original published sequence (Sutcliffe 1979) was 4362 bp long. Position O of the sequence was arbitrarily set between the A and T residues of the *Eco*RI recognition sequence (GAATTC). The sequence was revised by the inclusion of an additional CG base pair at position 526, thus increasing the size of the plasmid to 4363 bp (Backman & Boyer 1983, Peden 1983). More recently, Watson (1988) has revised the size yet again, this time to 4361 bp, by eliminating base pairs at coordinates 1893 and 1915. The most useful aspect of the DNA sequence is that it totally characterizes pBR322 in terms of its restriction sites, such that the exact length of every fragment can be calculated. These fragments can serve as DNA markers for sizing any other DNA fragment in the range of several base pairs up to the entire length of the plasmid.

There are over 40 enzymes with unique cleavage sites on the pBR322 genome (Fig. 4.8). The target sites of 11 of these enzymes lie within the Tc^R gene, and there are sites for a further two (*Cla*I and *Hin*dIII) within the promoter of that gene. There are unique sites for six enzymes within the Ap^R gene. Thus, cloning in pBR322 with the aid of any one of those 19 enzymes will result in insertional inactivation of either the Ap^R or the Tc^R markers. However, cloning in the other unique sites does not permit the easy selection of recombinants, because neither of the antibiotic resistance determinants is inactivated.

Following manipulation *in vitro*, *E. coli* cells transformed with plasmids with inserts in the Tc^R gene can be distinguished from those cells transformed with recircularized vector. The former are Ap^R and Tc^S, whereas the latter are both Ap^R and Tc^R. In practice, transformants are selected on the basis of their Ap resistance and then replica-plated on to Tc-containing media to identify those that are Tc^S. Cells transformed with pBR322 derivatives carrying inserts in the Ap^R gene can be identified more readily (Boyko & Ganschow 1982). Detection is based upon the ability of the β-lactamase produced by Ap^R cells to convert penicillin to penicilloic acid, which in turn binds iodine. Transformants are selected on rich medium containing soluble starch and Tc. When colonized plates are flooded with an indicator solution of iodine and penicillin, β-lactamase-producing (Ap^R) colonies clear the indicator solution whereas Ap^S colonies do not.

The *Pst*I site in the Ap^R gene is particularly useful, because the 3′ tetranucleotide extensions formed on digestion are ideal substrates for terminal transferase. Thus this site is excellent for cloning by the homopolymer tailing method described in the previous chapter (see p. 40). If oligo(dG.dC) tailing is used, the *Pst*I site is regenerated (see Fig. 3.11) and the insert may be cut out with that enzyme.

Plasmid pBR322 has been a widely used cloning vehicle. In addition, it has been widely used as a model system for the study of prokaryotic transcription and translation, as well as investigation of the effects of topological changes on DNA conformation. The popularity of pBR322 is a direct result of the availability of an extensive body of information on its structure and function. This in turn is increased with each new study. The reader wishing more detail on the structural features, transcriptional signals, replication, amplification, stability and conjugal mobility of pBR322 should consult the review of Balbás *et al.* (1986).

Example of the use of plasmid pBR322 as a vector: isolation of DNA fragments which carry promoters

Cloning into the *Hin*dIII site of pBR322 generally results in loss of tetracycline resistance. However, in some recombinants, Tc^R is retained or even increased. This is because the *Hin*dIII site lies within the promoter rather than the coding sequence. Thus whether or not insertional inactivation occurs depends on whether the cloned DNA carries a promoter-like sequence able to initiate transcription of the Tc^R gene. Widera *et al.* (1978) have used this technique to search for promoter-containing fragments.

Four structural domains can be recognized within *E. coli* promoters. These are:
• position 1, the purine initiation nucleotide from which RNA synthesis begins;
• position −6 to −12, the Pribnow box;
• the region around base pair −35;
• the sequence between base pairs −12 and −35.

Although the *Hin*dIII site lies within the Pribnow box (Rodriguez *et al.* 1979) the box is re-created on insertion of a foreign DNA fragment. Thus when insertional inactivation occurs it must be the region from −13 to −40 which is modified.

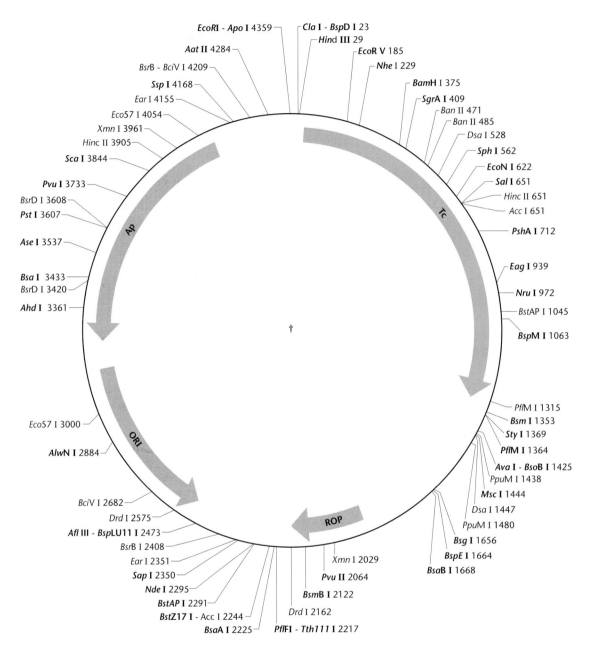

Fig. 4.8 Restriction map of plasmid pBR322 showing the location and direction of transcription of the ampicillin (Ap) and tetracycline (Tc) resistance loci, the origin of replication (*ori*) and the Col E1-derived Rop gene. The map shows the restriction sites of those enzymes that cut the molecule once or twice. The unique sites are shown in bold type. The coordinates refer to the position of the 5′ base in each recognition sequence with the first T in the *Eco*RI site being designated as nucleotide number 1. The exact positions of the loci are: Tc, 86–1268; Ap, 4084–3296; Rop, 1918–2105 and the origin of replication, 2535.

Improved vectors derived from pBR322

Over the years, numerous different derivatives of pBR322 have been constructed, many to fulfil special-purpose cloning needs. A compilation of the properties of some of these plasmids has been provided by Balbás *et al.* (1986).

Much of the early work on the improvement of pBR322 centred on the insertion of additional unique restriction sites and selectable markers, e.g. pBR325 encodes chloramphenicol resistance in addition to ampicillin and tetracycline resistance and has a unique *Eco*RI site in the Cm^R gene. Initially, each new vector was constructed in a series of steps analogous to those used in the generation of pBR322 itself (Fig. 4.7). Then the construction of improved vectors was simplified (Vieira & Messing 1982, 1987, Yanisch-Perron *et al.* 1985) by the use of *polylinkers* or *multiple cloning sites* (MCS), as exemplified by the pUC vectors (Fig. 4.9). An MCS is a short DNA sequence, 2.8 kb in the case of pUC19, carrying sites for many different restriction endonucleases. An MCS increases the number of potential cloning strategies available by extending the range of enzymes that can be used to generate a restriction fragment suitable for cloning. By combining them within an MCS, the sites are made contiguous, so that any two sites within it can be cleaved simultaneously without excising vector sequences.

The pUC vectors also incorporate a DNA sequence that permits rapid visual detection of an insert. The MCS is inserted into the *lacZ'* sequence, which encodes the promoter and the α-peptide of β-galactosidase. The insertion of the MCS into the lacZ' fragment does not affect the ability of the α-peptide to mediate complementation, but cloning DNA fragments into the MCS does. Therefore, recombinants can be detected by blue/white screening on growth medium containing Xgal (see Box 3.2 on p. 35). The usual site for insertion of the MCS is between the iniator ATG codon and codon 7, a region that encodes a functionally non-essential part of the α-complementation peptide. Recently, Slilaty and Lebel (1998) have reported that blue/white colour selection can be variable. They have found that reliable inactivation of complementation occurs only when the insert is made between codons 11 and 36.

Bacteriophage λ

Essential features

Bacteriophage λ is a genetically complex but very extensively studied virus of *E. coli* (Box 4.1). Because it has been the object of so much molecular-genetical research, it was natural that, right from the beginnings of gene manipulation, it should have been investigated and developed as a vector. The DNA of phage λ, in the form in which it is isolated from the phage particle, is a linear duplex molecule of about 48.5 kbp. The entire DNA sequence has been determined (Sanger *et al.* 1982). At each end are short single-stranded 5' projections of 12 nucleotides, which are complementary in sequence and by which the DNA adopts a circular structure when it is injected into its host cell, i.e. λ DNA naturally has cohesive termini, which associate to form the *cos* site.

Functionally related genes of phage λ are clustered together on the map, except for the two positive regulatory genes *N* and *Q*. Genes on the left of the conventional linear map (Fig. 4.10) code for head and tail proteins of the phage particle. Genes of the central region are concerned with recombination (e.g. *red*) and the process of lysogenization, in which the circularized chromosome is inserted into its host chromosome and stably replicated along with it as a prophage. Much of this central region, including these genes, is not essential for phage growth and can be deleted or replaced without seriously impairing the infectious growth cycle. Its dispensability is crucially important, as will become apparent later, in the construction of vector derivatives of the phage. To the right of the central region are genes concerned with regulation and prophage immunity to superinfection (*N*, *cro*, *cI*), followed by DNA synthesis (*O*, *P*), late function regulation (*Q*) and host cell lysis (*S*, *R*). Figure 4.11 illustrates the λ life cycle.

Promoters and control circuits

As we shall see, it is possible to insert foreign DNA into the chromosome of phage-λ derivative and, in some cases, foreign genes can be expressed efficiently via λ promoters. We must therefore briefly

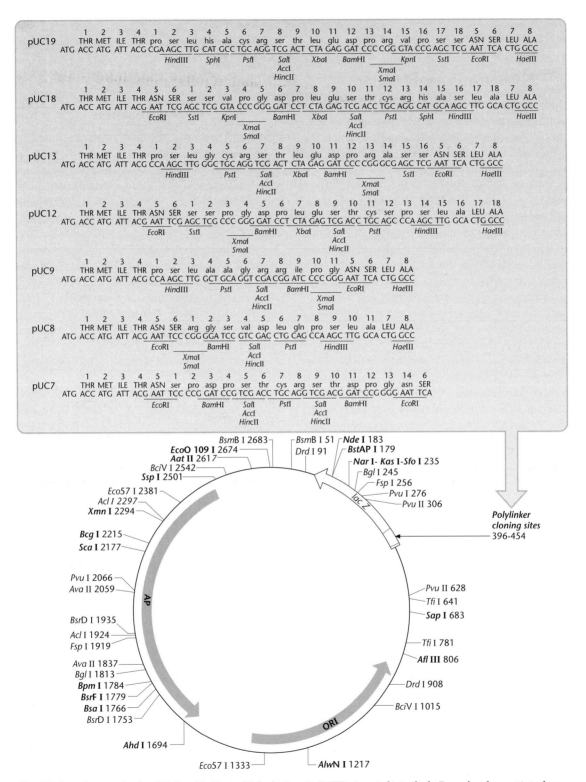

Fig. 4.9 Genetic maps of some pUC plasmids. The multiple cloning site (MCS) is inserted into the *lacZ* gene but does not interfere with gene function. The additional codons present in the *lacZ* gene as a result of the polylinker are labelled with lower-case letters. These polylinker regions (MCS) are identical to those of the M13 mp series of vectors (see p. 63).

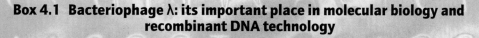

Box 4.1 Bacteriophage λ: its important place in molecular biology and recombinant DNA technology

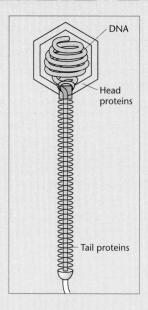

In the early 1950s, following some initial studies on *Bacillus megaterium*, André Lwoff and his colleagues at the Institut Pasteur described the phenomenon of lysogeny in *E. coli*. It became clear that certain strains of *E. coli* were lysogenized by phage, that is to say, these bacteria harboured phage λ in a dormant form, called a prophage. The lysogenic bacteria grew normally and might easily not have been recognized as lysogenic. However, when Lwoff exposed the bacteria to a moderate dose of ultraviolet light, the bacteria stopped growing, and after about 90 min of incubation the bacteria lysed, releasing a crop of phage into the medium.

The released phage were incapable of infecting more *E. coli* than had been lysogenized by phage λ (this is called immunity to superinfection), but non-lysogenic bacteria could be infected to yield another crop of virus. Not every non-lysogenic bacterium yielded virus; some bacteria were converted into lysogens because the phage switched to the dormant lifestyle – becoming prophage – rather than causing a lytic infection.

By the mid-1950s it was realized that the prophage consisted of a phage λ genome that had become integrated into the *E. coli* chromosome. It was also apparent to Lwoff's colleagues, Jacob and Monod, that the switching between the two states of the virus – the lytic and lysogenic lifestyles – was an example of a fundamental aspect of genetics that was gaining increasing attention, gene regulation.

Intensive genetic and molecular biological analysis of the phage, mainly in the 1960s and 1970s, led to a good understanding of the virus. The key molecule in maintaining the dormancy of the prophage and in conferring immunity to superinfection is the phage repressor, which is the product of the phage *cI* gene. In 1967 the phage repressor was isolated by Mark Ptashne (Ptashne 1967a,b). The advanced molecular genetics of the phage made it a good candidate for development as a vector, beginning in the 1970s and continuing to the present day, as described in the text. The development of vectors exploited the fact that a considerable portion of the phage genome encodes functions that are not needed for the infectious cycle. The ability to package recombinant phage DNA into virus particles *in vitro* was an important development for library construction (Hohn & Murray 1977).

A landmark in molecular biology was reached when the entire sequence of the phage λ genome, 48 502 nucleotide pairs, was determined by Fred Sanger and his colleagues (Sanger *et al.* 1982).

consider the promoters and control circuits affecting λ gene expression (see Ptashne (1992) for an excellent monograph on phage-λ control circuits).

In the lytic cycle, λ transcription occurs in three temporal stages: early, middle and late. Basically, early gene transcription establishes the lytic cycle (in competition with lysogeny), middle gene products replicate and recombine the DNA and late gene products package this DNA into mature phage particles. Following infection of a sensitive host, early transcription proceeds from major promoters situated immediately to the left (P_L) and right (P_R) of

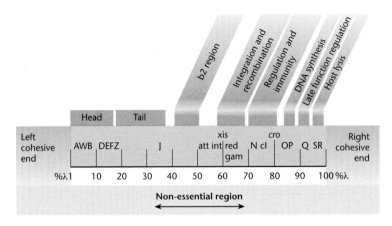

Fig. 4.10 Map of the λ chromosome, showing the physical position of some genes on the full-length DNA of wild-type bacteriophage λ. Clusters of functionally related genes are indicated.

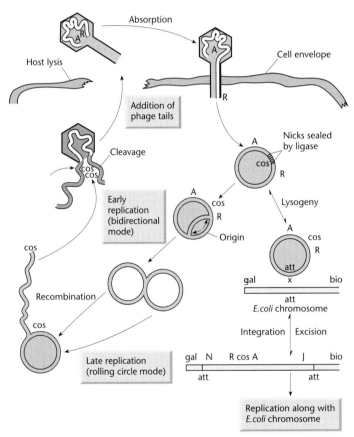

Fig. 4.11 Replication of phage-λ DNA in lytic and lysogenic cycles.

the repressor gene (*cI*) (Fig. 4.12). This transcription is subject to repression by the product of the *cI* gene and in a lysogen this repression is the basis of immunity to superinfecting λ. Early in infection, transcripts from P_L and P_R stop at termination sites t_L and t_{R1}. The site t_{R2} stops any transcripts that escape beyond t_{R1}. Lambda switches from early- to middle-

stage transcription by anti-termination. The *N* gene product, expressed from P_L, directs this switch. It interacts with RNA polymerase and, antagonizing the action of host termination protein ρ, permits it to ignore the stop signals so that P_L and P_R transcripts extend into genes such as *red*, *O* and *P* necessary for the middle stage. The early and middle transcripts

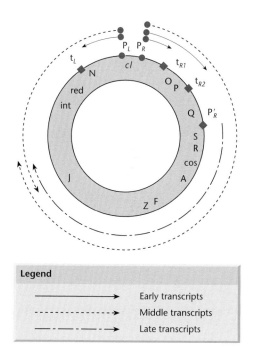

Fig. 4.12 Major promoters and transcriptional termination sites of phage λ. (See text for details.)

and patterns of expression therefore overlap. The *cro* product, when sufficient has accumulated, prevents transcription from P_L and P_R. The gene Q is expressed from the distal portion of the extended P_R transcript and is responsible for the middle-to-late switch. This also operates by anti-termination. The Q product specifically anti-terminates the short P_R transcript, extending it into the late genes, across the cohered *cos* region, so that many mature phage particles are ultimately produced.

Both N and Q play positive regulatory roles essential for phage growth and plaque formation; but an N^- phage can produce a small plaque if the termination site t_{R2} is removed by a small deletion termed *nin* (*N*-independent) as in λN^- *nin*.

Vector DNA

Wild-type λ DNA contains several target sites for most of the commonly used restriction endonucleases and so is not itself suitable as a vector. Derivatives of the wild-type phage have therefore been produced that either have a single target site at which foreign DNA can be inserted (*insertional* vectors) or have a pair of sites defining a fragment

that can be removed (*stuffer*) and replaced by foreign DNA (*replacement* vectors). Since phage λ can accommodate only about 5% more than its normal complement of DNA, vector derivatives are constructed with deletions to increase the space within the genome. The shortest λ DNA molecules that produce plaques of nearly normal size are 25% deleted. Apparently, if too much non-essential DNA is deleted from the genome, it cannot be packaged into phage particles efficiently. This can be turned to advantage for, if the replaceable fragment of a replacement-type vector is either removed by physical separation or effectively destroyed by treatment with a second restriction endonuclease that cuts it alone, then the deleted vector genome can give rise to plaques only if a new DNA segment is inserted into it. This amounts to positive selection for recombinant phage carrying foreign DNA.

Many vector derivatives of both the insertional and replacement types were produced by several groups of researchers early in the development of recombinant DNA technology (e.g. Thomas *et al.* 1974, Murray & Murray 1975, Blattner *et al.* 1977, Leder *et al.* 1977). Most of these vectors were constructed for use with *Eco*RI, *Bam*HI or *Hind*III, but their application could be extended to other endonucleases by the use of linker molecules. These early vectors have been largely superseded by improved vectors for rapid and efficient genomic or complementary DNA (cDNA) library construction (see Chapter 6).

Improved phage-λ vectors

As with plasmid vectors, improved phage-vector derivatives have been developed. There have been several aims, among which are the following.
• To increase the capacity for foreign DNA fragments, preferably for fragments generated by any one of several restriction enzymes (reviewed by Murray 1983).
• To devise methods for positively selecting recombinant formation.
• To allow RNA probes to be conveniently prepared by transcription of the foreign DNA insert; this facilitates the screening of libraries in chromosome walking procedures. An example of a vector with this property is λZAP (see p. 93).
• To develop vectors for the insertion of eukaryotic cDNA (p. 93) such that expression of the cDNA, in

the form of a fusion polypeptide with β-galactosidase, is driven in *E. coli*; this form of expression vector is useful in antibody screening. An example of such a vector is λgt11.

The first two points will be discussed here. The discussion of improved vectors in library construction and screening is deferred until Chapter 6.

The maximum capacity of phage-λ derivatives can only be attained with vectors of the replacement type, so that there has also been an accompanying incentive to devise methods for positively selecting recombinant formation without the need for prior removal of the stuffer fragment. Even when steps are taken to remove the stuffer fragment by physical purification of vector arms, small contaminating amounts may remain, so that genetic selection for recombinant formation remains desirable. The usual method of achieving this is to exploit the Spi⁻ phenotype.

Wild-type λ cannot grow on *E. coli* strains lysogenic for phage P2; in other words, the λ phage is Spi⁺ (sensitive to P2 inhibition). It has been shown that the products of λ genes *red* and *gam*, which lie in the region 64–69% on the physical map, are responsible for the inhibition of growth in a P2 lysogen (Herskowitz 1974, Sprague *et al.* 1978, Murray 1983). Hence vectors have been derived in which the stuffer fragment includes the region 64–69%, so that recombinants in which this has been replaced by foreign DNA are phenotypically Spi⁻ and can be positively selected by plating on a P2 lysogen (Karn *et al.* 1986, Loenen & Brammar 1980).

Deletion of the *gam* gene has other consequences. The *gam* product is necessary for the normal switch in λ DNA replication from the bidirectional mode to the rolling-circle mode (see Fig. 4.11). Gam⁻ phage cannot generate the concatemeric linear DNA which is normally the substrate for packaging into phage heads. However, *gam*⁻ phage do form plaques because the *rec* and *red* recombination systems act on circular DNA molecules to form multimers, which can be packaged. *gam*⁻ *red*⁻ phage are totally dependent upon *rec*-mediated exchange for plaque formation on *rec*⁺ bacteria. λ DNA is a poor substrate for this *rec*-mediated exchange. Therefore, such phage make vanishingly small plaques unless they contain one or more short DNA sequences called *chi* (crossover hot-spot instigator) sites, which stimulate *rec*-mediated exchange. Many of the current replacement vectors generate *red*⁻ *gam*⁻ clones and

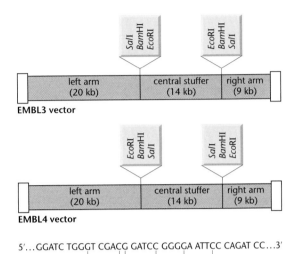

Fig. 4.13 The structure of bacteriophage λ cloning vectors EMBL3 and EMBL4. The polylinker sequence is present in opposite orientation in the two vectors.

so have been constructed with a *chi* site within the non-replaceable part of the phage.

The most recent generation of λ vectors, which are based on EMBL3 and EMBL4 (Frischauf *et al.* 1983; Fig. 4.13), have a capacity for DNA of size 9–23 kb. As well as being *chi*⁺, they have polylinkers flanking the stuffer fragment to facilitate library construction. Phages with inserts can be selected on the basis of their Spi⁻ phenotype, but there is an alternative. The vector can be digested with *Bam*HI and *Eco*RI prior to ligation with foreign DNA fragments produced with *Bam*HI. If the small *Bam*HI–*Eco*RI fragments from the polylinkers are removed, the stuffer fragment will not be reincorporated.

Packaging phage-λ DNA *in vitro*

So far, we have considered only one way of introducing manipulated phage DNA into the host bacterium, i.e. by transfection of competent bacteria (see Chapter 2). Using freshly prepared λ DNA that has not been subjected to any gene-manipulation procedures, transfection will result typically in about 10^5 plaques/μg of DNA. In a gene-manipulation experiment in which the vector DNA is restricted and then ligated with foreign DNA, this figure is reduced to about 10^4–10^3 plaques/μg of vector DNA. Even with perfectly efficient nucleic acid

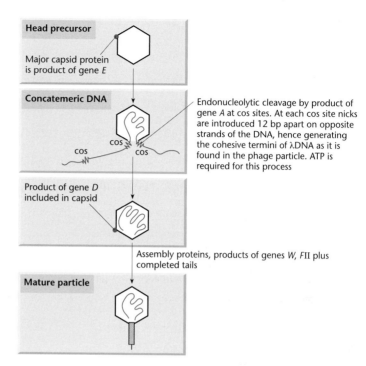

Head precursor

Major capsid protein is product of gene *E*

Concatemeric DNA

Endonucleolytic cleavage by product of gene *A* at cos sites. At each cos site nicks are introduced 12 bp apart on opposite strands of the DNA, hence generating the cohesive termini of λDNA as it is found in the phage particle. ATP is required for this process

cos cos cos cos

Product of gene *D* included in capsid

Assembly proteins, products of genes *W*, *FII* plus completed tails

Mature particle

Fig. 4.14 Simplified scheme showing packaging of phage-λ DNA into phage particles.

biochemistry, some of this reduction is inevitable. It is a consequence of the random association of fragments in the ligation reaction, which produces molecules with a variety of fragment combinations, many of which are inviable. Yet, in some contexts, 10^6 or more recombinants are required. The scale of such experiments can be kept within a reasonable limit by packaging the recombinant DNA into mature phage particles *in vitro*.

Placing the recombinant DNA in a phage coat allows it to be introduced into the host bacteria by the normal processes of phage infection, i.e. phage adsorption followed by DNA injection. Depending upon the details of the experimental design, packaging *in vitro* yields about 10^6 plaques/μg of vector DNA after the ligation reaction.

Figure 4.14 shows some of the events occurring during the packaging process that take place within the host during normal phage growth and which we now require to perform *in vitro*. Phage DNA in concatemeric form, produced by a rolling-circle replication mechanism (see Fig. 4.11), is the substrate for the packaging reaction. In the presence of phage head precursor (the product of gene *E* is the major capsid protein) and the product of gene *A*, the concatemeric DNA is cleaved into monomers and

encapsidated. Nicks are introduced in opposite strands of the DNA, 12 nucleotide pairs apart at each *cos* site, to produce the linear monomer with its cohesive termini. The product of gene *D* is then incorporated into what now becomes a completed phage head. The products of genes *W* and *FII*, among others, then unite the head with a separately assembled tail structure to form the mature particle.

The principle of packaging *in vitro* is to supply the ligated recombinant DNA with high concentrations of phage-head precursor, packaging proteins and phage tails. Practically, this is most efficiently performed in a very concentrated mixed lysate of two induced lysogens, one of which is blocked at the pre-head stage by an amber mutation in gene *D* and therefore accumulates this precursor, while the other is prevented from forming any head structure by an amber mutation in gene *E* (Hohn & Murray 1977). In the mixed lysate, genetic complementation occurs and exogenous DNA is packaged (Fig. 4.15). Although concatemeric DNA is the substrate for packaging (covalently joined concatemers are, of course, produced in the ligation reaction by association of the natural cohesive ends of λ), the *in vitro* system will package added monomeric DNA, which presumably first concatemerizes non-covalently.

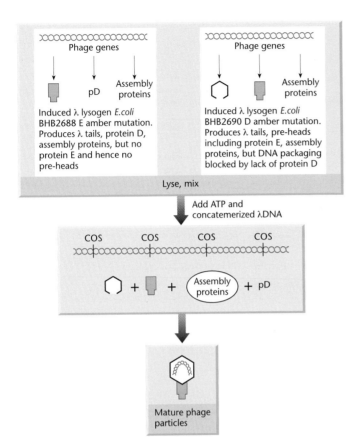

Fig. 4.15 *In vitro* packaging of concatemerized phage-λ DNA in a mixed lysate.

There are two potential problems associated with packaging *in vitro*. First, endogenous DNA derived from the induced prophages of the lysogens used to prepare the packaging lysate can itself be packaged. This can be overcome by choosing the appropriate genotype for these prophages, i.e. excision upon induction is inhibited by the *b2* deletion (Gottesmann & Yarmolinsky 1968) and *imm 434* immunity will prevent plaque formation if an *imm 434* lysogenic bacterium is used for plating the complex reaction mixture. Additionally, if the vector does not contain any amber mutation a non-suppressing host bacterium can be used so that endogenous DNA will not give rise to plaques. The second potential problem arises from recombination in the lysate between exogenous DNA and induced prophage markers. If troublesome, this can be overcome by using recombination-deficient (i.e. *red⁻ rec⁻*) lysogens and by UV-irradiating the cells used to prepare the lysate, so eliminating the biological activity of the endogenous DNA (Hohn & Murray 1977).

DNA cloning with single-stranded DNA vectors

M13, f1 and fd are filamentous coliphages containing a circular single-stranded DNA molecule. These coliphages have been developed as cloning vectors, for they have a number of advantages over other vectors, including the other two classes of vector for *E. coli*, plasmids and phage λ. However, in order to appreciate their advantages, it is essential to have a basic understanding of the biology of filamentous phages.

The biology of the filamentous coliphages

The phage particles have dimensions 900 nm × 9 nm and contain a single-stranded circular DNA molecule, which is 6407 (M13) or 6408 (fd) nucleotides long. The complete nucleotide sequences of fd and M13 are available and they are 97% homologous. The differences consist mainly of isolated nucleotides here and there, mostly affecting the redundant bases of codons, with no blocks of sequence divergence. Sequencing of f1 DNA indicates that it is very similar to M13 DNA.

The filamentous phages only infect strains of enteric bacteria harbouring F pili. The adsorption site appears to be the end of the F pilus, but exactly how the phage genome gets from the end of the F pilus to the inside of the cell is not known. Replication of phage DNA does not result in host-cell lysis. Rather, infected cells continue to grow and divide, albeit at a slower rate than uninfected cells, and extrude virus particles. Up to 1000 phage particles may be released into the medium per cell per generation (Fig. 4.16).

The single-stranded phage DNA enters the cell by a process in which decapsidation and replication are tightly coupled. The capsid proteins enter the cytoplasmic membrane as the viral DNA passes into the cell while being converted to a double-stranded replicative form (RF). The RF multiplies rapidly until about 100 RF molecules are formed inside the cell. Replication of the RF then becomes asymmetric, due to the accumulation of a viral-encoded single-stranded specific DNA-binding protein. This protein binds to the viral strand and prevents synthesis of the complementary strand. From this point on, only viral single strands are synthesized. These progeny single strands are released from the cell as filamentous particles following morphogenesis at the cell membrane. As the DNA passes through the membrane, the DNA-binding protein is stripped off and replaced with capsid protein.

Why use single-stranded vectors?

For several applications of cloned DNA, single-stranded DNA is required. Sequencing by the original dideoxy method required single-stranded DNA, as do techniques for oligonucleotide-directed mutagenesis and certain methods of probe preparation. The use of vectors that occur in single-stranded form is an attractive means of combining the cloning, amplification and strand separation of an originally double-stranded DNA fragment.

As single-stranded vectors, the filamentous phages have a number of advantages. First, the phage DNA is replicated via a double-stranded circular DNA (RF) intermediate. This RF can be purified and

manipulated *in vitro* just like a plasmid. Secondly, both RF and single-stranded DNA will transfect competent *E. coli* cells to yield either plaques or infected colonies, depending on the assay method. Thirdly, the size of the phage particle is governed by the size of the viral DNA and therefore there are no packaging constraints. Indeed, viral DNA up to six times the length of M13 DNA has been packaged (Messing *et al.* 1981). Finally, with these phages it is very easy to determine the orientation of an insert. Although the relative orientation can be determined from restriction analysis of RF, there is an easier method (Barnes 1980). If two clones carry the insert in opposite directions, the single-stranded DNA from them will hybridize and this can be detected by agarose gel electrophoresis. Phage from as little as 0.1 ml of culture can be used in assays of this sort, making mass screening of cultures very easy.

In summary, as vectors, filamentous phages possess all the advantages of plasmids while producing particles containing single-stranded DNA in an easily obtainable form.

Development of filamentous phage vectors

Unlike λ, the filamentous coliphages do not have any non-essential genes which can be used as cloning sites. However, in M13 there is a 507 bp intergenic region, from position 5498 to 6005 of the DNA sequence, which contains the origins of DNA replication for both the viral and the complementary strands. In most of the vectors developed so far, foreign DNA has been inserted at this site, although it is possible to clone at the carboxy-terminal end of gene IV (Boeke *et al.* 1979). The wild-type phages are not very promising as vectors because they contain very few unique sites within the intergenic region: *Asu*I in the case of fd, and *Asu*I and *Ava*I in the case of M13.

The first example of M13 cloning made use of one of 10 *Bsu*I sites in the genome, two of which are in the intergenic region (Messing *et al.* 1977). For cloning, M13 RF was partially digested with *Bsu*I and linear full-length molecules isolated by agarose gel electrophoresis. These linear monomers were blunt-end-ligated to a *Hin*dII restriction fragment comprising the *E. coli lac* regulatory region and the genetic information for the α-peptide of β-galactosidase. The complete ligation mixture was used to

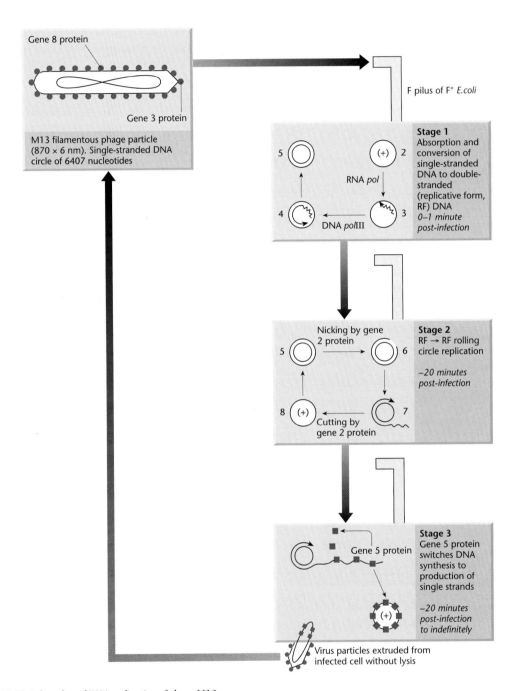

Fig. 4.16 Life cycle and DNA replication of phage M13.

transform a strain of *E. coli* with a deletion of the β-galactosidase α-fragment and recombinant phage detected by intragenic complementation on media containing IPTG and Xgal (see Box 3.2 on p. 35).

One of the blue plaques was selected and the virus in it designated M13 mp1.

Insertion of DNA fragments into the *lac* region of M13 mp1 destroys its ability to form blue plaques,

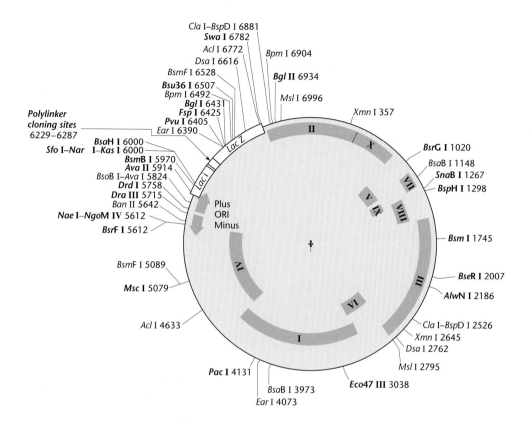

1

(Met) Thr Met Ile Thr Asn Ser Ser Ser Val Pro Gly Asp Pro Leu Glu Ser Thr Cys Arg His Ala Ser Leu Ala → LacZ´

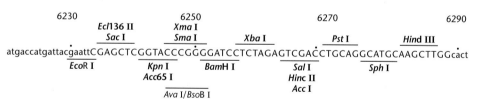

Fig. 4.17 Restriction map of cloning vector M13mp18. Phages M13mp18 and M13mp19 are 7249 bases in length and differ only in the orientation of the 54-base polylinker that they carry. The map shows the restriction sites of enzymes that cut the molecule once or twice. The unique sites are shown in bold type.

making detection of recombinants easy. However, the *lac* region only contains unique sites for *Ava*II, *Bgl*I and *Pvu*I and three sites for *Pvu*II, and there are no sites anywhere on the complete genome for the commonly used enzymes such as *Eco*RI or *Hind*III. To remedy this defect, Gronenborn and Messing (1978) used *in vitro* mutagenesis to change a single

base pair, thereby creating a unique *Eco*RI site within the *lac* fragment. This variant was designated M13 mp2. This phage derivative was further modified to generate derivatives with polylinkers upstream of the lac α-fragment (Fig. 4.17). These derivatives (mp7–mp11, mp18, mp19) are the exact M13 counterparts of the pUC plasmids shown in Fig. 4.9.

CHAPTER 5

Cosmids, phasmids and other advanced vectors

Introduction

In the 1970s, when recombinant DNA technology was first being developed, only a limited number of vectors were available and these were based on either high-copy-number plasmids or phage λ. Later, phage M13 was developed as a specialist vector to facilitate DNA sequencing. Gradually, a number of purpose-built vectors were developed, of which pBR322 is probably the best example, but the creation of each one was a major task. Over time, a series of specialist vectors was constructed, each for a particular purpose. During this period, there were many arguments about the relative benefits of plasmid and phage vectors. Today, the molecular biologist has available an enormous range of vectors and these are notable for three reasons. First, many of them combine elements from both plasmids and phages and are known as *phasmids* or, if they contain an M13 *ori* region, *phagemids*. Secondly, many different features that facilitate cloning and expression can be found combined in a single vector. Thirdly, purified vector DNA plus associated reagents can be purchased from molecular-biology suppliers. The hapless scientist who opens a molecular-biology catalogue is faced with a bewildering selection of vectors and each vender promotes different ones. Although the benefits of using each vector may be clear, the disadvantages are seldom obvious. The aim of this chapter is to provide the reader with a detailed explanation of the biological basis for the different designs of vector.

There are two general uses for cloning vectors: cloning large pieces of DNA and manipulating genes. When mapping and sequencing genomes, the first step is to subdivide the genome into manageable pieces. The larger these pieces, the easier it is to construct the final picture (see Chapter 7); hence the need to clone large fragments of DNA. Large fragments are also needed if it is necessary to 'walk'

along the genome to isolate a gene, and this topic is covered in Chapter 6. In many instances, the desired gene will be relatively easy to isolate and a simpler cloning vector can be used. Once isolated, the cloned gene may be expressed as a probe sequence or as a protein, it may be sequenced or it may be mutated *in vitro*. For all these applications, small specialist vectors are used.

Vectors for cloning large fragments of DNA

Cosmid vectors

As we have seen, concatemers of unit-length λ DNA molecules can be efficiently packaged if the *cos* sites, substrates for the packaging-dependent cleavage, are 37–52 kb apart (75–105% the size of λ$^+$ DNA). In fact, only a small region in the proximity of the *cos* site is required for recognition by the packaging system (Hohn 1975).

Plasmids have been constructed which contain a fragment of λ DNA including the *cos* site (Collins & Brüning 1978, Collins & Hohn 1979, Wahl *et al.* 1987, Evans *et al.* 1989). These plasmids have been termed *cosmids* and can be used as gene-cloning vectors in conjunction with the *in vitro* packaging system. Figure 5.1 shows a gene-cloning scheme employing a cosmid. Packaging the cosmid recombinants into phage coats imposes a desirable selection upon their size. With a cosmid vector of 5 kb, we demand the insertion of 32–47 kb of foreign DNA – much more than a phage-λ vector can accommodate. Note that, after packaging *in vitro*, the particle is used to infect a suitable host. The recombinant cosmid DNA is injected and circularizes like phage DNA but replicates as a normal plasmid without the expression of any phage functions. Transformed cells are selected on the basis of a vector drug-resistance marker.

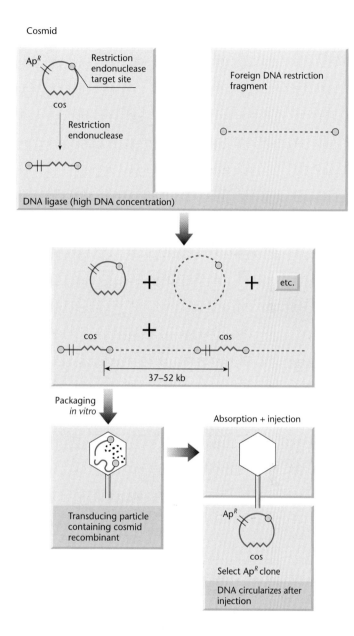

Fig. 5.1 Simple scheme for cloning in a cosmid vector. (See text for details.)

Cosmids provide an efficient means of cloning large pieces of foreign DNA. Because of their capacity for large fragments of DNA, cosmids are particularly attractive vectors for constructing libraries of eukaryotic genome fragments. Partial digestion with a restriction endonuclease provides suitably large fragments. However, there is a potential problem associated with the use of partial digests in this way. This is due to the possibility of two or more genome fragments joining together in the ligation reaction, hence creating a clone containing fragments that were not initially adjacent in the genome. This would give an incorrect picture of their chromosomal organization. The problem can be overcome by size fractionation of the partial digest.

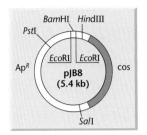

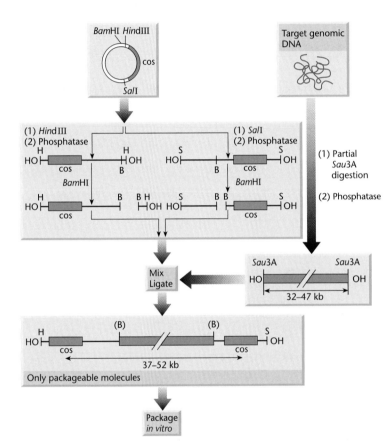

Fig. 5.2 Cosmid cloning scheme of Ish-Horowicz and Burke (1981). (a) Map of cosmid pJB8. (b) Application to the construction of a genomic library of fragments obtained by partial digestion with *Sau*3A. This restriction endonuclease has a tetranucleotide recognition site and generates fragments with the same cohesive termini as *Bam*HI (see p. 31).

Even with sized foreign DNA, in practice cosmid clones may be produced that contain non-contiguous DNA fragments ligated to form a single insert. The problem can be solved by dephosphorylating the foreign DNA fragments so as to prevent their ligation together. This method is very sensitive to the exact ratio of target-to-vector DNAs (Collins & Brüning 1978) because vector-to-vector ligation can occur. Furthermore, recombinants with a duplicated vector are unstable and break down in the host by recombination, resulting in the propagation of a non-recombinant cosmid vector.

Such difficulties have been overcome in a cosmid-cloning procedure devised by Ish-Horowicz and Burke (1981). By appropriate treatment of the cosmid vector pJB8 (Fig. 5.2), left-hand and right-hand vector ends are purified which are incapable of self-ligation but which accept dephosphorylated foreign DNA. Thus the method eliminates the need to size the foreign DNA fragments and prevents formation

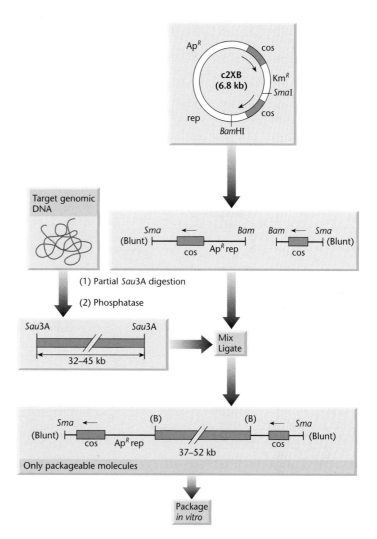

Fig. 5.3 Cosmid cloning scheme of Bates and Swift (1983). The cosmid c2XB contains two *cos* sites, separated by a site for the restriction endonuclease *Sma*I which creates blunt ends. These blunt ends ligate only very inefficiently under the conditions used and effectively prevent the formation of recombinants containing multiple copies of the vector.

of clones containing short foreign DNA or multiple vector sequences.

An alternative solution to these problems has been devised by Bates and Swift (1983) who have constructed cosmid c2XB. This cosmid carries a *Bam*HI insertion site and two *cos* sites separated by a blunt-end restriction site (Fig. 5.3). The creation of these blunt ends, which ligate only very inefficiently under the conditions used, effectively prevents vector self-ligation in the ligation reaction.

Modern cosmids of the pWE and sCos series (Wahl *et al.* 1987, Evans *et al.* 1989) contain features such as: (i) multiple cloning sites (Bates & Swift 1983, Pirrotta *et al.* 1983, Breter *et al.* 1987) for simple cloning using non-size-selected DNA; (ii) phage promoters flanking the cloning site; and (iii) unique *Not*I, *Sac*II or *Sfi*I sites (rare cutters, see Chapter 6) flanking the cloning site to permit removal of the insert from the vector as single fragments. Mammalian expression modules encoding dominant selectable markers (Chapter 10) may also be present, for gene transfer to mammalian cells if required.

BACs and PACs as alternatives to cosmids

Phage P1 is a temperate bacteriophage which has been extensively used for genetic analysis of *Escherichia coli* because it can mediate generalized transduction. Sternberg and co-workers have developed a P1 vector system which has a capacity for

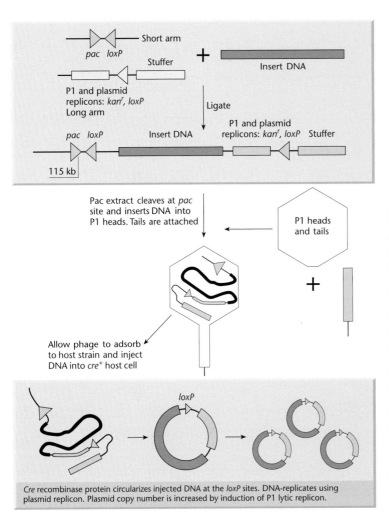

Pac extract cleaves at *pac* site and inserts DNA into P1 heads. Tails are attached

P1 heads and tails

Allow phage to adsorb to host strain and inject DNA into *cre*⁺ host cell

loxP

Cre recombinase protein circularizes injected DNA at the *loxP* sites. DNA-replicates using plasmid replicon. Plasmid copy number is increased by induction of P1 lytic replicon.

Fig. 5.4 The phage P1 vector system. The P1 vector Ad10 (Sternberg 1990) is digested to generate short and long vector arms. These are dephosphorylated to prevent self-ligation. Size-selected insert DNA (85–100 kb) is ligated with vector arms, ready for a two-stage processing by packaging extracts. First, the recombinant DNA is cleaved at the *pac* site by pacase in the packaging extract. Then the pacase works in concert with head/tail extract to insert DNA into phage heads, *pac* site first, cleaving off a headful of DNA at 115 kb. Heads and tails then unite. The resulting phage particle can inject recombinant DNA into host *E. coli*. The host is *cre*⁺. The *cre* recombinase acts on *loxP* sites to produce a circular plasmid. The plasmid is maintained at low copy number, but can be amplified by inducing the P1 lytic operon.

DNA fragments as large as 100 kb (Sternberg 1990, Pierce *et al.* 1992). Thus the capacity is about twice that of cosmid clones but less than that of yeast artificial chromosome (YAC) clones (see p. 159). The P1 vector contains a packaging site (*pac*) which is necessary for *in vitro* packaging of recombinant molecules into phage particles. The vectors contain two *loxP* sites. These are the sites recognized by the phage recombinase, the product of the phage *cre* gene, and which lead to circularization of the packaged DNA after it has been injected into an *E. coli* host expressing the recombinase (Fig. 5.4). Clones are maintained in *E. coli* as low-copy-number plasmids by selection for a vector kanamycin-resistance marker. A high copy number can be induced by

exploitation of the P1 lytic replicon (Sternberg 1990). This P1 system has been used to construct genomic libraries of mouse, human, fission yeast and *Drosophila* DNA (Hoheisel *et al.* 1993, Hartl *et al.* 1994).

Shizuya *et al.* (1992) have developed a bacterial cloning system for mapping and analysis of complex genomes. This BAC system (bacterial artificial chromosome) is based on the single-copy sex factor F of *E. coli*. This vector (Fig. 5.5) includes the λ *cos* N and P1 *loxP* sites, two cloning sites (*Hind*III and *Bam*HI) and several G+C restriction enzyme sites (e.g. *Sfi*I, *Not*I, etc.) for potential excision of the inserts. The cloning site is also flanked by T7 and SP6 promoters for generating RNA probes. This BAC can be transformed into *E. coli* very efficiently, thus avoiding the

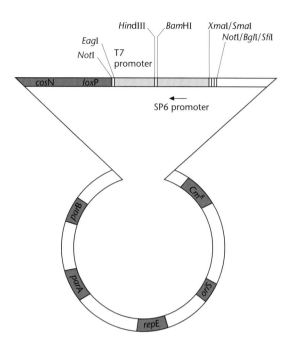

Fig. 5.5 Structure of a BAC vector derived from a mini-F plasmid. The *ori*S and *rep*E genes mediate the unidirectional replication of the F factor, while *par*A and *par*B maintain the copy number at a level of one or two per genome. Cm^R is a chloramphenicol-resistance marker. *cos*N and *lox*P are the cleavage sites for λ terminase and P1 *cre* protein, respectively. *Hind*III and *Bam*HI are unique cleavage sites for inserting foreign DNA. (Adapted from Shizuya *et al.* 1992.)

packaging extracts that are required with the P1 system. BACs are capable of maintaining human and plant genomic fragments of greater than 300 kb for over 100 generations with a high degree of stability (Woo *et al.* 1994) and have been used to construct genome libraries with an average insert size of 125 kb (Wang *et al.* 1995a). Subsequently, Ioannou *et al.* (1994) have developed a P1-derived artificial chromosome (PAC), by combining features of both the P1 and the F-factor systems. Such PAC vectors are able to handle inserts in the 100–300 kb range.

The first BAC vector, pBAC108L, lacked a selectable marker for recombinants. Thus, clones with inserts had to be identified by colony hybridization. While this once was standard practice in gene manipulation work, today it is considered to be inconvenient! Two widely used BAC vectors, pBeloBAC11 and pECBAC1, are derivatives of pBAC108L in

which the original cloning site is replaced with a *lac*Z gene carrying a multiple cloning site (Kim *et al.* 1996, Frijters *et al.* 1997). pBeloBAC11 has two *Eco*RI sites, one in the *lac*Z gene and one in the CM^R gene, wheras pECBAC1 has only the *Eco*RI site in the *lac*Z gene. Further improvements to BACs have been made by replacing the *lac*Z gene with the *sac*B gene (Hamilton *et al.* 1996). Insertional inactivation of *sac*B permits growth of the host cell on sucrose-containing media, i.e. positive selection for inserts. Frengen *et al.* (1999) have further improved these BACs by including a site for the insertion of a transposon. This enables genomic inserts to be modified after cloning in bacteria, a procedure known as *retrofitting*. The principal uses of retrofitting are the simplified introduction of deletions (Chatterjee & Coren 1997) and the introduction of reporter genes for use in the original host of the genomic DNA (Wang *et al.* 2001).

Choice of vector

The maximum size of insert that the different vectors will accommodate is shown in Table 5.1. The size of insert is not the only feature of importance. The absence of chimeras and deletions is even more important. In practice, some 50% of YACs show structural instability of inserts or are chimeras in which two or more DNA fragments have become incorporated into one clone. These defective YACs are unsuitable for use as mapping and sequencing reagents and a great deal of effort is required to identify them. Cosmid inserts sometimes contain

Table 5.1 Maximum DNA insert possible with different cloning vectors. YACs are discussed on p. 159.

Vector	Host	Insert size
λ phage	*E. coli*	5–25 kb
λ cosmids	*E. coli*	35–45 kb
P1 phage	*E. coli*	70–100 kb
PACs	*E. coli*	100–300 kb
BACs	*E coli*	≤ 300 kb
YACs	*Saccharomyces cerevisiae*	200–2000 kb

the same aberrations and the greatest problem with them arises when the DNA being cloned contains tandem arrays of repeated sequences. The problem is particularly acute when the tandem array is several times larger than the allowable size of a cosmid insert. Potential advantages of the BAC and PAC systems over YACs include lower levels of chimerism (Hartl *et al.* 1994, Sternberg 1994), ease of library generation and ease of manipulation and isolation of insert DNA. BAC clones seem to represent human DNA far more faithfully than their YAC or cosmid counterparts and appear to be excellent substrates for shotgun sequence analysis, resulting in accurate contiguous sequence data (Venter *et al.* 1996).

Specialist-purpose vectors

Vectors that can be used to make single-stranded DNA for sequencing

Whenever a new gene is cloned or a novel genetic construct is made, it is usual practice to sequence all or part of the chimeric molecule. As will be seen later (p. 120), the Sanger method of sequencing requires single-stranded DNA as the starting material. Originally, single-stranded DNA was obtained by cloning the sequence of interest in an M13 vector (see p. 60). Today, it is more usual to clone the sequence into a pUC-based phagemid vector which contains the M13 *ori* region as well as the pUC (Col E1) origin of replication. Such vectors normally replicate inside the cell as double-stranded molecules. Single-stranded DNA for sequencing can be produced by infecting cultures with a helper phage such as M13K07. This helper phage has the origin of replication of P15A and a kanamycin-resistance gene inserted into the M13 *ori* region and carries a mutation in the *gII* gene (Vieira & Messing 1987). M13K07 can replicate on its own. However, in the presence of a phagemid bearing a wild-type origin of replication, single-stranded phagemid is packaged preferentially and secreted into the culture medium. DNA purified from the phagemids can be used directly for sequencing.

Expression vectors

Expression vectors are required if one wants to prepare RNA probes from the cloned gene or to purify large amounts of the gene product. In either case, transcription of the cloned gene is required. Although it is possible to have the cloned gene under the control of its own promoter, it is more usual to utilize a promoter specific to the vector. Such vector-carried promoters have been optimized for binding of the *E. coli* RNA polymerase and many of them can be regulated easily by changes in the growth conditions of the host cell.

E. coli RNA polymerase is a multi-subunit enzyme. The core enzyme consists of two identical α subunits and one each of the β and β' subunits. The core enzyme is not active unless an additional subunit, the σ factor, is present. RNA polymerase recognizes different types of promoters depending on which type of σ factor is attached. The most common promoters are those recognized by the RNA polymerase with σ^{70}. A large number of σ^{70} promoters from *E. coli* have been analysed and a compilation of over 300 of them can be found in Lisser and Margalit (1993). A comparison of these promoters has led to the formulation of a consensus sequence (Fig. 5.6). If the transcription start point is assigned the position +1 then

		-35 Region					-10 Region	
			1 2 3 4 5 6 7 8 9 10 11 12 13 14 15 16 17					
CONSENSUS	• • •	TTGACA	• • • • • • • • •	• • • • • • • • •	TATAAT	• •		
lac		GGCTTTACACTTTATGCTTCCGGCTCGTATATTGT						
trp		CTGTTGACAATTAATCAT CGAACTAGTTAACTAG						
λP$_L$		GTGTTGACATAAATACCA CTGGCGGTGATACTGA						
rec A		CACTTGATACTGTATGAA GCATACAGTATAATTG						
*tac*I		CTGTTGACAATTAATCAT CGGCTCGTATAATGT						
*tac*II		CTGTTGACAATTAATCAT CGAACTAGTTTAATGT						

Fig. 5.6 The base sequence of the −10 and −35 regions of four natural promoters, two hybrid promoters and the consensus promoter.

this consensus sequence consists of the −35 region (5′-TTGACA-) and the −10 region, or Pribnow box (5′-TATAAT). RNA polymerase must bind to both sequences to initiate transcription. The strength of a promoter, i.e. how many RNA copies are synthesized per unit time per enzyme molecule, depends on how close its sequence is to the consensus. While the −35 and −10 regions are the sites of nearly all mutations affecting promoter strength, other bases flanking these regions can affect promoter activity (Hawley & McClure 1983, Dueschle *et al.* 1986, Keilty & Rosenberg 1987). The distance between the −35 and −10 regions is also important. In all cases examined, the promoter was weaker when the spacing was increased or decreased from 17 bp.

Upstream (UP) elements located 5′ of the −35 hexamer in certain bacterial promoters are A+T-rich sequences that increase transcription by interacting with the α subunit of RNA polymerase. Gourse *et al.* (1998) have identified UP sequences conferring increased activity to the *rrn* core promoter. The best UP sequence was portable and increased heterologous protein expression from the *lac* promoter by a factor of 100.

Once RNA polymerase has initiated transcription at a promoter, it will polymerize ribonucleotides until it encounters a transcription-termination site in the DNA. Bacterial DNA has two types of transcription-termination site: factor-independent and factor-dependent. As their names imply, these types are distinguished by whether they work with just RNA polymerase and DNA alone or need other factors before they can terminate transcription. The factor-independent transcription terminators are easy to recognize because they have similar sequences: an inverted repeat followed by a string of A residues (Fig. 5.7). Transcription is terminated in the string of A residues, resulting in a string of

U residues at the 3′ end of the mRNA. The factor-dependent transcription terminators have very little sequence in common with each other. Rather, termination involves interaction with one of the three known *E. coli* termination factors, Rho (ρ), Tau (τ) and NusA. Most expression vectors incorporate a factor-independent termination sequence downstream from the site of insertion of the cloned gene.

Vectors for making RNA probes

Although single-stranded DNA can be used as a sequence probe in hybridization experiments, RNA probes are preferred. The reasons for this are that the rate of hybridization and the stability are far greater for RNA–DNA hybrids compared with DNA–DNA hybrids. To make an RNA probe, the relevant gene sequence is cloned in a plasmid vector such that it is under the control of a phage promoter. After purification, the plasmid is linearized with a suitable restriction enzyme and then incubated with the phage RNA polymerase and the four ribonucleoside triphosphates (Fig. 5.8). No transcription terminator is required because the RNA polymerase will fall off the end of the linearized plasmid.

There are three reasons for using a phage promoter. First, such promoters are very strong, enabling large amounts of RNA to be made *in vitro*. Secondly, the phage promoter is not recognized by the *E. coli* RNA polymerase and so no transcription will occur inside the cell. This minimizes any selection of variant inserts. Thirdly, the RNA polymerases encoded by phages such as SP6, T7 and T3 are much simpler molecules to handle than the *E. coli* enzyme, since the active enzyme is a single polypeptide.

If it is planned to probe RNA or single-stranded DNA sequences, then it is essential to prepare RNA probes corresponding to both strands of the insert. One way of doing this is to have two different clones corresponding to the two orientations of the insert. An alternative method is to use a cloning vector in which the insert is placed between two different, opposing phage promoters (e.g. T7/T3 or T7/SP6) that flank a multiple cloning sequence (see Fig. 5.5). Since each of the two promoters is recognized by a different RNA polymerase, the direction of transcription is determined by which polymerase is used.

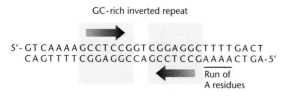

GC-rich inverted repeat

```
5′- GT CAAAAGCCT CCGGT CGGAGGCT T T T GACT
    CA GT T T T CGGAGGCCA GCCT CCGAAAA CT GA-5′
```
Run of
A residues

Fig. 5.7 Structure of a factor-independent transcriptional terminator.

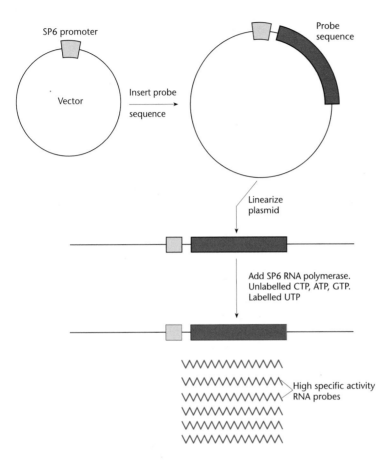

Fig. 5.8 Method for preparing RNA probes from a cloned DNA molecule using a phage SP6 promoter and SP6 RNA polymerase.

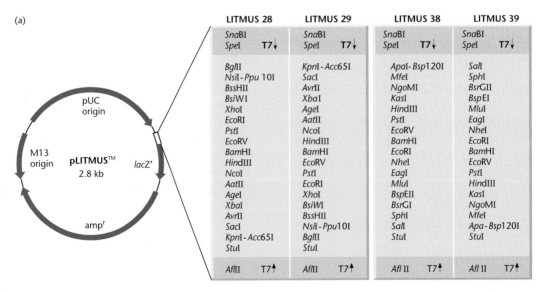

	LITMUS 28	LITMUS 29	LITMUS 38	LITMUS 39
	*Sna*BI	*Sna*BI	*Sna*BI	*Sna*BI
	*Spe*I T7↓	*Spe*I T7↓	*Spe*I T7↓	*Spe*I T7↓
	*Bgl*II	*Kpn*I - *Acc*65I	*Apa*I - *Bsp*120I	*Sal*I
	*Nsi*I - *Ppu*10I	*Sac*I	*Mfe*I	*Sph*I
	*Bss*HII	*Avr*II	*Ngo*MI	*Bsr*GII
	*Bsi*WI	*Xba*I	*Kas*I	*Bsp*EI
	*Xho*I	*Age*I	*Hind*III	*Mlu*I
	*Eco*RI	*Aat*II	*Pst*I	*Eag*I
	*Pst*I	*Nco*I	*Eco*RV	*Nhe*I
	*Eco*RV	*Hind*III	*Bam*HI	*Eco*RI
	*Bam*HI	*Bam*HI	*Eco*RI	*Bam*HI
	*Hind*III	*Eco*RV	*Nhe*I	*Eco*RV
	*Nco*I	*Pst*I	*Eag*I	*Pst*I
	*Aat*II	*Eco*RI	*Mlu*I	*Hind*III
	*Age*I	*Xho*I	*Bsp*EII	*Kas*I
	*Xba*I	*Bsi*WI	*Bsr*GI	*Ngo*MI
	*Avr*II	*Bss*HII	*Sph*I	*Mfe*I
	*Sac*I	*Nsi*I - *Ppu*10I	*Sal*I	*Apa* - *Bsp*120I
	*Kpn*I - *Acc*65I	*Bgl*II	*Stu*I	*Stu*I
	*Stu*I	*Stu*I		
	*Afl*II T7↑	*Afl*II T7↑	*Afl* II T7↑	*Afl* II T7↑

(a)

Fig. 5.9 Structure and use of the LITMUS vectors for making RNA probes. (a) Structure of the LITMUS vectors showing the orientation and restriction sites of the four polylinkers.

(b)

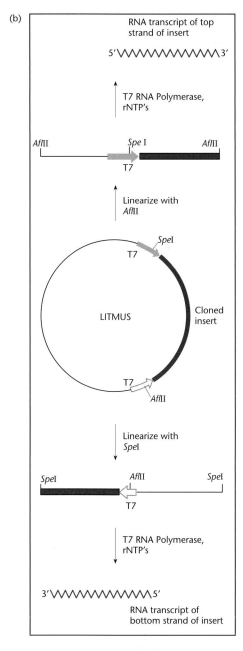

RNA transcript of top strand of insert

5′ \/\/\/\/\/\/\/\/\/\ 3′

T7 RNA Polymerase, rNTP's

*Afl*II — *Spe* I — *Afl*II

T7

Linearize with *Afl*II

*Spe*I

T7

LITMUS

Cloned insert

T7

*Afl*II

Linearize with *Spe*I

*Spe*I — *Afl*II — *Spe*I

T7

T7 RNA Polymerase, rNTP's

3′ \/\/\/\/\/\/\/\/\/\ 5′

RNA transcript of bottom strand of insert

Fig. 5.9 (*cont'd*) (b) Method of using the LITMUS vectors to selectively synthesize RNA probes from each strand of a cloned insert. (Figure reproduced courtesy of New England Biolabs.)

A further improvement has been introduced by Evans *et al.* (1995). In their LITMUS vectors, the polylinker regions are flanked by two modified T7 RNA polymerase promoters. Each contains a

Table 5.2 Factors affecting the expression of cloned genes.

Factor	Text
Promoter strength	Page 74
Transcriptional termination	Page 71
Plasmid copy number	Page 45, Chapter 4
Plasmid stability	Page 47, Chapter 4
Host-cell physiology	Chapters 4 & 5
Translational initiation sequences	Box 5.1, page 77
Codon choice	Box 5.1, page 77
mRNA structure	Box 5.1, page 77

unique restriction site (*Spe*I or *Afl*II) that has been engineered into the T7 promoter consensus sequence such that cleavage with the corresponding endonuclease inactivates that promoter. Both promoters are active despite the presence of engineered sites. Selective unidirectional transcription is achieved by simply inactivating the other promoter by digestion with *Spe*I or *Afl*II prior to *in vitro* transcription (Fig. 5.9). Since efficient labelling of RNA probes demands that the template be linearized prior to transcription, at a site downstream from the insert, cutting at the site within the undesired promoter performs both functions in one step. Should the cloned insert contain either an *Spe*I or an *Afl*II site, the unwanted promoter can be inactivated by cutting at one of the unique sites within the polylinker.

Vectors for maximizing protein synthesis

Provided that a cloned gene is preceded by a promoter recognized by the host cell, then there is a high probability that there will be *detectable* synthesis of the cloned gene product. However, much of the interest in the application of recombinant DNA technology is the possibility of facile synthesis of large quantities of protein, either to study its properties or because it has commercial value. In such instances, detectable synthesis is not sufficient: rather, it must be maximized. The factors affecting the level of expression of a cloned gene are shown in Table 5.2 and are reviewed by Baneyx (1999). Of these factors, only promoter strength is considered here.

Promoter	Uninduced level of CAT	Induced level of CAT	Ratio
λ P_L	0.0275	28.18	1025
trc	1.10	5.15	4.7
T7	1.14	15.40	13.5

Table 5.3 Control of expression of chloramphenicol acetyltransferase (CAT) in *E. coli* by three different promoters. The levels of CAT are expressed as μg/mg total protein.

When maximizing gene expression it is not enough to select the strongest promoter possible: the effects of overexpression on the host cell also need to be considered. Many gene products can be toxic to the host cell even when synthesized in small amounts. Examples include surface structural proteins (Beck & Bremer 1980), proteins, such as the PolA gene product, that regulate basic cellular metabolism (Murray & Kelley 1979), the cystic fibrosis transmembrane conductance regulator (Gregory *et al.* 1990) and lentivirus envelope sequences (Cunningham *et al.* 1993). If such cloned genes are allowed to be expressed there will be a rapid selection for mutants that no longer synthesize the toxic protein. Even when overexpression of a protein is not toxic to the host cell, high-level synthesis exerts a metabolic drain on the cell. This leads to slower growth and hence in culture there is selection for variants with lower or no expression of the cloned gene because these will grow faster. To minimize the problems associated with high-level expression, it is usual to use a vector in which the cloned gene is under the control of a *regulated* promoter.

Many different vectors have been constructed for regulated expression of gene inserts but most of those in current use contain one of the following controllable promoters: λ P_L, T7, *trc* (*tac*) or BAD. Table 5.3 shows the different levels of expression that can be achieved when the gene for chloramphenicol transacetylase (CAT) is placed under the control of three of these promoters.

The *trc* and *tac* promoters are hybrid promoters derived from the *lac* and *trp* promoters (Brosius 1984). They are stronger than either of the two parental promoters because their sequences are more like the consensus sequence. Like *lac*, the *trc* and *tac* promoters are inducibile by lactose and isopropyl-β-D-thiogalactoside (IPTG). Vectors using these promoters also carry the *lacO* operator and the *lacI* gene, which encodes the repressor.

The pET vectors are a family of expression vectors that utilize phage T7 promoters to regulate synthesis of cloned gene products (Studier *et al.* 1990). The general strategy for using a pET vector is shown in Fig. 5.10. To provide a source of phage-T7 RNA polymerase, *E. coli* strains that contain gene 1 of the phage have been constructed. This gene is cloned downstream of the *lac* promoter, in the chromosome, so that the phage polymerase will only be synthesized following IPTG induction. The newly synthesized T7 RNA polymerase will then transcribe the foreign gene in the pET plasmid. If the protein product of the cloned gene is toxic, it is possible to minimize the uninduced level of T7 RNA polymerase. First, a plasmid compatible with pET vectors is selected and the T7 *lysS* gene is cloned in it. When introduced into a host cell carrying a pET plasmid, the *lysS* gene will bind any residual T7 RNA polymerase (Studier 1991, Zhang & Studier 1997). Also, if a *lac* operator is placed between the T7 promoter and the cloned gene, this will further reduce transcription of the insert in the absence of IPTG (Dubendorff & Studier 1991). Improvements in the yield of heterologous proteins can sometimes be achieved by use of selected host cells (Miroux & Walker 1996).

The λ P_L promoter system combines very tight transcriptional control with high levels of gene expression. This is achieved by putting the cloned gene under the control of the P_L promoter carried on a vector, while the P_L promoter is controlled by a *cI* repressor gene in the *E. coli* host. This *cI* gene is itself under the control of the tryptophan (*trp*) promoter (Fig. 5.11). In the absence of exogenous tryptophan, the *cI* gene is transcribed and the *cI* repressor binds to the P_L promoter, preventing expression of the

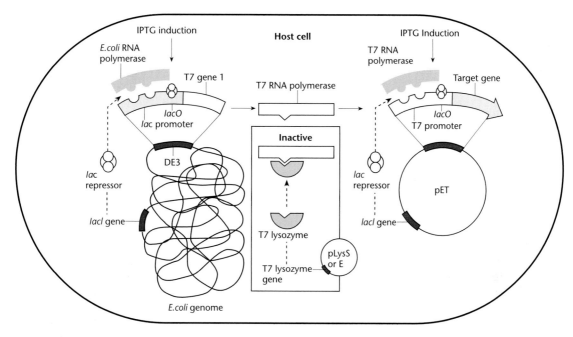

Fig. 5.10 Strategy for regulating the expression of genes cloned into a pET vector. The gene for T7 RNA polymerase (gene 1) is inserted into the chromosome of *E. coli* and transcribed from the *lac* promoter; therefore, it will be expressed only if the inducer IPTG is added. The T7 RNA polymerase will then transcribe the gene cloned into the pET vector. If the protein product of the cloned gene is toxic, it may be necessary to further reduce the transcription of the cloned gene before induction. The T7 lysozyme encoded by a compatible plasmid, pLysS, will bind to any residual T7 RNA polymerase made in the absence of induction and inactivate it. Also, the presence of *lac* operators between the T7 promoter and the cloned gene will further reduce transcription of the cloned gene in the absence of the inducer IPTG. (Reprinted with permission from the *Novagen Catalog*, p. 24, Novagen, Madison, Wisconsin, 1995.)

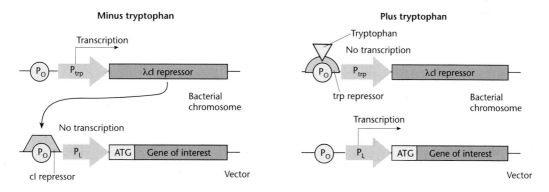

Fig. 5.11 Control of cloned gene expression using the λcI promoter. See text for details. (Diagram reproduced courtesy of In Vitrogen Corporation.)

cloned gene. Upon addition of tryptophan, the *trp* repressor binds to the *cI* gene, preventing synthesis of the cI repressor. In the absence of *cI* repressor, there is a high level of expression from the very strong P_L promoter.

The pBAD vectors, like the ones based on P_L promoter, offer extremely tight control of expression of cloned genes (Guzman *et al.* 1995). The pBAD vectors carry the promoter of the *ara*BAD (arabinose) operon and the gene encoding the positive and

(a)

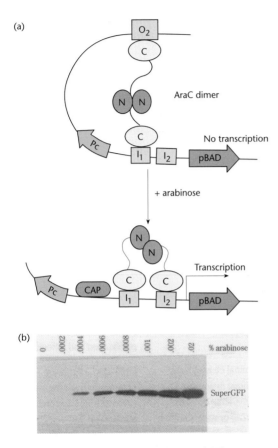

(b)

Fig. 5.12 Regulation of the pBAD promoter. (a) The conformational changes that take place on addition of arabinose. (b) Western blot showing the increase in synthesis of a cloned gene product when different levels of arabinose are added to a culture of the host cell.

negative regulator of this promoter, *ara*C. AraC is a transcriptional regulator that forms a complex with L-arabinose. In the absence of arabinose, AraC binds to the O_2 and I_1 half-sites of the *ara*BAD operon, forming a 210 bp DNA loop and thereby blocking transcription (Fig. 5.12). As arabinose is added to the growth medium, it binds to AraC, thereby releasing the O_2 site. This in turn causes AraC to bind to the I_2 site adjacent to the I_1 site. This releases the DNA loop and allows transcription to begin. Binding of AraC to I_1 and I_2 is activated in the presence of cAMP activator protein (CAP) plus cAMP. If glucose is added to the growth medium, this will lead to a repression of cAMP synthesis, thereby decreasing expression from the *ara*BAD promoter. Thus one can titrate the level of cloned gene product by varying

the glucose and arabinose content of the growth medium (Fig. 5.12). According to Guzman *et al.* (1995), the pBAD vectors permit fine-tuning of gene expression. All that is required is to change the sugar composition of the medium. However, this is disputed by others (Siegele & Hu 1997, Hashemzadeh-Bonehi *et al.* 1998).

Many of the vectors designed for high-level expression also contain translation-initiation signals optimized for *E. coli* expression (see Box 5.1).

Vectors to facilitate protein purification

Many cloning vectors have been engineered so that the protein being expressed will be fused to another protein, called a *tag*, which can be used to facilitate protein purification. Examples of tags include glutathione-*S*-transferase, the MalE (maltose-binding) protein and multiple histidine residues, which can easily be purified by affinity chromatography. The tag vectors are usually constructed so that the coding sequence for an amino acid sequence cleaved by a specific protease is inserted between the coding sequence for the tag and the gene being expressed. After purification, the tag protein can be cleaved off with the specific protease to leave a normal or nearly normal protein. It is also possible to include in the tag a protein sequence that can be assayed easily. This permits assay of the cloned gene product when its activity is not known or when the usual assay is inconvenient. Three different examples of tags are given below. The reader requiring a more detailed insight should consult the review by LaVallie and McCoy (1995).

To use a polyhistidine fusion for purification, the gene of interest is first engineered into a vector in which there is a polylinker downstream of six histidine residues and a proteolytic cleavage site. In the example shown in Fig. 5.13, the cleavage site is that for enterokinase. After induction of synthesis of the fusion protein, the cells are lysed and the viscosity of the lysate is reduced by nuclease treatment. The lysate is then applied to a column containing immobilized divalent nickel, which selectively binds the polyhistidine tag. After washing away any contaminating proteins, the fusion protein is eluted from the column and treated with enterokinase to release the cloned gene product.

Box 5.1 Optimizing translation

High-level expression of a cloned gene requires more than a strong promoter. The mRNA produced during transcription needs to be effectively translated into protein. Although many factors can influence the rate of translation, the most important is the interaction of the ribosome with the bases immediately upstream from the initiation codon of the gene. In bacteria, a key sequence is the ribosome-binding site or Shine–Dalgarno (S-D) sequence. The degree of complementarity of this sequence with the 16S rRNA can affect the rate of translation (De Boer & Hui 1990). Maximum complementarity occurs with the sequence 5′-UAAGGAGG-3′ (Ringquist *et al.* 1992). The spacing between the S-D sequence and the initiation codon is also important. Usually there are five to 10 bases, with eight being optimal. Decreasing the distance below 4 bp or increasing it beyond 14 bp can reduce translation by several orders of magnitude.

Translation is affected by the sequence of bases that follow the S-D site (De Boer *et al.* 1983b). The presence of four A residues or four T residues in this position gave the highest translational efficiency. Translational efficiency was 50% or 25% of maximum when the region contained, respectively, four C residues or four G residues.

The composition of the triplet immediately preceding the AUG start codon also affects the efficiency of translation. For translation of β-galactosidase mRNA, the most favourable combinations of bases in this triplet are UAU and CUU. If UUC, UCA or AGG replaced UAU or CUU, the level of expression was 20-fold less (Hui *et al.* 1984).

The codon composition following the AUG start codon can also affect the rate of translation. For example, a change in the third position of the fourth codon of a human γ-interferon gene resulted in a 30-fold change in the level of expression (De Boer & Hui 1990). Also, there is a strong bias in the second codon of many natural mRNAs, which is quite different from the general bias in codon usage. Highly expressed genes have AAA (Lys) or GCU (Ala) as the second codon. Devlin *et al.* (1988) changed all the G and C nucleotides for the first four codons of a granulocyte colony-stimulating factor gene and expression increased from undetectable to 17% of total cell protein.

Sequences upstream from the S-D site can affect the efficiency of translation of certain genes. In the *E. coli rnd* gene there is a run of eight uracil residues. Changing two to five of these residues has no effect on mRNA levels but reduces translation by up to 95% (Zhang & Deutscher 1992). Etchegaray and Inouye (1999) have identified an element downstream of the initiation codon, the downstream box, which facilitates formation of the translation-initiation complex. The sequence of the 3′ untranslated region of the mRNA can also be important. If this region is complementary to sequences within the gene, hairpin loops can form and hinder ribosome movement along the messenger.

The genetic code is degenerate, and hence for most of the amino acids there is more than one codon. However, in all genes, whatever their origin, the selection of synonymous codons is distinctly non-random (for reviews, see Kurland 1987, Ernst 1988 and McPherson 1988). The bias in codon usage has two components: correlation with tRNA availability in the cell, and non-random choices between pyrimidine-ending codons. Ikemura (1981a) measured the relative abundance of the 26 known tRNAs of *E. coli* and found a strong positive correlation between tRNA abundance and codon choice. Later, Ikemura (1981b) noted that the most highly expressed genes in *E. coli* contain mostly those codons corresponding to major tRNAs but few codons of minor tRNAs. In contrast, genes that are expressed less well use more suboptimal codons. Forman *et al.* (1998) noted significant misincorporation of lysine, in place of arginine, when the rare AGA codon was included in a gene overexpressed in *E. coli*. It should be noted that the bias in codon usage even extends to the stop codons (Sharp & Bulmer 1988). UAA is favoured in genes expressed at high levels, whereas UAG and UGA are used more frequently in genes expressed at a lower level.

For a review of translation the reader should consult Kozak (1999).

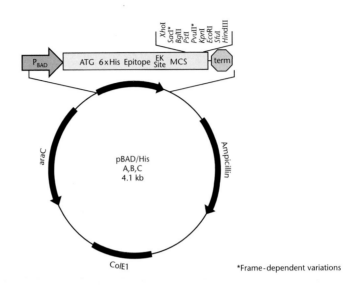

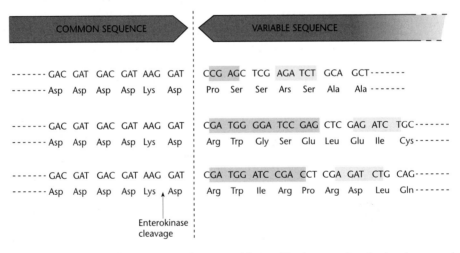

Fig. 5.13 Structure of a vector (pBAD/His, In Vitrogen Corporation) designed for the expression of a cloned gene as a fusion protein containing a polyhistidine sequence. Three different variants (A, B, C) allow the insert to be placed in each of the three translational reading frames. The sequence shaded pink shows the base sequence which is altered in each of the three vectors. The lightly-shaded box (AGATCT) is the Bgl II site of the polylinker. Note that the initial A residue of the restriction site is at a different point in the triplet codon in each of the three sequences.

For the cloned gene to be expressed correctly, it has to be in the correct translational reading frame. This is achieved by having three different vectors, each with a polylinker in a different reading frame (see Fig. 5.13). Enterokinase recognizes the sequence (Asp)$_4$Lys and cleaves immediately after the lysine residue. Therefore, after enterokinase cleavage, the cloned gene protein will have a few extra amino acids at the N terminus. If desired, the cleavage site and polyhistidines can be synthesized at the C terminus. If the cloned gene product itself contains an enterokinase cleavage site, then an alternative protease, such as thrombin or factor Xa, with a different cleavage site can be used.

To facilitate assay of the fusion proteins, short antibody recognition sequences can be incorporated into the tag between the affinity label and the protease cleavage site. Some examples of recognizable

Table 5.4 Peptide epitopes, and the antibodies that recognize them, for use in assaying fusion proteins.

Peptide sequence	Antibody recognition
-Glu-Gln-Lys-Leu-Ile Ser-Glu-Glu-Asp-Leu-	Anti-*myc* antibody
-His-His-His-His-His-His-COOH	Anti-His (C-terminal) antibody
-Gly-Lys-Pro-Ile-Pro-Asn-Pro-Leu-Leu-Gly-Leu- Asp-Ser-Thr-	Anti-V5 antibody

epitopes are given in Table 5.4. These antibodies can be used to detect, by western blotting, fusion proteins carrying the appropriate epitope. Note that a polyhistidine tag at the C terminus can function for both assay and purification.

Biotin is an essential cofactor for a number of carboxylases important in cell metabolism. The biotin in these enzyme complexes is covalently attached at a specific lysine residue of the biotin carboxylase carrier protein. Fusions made to a segment of the carrier protein are recognized in *E. coli* by biotin ligase, the product of the *bir*A gene, and biotin is covalently attached in an ATP-dependent reaction. The expressed fusion protein can be purified using streptavidin affinity chromatography (Fig. 5.14). *E. coli* expresses a single endogenous biotinylated protein, but it does not bind to streptavidin in its native configuration, making the affinity purification highly specific for the recombinant fusion protein. The presence of biotin on the fusion protein has an additional advantage: its presence can be detected with enzymes coupled to streptavidin.

The affinity purification systems described above suffer from the disadvantage that a protease is required to separate the target protein from the affinity tag. Also, the protease has to be separated from the protein of interest. Chong *et al.* (1997, 1998) have described a unique purification system that has neither of these disadvantages. The system utilizes a protein splicing element, an intein, from the *Saccharomyces cerevisiae VMA1* gene (see Box 5.2). The intein is modified such that it undergoes a self-cleavage reaction at its N terminus at low temperatures in the presence of thiols, such as cysteine, dithiothreitol or β-mercaptoethanol. The gene encoding the target protein is inserted into a multiple cloning site (MCS) of a vector to create a fusion between the C terminus of the target gene and

the N terminus of the gene encoding the intein. DNA encoding a small (5 kDa) chitin-binding domain from *Bacillus circulans* was added to the C terminus of the intein for affinity purification (Fig. 5.15).

The above construct is placed under the control of an IPTG-inducible T7 promoter. When crude extracts from induced cells are passed through a chitin column, the fusion protein binds and all contaminating proteins are washed through. The fusion is then induced to undergo intein-mediated self-cleavage on the column by incubation with a thiol. This releases the target protein, while the intein chitin-binding domain remains bound to the column.

Vectors to promote solubilization of expressed proteins

One of the problems associated with the overproduction of proteins in *E. coli* is the sequestration of the product into insoluble aggregates or 'inclusion bodies' (Fig. 5.16). They were first reported in strains overproducing insulin A and B chains (Williams *et al.* 1982). At first, their formation was thought to be restricted to the overexpression of heterologous proteins in *E. coli*, but they can form in the presence of high levels of normal *E. coli* proteins, e.g. subunits of RNA polymerase (Gribskov & Burgess 1983). Two parameters that can be manipulated to reduce inclusion-body formation are temperature and growth rate. There are a number of reports which show that lowering the temperature of growth increases the yield of correctly folded, soluble protein (Schein & Noteborn 1988, Takagi *et al.* 1988, Schein 1991). Media compositions and pH values that reduce the growth rate also reduce inclusion-body formation. Renaturation of misfolded proteins can sometimes be achieved following solubilization in guanidinium hydrochloride (Lilie *et al.* 1998).

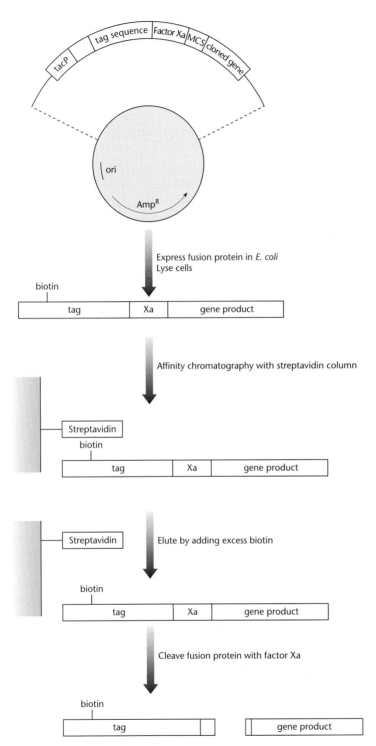

Fig. 5.14 Purification of a cloned gene product synthesized as a fusion to the biotin carboxylase carrier protein (tag). See text for details.

Box 5.2 Inteins, exeins and protein splicing

Protein splicing is defined as the excision of an intervening protein sequence (the *intein*) from a protein precursor. Splicing involves ligation of the flanking protein fragments (the *exteins*) to form a mature extein protein and the free intein. Protein splicing results in a native peptide bond between the ligated exteins and this differentiates it from other forms of autoproteolysis.

N-extein	Intein	C-extein

Precursor protein

↓ Splicing

N-extein	C-extein		Intein

Mature extein protein

Sequence comparison and structural analysis have indicated that the residues responsible for splicing are ~100 amino acids at the N terminus of the intein and ~50 amino acids at the C terminus. These two splicing regions are separated by a linker or a gene encoding a *homing endonuclease*. If present in a cell, the homing endonuclease makes a double-stranded break in the DNA at or near the insertion site (the home) of the intein encoding it. This endonuclease activity initiates the movement of the intein into another allele of the same gene if that allele lacks the intein.

N-extein	N-terminal splicing region	Linker or homing endonuclease gene	C-terminal splicing region	C-extein

Three 'genetic' methods of preventing inclusion-body formation have been described. In the first of these, the host cell is engineered to overproduce a chaperon (e.g. DnaK, GroEL or GroES proteins) in addition to the protein of interest (Van Dyk *et al.* 1989, Blum *et al.* 1992, Thomas *et al.* 1997). Castanie *et al.* (1997) have developed a series of vectors which are compatible with pBR322-type plasmids and which encode the overproduction of chaperons. These vectors can be used to test the effect of chaperons on the solubilization of heterologous gene products. Even with excess chaperon there is no guarantee of proper folding. The second method involves making minor changes to the amino acid sequence of the target protein. For example, cysteine-to-serine changes in fibroblast growth factor minimized inclusion-body formation (Rinas *et al.* 1992). The third method is derived from the observation that many proteins produced as insoluble aggregates in their native state are synthesized in soluble form as thioredoxin fusion proteins (LaVallie *et al.* 1993). More recently, Davis *et al.* (1999) have shown that the NusA and GrpE proteins, as well as bacterioferritin, are even better than thioredoxin at solubilizing proteins expressed at a high level. Kapust and Waugh (1999) have reported that the maltose-binding protein is also much better than thioredoxin.

Building on the work of LaVallie *et al.* (1993), a series of vectors has been developed in which the gene of interest is cloned into an MCS and the gene product is produced as a thioredoxin fusion protein with an enterokinase cleavage site at the fusion point. After synthesis, the fusion protein is released from the producing cells by osmotic shock and purified. The desired protein is then released by enterokinase cleavage. To simplify the purification of thioredoxin fusion proteins, Lu *et al.* (1996)

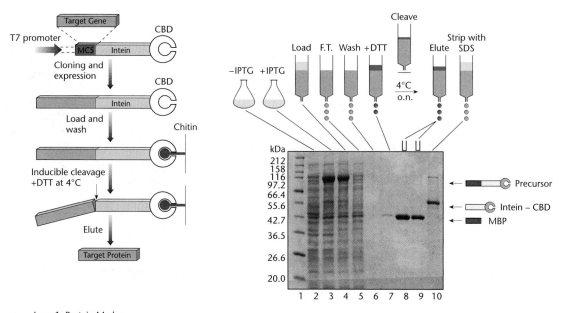

Lane 1: Protein Marker.
Lane 2: Crude extract from uninduced cells.
Lane 3: Crude extract from cells, induced at 15°C for 16 hours.
Lane 4: Clarified crude extract from induced cells.
Lane 5: Chitin column flow through (F.T.).
Lane 6: Chitin column wash.
Lane 7: Quick DTT wash to distribute DTT evenly throughout the chitin column.
Lanes 8-9: Fraction of eluted MBP after stopping column flow and inducing a self-cleavage reaction at 4°C overnight.
Lane 10: SDS stripping of remaining proteins bound to chitin column (mostly the cleaved intein-CBD fusion).

Fig. 5.15 Purification of a cloned gene product synthesized as a fusion with an intein protein. (Figure reproduced courtesy of New England Biolabs.)

systematically mutated a cluster of surface amino acid residues. Residues 30 and 62 were converted to histidine and the modified ('histidine patch') thioredoxin could now be purified by affinity chromatography on immobilized divalent nickel. An alternative purification method was developed by Smith *et al.* (1998). They synthesized a gene in which a short biotinylation peptide is fused to the N terminus of the thioredoxin gene to generate a new protein called BIOTRX. They constructed a vector carrying the BIOTRX gene, with an MCS at the C terminus, and the *bir*A gene. After cloning a gene in the MCS, a fused protein is produced which can be purified by affinity chromatography on streptavidin columns.

An alternative way of keeping recombinant proteins soluble is to export them to the periplasmic space (see next section). However, even here they may still be insoluble. Barth *et al.* (2000) solved this problem by growing the producing bacteria under osmotic stress (4% NaCl plus 0.5 mol/l sorbitol) in the presence of compatible solutes. Compatible solutes are low-molecular-weight osmolytes, such as glycine betaine, that occur naturally in halophilic bacteria and are known to protect proteins at high salt concentrations. Adding glycine betaine for the cultivation of *E. coli* under osmotic stress not only allowed the bacteria to grow under these otherwise inhibitory conditions but also produced a periplasmic environment for the generation of correctly folded recombinant proteins.

Vectors to promote protein export

Gram-negative bacteria such as *E. coli* have a complex wall–membrane structure comprising an inner, cytoplasmic membrane separated from an outer membrane by a cell wall and periplasmic space.

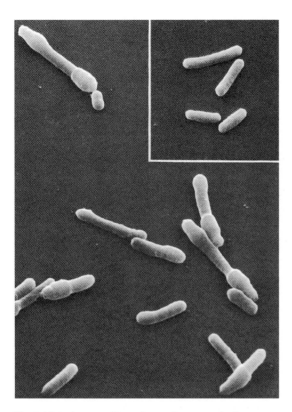
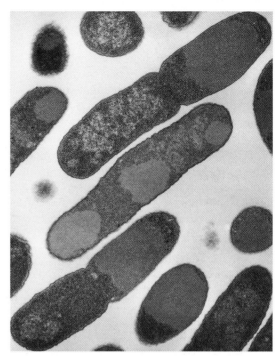

Fig. 5.16 Inclusions of Trp polypeptide–proinsulin fusion protein in *E. coli*. (*Left*) Scanning electron micrograph of cells fixed in the late logarithmic phase of growth; the inset shows normal *E. coli* cells. (*Right*) Thin section through *E. coli* cells producing Trp polypeptide–insulin A chain fusion protein. (Photographs reproduced from *Science* courtesy of Dr D.C. Williams (Eli Lilly & Co.) and the American Association for the Advancement of Science.)

Secreted proteins may be released into the periplasm or integrated into or transported across the outer membrane. In *E. coli* it has been established that protein export through the inner membrane to the periplasm or to the outer membrane is achieved by a universal mechanism known as the general export pathway (GEP). This involves the *sec* gene products (for review see Lory 1998). Proteins that enter the GEP are synthesized in the cytoplasm with a signal sequence at the N terminus. This sequence is cleaved by a signal or leader peptidase during transport. A signal sequence has three domains: a positively charged amino-terminal region, a hydrophobic core, consisting of five to 15 hydrophobic amino acids, and a leader peptidase cleavage site. A signal sequence attached to a normally cytoplasmic protein will direct it to the export pathway.

Many signal sequences derived from naturally occurring secretory proteins (e.g. OmpA, OmpT, PelB, β-lactamase, alkaline phosphatase and phage M13 gIII) support the efficient translocation of heterologous peptides across the inner membrane when fused to their amino termini. In some cases, however, the preproteins are not readily exported and either become 'jammed' in the inner membrane, accumulate in precursor inclusion bodies or are rapidly degraded within the cytoplasm. In practice, it may be necessary to try several signal sequences (Berges *et al.* 1996) and/or overproduce different chaperons to optimize the translocation of a particular heterologous protein. A first step would be to try the secretion vectors offered by a number of molecular-biology suppliers and which are variants of the vectors described above.

It is possible to engineer proteins such that they are transported through the outer membrane and are secreted into the growth medium. This is achieved by making use of the type I, Sec-independent secretion system. The prototype type I system is the haemolysin transport system, which requires a short carboxy-terminal secretion signal, two translocators (HlyB and D), and the outer-membrane protein TolC. If the protein of interest is fused to the carboxy-terminal secretion signal of a type I-secreted protein, it will be secreted into the medium provided HlyB, HlyD and TolC are expressed as well (Spreng *et al.* 2000). An alternative presentation of recombinant proteins is to express them on the surface of the bacterial cell using any one of a number of carrier proteins (for review, see Cornelis 2000).

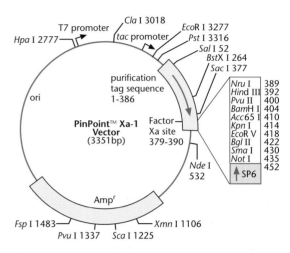

Fig. 5.17 Structural features of a PinPoint™ vector. (Figure reproduced courtesy of Promega Corporation.)

Putting it all together: vectors with combinations of features

Many of the vectors in current use, particularly those that are commercially available, have combinations of the features described in previous sections. Two examples are described here to show the connection between the different features. The first example is the LITMUS vectors that were described earlier (p. 73) and which are used for the generation of RNA probes. They exhibit the following features:
• The polylinkers are located in the *lacZ*α gene and inserts in the polylinker prevent α-complementation. Thus blue/white screening (see Box 3.2 on p. 35) can be used to distinguish clones with inserts from those containing vector only.
• The LITMUS polylinkers contain a total of 32 unique restriction sites. Twenty-nine of these enzymes leave four-base overhangs and three leave blunt ends. The three blunt cutting enzymes have been placed at either end of the polylinker and in the middle of it.
• The vectors carry both the pUC and the M13 *ori* regions. Under normal conditions the vector replicates as a double-stranded plasmid but, on infection with helper phage (M13KO7), single-stranded molecules are produced and packaged in phage protein.
• The single-stranded molecules produced on helper-phage addition have all the features necessary for DNA sequencing (see p. 123).

• The vectors are small (< 3 kb) and with a pUC *ori* have a high copy number.

The second example is the PinPoint series of expression vectors (Fig. 5.17). These vectors have the following features:
• Expression is under the control of either the T7 or the *tac* promoter, allowing the user great flexibility of control over the synthesis of the cloned gene product.
• Some of them carry a DNA sequence specifying synthesis of a signal peptide.
• Presence of an MCS adjacent to a factor-Xa cleavage site.
• Synthesis of an N-terminal biotinylated sequence to facilitate purification.
• Three different vectors of each type permitting translation of the cloned gene insert in each of the three reading frames.
• Presence of a phage SP6 promoter distal to the MCS to permit the synthesis of RNA probes complementary to the cloned gene. Note that the orientation of the cloned gene is known and so the RNA probe need only be synthesized from one strand.
What is absent from these vectors is an M13 origin of replication to facilitate synthesis of single strands for DNA sequencing.

CHAPTER 6

Cloning strategies

Introduction

In the early chapters of this book, we discussed DNA cutting and joining techniques, and introduced the different types of vectors that are available for cloning DNA molecules. Any cell-based cloning procedure has four essential parts: (i) a method for generating the DNA fragment for cloning; (ii) a reaction that inserts that fragment into the chosen cloning vector; (iii) a means for introducing that recombinant vector into a host cell wherein it is replicated; and (iv) a method for selecting recipient cells that have acquired the recombinant (Fig. 6.1). To simplify the description of such procedures, the assumption is made that we know exactly what we are cloning. This is indeed the case with simple *subcloning* strategies, where a defined restriction fragment is isolated from one cloning vector and inserted into another. However, we also need to consider what happens in cases where the source of donor DNA is very complex. We may wish, for example, to isolate a single gene from the human

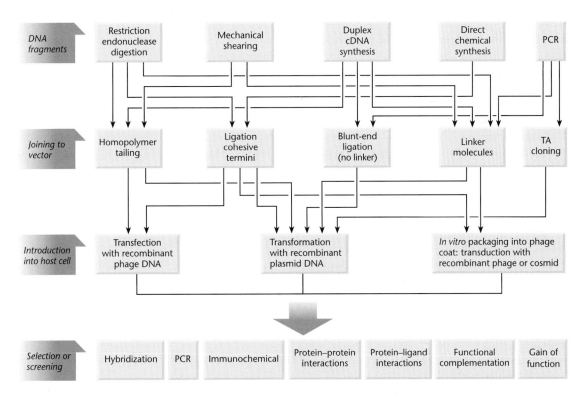

Fig. 6.1 Generalized overview of cloning strategies, with favoured routes shown by arrows. Note that in cell-based cloning strategies, DNA fragments are initially generated and cloned in a non-specific manner, so that screening for the desired clone is carried out at the end of the process. Conversely, when specific DNA fragments are obtained by PCR or direct chemical synthesis, there is no need for subsequent screening.

genome, in which case the target sequence could be diluted over a millionfold by unwanted genomic DNA. We need to find some way of rapidly sifting through, or *screening*, large numbers of unwanted sequences to identify our particular target.

There are two major strategies for isolating sequences from complex sources such as genomic DNA. The first, a cell-based cloning strategy, is to divide the source DNA into manageable fragments and *clone everything*. Such a collection of clones, representative of the entire starting population, is known as a *library*. We must then *screen the library* to identify our clone of interest, using a procedure that discriminates between the desired clone and all the others. A number of such procedures are discussed later in the chapter. The second strategy is to selectively amplify the target sequence directly from the source DNA using the polymerase chain reaction (PCR), and then clone this individual fragment. Each strategy has its advantages and disadvantages. Note that, in the library approach, screening is carried out after the entire source DNA population has been cloned indiscriminately. Conversely, in the PCR approach, the screening step is built into the first stage of the procedure, when the fragments are generated, so that only selected fragments are actually cloned. In this chapter we consider principles for the construction and screening of genomic and complementary DNA (cDNA) libraries, and compare the library-based route of gene isolation to equivalent PCR-based techniques.

Cloning genomic DNA

Genomic DNA libraries

Producing representative genomic libraries in λ cloning vectors

Following the example above, let us suppose that we wish to clone a single-copy gene from the human genome. How might this be achieved? We could simply digest total human DNA with a restriction endonuclease, such as *Eco*RI, insert the fragments into a suitable phage-λ vector and then attempt to isolate the desired clone. How many recombinants would we have to screen in order to isolate the right one? Assuming *Eco*RI gives, on average, fragments of about 4 kb, and given that the size of the human haploid genome is 2.8×10^6 kb, we can see that over 7×10^5 independent recombinants must be prepared and screened in order to have a reasonable chance of including the desired sequence. In other words we have to obtain a very large number of recombinants, which together contain a complete collection of all of the DNA sequences in the entire human genome, a human *genomic library*. The sizes of some other genomes are listed in Table 6.1.

There are two problems with the above approach. First, the gene may be cut internally one or more times by *Eco*RI so that it is not obtained as a single fragment. This is likely if the gene is large. Also, it may be desirable to obtain extensive regions flanking

Table 6.1 Genome sizes of selected organisms.

Organism	Genome size (kb) (haploid where appropriate)
Escherichia coli	4.0×10^3
Yeast (*Saccharomyces cerevisiae*)	1.35×10^4
Arabidopsis thaliana (higher plant)	1.25×10^5
Tobacco	1.6×10^6
Wheat	5.9×10^6
Zea mays	1.5×10^7
Drosophila melanogaster	1.8×10^5
Mouse	2.3×10^6
Human	2.8×10^6
Xenopus laevis	3.0×10^6

the gene or whole gene clusters. Fragments averaging about 4 kb are likely to be inconveniently short. Alternatively, the gene may be contained on an *Eco*RI fragment that is larger than the vector can accept. In this case the appropriate gene would not be cloned at all.

These problems can be overcome by cloning *random* DNA fragments of a large size (for λ replacement vectors, approximately 20 kb). Since the DNA is randomly fragmented, there will be no systematic exclusion of any sequence. Furthermore, clones will overlap one another, allowing the sequence of very large genes to be assembled and giving an opportunity to 'walk' from one clone to an adjacent one (p. 107). Because of the larger size of each cloned DNA fragment, fewer clones are required for a complete or nearly complete library. How many clones are required? Let n be the size of the genome relative to a single cloned fragment. Thus, for the human genome (2.8×10^6 kb) and an average cloned fragment size of 20 kb, $n = 1.4 \times 10^5$. The number of independent recombinants required in the library must be greater than n, because sampling variation will lead to the inclusion of some sequences several times and the exclusion of other sequences in a library of just n recombinants. Clarke and Carbon (1976) have derived a formula that relates the probability (P) of including any DNA sequence in a random library of N independent recombinants:

$$N = \frac{\ln(1 - P)}{\ln\left(1 - \dfrac{1}{n}\right)}$$

Therefore, to achieve a 95% probability ($P = 0.95$) of including any particular sequence in a random human genomic DNA library of 20 kb fragment size:

$$N = \frac{\ln(1 - 0.95)}{\ln\left(1 - \dfrac{1}{1.4 \times 10^5}\right)} = 4.2 \times 10^5$$

Notice that a considerably higher number of recombinants is required to achieve a 99% probability, for here $N = 6.5 \times 10^5$.

How can appropriately sized random fragments be produced? Various methods are available. Random breakage by mechanical shearing is appropriate because the average fragment size can be controlled,

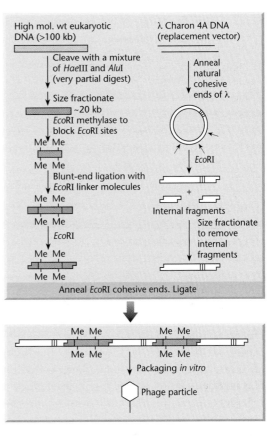

Fig. 6.2 Maniatis' strategy for producing a representative gene library.

but insertion of the resulting fragments into vectors requires additional steps. The more commonly used procedure involves restriction endonucleases. In the strategy devised by Maniatis *et al.* (1978) (Fig. 6.2), the target DNA is digested with a mixture of *two* restriction enzymes. These enzymes have tetranucleotide recognition sites, which therefore occur frequently in the target DNA and in a complete double-digest would produce fragments averaging less than 1 kb. However, only a partial restriction digest is carried out, and therefore the majority of the fragments are large (in the range 10–30 kb). Given that the chances of cutting at each of the available restriction sites are more or less equivalent, such a reaction effectively produces a random set of overlapping fragments. These can be size-fractionated, e.g. by gel electrophoresis, so as to give a random population of fragments of about

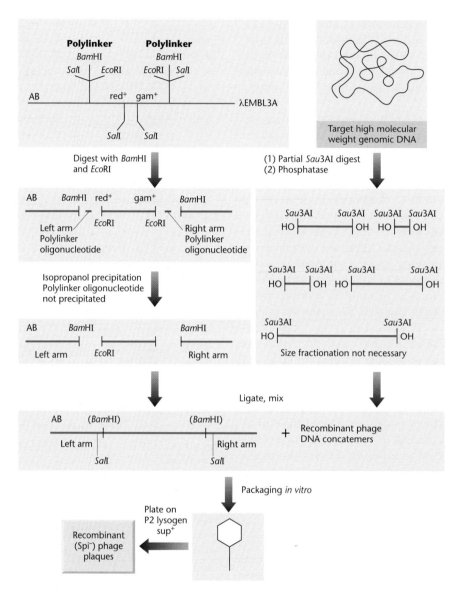

Fig. 6.3 Creation of a genomic DNA library using the phage-λ vector EMBL3A. High-molecular-weight genomic DNA is partially digested with *Sau*3AI. The fragments are treated with phosphatase to remove their 5′ phosphate groups. The vector is digested with *Bam*HI and *Eco*RI, which cut within the polylinker sites. The tiny *Bam*HI/*Eco*RI polylinker fragments are discarded in the isopropanol precipitation, or alternatively the vector arms may be purified by preparative agarose gel electrophoresis. The vector arms are then ligated with the partially digested genomic DNA. The phosphatase treatment prevents the genomic DNA fragments from ligating together. Non-recombinant vector cannot re-form because the small polylinker fragments have been discarded. The only packageable molecules are recombinant phages. These are obtained as plaques on a P2 lysogen of *sup*⁺ *E. coli*. The Spi⁻ selection ensures recovery of phage lacking *red* and *gam* genes. A *sup*⁺ host is necessary because, in this example, the vector carries amber mutations in genes *A* and *B*. These mutations increase biological containment, and can be applied to selection procedures, such as recombinational selection, or tagging DNA with a *sup*⁺ gene. Ultimately, the foreign DNA can be excised from the vector by virtue of the *Sal*I sites in the polylinker. (*Note*: Rogers *et al*. (1988) have shown that the EMBL3 polylinker sequence is not exactly as originally described. It contains an extra sequence with a previously unreported *Pst*I site. This does not affect most applications as a vector.)

20 kb, which are suitable for insertion into a λ replacement vector. Packaging *in vitro* (p. 58) ensures that an appropriately large number of independent recombinants can be recovered, which will give an almost completely representative library.

Development of λ replacement vectors for genomic library construction

In the Maniatis strategy, the use of two different restriction endonucleases with completely unrelated recognition sites, *Hae*III and *Alu*I, assists in obtaining fragmentation that is nearly random. These enzymes both produce blunt ends, and the cloning strategy requires linkers (see Fig. 6.2). Therefore, in the early days of vector development, a large number of different vectors became available with alternative restriction sites and genetic markers suitable for varied cloning strategies. A good example of this diversity is the Charon series, which included both insertion- and replacement-type vectors (Blattner *et al.* 1977, Williams & Blattner 1979).

A convenient simplification can be achieved by using a *single* restriction endonuclease that cuts frequently, such as *Sau*3AI. This will create a partial digest that is slightly less random than that achieved with a pair of enzymes. However, it has the great advantage that the *Sau*3AI fragments can be readily inserted into λ replacement vectors, such as λEMBL3 (Frischauf *et al.* 1983), which have been digested with *Bam*HI (Fig. 6.3). This is because *Sau*3AI and *Bam*HI create the same cohesive ends (see p. 32). Due to the convenience and efficiency of this strategy, the λEMBL series of vectors have been very widely used for genomic library construction (p. 58). Note that λEMBL vectors also carry the *red* and *gam* genes on the stuffer fragment and a *chi* site on one of the vector arms, allowing convenient positive selection on the basis of the Spi phenotype (see p. 58). Most λ vectors currently used for genomic library construction are positively selected on this basis, including λ2001 (Karn *et al.* 1984), λDASH and λFIX (Sorge 1988). λDASH and λFIX (and recently improved versions, λDASHII and λFIXII) are particularly versatile because the multiple cloning sites flanking the stuffer fragment contain opposed promoters for the T3 and T7 RNA polymerases. If the recombinant vector is digested with a restriction endonuclease that cuts frequently, only short fragments of insert DNA are left attached to these promoters. This allows RNA probes to be generated corresponding to the *ends* of any genomic insert. These are ideal for probing the library to identify overlapping clones as part of a chromosome walk (p. 107) and have the great advantage that they can be made conveniently, directly from the vector, without recourse to subcloning. Vector maps of λDASH and λFIX are shown in Fig. 6.4. λFIX is similar to λDASH, except that it incorporates additional *Xho*I sites flanking the stuffer fragment. Digestion of the vector with *Xho*I followed by partial filling of the sticky ends prevents vector religation. However, partially filled *Sau*3AI sticky ends are compatible with the partially filled *Xho*I ends, although not with each other. This strategy prevents the ligation of vector arms without genomic DNA, and also prevents the insertion of multiple fragments.

Genomic libraries in high-capacity vectors

In place of phage-λ derivatives, a number of higher-capacity cloning vectors such as cosmids, bacterial artificial chromosomes (BACs), P1-derived artificial chromosomes (PACs) and yeast artificial chromosomes (YACs) are available for the construction of genomic libraries. The advantage of such vectors is that the average insert size is much larger than for λ replacement vectors. Thus, the number of recombinants that need to be screened to identify a particular gene of interest is correspondingly lower, large genes are more likely to be contained within a single clone and fewer steps are needed for a chromosome walk (p. 107). Generally, strategies similar to the Maniatis method discussed above are used to construct such libraries, except that the partial restriction digest conditions are optimized for larger fragment sizes, and size fractionation must be performed by specialized electrophoresis methods that can separate fragments over 30 kb in length. Pulsed-field gel electrophoresis (PFGE) and field-inversion gel electrophoresis (FIGE) have been devised for this purpose (p. 10). High-molecular-weight donor DNA fragments can also be prepared using restriction enzymes that cut very rarely.

Cosmids may be favoured over λ vectors because they accept inserts of up to 45 kb. However, since

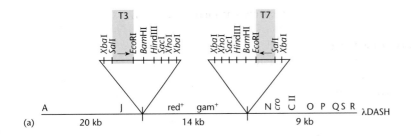

(a)

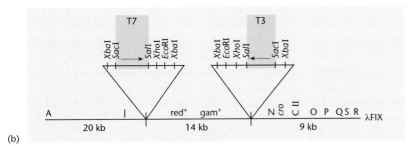

(b)

Fig. 6.4 The replacement vectors λDASH and λFIX. Promoters specific for the bacteriophage T3 and T7 RNA polymerases are located adjacent to the cloning sites, allowing RNA probes to be generated that correspond to each end of the insert.

these vectors are maintained as high-copy-number plasmids in *Escherichia coli*, they have a tendency to be unstable, undergoing deletions that favour increased replication. Most workers also find that plaques give less background hybridization than colonies, so screening libraries of phage-λ recombinants by plaque hybridization gives cleaner results than screening cosmid libraries by colony hybridization (see p. 104). It may also be desired to retain and store an *amplified genomic library*. With phage, the initial recombinant DNA population is packaged and plated out, and can be screened at this stage. Alternatively, the plates containing the recombinant plaques can be washed to give an amplified library of recombinant phage. The amplified library can then be stored almost indefinitely, because phage have a long shelf-life. The amplification is so great that samples of the amplified library can be plated out and screened with different probes on hundreds of occasions. With bacterial colonies containing cosmids, it is also possible to store an amplified library, but bacterial populations cannot be stored as readily as phage populations, and there is often an unacceptable loss of viability (Hanahan & Meselson 1980). A word of caution is necessary, however, when considering the use of any amplified library. This is due to the possibility of *distortion*. Not all recombinants in a population will propagate equally well, e.g. variations in target DNA size or sequence may affect replication of a recombinant phage, plasmid or cosmid. Therefore, when a library is put through an amplification step, particular recombinants may be increased in frequency, decreased in frequency or lost altogether. The development of modern vectors and cloning strategies has simplified library construction to the point where many workers now prefer to create a new library for each screening, rather than risk using a previously amplified one. Furthermore, pre-made libraries are available from many commercial sources and the same companies often offer custom library services. These libraries are often of high quality and such services are becoming increasingly popular.

The highest-capacity vectors – BACs, PACs and YACs – would seem to be ideal for library construction because of the very large insert sizes. However, such libraries are generally more difficult to prepare, and the larger inserts can be less than straightforward to work with. Unless the genomic target sequence is known to be very large and needs to be isolated as a single clone, λ replacement vectors or cosmids remain the most appropriate choice for many experiments. The main application of BAC, PAC and YAC libraries is for genome mapping, sequencing and the assembly of clone contigs.

Subgenomic libraries

Genomic libraries have been prepared from single human chromosomes separated by flow cytometry (e.g. Davies *et al*. 1981). Even greater enrichment for particular regions of the genome is possible using the technically demanding technique of *chromosome microdissection*. In *Drosophila*, it has been possible to physically excise a region of the salivary gland chromosome by micromanipulation, and then digest the DNA and clone it in phage-λ vectors, all within a microdrop under oil (Scalenghe *et al*. 1981). Similarly, this has been achieved with specific bands of human chromosomes using either extremely fine needles or a finely focused laser beam (Ludecke *et al*. 1989, 1990). Regardless of the species, this is a laborious and difficult technique and is prone to contamination with inappropriate DNA fragments. It has been rendered obsolete with the advent of high-capacity vectors, such as YACs.

PCR as an alternative to genomic DNA cloning

The PCR is a robust technique for amplifying specific DNA sequences from complex sources. In principle, therefore, PCR with specific primers could be used to isolate genes directly from genomic DNA, obviating the need for the production of genomic libraries. However, a serious limitation is that standard PCR conditions are suitable only for the amplification of short products. The maximum product size that can be obtained is about 5 kb, although the typical size is more likely to be 1–2 kb. This reflects the poor processivity of PCR enzymes such as *Taq* polymerase, and their lack of proofreading activity. Both of these deficiencies increase the likelihood of the enzyme detaching from the template, especially if the template is long. The extreme reaction conditions required for the PCR are also thought to cause damage to bases and generate nicks in DNA strands, which increases the probability of premature termination on long templates.

Long PCR

Modifications to reaction conditions can improve polymerase processivity by lowering the reaction temperature and increasing the pH, thereby protecting the template from damage (Foord & Rose 1994). The use of such conditions in combination with two DNA polymerases, one of which is a proofreading enzyme, has been shown to dramatically improve the performance of PCR using long templates (Barnes 1994, Cheng *et al*. 1994a). Essentially, the improvements come about because the proofreading enzyme removes mismatched bases that are often incorporated into growing DNA strands by enzymes such as *Taq* polymerase. Under normal conditions, *Taq* polymerase would stall at these obstructions and, lacking the intrinsic proofreading activity to correct them, the reaction would most likely be aborted.

Using such polymerase mixtures, it has been possible to amplify DNA fragments of up to 22 kb directly from human genomic DNA, almost the entire 16.6 kb human mitochondrial genome and the complete or near-complete genomes of several viruses, including 42 kb of the 45 kb phage-λ genome (Cheng *et al*. 1994a,b). Several commercial companies now provide cocktails of enzymes suitable for long PCR, e.g. TaqPlus Long PCR system, marketed by Stratagene, which is essentially a mixture of *Taq* polymerase and the thermostable proofreading enzyme *Pfu* polymerase. The technique has been applied to the structural analysis of human genes (e.g. Ruzzo *et al*. 1998, Bochmann *et al*. 1999) and viral genomes, including human immunodeficiency virus (HIV) (Dittmar *et al*. 1997). Long PCR has particular diagnostic value for the analysis of human triplet-repeat disorders, such as Friedreich's ataxia (Lamont *et al*. 1997). However, while long PCR is useful for the isolation of genes where sequence information is already available, it is unlikely to replace the use of genomic libraries, since the latter represent a permanent, full-genome resource that can be shared by numerous laboratories. Indeed, genomic libraries may be used in preference to total genomic DNA as the starting-point for gene isolation by long PCR.

Fragment libraries

Traditional genomic libraries cannot be prepared from small amounts of starting material, e.g. single cells, or from problematical sources, such as fixed tissue. In these cases, PCR is the only available strategy for gene isolation. However, as well as being useful for the isolation of specific fragments, the PCR can be used to generate libraries, i.e. by amplifying a representative collection of random genomic fragments. This can be achieved using either random primers followed by size selection for suitable PCR products or a strategy in which genomic DNA is digested with restriction enzymes and then linkers are ligated to the ends of the DNA fragments, providing annealing sites for one specific type of primer (e.g. Zhang *et al.* 1992, Cheung & Nelson 1996). These techniques are powerful because they allow genomic fragment libraries to be prepared from material that could not yield DNA of suitable quality or quantity for conventional library construction, but competition among the templates generally does not allow the production of a truly representative library.

cDNA cloning

Properties of cDNA

cDNA is prepared by reverse-transcribing cellular RNA. Cloned eukaryotic cDNAs have their own special uses, which derive from the fact that they lack introns and other non-coding sequences present in the corresponding genomic DNA. Introns are rare in bacteria but occur in most genes of higher eukaryotes. They can be situated within the coding sequence itself, where they then interrupt the collinear relationship of the gene and its encoded polypeptide, or they may occur in the 5′ or 3′ untranslated regions. In any event, they are copied by RNA polymerase when the gene is transcribed. The primary transcript goes through a series of processing events in the nucleus before appearing in the cytoplasm as mature mRNA. These events include the removal of intron sequences by a process called *splicing*. In mammals, some genes contain numerous large introns that represent the vast majority of the sequence. For example, the human dystrophin gene contains 79 introns, representing over 99% of the sequence. The gene is nearly 2.5 Mb in length and yet the corresponding cDNA is only just over 11 kb. Thus, one advantage of cDNA cloning is that in many cases the size of the cDNA clone is significantly lower than that of the corresponding genomic clone. Since removal of eukaryotic intron transcripts by

splicing does not occur in bacteria, eukaryotic cDNA clones find application where bacterial expression of the foreign DNA is necessary, either as a prerequisite for detecting the clone (see p. 109) or because expression of the polypeptide product is the primary objective. Also, where the sequence of the genomic DNA is available, the position of intron/exon boundaries can be assigned by comparison with the cDNA sequence.

cDNA libraries

Under some circumstances, it may be possible to prepare cDNA directly from a purified mRNA species. Much more commonly, a *cDNA library* is prepared by reverse-transcribing a population of mRNAs and then screened for particular clones. An important concept is that the cDNA library is representative of the RNA population from which it was derived. Thus, whereas genomic libraries are essentially the same, regardless of the cell type or developmental stage from which the DNA was isolated, the contents of cDNA libraries will vary widely according to these parameters. A given cDNA library will also be enriched for abundant mRNAs but may contain only a few clones representing rare mRNAs. Furthermore, where a gene is differentially spliced, a cDNA library will contain different clones representing alternative splice variants.

Table 6.2 shows the abundances of different classes of mRNAs in two representative tissues. Generally, mRNAs can be described as abundant, moderately abundant or rare. Notice that, in the chicken oviduct, one mRNA type is superabundant. This encodes ovalbumin, the major egg-white protein. Therefore, the starting population is naturally so enriched in ovalbumin mRNA that isolating the ovalbumin cDNA can be achieved without the use of a library. An appropriate strategy for obtaining such abundant cDNAs is to clone them directly in an M13 vector, such as M13 mp8. A set of clones can then be sequenced immediately and identified on the basis of the polypeptide that each encodes. A successful demonstration of this strategy was reported by Putney *et al.* (1983), who determined DNA sequences of 178 randomly chosen muscle cDNA clones. Based on the amino acid sequences available for 19 abundant muscle-specific proteins, they were able to

Table 6.2 Abundance classes of typical mRNA populations.

Source	Number of different mRNAs	Abundance (molecules/cell)
Mouse liver cytoplasmic poly(A)⁺	9	12 000
	700	300
	11 500	15
Chick oviduct polysomal poly(A)⁺	1	100 000
	7	4 000
	12 500	5

References: mouse (Young *et al.* 1976); chick oviduct (Axel *et al.* 1976).

identify clones corresponding to 13 of these 19 proteins, including several protein variants.

For the isolation of cDNA clones in the moderate- and low-abundance classes, it is usually necessary to construct a cDNA library. Once again, the high efficiency obtained by packaging *in vitro* makes phage-λ vectors attractive for obtaining large numbers of cDNA clones. Phage-λ insertion vectors are particularly well suited for cDNA cloning and some of the most widely used vectors are discussed in Box 6.1.

Typically, 10^5–10^6 clones are sufficient for the isolation of low-abundance mRNAs from most cell types, i.e. those present at 15 molecules per cell or above. However, some mRNAs are even less abundant than this, and may be further diluted if they are expressed in only a few specific cells in a particular tissue. Under these circumstances, it may be worth enriching the mRNA preparation prior to library construction, e.g. by size fractionation, and testing the fractions for the presence of the desired molecule. One way in which

---

Box 6.1 Phage-λ vectors for cDNA cloning and expression

λgt10 and λgt11

Most early cDNA libraries were constructed using plasmid vectors, and were difficult to store and maintain for long periods. They were largely replaced by phage-λ libraries, which can be stored indefinitely and can also be prepared to much higher titres. λgt10 and λgt11 were the standard vectors for cDNA cloning until about 1990. Both λgt10 and λgt11 are insertion vectors, and they can accept approximately 7.6 kb and 7.2 kb of foreign DNA, respectively. In each case, the foreign DNA is introduced at a unique *Eco*RI cloning site. λgt10 is used to make libraries that are screened by hybridization. The *Eco*RI site interrupts the phage *cI* gene, allowing selection on the basis of plaque

morphology. λgt11 contains an *E. coli lacZ* gene driven by the *lac* promoter. If inserted in the correct orientation and reading frame, cDNA sequences cloned in this vector can be expressed as β-galactosidase fusion proteins, and can be detected by immunological screening or screening with other ligands (see p. 109). λgt11 libraries can also be screened by hybridization, although λgt10 is more appropriate for this screening strategy because higher titres are possible.

λZAP series

While phage-λ vectors generate better libraries, they cannot be manipulated *in vitro* with the convenience of plasmid vectors. Therefore, phage clones have to

continued

Box 6.1 *continued*

be laboriously subcloned back into plasmids for further analysis. This limitation of conventional phage-λ vectors has been addressed by the development of hybrids, sometimes called *phasmids*, which possess the most attractive features of both bacteriophage λ and plasmids (see Chapter 5). The most popular current vectors for cDNA cloning are undoubtedly those of the λZAP series marketed by Stratagene (Short *et al*. 1988). A map of the original λZAP vector is shown below. The advantageous features of this vector are: (i) the high capacity – up to 10 kb of foreign DNA can be cloned, which is large enough to encompass most cDNAs; (ii) the presence of a polylinker with six unique restriction sites, which increases cloning versatility and also allows directional cloning; and (iii) the availability of T3 and T7 RNA polymerase sites flanking the polylinker,

allowing sense and antisense RNA to be prepared from the insert. Most importantly, all these features are included within a plasmid vector called pBluescript, which is itself inserted into the phage genome. Thus the cDNA clone can be recovered from the phage and propagated as a high-copy-number plasmid without any subcloning, simply by coinfecting the bacteria with a helper f1 phage that nicks the λZAP vector at the flanks of the plasmid and facilitates excision. Another member of this series, λZAP Express, also includes the human cytomegalovirus promoter and SV40 terminator, so that fusion proteins can be expressed in mammalian cells as well as bacteria. Thus, cDNA libraries can be cloned in the phage vector in *E. coli*, rescued as plasmids and then transfected into mammalian cells for expression cloning.

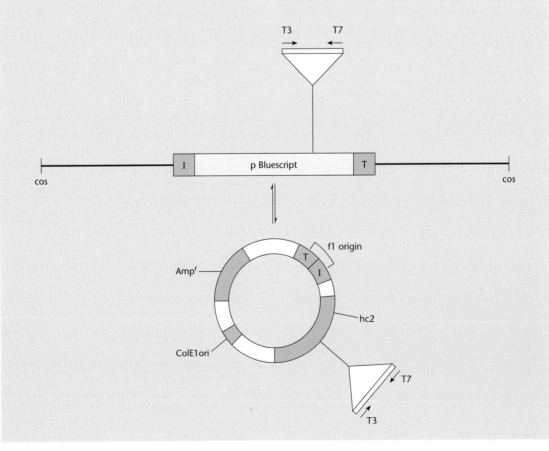

this can be achieved is to inject mRNA fractions into *Xenopus* oocytes (p. 215) and test them for production of the corresponding protein (Melton 1987). See also the discussion of subtraction cloning on p. 115.

Preparation of cDNA for library construction

The cDNA synthesis reaction

The synthesis of double-stranded cDNA suitable for insertion into a cloning vector involves three major steps: (i) first-strand DNA synthesis on the mRNA template, carried out with a reverse transcriptase; (ii) removal of the RNA template; and (iii) second-strand DNA synthesis using the first DNA strand as a template, carried out with a DNA-dependent DNA polymerase, such as *E. coli* DNA polymerase I. All DNA polymerases, whether they use RNA or DNA as the template, require a primer to initiate strand synthesis.

Development of cDNA cloning strategies

The first reports of cDNA cloning were published in the mid-1970s and were all based on the homopolymer tailing technique, which is described briefly in Chapter 3. Of several alternative methods, the one that became the most popular was that of Maniatis *et al.* (1976). This involved the use of an oligo-dT primer annealing at the polyadenylate tail of the mRNA to prime first-strand cDNA synthesis, and took advantage of the fact that the first cDNA strand has the tendency to transiently fold back on itself, forming a hairpin loop, resulting in self-priming of the second strand (Efstratiadis *et al.* 1976). After the synthesis of the second DNA strand, this loop must be cleaved with a single-strand-specific nuclease, e.g. S1 nuclease, to allow insertion into the cloning vector (Fig. 6.5).

A serious disadvantage of the hairpin method is that cleavage with S1 nuclease results in the loss of a certain amount of sequence at the 5′ end of the clone. This strategy has therefore been superseded by other methods in which the second strand is primed in a separate reaction. One of the simplest strategies is shown in Fig. 6.6 (Land *et al.* 1981). After first-strand synthesis, which is primed with an

oligo-dT primer as usual, the cDNA is tailed with a string of cytidine residues using the enzyme terminal transferase. This artificial oligo-dC tail is then used as an annealing site for a synthetic oligo-dG primer, allowing synthesis of the second strand. Using this method, Land *et al.* (1981) were able to isolate a full-length cDNA corresponding to the chicken lysozyme gene. However, the efficiency can be lower for other cDNAs (e.g. Cooke *et al.* 1980).

For cDNA expression libraries, it is advantageous if the cDNA can be inserted into the vector in the correct orientation. With the self-priming method, this can be achieved by adding a synthetic linker to the double-stranded cDNA molecule before the hairpin loop is cleaved (e.g. Kurtz & Nicodemus 1981; Fig. 6.7a). Where the second strand is primed separately, direction cloning can be achieved using an oligo-dT primer containing a linker sequence (e.g. Coleclough & Erlitz 1985; Fig. 6.7b). An alternative is to use primers for cDNA synthesis that are already linked to a plasmid (Fig. 6.7c). This strategy was devised by Okayama and Berg (1982) and has two further notable characteristics. First, full-length cDNAs are *preferentially obtained* because an RNA–DNA hybrid molecule, the result of first-strand synthesis, is the substrate for a terminal transferase reaction. A cDNA that does not extend to the end of the mRNA will present a shielded 3-hydroxyl group, which is a poor substrate for tailing. Secondly, the second-strand synthesis step is primed by nicking the RNA at multiple sites with RNase H. Second-strand synthesis therefore occurs by a nick-translation type of reaction, which is highly efficient. Simpler cDNA cloning strategies incorporating replacement synthesis of the second strand are widely used (e.g. Gubler & Hoffman 1983, Lapeyre & Amalric 1985). The Gubler–Hoffman reaction, as it is commonly known, is show in Fig. 6.8.

Full-length cDNA cloning

Limitations of conventional cloning strategies

Conventional approaches to the production of cDNA libraries have two major drawbacks. First, where oligo-dT primers are used to initiate first-strand synthesis, there is generally a 3′-end bias (preferential recovery of clones representing the 3′ end of cDNA

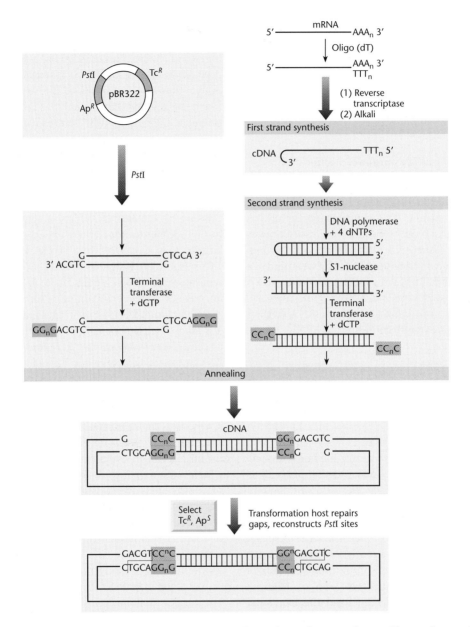

Fig. 6.5 An early cDNA cloning strategy, involving hairpin-primed second-strand DNA synthesis and homopolymer tailing to insert the cDNA into the vector.

sequences) in the resulting library. This can be addressed through the use of random oligonucleotide primers, usually hexamers, for both first- and second-strand cDNA synthesis. However, while this eliminates 3′-end bias in library construction, the resulting clones are much smaller, such that full-

length cDNAs must be assembled from several shorter fragments. Secondly, as the size of a cDNA increases, it becomes progressively more difficult to isolate full-length clones. This is partly due to deficiencies in the reverse-transcriptase enzymes used for first-strand cDNA synthesis. The enzymes

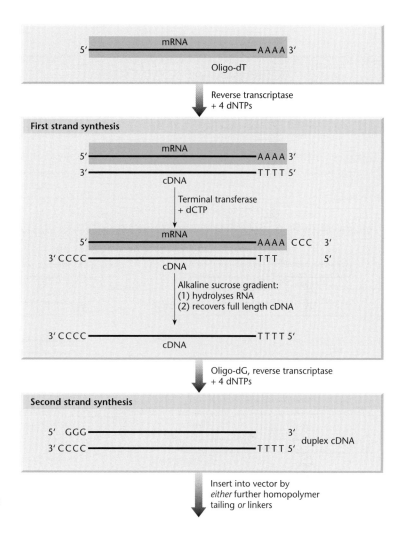

Fig. 6.6 Improved method for cDNA cloning. The first strand is tailed with oligo(dC) allowing the second strand to be initiated using an oligo(dG) primer.

are usually purified from avian myeloblastosis virus (AMV) or produced from a cloned Moloney murine leukaemia virus (MMLV) gene in *E. coli*. Native enzymes have poor processivity and intrinsic RNase activity, which leads to degradation of the RNA template (Champoux 1995). Several companies produce engineered murine reverse transcriptases that lack RNase H activity, and these are more efficient in the production of full-length cDNAs (Gerard & D'Allesio 1993). An example is the enzyme Super-Script II, marketed by Life Technologies (Kotewicz *et al.* 1988). This enzyme can also carry out reverse transcription at temperatures of up to 50°C. The native enzymes function optimally at 37°C and therefore tend to stall at sequences that are rich in secondary

structure, as often found in 5′ and 3′ untranslated regions.

Selection of 5′ mRNA ends

Despite improvements in reverse transcriptases, the generation of full-length clones corresponding to large mRNAs remains a problem. This has been addressed by the development of cDNA cloning strategies involving the selection of mRNAs with intact 5′ ends. Nearly all eukaryotic mRNAs have a 5′ end cap, a specialized, methylated guanidine residue that is inverted with respect to the rest of the strand and is recognized by the ribosome prior to the initiation of protein synthesis. Using a combination

(a)

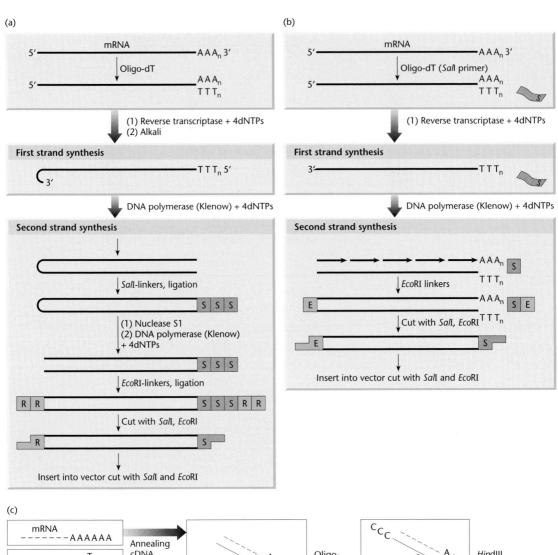

(c)

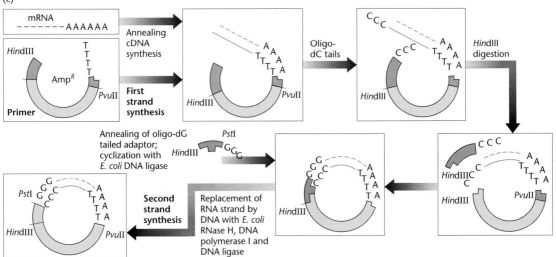

(b)

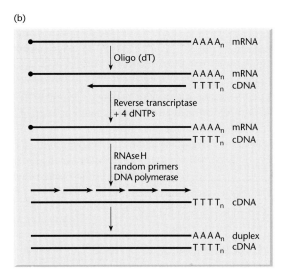

Fig. 6.8 The Gubler–Hoffman method, a simple and general method for non-directional cDNA cloning. First-strand synthesis is primed using an oligo(dT) primer. When the first strand is complete, the RNA is removed with RNase H and the second strand is random-primed and synthesized with DNA polymerase I. T4 DNA polymerase is used to ensure that the molecule is blunt-ended prior to insertion into the vector.

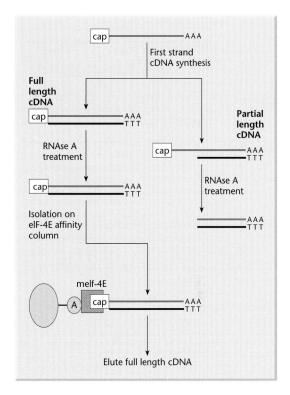

Fig. 6.9 The CAPture method of full-length cDNA cloning, using the eukaryotic initiation factor eIF-4E to select mRNAs with caps protected from RNase digestion by a complementary DNA strand.

of cap selection and nuclease treatment, it is possible to select for full-length first-strand cDNAs and thus generate libraries highly enriched in full-length clones.

An example is the method described by Edery *et al.* (1995) (Fig. 6.9). In this strategy, first-strand cDNA synthesis is initiated as usual, using an oligo-dT primer. Following the synthesis reaction, the hybrid molecules are treated with RNase A, which only

digests single-stranded RNA. DNA–RNA hybrids therefore remain intact. If the first-strand cDNA is full-length, it reaches all the way to the 5′ cap of the mRNA, which is therefore protected from cleavage by RNase A. However, part-length cDNAs will leave a stretch of unprotected single-stranded RNA between the end of the double-stranded region and the cap, which is digested away with the enzyme. In the next stage of the procedure, the eukaryotic translational initiation factor eIF-4E is used to isolate full-length molecules by affinity capture. Incomplete cDNAs and cDNAs synthesized on broken templates will lack the cap and will not be retained. A similar method based on the biotinylation of mRNA has also been reported (Caminci *et al.* 1996). Both methods, however, also co-purify cDNAs resulting from the mispriming of first-strand synthesis, which can account for up to 10% of the clones in a library. An alternative method, *oligo-capping*, addresses this

---

Fig. 6.7 (*opposite*) Methods for directional cDNA cloning. (a) An early strategy in which the formation of a loop is exploited to place a specific linker (in this example, for *Sal*I) at the open end of the duplex cDNA. Following this ligation, the loop is cleaved and trimmed with S1 nuclease and *Eco*R1 linkers are added to both ends. Cleavage with *Eco*R1 and *Sal*I generates a restriction fragment that can be unidirectionally inserted into a vector cleaved with the same enzymes. (b) A similar strategy, but second-strand cDNA synthesis is random-primed. The oligo(dT) primer carries an extension forming a *Sal*I site. During second-strand synthesis, this forms a double-stranded *Sal*I linker. The addition of further *Eco*RI linkers to both ends allows the cDNA to be unidirectionally cloned, as above. (c) The strategy of Okayama and Berg (1982), where the mRNA is linked unidirectionally to the plasmid cloning vector prior to cDNA synthesis, by virtue of a cDNA tail.

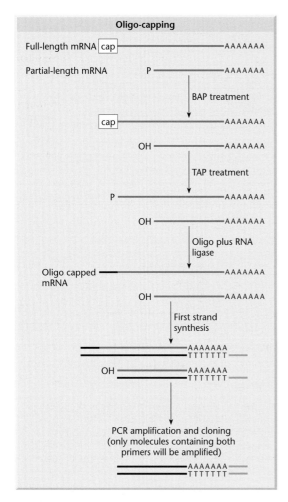

Fig. 6.10 Oligo-capping, the addition of specific oligonucleotide primers to full-length RNAs by sequential treatment with alkaline phosphatase and acid pyrophosphatase. Once the oligo cap has annealed to the 5′ end of the mRNA, it can serve as a primer binding site for PCR amplification.

group. Full-length molecules can be ligated to a specific oligonucleotide, while broken and degraded molecules cannot. The result is an oligo-capped population of full-length mRNAs. This selected population is then reverse-transcribed using an oligo-dT primer. Second-strand synthesis and cloning is then carried out by PCR using the oligo-dT primer and a primer annealing to the oligonucleotide cap. Only full-length cDNAs annealing to both primers will be amplified, thus eliminating broken or degraded RNAs, incomplete first cDNA strands (which lack a 5′ primer annealing site) and misprimed cDNAs (which lack a 3′ primer annealing site).

PCR as an alternative to cDNA cloning

Reverse transcription followed by the PCR (RT-PCR) leads to the amplification of RNA sequences in cDNA form. No modification to the basic PCR strategy (p. 19) is required, except that the template for PCR amplification is generated in the same reaction tube in a prior reverse-transcription reaction (see Kawasaki 1990, Dieffenbach & Dvesler 1995). Using gene-specific primers, RT-PCR is a sensitive means for detecting, quantifying and cloning specific cDNA molecules. Reverse transcription is carried out using a specific 3′ primer that generates the first cDNA strand, and then PCR amplification is initiated following the addition of a 5′ primer to the reaction mix. The sensitivity is such that total RNA can be used as the starting material, rather than the poly(A)$^+$ RNA which is used for conventional cDNA cloning. Total RNA also contains ribosomal RNA (rRNA) and transfer RNA (tRNA), which can be present in a great excess to mRNA.

Due to the speed with which RT-PCR can be carried out, it is an attractive approach for obtaining a specific cDNA sequence for cloning. In contrast, screening a cDNA library is laborious, even presuming that a suitable cDNA library is already available and does not have to be constructed for the purpose. Quite apart from the labour involved, a cDNA library may not yield a cDNA clone with a full-length coding region, because, as described above, generating a full-length cDNA clone may be technically challenging, particularly with respect to long mRNAs. Furthermore, the sought-after cDNA may be very rare even in specialized libraries. Does this mean

problem by performing selection at the RNA stage (Maruyama & Sugano 1994, Suzuki *et al.* 1997, 2000; Fig. 6.10). The basis of the method is that RNA is sequentially treated with the enzymes alkaline phosphatase and acid pyrophosphatase. The first enzyme removes phosphate groups from the 5′ ends of uncapped RNA molecules, but does not affect full-length molecules with a 5′ cap. The second treatment removes the cap from full-length RNAs, leaving a 5′-terminal residue with a phosphate that cDNA libraries have been superseded? Despite the advantages of RT-PCR, there are still reasons for constructing cDNA libraries. The first reflects the availability of starting material and the permanence of the library. A sought-after mRNA may occur in a source that is not readily available, perhaps a small number of cells in a particular human tissue. A good-quality cDNA library has only to be constructed once from this tissue to give a virtually infinite resource for future use. The specialized library is permanently available for screening. Indeed, the library may be used as a source from which a specific cDNA can be obtained by PCR amplification. The second reason concerns screening strategies. The PCR-based approaches are dependent upon specific primers. However, with cDNA libraries, screening strategies are possible that are based upon expression, e.g. immunochemical screening, rather than nucleic acid hybridization (see below).

As discussed above for genomic libraries, PCR can be used to provide the DNA for library construction when the source is unsuitable for conventional approaches, e.g. a very small amount of starting material or fixed tissue. Instead of gene-specific primers, universal primers can be used that lead to the amplification of all mRNAs, which can then be subcloned into suitable vectors. A disadvantage of PCR-based strategies for cDNA library construction is that the DNA polymerases used for PCR are more error-prone than those used conventionally for second-strand synthesis, so the library may contain a large number of mutations. There is also likely to be a certain amount of distortion due to competition among templates, and a bias towards shorter cDNAs.

A potential problem with RT-PCR is false results resulting from the amplification of contaminating genomic sequences in the RNA preparation. Even trace amounts of genomic DNA may be amplified. In the study of eukaryotic mRNAs, it is therefore desirable to choose primers that anneal in different exons, such that the products expected from the amplification of cDNA and genomic DNA would be different sizes or, if the intron is suitably large, so that genomic DNA would not be amplified at all. Where this is not possible (e.g. when bacterial RNA is used as the template), the RNA can be treated with DNase prior to amplification to destroy any contaminating DNA.

Rapid amplification of cDNA ends (RACE)

Another way to address the problem of incomplete cDNA sequences in libraries is to use a PCR-based technique for the *rapid amplification of cDNA ends* (RACE) (Frohman *et al.* 1988). Both 5′ RACE and 3′ RACE protocols are available, although 3′ RACE is usually only required if cDNAs have been generated using random primers. In each case, only limited knowledge of the mRNA sequence is required. A single stretch of sequence within the mRNA is sufficient, so an incomplete clone from a cDNA library is a good starting-point. From this sequence, specific primers are chosen which face *outwards* and which produce overlapping cDNA fragments. In the two RACE protocols, extension of the cDNAs from the ends of the transcript to the specific primers is accomplished by using primers that hybridize either at the natural 3′ poly(A) tail of the mRNA, or at a synthetic poly(dA) tract added to the 5′ end of the first-strand cDNA (Fig. 6.11). Finally, after amplification, the overlapping RACE products can be combined if desired, to produce an intact full-length cDNA.

Although simple in principle, RACE suffers from the same limitations that affect conventional cDNA cloning procedures. In 5′ RACE, for example, the reverse transcriptase may not, in many cases, reach the authentic 5′ end of the mRNA, but all first-strand cDNAs, whether full length or truncated, are tailed in the subsequent reaction, leading to the amplification of a population of variable-length products. Furthermore, as might be anticipated, since only a single *specific* primer is used in each of the RACE protocols, the specificity of amplification may not be very high. This is especially problematical where the specific primer is degenerate. In order to overcome this problem, a modification of the RACE method has been devised which is based on using nested primers to increase specificity (Frohman & Martin 1989). Strategies for improving the specificity of RACE have been reviewed (Schaefer 1995, Chen 1996).

Screening strategies

The identification of a specific clone from a DNA library can be carried out by exploiting either the sequence of the clone or the structure/function of its

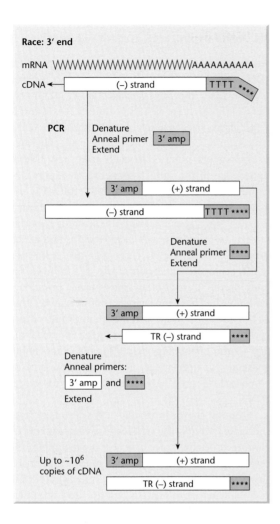

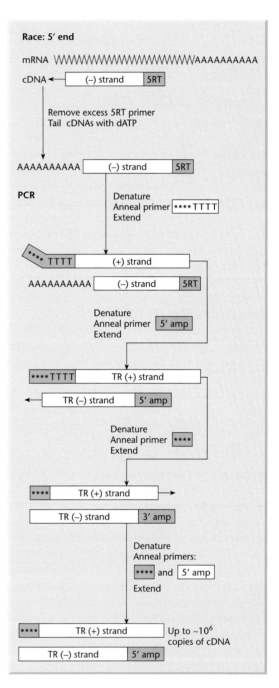

Fig. 6.11 Rapid amplification of cDNA ends (RACE) (Frohman *et al.* 1988). *3′ Protocol.* The mRNA is reverse transcribed using an oligo(dT$_{17}$) primer which has a 17 nucleotide extension at its 5′-end. This extension, the anchor sequence, is designed to contain restriction sites for subsequent cloning. Amplification is performed using the anchor 17-mer (which has a T_{m} higher than oligo(dT$_{17}$)) and a primer specific for the sought-after cDNA. *5′ Protocol.* The mRNA is reverse transcribed from a specific primer. The resultant cDNA is then extended by terminal transferase to create a poly(dA) tail at the 3′-end of the cDNA. Amplification is performed with the oligo(dT$_{17}$)/anchor system as used for the 3′ protocol, and the specific primer. Open boxes represent DNA strands being synthesized; coloured boxes represent DNA from a previous step. The diagram is simplified to show only how the *new* product from a previous step is used. Molecules designated TR, truncated, are shorter than full-length (+) or (−) strands.

Box 6.2 Expressed sequence tags (ESTs) for high-throughput genome research

The gold standard in the analysis of individual genes is a full-length cDNA clone that has been independently sequenced several times to ensure accuracy. Such clones are desirable for accurate archiving and for the detailed mapping of genomic transcription units, i.e. to determine the transcriptional start and stop sites, and all intron/exon boundaries. However, as discussed in the main text, such clones can be difficult and expensive to obtain. Technology has not yet advanced to the stage where full-length cDNAs can be produced and sequenced in a high-throughput manner.

Fortunately, full-length cDNA clones are not required for many types of analysis. Even short cDNA sequence fragments can be used to unambiguously identify specific genes and therefore map them on to physical gene maps or provide information about their expression patterns. The development of high-throughput sequencing technology has allowed thousands of clones to be picked randomly from cDNA libraries and subjected to single-pass sequencing to generate 200–300 bp cDNA signatures called *expressed sequence tags* (ESTs) (Wilcox *et al.* 1991, Okubo *et al.* 1992). Although short and somewhat inaccurate, very large numbers of sequences can be collected rapidly and inexpensively and deposited into public databases that can be searched using the Internet. The vast majority of database sequences are now ESTs rather than full cDNA or genomic clones. ESTs have been used for gene discovery, as physical markers on genomic maps and for the identification of genes in genomic clones (e.g. Adams *et al.* 1991, 1992, Banfi *et al.* 1996). Over two million ESTs from numerous species are currently searchable using the major public EST database, dbEST (Boguski *et al.* 1993). The development of EST informatics has been reviewed (Boguski 1995, Gerhold & Caskey 1996, Hartl 1996, Okubo & Matsubara 1997). The advent of ESTs prompted a wide public debate on the concept of patenting genes. The National Institutes of Health attempted to patent more than 1000 of the first EST sequences to be generated, but the patent application was rejected, predominantly on the grounds that ESTs are incomplete sequences and lack precise functional applications (for discussion, see Roberts 1992).

As well as their use for mapping, ESTs are useful for expression analysis. PCR primers designed around ESTs have been used to generate large numbers of target sequences for cDNA microarrays (see Box 6.5), and the partial cDNA fragments used for techniques such as differential-display PCR are also essentially ESTs (p. 116). The ultimate EST approach to expression analysis is *serial analysis of gene expression* (SAGE). In this technique, the size of the sequence tag is only 9–10 bp (the minimum that is sufficient to uniquely identify specific transcripts) and multiple tags are ligated into a large concatemer allowing expressed genes to be 'read' by cloning and sequencing the tags serially arranged in each clone (for details see Velculescu *et al.* 1995).

---

expressed product. The former applies to any type of library, genomic or cDNA, and can involve either nucleic acid hybridization or the PCR. In each case, the design of the probe or primers can be used to home in on one specific clone or a group of structurally related clones. Note that PCR screening can also be used to isolate DNA sequences from uncloned genomic DNA and cDNA. Screening the product of a clone applies only to expression libraries, i.e. libraries where the DNA fragment is expressed to yield a protein. In this case, the clone can be identified because its product is recognized by an antibody or a ligand of some nature, or because the biological activity of the protein is preserved and can be assayed in an appropriate test system.

Sequence-dependent screening

Screening by hybridization

Nucleic acid hybridization is the most commonly used method of library screening because it is rapid,

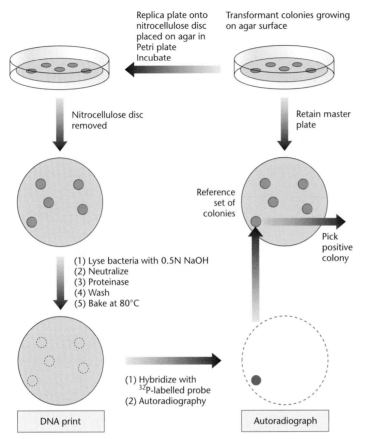

Replica plate onto
nitrocellulose disc
placed on agar in
Petri plate
Incubate

Transformant colonies growing
on agar surface

Nitrocellulose disc
removed

Retain master
plate

Reference
set of
colonies

Pick
positive
colony

(1) Lyse bacteria with 0.5N NaOH
(2) Neutralize
(3) Proteinase
(4) Wash
(5) Bake at 80°C

(1) Hybridize with
^{32}P-labelled probe
(2) Autoradiography

DNA print

Autoradiograph

Fig. 6.12 Grunstein–Hogness method for detection of recombinant clones by colony hybridization.

it can be applied to very large numbers of clones and, in the case of cDNA libraries, can be used to identify clones that are not full-length (and therefore cannot be expressed).

Grunstein and Hogness (1975) developed a screening procedure to detect DNA sequences in transformed colonies by hybridization *in situ* with radioactive RNA probes. Their procedure can rapidly determine which colony among thousands contains the target sequence. A modification of the method allows screening of colonies plated at a very high density (Hanahan & Meselson 1980). The colonies to be screened are first replica-plated on to a nitrocellulose filter disc that has been placed on the surface of an agar plate prior to inoculation (Fig. 6.12). A reference set of these colonies on the master plate is retained. The filter bearing the colonies is removed and treated with alkali so that the bacterial colonies are lysed and the DNA they contain is denatured.

The filter is then treated with proteinase K to remove protein and leave denatured DNA bound to the nitrocellulose, for which it has a high affinity, in the form of a 'DNA print' of the colonies. The DNA is fixed firmly by baking the filter at 80°C. The defining, labelled RNA is hybridized to this DNA and the result of this hybridization is monitored by autoradiography. A colony whose DNA print gives a positive autoradiographic result can then be picked from the reference plate.

Variations of this procedure can be applied to phage plaques (Jones & Murray 1975, Kramer *et al.* 1976). Benton and Davis (1977) devised a method called *plaque lift*, in which the nitrocellulose filter is applied to the upper surface of agar plates, making direct contact between plaques and filter. The plaques contain phage particles, as well as a considerable amount of unpackaged recombinant DNA. Both phage and unpackaged DNA bind to the filter and

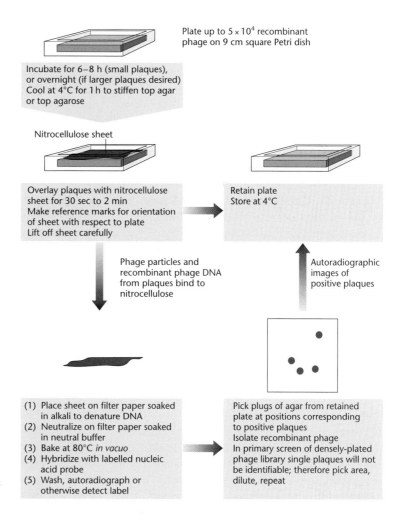

Plate up to 5×10^4 recombinant phage on 9 cm square Petri dish

Incubate for 6–8 h (small plaques), or overnight (if larger plaques desired) Cool at 4°C for 1 h to stiffen top agar or top agarose

Nitrocellulose sheet

Overlay plaques with nitrocellulose sheet for 30 sec to 2 min Make reference marks for orientation of sheet with respect to plate Lift off sheet carefully

Retain plate Store at 4°C

Phage particles and recombinant phage DNA from plaques bind to nitrocellulose

Autoradiographic images of positive plaques

(1) Place sheet on filter paper soaked in alkali to denature DNA
(2) Neutralize on filter paper soaked in neutral buffer
(3) Bake at 80°C *in vacuo*
(4) Hybridize with labelled nucleic acid probe
(5) Wash, autoradiograph or otherwise detect label

Pick plugs of agar from retained plate at positions corresponding to positive plaques Isolate recombinant phage In primary screen of densely-plated phage library single plaques will not be identifiable; therefore pick area, dilute, repeat

Fig. 6.13 Benton and Davis' plaque-lift procedure.

can be denatured, fixed and hybridized. This method has the advantage that several identical DNA prints can easily be made from a single-phage plate: this allows the screening to be performed in duplicate, and hence with increased reliability, and also allows a single set of recombinants to be screened with two or more probes. The Benton and Davis (1977) procedure is probably the most widely applied method of library screening, successfully applied in thousands of laboratories to the isolation of recombinant phage by nucleic acid hybridization (Fig. 6.13). More recently, however, library presentation and screening have become increasingly automated. Box 6.3 considers the advantages of gridded reference libraries.

In place of RNA probes, DNA or synthetic oligonucleotide probes can be used. A number of alternative labelling methods are also available that avoid the use

of radioactivity. These methods involve the incorporation of chemical labels into the probe, such as digoxigenin or biotin, which can be detected with a specific antibody or the ligand streptavidin, respectively.

Probe design

A great advantage of hybridization for library screening is that it is extremely versatile. Conditions can be used in which hybridization is very stringent, so that only sequences identical to the probe are identified. This is necessary, for example, to identify genomic clones corresponding to a specific cDNA or to identify overlapping clones in a chromosome walk (see below). Alternatively, less stringent conditions can be used to identify both identical and related sequences. This is appropriate where a probe from one species

---

Box 6.3 Gridded (arrayed) hybridization reference libraries

Traditionally, library screening by hybridization involves taking a plaque lift or colony blot, which generates a replica of the distribution of clones on an agar plate. However, an alternative is to individually pick clones and arrange them on the membrane in the form of a regular grid. Once a laborious process, gridding or arraying has been considerably simplified through the use of robotics. Machines can be programmed to pick clones from microtitre dishes and spot them onto membranes at a high density; then the membrane can be hybridized with a radioactive probe as normal. Using traditional libraries, positive clones are detected by autoradiography and the X-ray film must be aligned with the original plates in order to identify the corresponding plaques. With gridded libraries, however, positive signals can be used to obtain sets of coordinates, which then identify the corresponding clone from the original microtitre dishes. Since identical sets of membranes can be easily prepared,

duplicates can be distributed to other laboratories for screening. These laboratories can then determine the coordinates of their positive signals and order the corresponding clone from the source laboratory. Thus, one library can serve a number of different users and all data can be centralized (Zehetner & Lehrach 1994). Gridded libraries, while convenient for screening and data sharing, are more expensive to prepare than traditional libraries. Therefore, they are often prepared for high-value libraries with wide applications, such as genomic libraries cloned in high-capacity P1, BAC or YAC vectors (Bentley *et al.* 1992) and also for valuable cDNA libraries (Lennon & Lehrach 1991). It is possible to plate libraries at a density of one clone per well, although for PACs and BACs it is more common to pool clones in a hierarchical manner, so that individual clones may be identified by successive rounds of screening on smaller subpools (e.g. Shepherd *et al.* 1994, Shepherd & Smoller 1994).

---

is being used to isolate a homologous clone from another species (e.g. see Old *et al.* 1982). Probes corresponding to a conserved functional domain of a gene may also cross-hybridize with several different clones in the same species at lower stringency, and this can be exploited to identify members of a gene family. The identification of the vertebrate *Hox* genes provides an example in which cross-species hybridization was used to identify a family of related clones (Levine *et al.* 1984). In this case a DNA sequence was identified that was conserved between the *Drosophila* developmental genes *fushi tarazu* and *Antennapedia*. When this sequence, the homoeobox, was used to screen a Southern blot of DNA from other species, including frogs and humans, several hybridizing bands were revealed. This led to the isolation of a number of clones from vertebrate cDNA libraries representing the large family of *Hox* genes that play a central role in animal development.

Hybridization thus has the potential to isolate any sequence from any library *if a probe is available*. If a suitable DNA or RNA probe cannot be obtained from an existing cloned DNA, an alternative strategy is to

make an oligonucleotide probe by chemical synthesis. This requires some knowledge of the amino acid sequence of the protein encoded by the target clone. However, since the genetic code is degenerate (i.e. most amino acids are specified by more than one codon), degeneracy must be incorporated into probe design so that a mixture of probes is made, at least one variant of which will specifically match the target clone. Amino acid sequences known to include methionine and tryptophan are particularly valuable because these amino acids are each specified by a single codon, hence reducing the degeneracy of the resulting probe. Thus, for example, the oligopeptide His-Phe-Pro-Phe-Met may be identified and chosen to provide a probe sequence, in which case 32 different oligonucleotides would be required:

$$5' \quad CA_{C}^{T}TT_{C}^{T}CCCTT_{C}^{T}ATG \quad 3'$$
$$A$$
$$G$$

These 32 different sequences do not have to be synthesized individually because it is possible to perform a mixed addition reaction for each polymerization step. This mixture is then end-labelled with a single isotopic or alternatively labelled nucleotide, using an exchange reaction. This mixed-probe method was originally devised by Wallace and co-workers (Suggs *et al.* 1981). To cover all codon possibilities, degeneracies of 64-fold (Orkin *et al.* 1983) or even 256-fold (Bell *et al.* 1984) have been employed successfully. What length of oligonucleotide is required for reliable hybridization? Even though 11-mers can be adequate for Southern blot hybridization (Singer-Sam *et al.* 1983), longer probes are necessary for good colony and plaque hybridization. Mixed probes of 14 nucleotides have been successful, although 16-mers are typical (Singer-Sam *et al.* 1983).

An alternative strategy is to use a single longer probe of 40–60 nucleotides. Here the uncertainty at each codon is largely ignored and instead increased probe length confers specificity. Such probes are usually designed to incorporate the most commonly used codons in the target species, and they may include the non-standard base inosine at positions of high uncertainty because this can pair with all four conventional bases. Such probes are sometimes termed *guessmers*. Hybridization is carried out under low stringency to allow for the presence of mismatches. This strategy is examined theoretically by Lathe (1985) and has been applied to sequences coding for human coagulation factor VIII (Toole *et al.* 1984, Wood *et al.* 1984) and the human insulin receptor (Ullrich *et al.* 1985).

Chromosome walking

Earlier in this chapter, we discussed the advantages of making genomic libraries from random DNA fragments. One of these advantages is that the resulting fragments overlap, which allows genes to be cloned by *chromosome walking*. The principle of chromosome walking is that overlapping clones will hybridize to each other, allowing them to be assembled into a contiguous sequence. This can be used to isolate genes whose function is unknown but whose genetic location is known, a technique known as *positional cloning*.

To begin a chromosome walk, it is necessary to have in hand a genomic clone that is known to lie very close to the suspected location of the target gene. In humans, for example, this could be a restriction fragment length polymorphism that has been genetically mapped to the same region. This clone is then used to screen a genomic library by hybridization, which should reveal any overlapping clones. These overlapping clones are then isolated, labelled and used in a second round of screening to identify further overlapping clones, and the process is repeated to build up a contiguous map. If the same library is used for each round of screening, previously identified clones can be distinguished from new ones, so that walking back and forth along the same section of DNA is prevented. Furthermore, modern vectors, such as λDASH and λFIX, allow probes to be generated from the end-points of a given genomic clone by *in vitro* transcription (see Fig. 6.4), which makes it possible to walk specifically in one direction. In *Drosophila*, the progress of a walk can also be monitored by using such probes for *in situ* hybridization against polytene chromosomes. Monitoring is necessary due to the dangers posed by repetitive DNA. Certain DNA sequences are highly repetitive and are dispersed throughout the genome. Hybridization with such a sequence could disrupt the orderly progress of a walk, in the worst cases causing a 'warp' to another chromosome. The probe used for stepping from one genomic clone to the next must be a unique sequence clone, or a subclone that has been shown to contain only a unique sequence.

Chromosome walking is simple in principle, but technically demanding. For large distances, it is advisable to use libraries based on high-capacity vectors, such as BACs and YACs, to reduce the number of steps involved. Before such libraries were available, some ingenious strategies were used to reduce the number of steps needed in a walk. In one of the first applications of this technology, Hogness and his co-workers (Bender *et al.* 1983) cloned DNA from the *Ace* and *rosy* loci and the homoeotic *Bithorax* gene complex in *Drosophila*. The number of steps was minimized by exploiting the numerous strains carrying well-characterized inversions and translocations of specific chromosome regions. A different strategy, called *chromosome jumping*, has been used for human DNA (Collins & Weissman 1984, Poustka & Lehrach 1986). This involves the circularization of very large genomic fragments

---

Box 6.4 A landmark publication. Identification of the cystic fibrosis gene by chromosome walking and jumping

Cystic fibrosis (CF) is a relatively common severe autosomal recessive disorder. Until the CF gene was cloned, there was little definite information about the primary genetic defect. The cloning of the CF gene was a breakthrough for studying the biochemistry of the disorder (abnormal chloride-channel function), for providing probes for prenatal diagnosis and for potential treatment by somatic gene therapy or other means. The publication is especially notable for the generality of the cloning strategy. In the absence of any direct functional information about the CF gene, the chromosomal location of the gene was used as the basis of the cloning strategy. Starting from markers identified by linkage analysis as being close to the CF locus on chromosome 7, a total of about 500 kb was encompassed by a combination of chromosome walking and jumping. Jumping was found to be very important to overcome problems caused by 'unclonable' regions which halted the sequential walks, and in one case achieved a distance of 100 kb (Collins *et al.* 1987). In this work, large numbers of clones were involved, obtained from several different phage and cosmid genomic libraries. Among these libraries, one was prepared using the Maniatis strategy using the λCharon 4A vector, and several were prepared using the λDASH and λFIX vectors (Fig. 6.4) after partial digestion of human genomic DNA with *Sau*3AI. Cloned regions were aligned with a map of the genome in the CF region, obtained by long-range restriction mapping using rare-cutting enzymes, such as *Not*I, in combination with pulsed-field gel electrophoresis (p. 10). The actual CF gene was detected in this cloned region by a number of criteria, such as the identification of open reading frames, the detection of cDNAs hybridizing to the genomic clones, the detection of cross-hybridizing sequences in other species and the presence of CpG islands, which are known to be associated with the 5′ ends of many genes in mammals.

From: Rommens *et al.* (1989) *Science* 245: 1059–65.

---

generated by digestion with endonucleases, such as *Not*I, which cut at very rare target sites. This is followed by subcloning of the region covering the closure of the fragment, thus bringing together sequences that were located a considerable distance apart. In this way a *jumping library* is constructed, which can be used for long-distance chromosome walks (Collins *et al.* 1987, Richards *et al.* 1988). The application of chromosome walking and jumping to the cloning of the human cystic fibrosis gene is discussed in Box 6.4.

Screening by PCR

The PCR is widely used to isolate specific DNA sequences from uncloned genomic DNA or cDNA, but it also a useful technique for library screening (Takumi, 1997). As a screening method, PCR has the same versatility as hybridization, and the same limitations. It is possible to identify any clone by PCR but only if there is sufficient information about its sequence to make suitable primers.*

To isolate a specific clone, PCR is carried out with gene-specific primers that flank a unique sequence in the target. A typical strategy for library screening by PCR is demonstrated by Takumi and Lodish (1994). Instead of plating the library out on agar, as would be necessary for screening by hybridization, pools of clones are maintained in multiwell plates. Each well is screened by PCR and positive wells are identified. The clones in each positive well are then

---

* Note that, in certain situations, clever experimental design can allow the PCR to be used to isolate specific but unknown DNA sequences. One example of this is 5′ RACE, which is discussed on p. 101. Another is inverse PCR (p. 267), which can be used to isolate unknown flanking DNA surrounding the insertion site of an integrating vector. In each case, primers are designed to bind to known sequences that are joined to the DNA fragment of interest, e.g. synthetic homopolymer tails, linkers or parts of the cloning vector.

diluted into a series in a secondary set of plates and screened again. The process is repeated until wells carrying homogeneous clones corresponding to the gene of interest have been identified.

There are also several applications where the use of *degenerate primers* is favourable. A degenerate primer is a mixture of primers, all of similar sequence but with variations at one or more positions. This is analogous to the use of degenerate oligonucleotides as hybridization probes, and the primers are synthesized in the same way. A common circumstance requiring the use of degenerate primers is when the primer sequences have to be deduced from amino acid sequences (Lee *et al.* 1988). Degenerate primers may also be employed to search for novel members of a known family of genes (Wilks 1989) or to search for homologous genes between species (Nunberg *et al.* 1989). As with oligonucleotide probes, the selection of amino acids with low codon degeneracy is desirable. However, a 128-fold degeneracy in each primer can be successful in amplifying a single-copy target from the human genome (Girgis *et al.* 1988). Under such circumstances, the concentration of any individual primer sequence is very low, so mismatching between primer and template must occur under the annealing conditions chosen. Since mismatching of the 3′-terminal nucleotide of the primer may prevent efficient extension, degeneracy at this position is to be avoided.

Screening expression libraries (expression cloning)

If a DNA library is established using expression vectors (see Chapter 5), each individual clone can be expressed to yield a polypeptide. While all libraries can be screened by hybridization or PCR, as discussed above, expression libraries are useful because they allow a range of alternative techniques to be employed, each of which exploits some structural or functional property of the gene product. This can be important in cases where the DNA sequence of the target clone is completely unknown and there is no strategy available to design a suitable probe or set of primers.

For higher eukaryotes, all expression libraries are cDNA libraries, since these lack introns and the clones are in most cases of a reasonable size. Gener-

ally, a random primer method is used for cDNA synthesis, so there is a greater representation of 5′ sequences. As discussed above, such libraries are representative of their source, so certain cDNAs are abundant and others rare. However, it should be noted that bacterial expression libraries and many yeast expression libraries are usually genomic, since there are few introns in bacteria and some yeasts and very little intergenic DNA. Efficient expression libraries can be generated by cloning randomly sheared genomic DNA or partially digested DNA, and therefore all genes are represented at the same frequency (Young *et al.* 1985). A potential problem with such libraries is that clones corresponding to a specific gene may carry termination sequences from the gene lying immediately upstream, which can prevent efficient expression. For this reason, conditions are imposed so that the size of the fragments for cloning is smaller than that of the target gene, and enough recombinants are generated for there to be a reasonable chance that each gene fragment will be cloned in all six possible reading frames (three in each orientation). Considerations for cloning DNA in *E. coli* expression vectors are discussed further in Chapter 8.

Immunological screening

Immunological screening involves the use of antibodies that specifically recognize antigenic determinants on the polypeptide synthesized by a target clone. This is one of the most versatile expression-cloning strategies, because it can be applied to any protein for which an antibody is available. Unlike the screening strategies discussed below, there is also no need for that protein to be functional. The molecular target for recognition is generally an *epitope*, a short sequence of amino acids that folds into a particular three-dimensional conformation on the surface of the protein. Epitopes can fold independently of the rest of the protein and therefore often form even when the polypeptide chain is incomplete or when expressed as a fusion with another protein. Importantly, many epitopes can form under denaturing conditions, when the overall conformation of the protein is abnormal.

The first immunological screening techniques were developed in the late 1970s, when expression libraries were generally constructed using plasmid

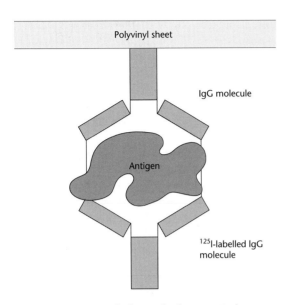

Fig. 6.14 Antigen–antibody complex formation in the immunochemical detection method of Broome and Gilbert. (See text for details.)

vectors. The method of Broome and Gilbert (1978) was widely used at the time. This method exploited the fact that antibodies adsorb very strongly to certain types of plastic, such as polyvinyl, and that IgG antibodies can be readily labelled with ^{125}I by iodination *in vitro*. As usual, transformed cells were plated out on Petri dishes and allowed to form colonies. In order to release the antigen from positive clones, the colonies were lysed, e.g. using chloroform vapour or by spraying with an aerosol of virulent phage (a replica plate is required because this procedure kills the bacteria). A sheet of polyvinyl that had been coated with the appropriate antibody was then applied to the surface of the plate, allowing antigen–antibody complexes to form. The sheet was then removed and exposed to ^{125}I-labelled IgG specific to a *different* determinant on the surface of the antigen (i.e. a determinant not involved in the initial binding of the antigen to the antibody-coated sheet (Fig. 6.14)). The sheet was then washed and exposed to X-ray film. The clones identified by this procedure could then be isolated from the replica plate. Note that this 'sandwich' technique is applicable only where two antibodies recognizing different determinants of the same protein are available. However, if the protein is expressed as a fusion, antibodies that bind to each component of the fusion can be used, efficiently selecting for recombinant molecules.

While plasmid libraries have been useful for expression screening (Helfman *et al.* 1983, Helfman & Hughes 1987), it is much more convenient to use bacteriophage-λ insertion vectors, because these have a higher capacity and the efficiency of *in vitro* packaging allows large numbers of recombinants to be prepared and screened. Immunological screening with phage-λ cDNA libraries was introduced by Young and Davies (1983) using the expression vector λgt11, which generates fusion proteins with β-galactosidase under the control of the *lac* promoter (see Box 6.1 for a discussion of λgt11 and similar fusion vectors, such as λZAP). In the original technique, screening was carried out using colonies of induced lysogenic bacteria, which required the production of replica plates, as above. A simplification of the method is possible by directly screening plaques of recombinant phage. In this procedure (Fig. 6.15), the library is plated out at moderately high density (up to 5×10^4 plaques/9 cm^2 plate), with *E. coli* strain Y1090 as the host. This *E. coli* strain overproduces the *lac* repressor and ensures that no expression of cloned sequences (which may be deleterious to the host) takes place until the inducer isopropyl-β-D-thiogalactoside (IPTG) is presented to the infected cells. Y1090 is also deficient in the *lon* protease, hence increasing the stability of recombinant fusion proteins. Fusion proteins expressed in plaques are absorbed on to a nitrocellulose membrane overlay, and this membrane is processed for antibody screening. When a positive signal is identified on the membrane, the positive plaque can be picked from the original agar plate (a replica is not necessary) and the recombinant phage can be isolated.

The original detection method using iodinated antibodies has been superseded by more convenient methods using non-isotopic labels, which are also more sensitive and have a lower background of non-specific signal. Generally, these involve the use of unlabelled primary antibodies directed against the polypeptide of interest, which are in turn recognized by secondary antibodies carrying an enzymatic label. As well as eliminating the need for isotopes, such methods also incorporate an amplification step, since two or more secondary antibodies bind to the primary antibody. Typically, the secondary antibody

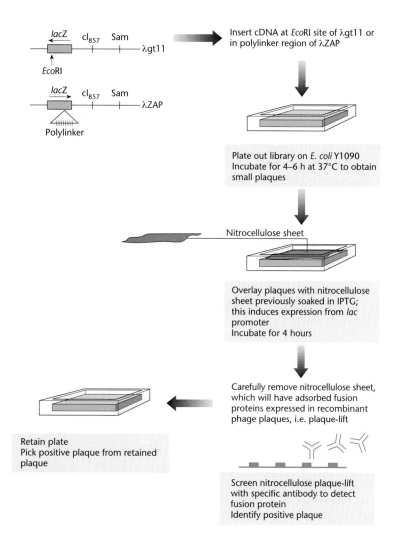

Insert cDNA at *Eco*RI site of λgt11 or in polylinker region of λZAP

Plate out library on *E. coli* Y1090
Incubate for 4–6 h at 37°C to obtain small plaques

Nitrocellulose sheet

Overlay plaques with nitrocellulose sheet previously soaked in IPTG; this induces expression from *lac* promoter
Incubate for 4 hours

Carefully remove nitrocellulose sheet, which will have adsorbed fusion proteins expressed in recombinant phage plaques, i.e. plaque-lift

Retain plate
Pick positive plaque from retained plaque

Screen nitrocellulose plaque-lift with specific antibody to detect fusion protein
Identify positive plaque

Fig. 6.15 Immunochemical screening of λgt11 or λZAP recombinant plaques.

recognizes the species-specific constant region of the primary antibody and is conjugated to either horse-radish peroxidase (De Wet *et al.* 1984) or alkaline phosphatase (Mierendorf *et al.* 1987), each of which can in turn be detected using a simple colorimetric assay carried out directly on the nitrocellulose filter. Polyclonal antibodies, which recognize many different epitopes, provide a very sensitive probe for immunological screening, although they may also cross-react to proteins in the expression host. Monoclonal antibodies and cloned antibody fragments can also be used, although the sensitivity of such reagents is reduced because only a single epitope is recognized.

South-western and north-western screening

We have seen how fusion proteins expressed in plaques produced by recombinant λgt11 or λZAP vectors may be detected by immunochemical screening. A closely related approach has been used for the screening and isolation of clones expressing sequence-specific DNA-binding proteins. As above, a plaque lift is carried out to transfer a print of the library on to nitrocellulose membranes. However, the screening is carried out, without using an antibody, by incubating the membranes with a radiolabelled *double-stranded* DNA oligonucleotide probe, containing the recognition sequence for the target DNA-binding protein. This technique is called *south-western* screening, because it combines the principles of Southern and western blots. It has been particularly successful in the isolation of clones expressing cDNA sequences corresponding to certain mammalian transcription factors (Singh *et al.* 1988, Staudt *et al.* 1988, Vinson *et al.* 1988, Katagiri *et al.* 1989, Williams *et al.* 1991, Xiao *et al.* 1991). A limitation of this technique is that, since individual plaques contain only single cDNA clones, transcription factors that function only in the form of heterodimers or as part of a multimeric complex do not recognize the DNA probe and the corresponding cDNAs cannot be isolated. Clearly the procedure can also be successful only in cases where the transcription factor remains functional when expressed as a fusion polypeptide. It is also clear that the affinity of the polypeptide for the specific DNA sequence must be high, and this has led to the preferential isolation of certain types of transcription factor (reviewed by Singh 1993). More recently, a similar technique has been used to isolate sequence-specific RNA-binding proteins, in this case using a single-stranded RNA probe. By analogy to the above, this is termed *north-western screening* and has been successful in a number of cases (e.g. see Qian & Wilusz 1993; reviewed by Bagga & Wilusz 1999). Both south-western and north-western screening are most efficient when the oligonucleotide contains the binding sequence in multimeric form. This may mean that several fusion polypeptides on the filter bind to each probe, hence greatly increasing the average dissociation time. To minimize non-specific binding, a large excess of unlabelled double-stranded DNA (or single-stranded RNA) is mixed with the specific probe. However, it is usually necessary to confirm the specificity of binding in a second round of screening, using the specific oligonucleotide probe and one or more alternative probes containing a similar sequences that are not expected to be recognized.

Screening with alternative ligands

As well as DNA and RNA, a whole range of alternative 'ligands' can be used to identify polypeptides that specifically bind certain molecules (for example, as an alternative to south-western screening). Such techniques are not widely used because they generally have a low sensitivity and their success depends on the preservation of the appropriate interacting domain of the protein when exposed on the surface of a nitrocellulose filter. Furthermore, as discussed in Chapter 9, the yeast two-hybrid system and its derivatives now provide versatile assay formats for many specific types of protein–protein interaction, with the advantage that such interactions are tested in living cells, so the proteins involved are more likely to retain their functional interacting domains.

Functional cloning

Finally, we consider screening methods that depend on the full biological activity of the protein. This is often termed *functional cloning*. In contrast to positional cloning, described above, functional cloning is possible in complete ignorance of the whereabouts of the gene in the genome and requires no prior knowledge of the nucleotide sequence of the clone or the amino acid sequence of its product. As long as the expressed protein is functional and that function can be exploited to screen an expression library, the corresponding clone can be identified.

Screening by functional complementation

Functional complementation is the process by which a particular DNA sequence compensates for a missing function in a mutant cell, and thus restores the wild-type phenotype. This can be a very powerful method of expression cloning, because, if the mutant cells are non-viable under particular growth conditions, cells carrying the clone of interest can be positively selected, allowing the corresponding clones to be isolated.

Ratzkin and Carbon (1977) provide an early example of how certain eukaryotic genes can be cloned on the basis of their ability to complement auxotrophic mutations in *E. coli*. These investigators inserted fragments of yeast DNA, obtained by mechanical shearing, into the plasmid ColEl, using a homopolymer-tailing procedure. They transformed *E. coli hisB* mutants, which are unable to synthesize histidine, with the recombinant plasmids and plated the bacteria on minimal medium. In this way, they selected for complementation of the mutation and isolated clones carrying an expressed yeast *his* gene. If the function of the gene is highly conserved, it is

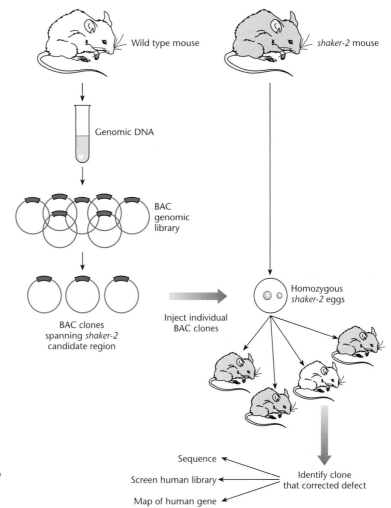

Fig. 6.16 Functional complementation in transgenic mice to isolate the *Shaker-2* gene. Homozygous *shaker-2* fertilized mouse eggs were injected with BAC clones derived from the *Shaker-2* candidate region of a wild-type mouse. Progeny were screened for restoration of the wild-type phenotype, thus identifying the BAC clone corresponding to the *Shaker-2* gene. This clone is then sequenced and used to isolate and map the corresponding human disease gene *DFNB3*.

quite possible to carry out functional cloning of, for example, mammalian proteins in bacteria and yeast. Thus, complementation in yeast has been used to isolate cDNAs for a number of mammalian metabolic enzymes (e.g. Botstein & Fink 1988) and certain highly conserved transcription factors (e.g. Becker *et al.* 1991), as well as regulators of meiosis in plants (Hirayama *et al.* 1997). This approach can also be used in mammalian cells, as demonstrated by Strathdee *et al.* (1992), who succeeded in isolating the *FACC* gene, corresponding to complementation group C of Fanconi's anaemia. Generally a pool system is employed, where cells are transfected with a complex mix of up to 100 000 clones. Pools that successfully complement the mutant phenotype are then subdivided for a further round of transfection, and the procedure repeated until the individual cDNA responsible is isolated.

Functional complementation is also possible in transgenic animals and plants. In this way, Probst *et al.* (1998) were able to clone the mouse deafness-associated gene *Shaker-2*, and from there its human homologue, *DFNB3* (Fig. 6.16). The *shaker-2* mutation had previously been mapped to a region of the mouse genome that is syntenic to the region involved in a human deafness disorder. BAC clones corresponding to this region were therefore prepared from wild-type mice and microinjected into the eggs of *shaker-2* mutants. The resulting transgenic mice were screened for restoration of a normal hearing phenotype, allowing a BAC clone corresponding to the functional *Shaker-2* gene to be identified. The gene was shown to encode a cytoskeletal myosin protein. This was then used to screen a human genomic library, resulting in the identification of the equivalent human gene. Note that no sequence information was required for this screening procedure, and without the functional assay there would have been no way to identify either the mouse or human gene except through a laborious chromosome walk from a linked marker. The recent development of high-capacity transformation vectors for plants (p. 236) has allowed similar methods to be used to identify plant genes (e.g. Sawa *et al.* 1999, Kubo & Kakimoto 2001).

Screening by 'gain of function'

Complementation analysis can be used only if an appropriate mutant expression host is available. In many cases, however, the function of the target gene is too specialized for such a technique to work in a bacterial or yeast expression host and, even in a higher eukaryotic system, loss of function in the host may be fully or partially compensated by one or more other genes. As an alternative, it may be possible to identify clones on the basis that they confer a gain of function on the host cell. In some cases, this gain of function is a selectable phenotype that allows cells containing the corresponding clone to be positively selected. For example, in an early example of the expression of a mammalian gene in *E. coli*, Chang *et al.* (1978) constructed a population of recombinant plasmids containing cDNA derived from unfractionated mouse mRNA. This population of mRNA molecules was expected to contain the transcript for dihydrofolate reductase (DHFR). Mouse DHFR is much less sensitive to inhibition by the drug trimethoprim than *E. coli* DHFR, so growing transformants in medium containing the drug allowed selection for those cells containing the mouse *Dhfr* cDNA.

In other cases, the phenotype conferred by the clone of interest is not selectable, but can be detected because it causes a visible change in phenotype. In mammalian cells, for example, clones corresponding to cellular oncogenes have been identified on the basis of their ability to stimulate the proliferation

of quiescent mouse 3T3 fibroblast cells either in culture or when transplanted into 'nude mice' (e.g. Brady *et al.* 1985). Many different specific assays have also been developed for the functional cloning of cDNAs encoding particular types of gene product. For example, *Xenopus* melanophores have been used for the functional cloning of G-protein-coupled receptors. Melanophores are dark cells containing many pigment organelles, called melanosomes. A useful characteristic of these organelles is that they disperse when adenyl cyclase or phospholipase C are active and aggregate when these enzymes are inhibited. Therefore, the expression of cDNAs encoding G-protein-coupled receptors and many types of receptor tyrosine kinases leads to redistribution of pigmentation within the cell, which can be used as an assay for the identification of receptor cDNAs (reviewed be Lerner 1994).

Difference cloning

Difference cloning refers to a range of techniques used to isolate sequences that are represented in one source of DNA but not another. Normally this means differentially expressed cDNAs, representing genes that are active in one tissue but inactive in another, but the technique can also be applied to genomic DNA to identify genes corresponding to deletion mutants. There are a number of cell-based differential cloning methods and also a range of PCR techniques. Each method follows one of two principles: either the differences between two sources are displayed, allowing differentially expressed clones to be visually identified, or the differences are exploited to generate a collection of clones that are enriched for differentially expressed sequences. The analysis of differential gene expression has taken on new importance recently with the advent of high-throughput techniques allowing the monitoring of many and, in some cases, all genes simultaneously.

Difference cloning with DNA libraries

Displaying differences – differential screening

An early approach to difference cloning was *differential screening*, a simple variation on normal hybridization-based library screening protocols that is useful for the identification of differentially expressed cDNAs that are also moderately abundant (e.g. Dworkin & Dawid 1980). Let us consider, for example, the isolation of cDNAs derived from mRNAs which are abundant in the gastrula embryo of the frog *Xenopus* but which are absent, or present at low abundance, in the egg. A cDNA library is prepared from gastrula mRNA. Replica filters carrying identical sets of recombinant clones are then prepared. One of these filters is then probed with ^{32}P-labelled mRNA (or cDNA) from gastrula embryos and one with ^{32}P-labelled mRNA (or cDNA) from the egg. Some colonies will give a positive signal with both probes; these represent cDNAs derived from mRNA types that are abundant at both stages of development. Some colonies will not give a positive signal with either probe; these correspond to mRNA types present at undetectably low abundance in both tissues. This is a feature of using *complex probes*, which are derived from mRNA populations rather than single molecules: only abundant or moderately abundant sequences in the probe carry a significant proportion of the label and are effective in hybridization. Importantly, some colonies give a positive signal with the gastrula probe, but not with the egg probe. These can be visually identified and should correspond to differentially expressed sequences.

A recent resurgence in the popularity of differential screening has come about through the development of DNA microarrays (Schena *et al.* 1995). In this technique, cDNA clones are transferred to a miniature solid support in a dense grid pattern and screened simultaneously with complex probes from two sources, which are labelled with different fluorophores. Clones that are expressed in both tissues will fluoresce in a colour that represents a mixture of fluorophores, while differentially expressed clones will fluoresce in a colour closer to the pure signal of one or other of the probes. A similar technique involves the use of DNA chips containing densely arrayed oligonucleotides. These methods are compared in Box 6.5.

Enrichment for differences – subtractive cloning

An alternative to differential screening is to generate a library that is enriched in differentially expressed

clones by removing sequences that are common to two sources. This is called a *subtracted cDNA library* and should greatly assist the isolation of rare cDNAs. If we use the same example as above, the aim of the experiment would be to generate a library enriched for cDNAs derived from gastrula-specific mRNAs. This would be achieved by hybridizing first-strand cDNAs prepared from gastrula mRNA with a large excess of mRNA from *Xenopus* oocytes. If this driver population is labelled in some way, allowing it to be removed from the mixed population, only gastrula-specific cDNAs would remain behind. A suitable labelling method would be to add biotin to all the oocyte mRNA, allowing oocyte/gastrula RNA/cDNA hybrids as well as excess oocyte mRNA to be subtracted by binding to streptavidin, for which biotin has great affinity (Duguid *et al.* 1988, Rubinstein *et al.* 1990). Highly enriched libraries can be prepared by several rounds of extraction with driver mRNA, resulting in highly enriched subtracted libraries (reviewed by Sagerstrom *et al.* 1997).

An example of subtractive cDNA cloning and differential screening is provided by Nedivi *et al.* (1993). These investigators were interested in the isolation of rat cDNAs that are induced in a particular region of the brain (the dentate gyrus (DG)) known to be involved in learning and memory. The inducing stimulus was kainate, a glutamate analogue that induces seizures and memory-related synaptic changes. Poly(A)$^+$ RNA was extracted from the DG of kainate-treated animals and used for first-strand cDNA synthesis. Ubiquitous sequences present in the activated DG cDNA preparation were hybridized with an excess of poly(A)$^+$ RNA from total uninduced rat brain. This RNA had previously been biotinylated (using a photobiotinylation procedure) and so hybrids and excess RNA could be removed using a streptavidin extraction method (Sive & St John 1988). The unhybridized cDNA was then converted into double-stranded form by conventional methods and used to construct the subtracted cDNA library in λZAP. This subtracted library was differentially screened using radiolabelled cDNA from activated and non-activated DG as the differential probes. A large number of activated DG clones were isolated, of which 52 were partially sequenced. One-third of these clones corresponded to known genes; the remainder were new.

Box 6.5 Differential screening with DNA chips

cDNA microarrays

Miniaturization and automation have facilitated the development of DNA microarrays, in which DNA sequences are displayed on the surface of a small 'chip' of either nylon or glass. In the initial description of this technology (Schena *et al.* 1995), up to 10 000 cDNA clones, each in the region of several hundred nucleotides in length, could be arrayed on a single microscope slide. The cDNAs were either obtained from an existing library or generated *de novo* by PCR. In each case, the machine transfers a small amount of liquid from a standard 96-well microtitre plate on to a poly-L-lysine-coated microscope slide, and the DNA is fixed in position by UV irradiation. Arrays are used predominantly for the multiplex analysis of gene expression profiles. Total RNA is used to prepare fluorescently labelled cDNA probes and signals are detected using a laser. Each hybridization experiment generates a large amount of data. Comparisons of expression profiles generated using probes from different sources can identify genes that are differentially expressed in various cell types, at different developmental stages or in response to induction (reviewed by Schena *et al.* 1998, Xiang & Chen 2000). There have been many successes with this relatively new technology, including the identification of genes involved in the development of the nervous system (Wen *et al.* 1998) and genes involved in inflammatory disease (Heller *et al.* 1997). Arrays have been constructed including every gene in the genome of *E. coli* (Tao *et al.* 1999)

and most genes of the yeast *Saccharomyces cerevisiae* (De Risi *et al.* 1997). This has allowed comprehensive parallel analysis of the expression of all genes simultaneously in a variety of experimental assays (Cho *et al.* 1998, Chu *et al.* 1998, Spellman *et al.* 1998, Jelinsky & Samson 1999). It is likely that complete genome arrays will be available for higher eukaryotes, including humans, within the next few years, offering an unprecedented ability to capture functional snapshots of the genome in action.

Oligonucleotide chips

An alternative to spotting presynthesized cDNAs or ESTs on to slides is to synthesize oligonucleotides *in situ* on silicon or glass wafers, using similar processes to those used in the manufacture of semiconductors (Lockhart *et al.* 1996, Shalon *et al.* 1996). Using current techniques, up to 1 000 000 oligonucleotides can by synthesized in tightly packed regular arrays on chips approximately 1 cm^2 (Lipshutz *et al.* 1999). Unlike cDNA arrays, a hybridizing probe sequence is recognized not by a single cognate cDNA, but by a combination of short oligonucleotides, from which its sequence can be deduced (reviewed by Southern 1996a,b). Chips are more versatile than arrays, because they can be used not only for expression analysis but also for DNA sequencing (resequencing) (Chee *et al.* 1996) and the analysis of differences between genomes at the level of single nucleotide polymorphisms (Hacia 1999).

Difference cloning by PCR

Displaying differences – differential-display PCR and arbitrarily primed PCR

As expected, PCR-based methods for difference cloning are more sensitive and rapid than library-based methods, and can be applied to small amounts of starting material. Two similar methods have been

described which use pairs of short arbitrary primers to amplify pools of partial cDNA sequences. If the same primer combinations are used to amplify cDNAs from two different tissues, the products can be fractionated side by side on a sequencing gel, and differences in the pattern of bands generated, the *mRNA fingerprint*, therefore reveal differentially expressed genes (Fig. 6.17). Essentially, the distinction between the two techniques concerns the

---

Box 6.6 A landmark publication. Subtraction cloning of the human Duchenne muscular dystrophy (*DMD*) gene

While most subtractive cloning experiments involve cDNAs, this publication reports one of the few successful attempts to isolate a gene using a subtracted *genomic* library. The study began with the identification of a young boy, known as 'BB', who suffered from four X-linked disorders, including DMD. Cytogenetic analysis showed that the boy had a chromosome deletion in the region Xp21, which was known to be the DMD locus. Kunkel's group then devised a subtraction-cloning procedure to isolate the DNA sequences that were deleted in BB. Genomic DNA was isolated from BB and randomly sheared, generating fragments with blunt ends and non-specific overhangs. DNA was also isolated from an aneuploid cell line with four (normal) X chromosomes. This DNA was digested with the restriction enzyme *Mbo*I, generating sticky ends suitable for cloning. The *Mbo*I fragments were mixed with a large excess of the randomly sheared DNA from BB, and the mixture was denatured and then persuaded to reanneal extensively, using phenol enhancement. The principle behind the strategy was

that, since the randomly fragmented DNA was present in a vast excess, most of the DNA from the cell line would be sequestered into hybrid DNA molecules that would be unclonable. However, those sequences present among the *Mbo*I fragments but absent from BB's DNA due to the deletion would only be able to reanneal to complementary strands from the cell line. Such strands would have intact *Mbo*I sticky ends and could therefore be ligated into an appropriate cloning vector. Using this strategy, Kunkel and colleagues generated a genomic library that was highly enriched for fragments corresponding to the deletion in BB. Subclones from the library were tested by hybridization against normal DNA and DNA from BB to confirm that they mapped to the deletion. To confirm that the genuine DMD gene had been isolated, the positive subclones were then tested against DNA from many other patients with DMD, revealing similar deletions in 6.5% of cases.

From Kunkel (1986) *Nature* 322: 73–77.

---

primer used for first-strand cDNA synthesis. In the differential-display PCR technique (Liang & Pardee 1992), the antisense primer is an oligo-dT primer with a specific two-base extension, which thus binds at the 3′ end of the mRNA. Conversely, in the arbitrarily primed PCR method (Welsh *et al.* 1992), the antisense primer is arbitrary and can in principle anneal anywhere in the message. In each case, an arbitrary sense primer is used, allowing the amplification of partial cDNAs from pools of several hundred mRNA molecules. Following electrophoresis, differentially expressed cDNAs can be excised from the gel and characterized further, usually to confirm its differential expression.

Despite the fact that these display techniques are problematical and appear to generate a large number of false-positive results, there have been remarkable successes. In the original report by

Liang and Pardee (1992), the technique was used to study differences between tumour cells and normal cells, resulting in the identification of a number of genes associated with the onset of cancer (Liang *et al.* 1992). Further cancer-related gene products have been discovered by other groups using differential display (Sager *et al.* 1993, Okamato & Beach 1994). The technique has also been used successfully to identify developmentally regulated genes (e.g. Adati *et al.* 1995) and genes that are induced by hormone treatment (Nitsche *et al.* 1996). An advantage of display techniques over subtracted libraries is that changes can be detected in related mRNAs representing the same gene family. In subtractive-cloning procedures, such differences are often overlooked because the excess of driver DNA can eliminate such sequences (see review by McClelland *et al.* 1995).

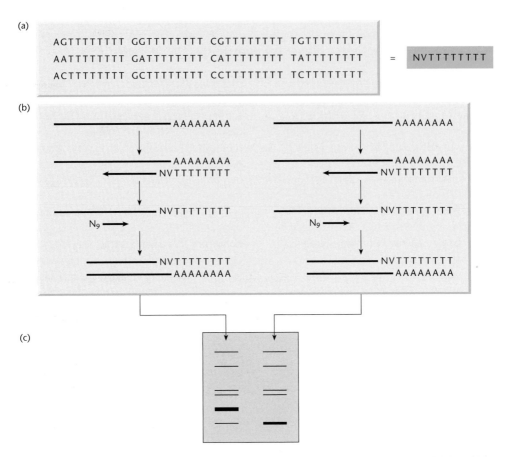

Fig. 6.17 Summary of the differential mRNA display technique, after Liang and Pardee (1992). (a) A set of 12 oligo(dT) primers is synthesized, each with a different two-base extension; the generic designation for this primer set is NVTTTTTTTTT, where N is any nucleotide and V is any nucleotide except T. (b) Messenger RNA from two sources is then converted into cDNA using these primers, generating 12 non-overlapping pools of first strand cDNA molecules for each source, The PCR is then carried out using the appropriate oligo(dT) primer and a set of arbitrary 9-mers (N_9), which may anneal anywhere within the cDNA sequence. This facilitates the amplification of pools of cDNA fragments, essentially the same as expressed sequence tags (ESTs). (c) Pools of PCR products, derived from alternative mRNA sources but amplified with the same pair of primers, are then compared side by side on a sequencing gel. Bands present in one lane but absent from the other are likely to represent differentially-expressed genes. The corresponding bands can be excised from the sequencing gel and the PCR products subcloned, allowing sequence annotation and expression analysis, e.g. by northern blot or *in situ* hybridization, to confirm differential expression.

Enrichment for differences – representational difference analysis

Representational difference analysis is a PCR subtraction technique, i.e. common sequences between two sources are eliminated prior to amplification. The method was developed for the comparative analysis of genomes (Lisitsyn *et al.* 1993) but has been modified for cloning differentially expressed genes (Hubank & Schatz 1994). Essentially, the technique involves the same principle as subtraction hybridization, in that a large excess of a DNA from one source, the driver, is used to make common sequences in the other source, the tester, unclonable (in this case unamplifiable). The general scheme is shown in Fig. 6.18. cDNA is prepared from two sources, digested with restriction enzymes and amplified. The amplified products from one source

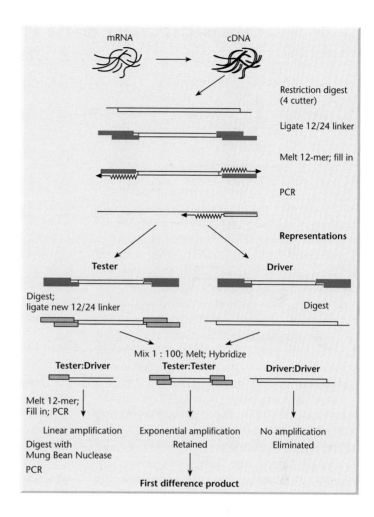

Fig. 6.18 Basic strategy for cDNA-based
representational difference analysis.
See text for details.

are then annealed to specific linkers that provide
annealing sites for a unique pair of PCR primers.
These linkers are not added to the driver cDNA. A
large excess of driver cDNA is then added to the
tester cDNA and the populations are mixed. Driver/
driver fragments possess no linkers and cannot be
amplified, while driver/tester fragments possess only
one primer annealing site and will only be amplified
in a linear fashion. However, cDNAs that are present
only in the tester will possess linkers on both strands
and will be amplified exponentially and can therefore be isolated and cloned.

CHAPTER 7

Sequencing and mutagenesis

Introduction

DNA sequencing is a fundamental requirement for modern gene manipulation. Knowledge of the sequence of a DNA region may be an end in its own right, perhaps in understanding an inherited human disorder. More importantly, sequence information is a prerequisite for planning any substantial manipulation of the DNA; for example, a computer search of the sequence for all known restriction-endonuclease target sites will provide a complete and precise restriction map. Similarly, mutants are an essential prerequisite for any genetic study and never more so than in the study of gene structure and function relationships.

Classically, mutants are generated by treating the test organism with chemical or physical agents that modify DNA (mutagens). This method of mutagenesis has been extremely successful, as witnessed by the growth of molecular biology, but suffers from a number of disadvantages. First, any gene in the organism can be mutated and the frequency with which mutants occur in the gene of interest can be very low. This means that selection strategies have to be developed. Second, even when mutants with the desired phenotype are isolated, there is no guarantee that the mutation has occurred in the gene of interest. Third, prior to the development of gene cloning and sequencing techniques, there was no way of knowing where in the gene the mutation had occurred and whether it arose by a single base change, an insertion of DNA or a deletion.

As techniques in molecular biology have developed, so that the isolation and study of a single gene is not just possible but routine, so mutagenesis has also been refined. Instead of crudely mutagenizing many cells or organisms and then analysing many thousands or millions of offspring to isolate a desired mutant, it is now possible to specifically change any given base in a cloned DNA sequence. This technique is known as *site-directed mutagenesis*. It has become a basic tool of gene manipulation, for it simplifies DNA manipulations that in the past required a great deal of ingenuity and hard work, e.g. the creation or elimination of cleavage sites for restriction endonucleases. The importance of site-directed mutagenesis goes beyond gene structure–function relationships, for the technique enables mutant proteins to be generated with very specific changes in particular amino acids (protein engineering). Such mutants facilitate the study of the mechanisms of catalysis, substrate specificity, stability, etc (see p. 299 *et seq.*).

Basic DNA sequencing

The first significant DNA sequence to be obtained was that of the cohesive ends of phage-λ DNA (Wu & Taylor 1971), which are only 12 bases long. The methodology used was derived from RNA sequencing and was not applicable to large-scale DNA sequencing. An improved method, plus and minus sequencing, was developed and used to sequence the 5386 bp phage ΦX 174 genome (Sanger *et al.* 1977a). This method was superseded in 1977 by two different methods, that of Maxam and Gilbert (1977) and the chain-termination or dideoxy method (Sanger *et al.* 1977b). For a while the Maxam and Gilbert (1977) method, which makes use of chemical reagents to bring about base-specific cleavage of DNA, was the favoured procedure. However, refinements to the chain-termination method meant that by the early 1980s it became the preferred procedure. To date, most large sequences have been determined using this technology, with the notable exception of bacteriophage T7 (Dunn & Studier 1983). For this reason, only the chain-termination method will be described here.

The chain-terminator or dideoxy procedure for DNA sequencing capitalizes on two properties of DNA polymerases: (i) their ability to synthesize faithfully a complementary copy of a single-stranded DNA template; and (ii) their ability to use 2′,3′-

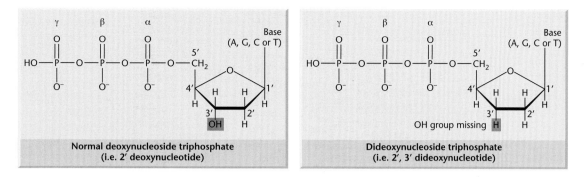

Fig. 7.1 Dideoxynucleoside triphosphates act as chain terminators because they lack a 3′ OH group. Numbering of the carbon atoms of the pentose is shown (primes distinguish these from atoms in the bases). The α, β and γ phosphorus atoms are indicated.

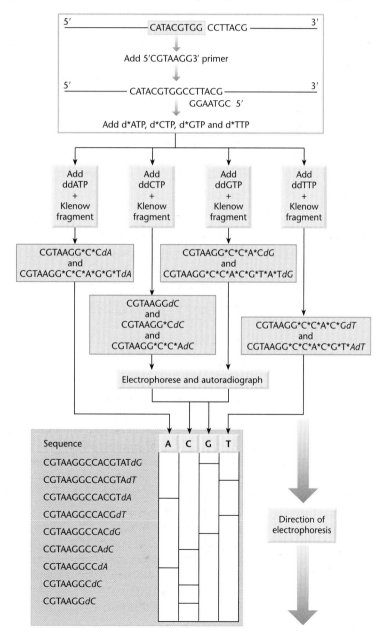

Fig. 7.2 DNA sequencing with dideoxynucleoside triphosphates as chain terminators. In this figure asterisks indicate the presence of ^{32}P and the prefix 'd' indicates the presence of a dideoxynucleoside. At the top of the figure the DNA to be sequenced is enclosed within the box. Note that unless the primer is also labelled with a radioisotope the smallest band with the sequence CGTAAGGdC will not be detected by autoradiography, as no labelled bases were incorporated.

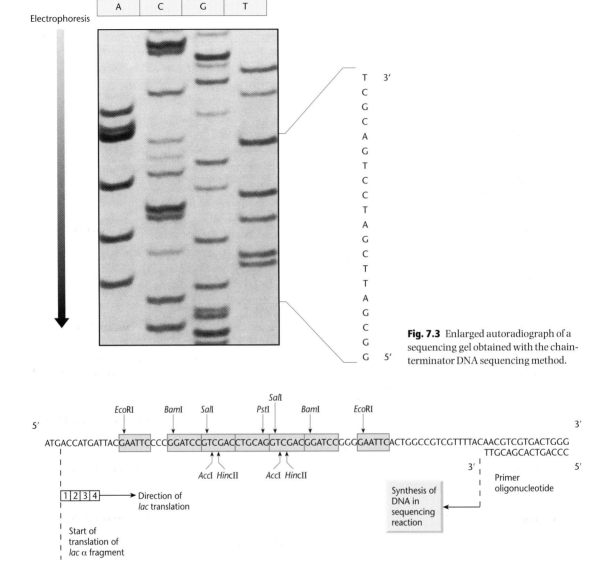

Fig. 7.3 Enlarged autoradiograph of a sequencing gel obtained with the chain-terminator DNA sequencing method.

Fig. 7.4 Sequence of M13 mp7 DNA in the vicinity of the multipurpose cloning region. The upper sequence is that of M13 mp7 from the ATG start codon of the β-galactosidase α fragment, through the multipurpose cloning region and back into the β-galactosidase gene. The short sequence at the right-hand side is that of the primer used to initiate DNA synthesis across the cloned insert. The numbered boxes correspond to the amino acids of the β-galactosidase fragment.

dideoxynucleotides as substrates (Fig. 7.1). Once the analogue is incorporated at the growing point of the DNA chain, the 3′ end lacks a hydroxyl group and is no longer a substrate for chain elongation. Thus, the growing DNA chain is terminated, i.e. dideoxynucleotides act as chain terminators. In practice,

the Klenow fragment of DNA polymerase is used because this lacks the 5′ → 3′ exonuclease activity associated with the intact enzyme. Initiation of DNA synthesis requires a primer and this is usually a chemically synthesized oligonucleotide which is annealed close to the sequence being analysed.

DNA synthesis is carried out in the presence of the four deoxynucleoside triphosphates, one or more of which is labelled with ^{32}P, and in four separate incubation mixes containing a low concentration of one each of the four dideoxynucleoside triphosphate analogues. Therefore, in each reaction there is a population of partially synthesized radioactive DNA molecules, each having a common 5′ end, but each varying in length to a base-specific 3′ end (Fig. 7.2). After a suitable incubation period, the DNA in each mixture is denatured and electrophoresed in a sequencing gel.

A sequencing gel is a high-resolution gel designed to fractionate single-stranded (denatured) DNA fragments on the basis of their size and which is capable of resolving fragments differing in length by a single base pair. They routinely contain 6–20% polyacrylamide and 7 mol/l urea. The function of the urea is to minimize DNA secondary structure, which affects electrophoretic mobility. The gel is run at sufficient power to heat up to about 70°C. This also minimizes DNA secondary structure. The labelled DNA bands obtained after such electrophoresis are revealed by autoradiography on large sheets of X-ray film and from these the sequence can be read (Fig. 7.3).

To facilitate the isolation of single strands, the DNA to be sequenced may be cloned into one of the clustered cloning sites in the *lac* region of the M13 mp series of vectors (Fig. 7.4). A feature of these vectors is that cloning into the same region can be mediated by any one of a large selection of restriction enzymes but still permits the use of a single sequencing primer.

Modifications of chain-terminator sequencing

The sharpness of the autoradiographic images can be improved by replacing the ^{32}P-radiolabel with the much lower-energy ^{33}P or ^{35}S. In the case of ^{35}S, this is achieved by including an α-^{35}S-deoxynucleoside triphosphate (Fig. 7.5) in the sequencing reaction. This modified nucleotide is accepted by DNA polymerase and incorporated into the growing DNA chain. Non-isotopic detection methods have also been developed with chemiluminescent, chromogenic or fluorogenic reporter systems. Although the sensitivity of these methods is not as great as with radiolabels, it is adequate for many purposes.

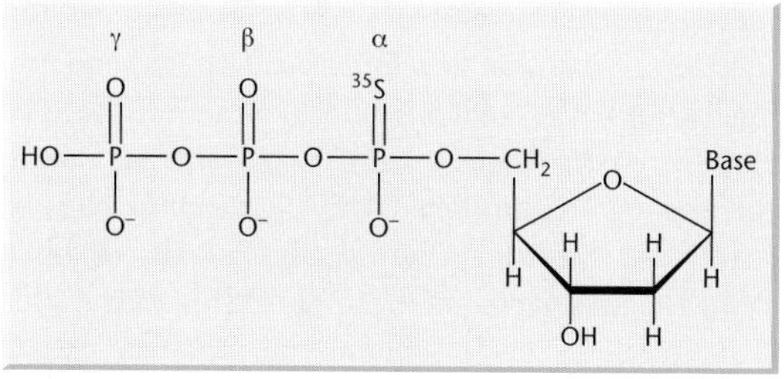

Fig. 7.5 Structure of an α-^{35}S-deoxynucleoside triphosphate.

Other technical improvements to Sanger's original method have been made by replacing the Klenow fragment of *Escherichia coli* DNA polymerase I. Natural or modified forms of the phage T7 DNA polymerase ('Sequenase') have found favour, as has the DNA polymerase of the thermophilic bacterium *Thermus aquaticus* (Taq DNA polymerase). The T7 DNA polymerase is more processive than Klenow polymerase, i.e. it is capable of polymerizing a longer run of nucleotides before releasing them from the template. Also, its incorporation of dideoxynucleotides is less affected by local nucleotide sequences and so the sequencing ladders comprise a series of bands with more even intensities. The Taq DNA polymerase can be used in a chain-termination reaction carried out at high temperatures (65–70°C) and this minimizes chain-termination artefacts caused by secondary structure in the DNA. Tabor and Richardson (1995) have shown that replacing a single phenylalanine residue of Taq DNA polymerase with a tyrosine residue results in a thermostable sequencing enzyme that no longer discriminates between dideoxy- and deoxynucleotides.

The combination of chain-terminator sequencing and M13 vectors to produce single-stranded DNA is very powerful. Very good-quality sequencing is obtainable with this technique, especially when the improvements given by ^{35}S-labelled precursors and T7 DNA polymerase are exploited. Further modifications allow sequencing of 'double-stranded' DNA, i.e. double-stranded input DNA is denatured by alkali and neutralized and then one strand is annealed with a specific primer for the actual chain-terminator sequencing reactions. This approach has gained in popularity as the convenience of having a universal primer has grown less important with the widespread availability of oligonucleotide synthesizers. With this development, Sanger sequencing has been liberated from its attachment to the M13 cloning system; for example, polymerase chain reaction (PCR)-amplified DNA segments can be sequenced directly. One variant of the double-stranded approach, often employed in automated sequencing, is 'cycle sequencing'. This involves a *linear* amplification of the sequencing reaction, using 25 cycles of denaturation, annealing of a specific primer to one strand only and extension in the presence of Taq DNA polymerase plus labelled dideoxynucleotides. Alternatively, labelled primers can be used with unlabelled dideoxynucleotides.

Automated DNA sequencing

In manual sequencing, the DNA fragments are radiolabelled in four chain-termination reactions, separated on the sequencing gel in four lanes and detected by autoradiography. This approach is not well suited to automation. To automate the process, it is desirable to acquire sequence data in real time by detecting the DNA bands within the gel during the electrophoretic separation. However, this is not trivial, as there are only about 10^{-15}–10^{-16} moles of DNA per band. The solution to the detection problem is to use fluorescence methods. In practice, the fluorescent tags are attached to the chain-terminating nucleotides. Each of the four dideoxynucleotides carries a spectrally different fluorophore. The tag is incorporated into the DNA molecule by the DNA polymerase and accomplishes two operations in one step: it terminates synthesis and it attaches the fluorophore to the end of the molecule. Alternatively, fluorescent primers can be used with non-labelled dideoxynucleotides. By using four different fluorescent dyes, it is possible to electrophorese all four chain-terminating reactions together in one lane of a sequencing gel. The DNA bands are detected by their fluorescence as they electrophorese past a detector (Fig. 7.6). If the detector is made to

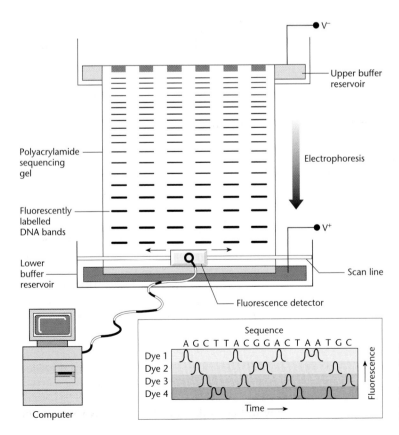

Fig. 7.6 Block diagram of an automated DNA sequencer and idealized representation of the correspondence between fluorescence in a single electrophoresis lane and nucleotide sequence.

scan horizontally across the base of a slab gel, many separate sequences can be scanned, one sequence per lane. Because the different fluorophores affect the mobility of fragments to different extents, sophisticated software is incorporated into the scanning step to ensure that bands are read in the correct order. A simpler method is to use only one fluorophore and to run the different chain-terminating reactions in different lanes.

For high-sensitivity DNA detection in four-colour sequencing and high-accuracy base calling, one would ideally like the following criteria to be met: each of the four dyes to exhibit strong absorption at a common laser wavelength; to have an emission maximum at a distinctly different wavelength; and to introduce the same relative mobility shift of the DNA sequencing fragments. Recently, dyes with these properties have been identified and successfully applied to automated sequencing (Glazer & Mathies 1997).

Automated DNA sequencers offer a number of advantages that are not particularly obvious. First, manual sequencing can generate excellent data, but even in the best sequencing laboratories poor autoradiographs are frequently produced that make sequence reading difficult or impossible. Usually the problem is related to the need to run different termination reactions in different tracks of the gel. Skilled DNA sequencers ignore bad sequencing tracks, but many laboratories do not. This leads to poor-quality sequence data. The use of a single-gel track for all four dideoxy reactions means that this problem is less acute in automated sequencing. Nevertheless, it is desirable to sequence a piece of DNA several times and on both strands, to eliminate errors caused by technical problems. It should be noted that long runs of the same nucleotide or a high G+C content can cause compression of the bands on a gel, necessitating manual reading of the data, even with an automated system. Note also that multiple, tandem short repeats, which are common in the DNA of higher eukaryotes, can reduce the fidelity of DNA copying, particularly with Taq DNA polymerase. The second advantage of automated DNA sequencers is that the output from them is in machine-readable form. This eliminates the errors that arise when DNA sequences are read and transcribed manually.

A third advantage derives from the new generation of sequencers that have been introduced recently. In these sequencers, the slab gel is replaced with 48 or 96 capillaries filled with the gel matrix. The key feature of this system is that the equipment has been designed for use with robotics, thereby minimizing hands-on time and increasing throughput. With a 96-capillary sequencer, it is possible to sequence up to 750 000 nucleotides per day.

Sequence accuracy

As part of a programme to sequence a 96 kb stretch of mouse DNA, Wilson *et al.* (1992) analysed 288 sequences containing part of the vector DNA. By comparing raw sequence data with known vector sequences, it was possible to calculate the error frequency. Sequences that were read beyond 400 bp contained an average of 3.2% error, while those less than 400 bp had 2.8% error. At least one-third of the errors were due to ambiguities in sequence reading. In those sequences longer than 400 bp that were read, most errors occurred late in the sequence and were often present as extra bases in a run of two or more of the same nucleotide. The remainder of the errors were due to secondary structure in the template DNA. However, because the complete sequence was analysed with an average 5.9-fold redundancy and most of it on both strands, the final error frequency is estimated to be less than 0.1%. In comparison, 35 different European laboratories were engaged in sequencing the *Saccharomyces cerevisiae* genome, with the attendant possibility of a very high error frequency. However, by using a DNA coordinator who implemented quality-control procedures (Table 7.1), the overall sequence accuracy for yeast chromosome XI (666 448 bp) was estimated to be 99.97% (Dujon *et al.* 1994), i.e. similar to that noted above. Lipshutz *et al.* (1994) have described a software program for estimating DNA sequence confidence. Fabret *et al.* (1995) have analysed the errors in finished sequences. They took advantage of the fact that the surfactin operon of *Bacillus subtilis* had been sequenced by three independent groups. This enabled the *actual* error rate to be calculated. It was found to range from 0.02 to 0.27%, the different error rates being ascribed to the detailed sequencing tactics used. Other studies of DNA sequencing

Table 7.1 Quality control of sequence data from 35 laboratories engaged in sequencing yeast chromosome XI. (Data from Dujon *et al.* 1994 with permission from *Nature*, © Macmillan Magazines Ltd.)

Method of verification	Total number of fragments	Total bp verified	Error % detected
Original overlap between cosmids	28	63 424	0.02
Resequencing of selected segments (3–5 kb long)	21	72 270	0.03
Resequencing of random segments (~300 bp long)	71	18 778	0.05
Resequencing of suspected segments (~300 bp long) from designed oligonucleotide pairs	60	17 035	0.03
Total	180	171 507	
Average error rate			0.03

accuracy have been summarized by Yager *et al.* (1997). In general, a sequence read of > 350 nucleotides at 99% accuracy can be expected using current ultrathin-slab gel technology. Reading lengths in excess of 1000 nucleotides have been reported (Noolandi *et al.* 1993; Voss *et al.* 1995).

Whole-genome sequencing

As noted in Chapter 1 (p. 2), many different genomes have been completely sequenced and the list includes viruses, bacteria, yeast, *Caenorhabditis*, *Drosophila*, *Arabidopsis* and humans. A detailed description of the methodology used for sequencing these genomes is outside the scope of this book and the interested reader is referred to the sister publication *Principles of Genome Analysis* (Primrose & Twyman 2002). Suffice is to say that the underlying principle is to subdivide the genome into small fragments of a size suitable for sequencing by the methods just described. Provided that the fragments overlap, the individual sequences can then be assembled into the complete genome sequence. However, the scale of the task of complete sequence assembly can be gauged from a comparison of the length of fragment that can be sequenced (600–1000 nucleotides) with the number of such fragments in the genome (> 3 million for the human genome)!

Analysing sequence data

DNA sequence databases

Since the current DNA sequencing technology was developed, a large amount of DNA sequence data has accumulated. These data are maintained in three databases: the National Center for Biotechnology Information in the USA, the DNA Databank of Japan and the European Bioinformatic Institute in the UK (Benson *et al.* 1996, 1997, Stoesser *et al.* 1997, Tateno & Gojobori 1997). Each of these three groups collects a portion of the total sequence data reported worldwide and all new and updated database entries are exchanged between the groups on a daily basis. In addition, several specialized genome databases exist, including seven for bacterial genomes: four for *E. coli*, two for *B. subtilis* and one at the Institute for Genome Research, an organization responsible for the complete sequencing of a number of genomes. Users worldwide can access these databases directly via the Worldwide Web or receive the information on CD-ROMs. The former option is the best because it ensures that an up-to-date database is being used. There are a number of different sequence-retrieval systems and the best of these are Network Entrez and DNA Workbench (Brenner 1995).

In working with databases, it is important to recognize their inherent deficiencies. As well as errors in the sequencing process itself, there can be transcriptional errors when the data are transferred from laboratory notebook to publications and databases. For example, when screening 300 human protein sequences in the SWISS-PROT database that had been published separately more than once, Bork (1996) found that 0.3% of the amino acids were different. This is a lower limit, for many corrections will already have been made and in many instances the sequences appearing in two different publications are not independent. Note that only stop codons and frame shifts can be detected unambiguously: point mutations are hard to verify, as natural polymorphisms or strain differences cannot easily be excluded. Sequencing by hybridization may be of great use here. Other database problems include misspelling of genes, resulting in confusion with ones of similar name or the generation of synonyms: different genes being given the same name and multiple synonyms for the same gene. Examples of the latter are the *E. coli* gene *hns*, which has eight synonyms, and the protein annexin V, which has five synonyms.

Analysing sequence data

Discovering new genes and their functions is a key step in analysing new sequence data. The process is facilitated by special-purpose gene-finding software, by searches in key databases and by programs for finding particular sites relevant to gene expression, e.g. splice sites and promoters. Unfortunately, no one software package contains all the necessary tools. Rather, optimal gene finding is dependent on combining evidence derived from use of multiple software tools (Table 7.2).

Fickett (1996) has described a framework for finding genes which makes use of a number of different software programs. Evidence for the presence or absence of genes in a sequence is gathered from a number of sources. These include sequence similarity to other features, such as repeats, which are unlikely to overlap protein-coding sequences, sequence similarity to other genes, statistical regularity evincing apparent codon bias over a region, and matches to template patterns for functional sites on the DNA,

such as consensus sequences for TATA boxes. All the information so gathered is integrated to make as coherent a picture as possible.

When analysing sequences from eukaryotes, it is best to locate and remove interspersed repeats before searching for genes. Not only can repeats confuse other analyses, such as database searches, but they provide important negative information on the location of gene features. For example, such repeats rarely overlap the promoters or coding portions of exons. Once this is done, the next step is to identify open reading frames (ORFs). Despite the availability of sophisticated software search routines, unambiguous assignment of ORFs is not easy. For example, the *Haemophilus influenzae* genome sequence submitted to the database included 1747 predicted protein-encoding genes (Fleischmann *et al.* 1995). When Tatusov *et al.* (1996) reanalysed the sequence data using different algorithms and different discrimination criteria, they identified a new set of 1703 putative protein-encoding genes. In addition to 1572 ORFs, which remained the same, they identified 23 new ORFs, modified 107 others and discarded the balance. Note that gene finding is relatively easy in compact and almost intron-free genomes, such as yeast. In higher plants and animals, the task is much greater, for a 2 kb ORF could be split into 15 exons spread over 30 kb of genomic DNA.

Database searches

Searching for a known homologue is the most widely used means of identifying genes in a new sequence. If a putative protein encoded by an uncharacterized ORF shows statistically significant similarity to another protein of known function, this simultaneously proves, beyond doubt, that the ORF in question is a bona fide new gene and predicts its likely function. Even if the homologue of the new protein has not been characterized, useful information is produced in the form of conserved motifs that may be important for protein function. In this way, Koonin *et al.* (1994) analysed the information contained in the complete sequence of yeast chromosome III and found that 61% of the probable gene products had significant similarities to sequences in the current databases. As many as 54% of them had

Table 7.2 Internet tools for gene discovery in DNA sequence data.

Important Home Pages

European Molecular Biology Laboratory (EMBL)
http://www.ebi.ac.uk/ebi_docs/embl_db/ebi/topembl.html
Cambridge, UK.

UK Human Genome Mapping Project – Resource Center (HGMP-RC)
http://www.hgmp.mrc.ac.uk/

SeqNet: UK Node of European Molecular Biology Network (EMBNet)
http://www.seqnet.dl.ac.uk/About/SEQNET/

GenBank
http://www.ncbi.nlm.nih.gov/Genbank/
GenBank at the National Center for Biotechnology (NCBI) of the National Library of Medicine (NLM) at the National Institutes for
Health (NIH) campus, USA.

DNA Databank of Japan (DDBJ)
http://www.ddbj.nig.ac.jp

Genome Sequence DataBase (GSDB)
http://seqsim.ncgr.org/
The National Center for Genome Resources, Genome Sequence Database.
The server is a supercomputer with genomic algorithm acceleration.

The Institute for Genomic Research (TIGR)
http://www.tigr.org/

The Sanger Centre
http://www.sanger.ac.uk/

Swiss Institute of Bioinformatics (Expasy)
http://www.expasy.ch/

Genomic Sequence Databases

These are the three primary sequence databases. They exchange sequence information daily.

European Molecular Biology Laboratory (EMBL)
http://www2.ebi.ac.uk/Help/General/general.html
Cambridge, UK.

GenBank
http://www.ncbi.nlm.nih.gov/Web/Search/index.html
GenBank at the National Center for Biotechnology (NCBI) of the National Library of Medicine (NLM) at the National Institutes for
Health (NIH) campus, USA.

DNA Databank of Japan (DDBJ)
http://www.ddbj.nig.ac.jp

Sequence Search and Retrieval

Sequence Retrieval System (SRS)
http://srs.hgmp.mrc.ac.uk/

Entrez
http://www.ncbi.nlm.nih.gov/Entrez/
Very comprehensive indeed!

Open Reading Frame (ORF) Finder

ORF Finder
http://www.ncbi.nlm.nih.gov/gorf/gorf.html
Finds likely open reading frames in a sequence.

continued

Table 7.2 *(cont'd)*

Sequence Translation

Nucleotide to Protein (EMBL)
5-4-1-DUW/UK
http://www.ebi.ac.uk/contrib/tommaso/translate.html
Translates a nucleotide sequence into a protein sequence.

Forward and Reverse Translation
5-5-2-DUPUFWE/UK
http://www.sanger.ac.uk/Software/Wise2/genewiseform.shtml
Translates a protein sequence into a genomic DNA sequence, and vice versa.

Protein and cDNA Translation
5-4-2-DcUPU/UK
http://www.sanger.ac.uk/Software/Wise2/genewiseform.shtml
Translates a protein sequence into a cDNA sequence, and vice versa.

Protein Sequence Databases

Translated EMBL (TrEMBL)
http://www.expasy.ch/sprot/sprot-top.html
Database of all the protein-coding regions stored in the EMBL database.
Comprehensive, but (generally) at the cost of poor annotation.

SWISS-PROT
http://expasy.hcuge.ch/sprot/sprot-top.html
A database of protein sequences, translated from the EMBL genomic database. Protein sequences have been checked and
annotated.

PIR
http://www.nbrf.georgetown.edu/pir/
Four databases: PIR1 is the most comprehensive with entries classified and annotated. PIR4 is the least comprehensive, with
unencoded or untranslated entries.

Protein Three-dimensional Structure Database

Protein Databank
http://pdb.pdb.bnl.gov/
Three-dimensional protein structures which can be downloaded and viewed locally (viewer required) or viewed in a hypertext
browser window (e.g. Netscape). The structures are experimentally determined by X-ray crystallography and nuclear magnetic
resonance (NMR) imaging.

Protein Analysis Utilities

Web Cutter – Restriction Enzyme Mapping Utility
http://www.medkem.gu.se/cutter/
Map restriction-enzyme sites on your sequence. Easy to use and comprehensive options.

Sequence Identification (BLAST), FASTA, etc.

The Sanger Centre Database Search Services
5-2-5-DDPDFWE/UK – Clean, simple design.
http://www.sanger.ac.uk/DataSearch/
BLAST and WU-BLAST 2.0 searches which can be refined to finished and/or unfinished genomic sequences.

BLAST 2 Similarity Search (EMBNet)
5-3-2-PUWE/Switzerland
http://www.ch.embnet.org/software/frameBLAST.html
WU-BLAST 2.0 similarity searches.

FASTA 3 (EMBL)
http://www2.ebi.ac.uk/fasta3/
FASTA 3 similarity search. Links to GenBank sequences returned. MolView presents alignment results well and in colour.

continued

Table 7.2 *(cont'd)*

FASTA
http://www2.igh.cnrs.fr/bin/fasta-guess.cgi
FASTA similarity search and a clean, basic and simple interface.

Blitz
http://www.ebi.ac.uk/searches/blitz.html
Blitz should only be used if BLAST and FASTA have failed to yield the desired results. It is a sensitive search, but also slow!

Beauty
http://dot.imgen.bcm.tmc.edu:9331/seq-search/protein-search.html
Beauty is an enhanced BLAST search, returning output which predicts the function of the protein being tested.

Sequence Alignment
Network Protein Sequence Analysis
5-4-1-PUW/France
http://pbil.ibcp.fr/NPSA/npsa_clustalw.html
ClustalW multiple sequence alignment.

Multiple Sequence Alignment with Hierarchical Clustering
5-5-3-PUW/France
http://www.toulouse.inra.fr/multalin.html
Sequence alignment with a colour output where differing or similar amino acids in the alignment can be highlighted.

ALIGN
5-1-1-DUPUWE/France
http://www2.igh.cnrs.fr/bin/align-guess.cgi
Applies the BLOSUM50 matrix to deduce the optimal alignment between two sequences.

New Sequence Submission
EMBL
http://www.ebi.ac.uk/Submissions/index.html

BankIt (GenBank)
http://www.ncbi.nlm.nih.gov/BankIt/

known functions or were related to functionally characterized proteins and 19% were similar to proteins of known three-dimensional structure. Other examples are given later. The methods of choice for screening databases in this way are those such as BLASTX that translate the query nucleotide sequence in all six reading frames and compare the resulting putative protein sequences to the protein-sequence database. Such methods allow the detection of frame-shift errors and will not miss even small ORFs if homologues are present in the database.

Significant matches of a novel ORF to another sequence may be in any of four classes (Oliver 1996). First, a match may predict both the biochemical and physiological function of the novel gene. An example is ORF YCR24c from yeast chromosome III and the *E. coli* Asn-tRNA synthetase.

Secondly, a match may define the biochemical function of a gene product without revealing its biochemical function. An example of this is the five protein-kinase genes found on yeast chromosome III whose biochemical function is clear (they phosphorylate proteins) but whose particular physiological function in yeast is unknown. Thirdly, a match may be to a gene from another organism whose function is unknown in that organism, e.g. ORF YCR63w from yeast, protein G10 from *Xenopus* and novel genes from *Caenorhabditis* and humans, none of whose functions is known. Finally, a match may occur to a gene of known function that merely reveals that our understanding of that function is very superficial, e.g. yeast ORF YCL17c and the Nif S protein of nitrogen-fixing bacteria. After similar sequences were found in a number of bacteria which do not fix nitrogen, it was shown that the Nif S protein is a pyridoxal phosphate-dependent aminotransferase.

Because the DNA and protein-sequence databases are updated daily, it could be that homologues of previously unidentified proteins have been found. Thus it could pay to repeat the search of the databases at regular intervals, as Robinson *et al.* (1994) found. Starting with more than 18 million bp of prokaryotic DNA sequence, they executed a systematic search for previously undetected protein-coding genes. They removed all DNA regions known to encode proteins or structural RNAs and used an algorithm to translate the remaining DNA in all six reading frames. A search of the resultant translations against the protein-sequence databases uncovered more than 450 genes which previously had escaped detection. Seven of these genes belonged to gene families not previously identified in prokaryotes. Others belonged to gene families with critical roles in metabolism. Clearly, periodic exhaustive reappraisal of databases can be very productive!

If homology with known proteins is to be used to analyse sequence data, then sequence accuracy is paramount. With the gene density and ORF size distribution of yeast (Dujon *et al.* 1994) and the nematode (Wilson *et al.* 1994), even relatively rare sequencing errors, many of which are missing nucleotides, will result in a large fraction of the protein-coding genes being affected by frame shifts. At 99% accuracy virtually all genes will contain errors and at 99.9% accuracy two-thirds of the genes will still contain an error, most probably a frame shift. With the 99.97% accuracy obtained in the yeast sequencing, about one-third of predicted genes will still contain errors (Fig. 7.7).

Because sequence-similarity search programs are so vital to the analysis of DNA sequence data, much attention has been paid to the precise algorithms used and their precise speed. However, the effectiveness of these searches is dependent on a number of factors. These include the choice of scoring system, the statistical significance of alignments and the nature and extent of sequence redundancy in the databases (Altschul *et al.* 1994; Coulson 1994). Most database search algorithms rank alignments by a score whose calculation is dependent upon a particular scoring system. Optimal strategies for detecting similarities between DNA protein-coding regions differ from those for non-coding regions.

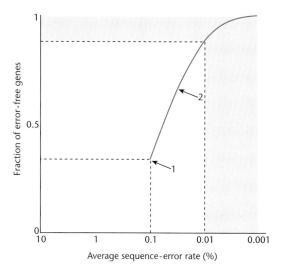

Fig. 7.7 Quantitative effect of sequence accuracy on gene accuracy. The figure indicates the fraction of error-free genes predicted from a DNA sequence, as a function of average sequence accuracy. The theoretical curve was computed using the size distribution of protein-coding genes in yeast and assuming random occurrence of sequencing errors. All types of errors are considered together. In practice, frame shifts, which have the most deleterious effects on gene interpretation, represent about two-thirds of all sequencing errors. The figure illustrates the difference between an average sequence accuracy of 99.9%, where only one-third of all genes are properly described, and an average sequence accuracy of 99.99%, where more than 85% of genes are properly described. It also illustrates the difference between average database entries (99.9% accuracy, arrow 1) and systematic sequencing programs with sequence verifications (99.97% accuracy, arrow 2). (Redrawn with permission from Dujon 1996.)

Special scoring systems have been developed for detecting frame-shift errors in the databases, a problem highlighted above. Thus a database search program should make use of a variety of scoring systems. Furthermore, given a query sequence, most database search programs will produce an ordered list of imperfectly matching database similarities, but none of them need have any biological significance. Many of the genes originally predicted by these statistical methods have subsequently proved to be homologous to newly described genes or have been confirmed experimentally, thus supporting the robustness of the predictive methods (Koonin *et al.* 1996).

Eventually, with the accumulation of new sequences, sequence conservation will become the definitive criterion for gene identification and the contribution of statistical methods will decrease. Nevertheless, it is still likely that some genes will not have identifiable homologues and statistical and experimental approaches will remain necessary for their detection. Furthermore, even for genes that have homologues, statistical methods of coding-potential analysis will remain useful for localizing frame shifts and choosing among the possible initiation codons.

Changing genes: site-directed mutagenesis

Three different methods of site-directed mutagenesis have been devised: cassette mutagenesis, primer extension and procedures based on the PCR. All three are described below but the reader wishing more detail should consult the review of Ling and Robinson (1998).

In some cases, the goal of protein engineering is to generate a molecule with an improvement in some operating parameter, but it is not known what amino acid changes to make. In this situation, a random mutagenesis strategy provides a route to the desired protein. However, methods based on gene manipulation differ from traditional mutagenesis in that the mutations are restricted to the gene of interest or a small portion of it. Genetic engineering also provides a number of simple methods of generating chimeric proteins where each domain is derived from a different protein.

It should not be forgotten that constructing the mutant DNA is only part of the task. The vector for expression, the expression system and strategies for purification and assay must also be considered before embarking on protein mutagenesis.

Cassette mutagenesis

In cassette mutagenesis, a synthetic DNA fragment containing the desired mutant sequence is used to replace the corresponding sequence in the wild-type gene. This method was originally used to generate improved variants of the enzyme subtilisin (Wells *et al.* 1985). It is a simple method for which the efficiency of mutagenesis is close to 100%. The disadvantages are the requirement for unique restriction sites flanking the region of interest and the limitation on the realistic number of different oligonucleotide replacements that can be synthesized. The latter problem can be minimized by the use of doped oligonucleotides (Fig. 7.8; Reidhaar-Olson and Sauer, 1988), as described on pp. 107–109.

Primer extension: the single-primer method

The simplest method of site-directed mutagenesis is the single-primer method (Gillam *et al.* 1980, Zoller & Smith 1983). The method involves priming *in vitro* DNA synthesis with a chemically synthesized oligonucleotide (7–20 nucleotides long) that carries a base mismatch with the complementary sequence. As shown in Fig. 7.9, the method requires that the DNA to be mutated is available in single-stranded form, and cloning the gene in M13-based vectors makes this easy. However, DNA cloned in a plasmid and obtained in duplex form can also be converted to a partially single-stranded molecule that is suitable (Dalbadie-McFarland *et al.* 1982).

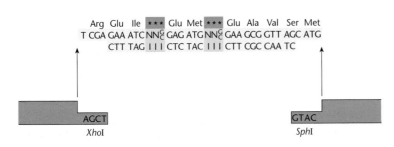

Arg Glu Ile *** Glu Met *** Glu Ala Val Ser Met
T CGA GAA ATC NNC GAG ATG NNC GAA GCG GTT AGC ATG
  CTT TAG I I I CTC TAC I I I CTT CGC CAA TC

AGCT
*Xho*I

GTAC
*Sph*I

Fig. 7.8 Mutagenesis by means of doped oligonucleotides. During synthesis of the upper strand of the oligonucleotide, a mixture of all four nucleotides is used at the positions indicated by the letter N. When the lower strand is synthesized, inosine (I) is inserted at the positions shown. The double-stranded oligonucleotide is inserted into the relevant position of the vector.

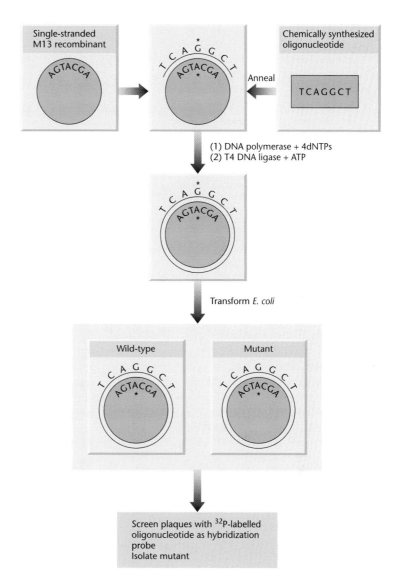

Fig. 7.9 Oligonucleotide-directed mutagenesis. Asterisks indicate mismatched bases. Originally the Klenow fragment of DNA polymerase was used, but now this has been largely replaced with T7 polymerase.

The synthetic oligonucleotide primes DNA synthesis and is itself incorporated into the resulting heteroduplex molecule. After transformation of the host *E. coli*, this heteroduplex gives rise to homoduplexes whose sequences are either that of the original wild-type DNA or that containing the mutated base. The frequency with which mutated clones arise, compared with wild-type clones, may be low. In order to pick out mutants, the clones can be screened by nucleic acid hybridization (see Chapter 6) with ^{32}P-labelled oligonucleotide as probe. Under suitable conditions of stringency, i.e. temperature and cation concentration, a positive signal will be obtained only with mutant clones. This allows ready detection of the desired mutant (Wallace *et al.* 1981, Traboni *et al.* 1983). In order to check that the procedure has not introduced other adventitious changes, it is prudent to check the sequence of the mutant directly by DNA sequencing. This was a particular necessity with early versions of the technique which made use of *E. coli* DNA polymerase. The more recent use of the high-fidelity DNA polymerases from phages T4 and T7 has minimized the problem of extraneous mutations, as well as shortening the time for copying

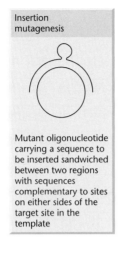

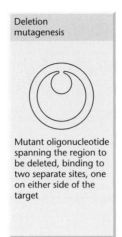

Multiple point mutations	Insertion mutagenesis	Deletion mutagenesis
Mutant oligonucleotide with multiple (four) single base pair mismatches	Mutant oligonucleotide carrying a sequence to be inserted sandwiched between two regions with sequences complementary to sites on either sides of the target site in the template	Mutant oligonucleotide spanning the region to be deleted, binding to two separate sites, one on either side of the target

Fig. 7.10 Oligonucleotide-directed mutagenesis used for multiple point mutation, insertion mutagenesis and deletion mutagenesis.

the second strand. Also, these polymerases do not 'strand-displace' the oligomer, a process which would eliminate the original mutant oligonucleotide.

A variation of the procedure (Fig. 7.10) outlined above involves oligonucleotides containing inserted or deleted sequences. As long as stable hybrids are formed with single-stranded wild-type DNA, priming of *in vitro* DNA synthesis can occur, ultimately giving rise to clones corresponding to the inserted or deleted sequence (Wallace *et al.* 1980, Norrander *et al.* 1983).

Deficiencies of the single-primer method

The efficiency with which the single-primer method yields mutants is dependent upon several factors. The double-stranded heteroduplex molecules that are generated will be contaminated both by any single-stranded non-mutant template DNA that has remained uncopied and by partially double-stranded molecules. The presence of these species considerably reduces the proportion of mutant progeny. They can be removed by sucrose gradient centrifugation or by agarose gel electrophoresis, but this is time-consuming and inconvenient.

Following transformation and *in vivo* DNA synthesis, segregation of the two strands of the heteroduplex molecule can occur, yielding a mixed population of mutant and non-mutant progeny. Mutant progeny have to be purified away from parental molecules, and this process is complicated by the cell's mismatch repair system. In theory, the mismatch repair system should yield equal numbers

of mutant and non-mutant progeny, but in practice mutants are counterselected. The major reason for this low yield of mutant progeny is that the methyl-directed mismatch repair system of *E. coli* favours the repair of non-methylated DNA. In the cell, newly synthesized DNA strands that have not yet been methylated are preferentially repaired at the position of the mismatch, thereby eliminating a mutation. In a similar way, the non-methylated *in vitro*-generated mutant strand is repaired by the cell so that the majority of progeny are wild type (Kramer, B. *et al.* 1984). The problems associated with the mismatch repair system can be overcome by using host strains carrying the *mut*L, *mut*S or *mut*H mutations, which prevent the methyl-directed repair of mismatches.

A heteroduplex molecule with one mutant and one non-mutant strand must inevitably give rise to both mutant and non-mutant progeny upon replication. It would be desirable to suppress the growth of non-mutants, and various strategies have been developed with this in mind (Kramer, B. 1984, Carter *et al.* 1985, Kunkel 1985, Sayers & Eckstein 1991).

Another disadvantage of all of the primer extension methods is that they require a single-stranded template. In contrast, with PCR-based mutagenesis (see below) the template can be single-stranded or double-stranded, circular or linear. In comparison with single-stranded DNAs, double-stranded DNAs are much easier to prepare. Also, gene inserts are in general more stable with double-stranded DNAs. These facts account for the observation that most commercial mutagenesis kits use double-stranded templates.

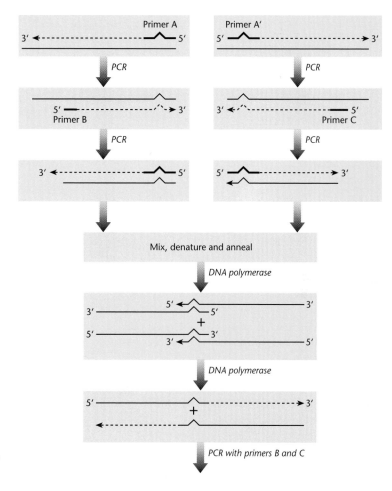

Fig. 7.11 Site-directed mutagenesis by means of the PCR. The steps shown in the top-left corner of the diagram show the basic PCR method of mutagenesis. The bottom half of the figure shows how the mutation can be moved to the middle of a DNA molecule. Primers are shown in bold and primers A and A' are complementary.

PCR methods of site-directed mutagenesis

Early work on the development of the PCR method of DNA amplification showed its potential for mutagenesis (Scharf *et al.* 1986). Single bases mismatched between the amplification primer and the template become incorporated into the template sequence as a result of amplification (Fig. 7.11). Higuchi *et al.* (1988) have described a variation of the basic method which enables a mutation in a PCR-produced DNA fragment to be introduced anywhere along its length. Two primary PCR reactions produce two overlapping DNA fragments, both bearing the same mutation in the overlap region. The overlap in sequence allows the fragments to hybridize (Fig. 7.11). One of the two possible hybrids is extended by DNA

polymerase to produce a duplex fragment. The other hybrid has recessed 5' ends and, since it is not a substrate for the polymerase, is effectively lost from the reaction mixture. As with conventional primer-extension mutagenesis, deletions and insertions can also be created.

The method of Higuchi *et al.* (1988) requires four primers and three PCRs (a pair of PCRs to amplify the overlapping segments and a third PCR to fuse the two segments). Sarkar and Sommer (1990) have described a simpler method, which utilizes three oligonucleotide primers to perform two rounds of PCR. In this method, the product of the first PCR is used as a *megaprimer* for the second PCR (Fig. 7.12).

The advantage of a PCR-based mutagenic protocol is that the desired mutation is obtained with 100% efficiency. There are two disadvantages. First,

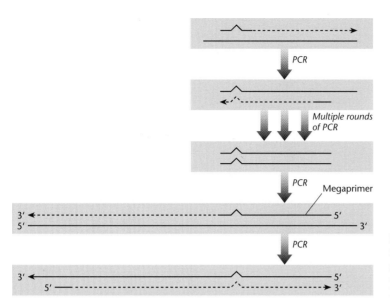

Fig. 7.12 The megaprimer method of mutagenesis. The mutant molecule produced in the early rounds of PCR acts as a primer ('megaprimer') for later rounds of PCR.

the PCR product usually needs to be ligated into a vector, although Sarkar and Sommer (1990) have generated the mutant protein directly, using coupled *in vitro* transcription and translation. Secondly, *Taq* polymerase copies DNA with low fidelity (see p. 23). Thus the sequence of the entire amplified segment generated by PCR mutagenesis must be determined to ensure that there are no extraneous mutations. Alternatively, thermostable polymerases with improved fidelity can be used (Cariello *et al.* 1991, Lundberg *et al.* 1991, Mattila *et al.* 1991).

Selection of mutant peptides by phage and phasmid display

In phage display, a segment of foreign DNA is inserted into either a phagemid or an infectious filamentous phage genome and expressed as a fusion product with a phage coat protein. It is a very powerful technique for selecting and engineering polypeptides with novel functions. The technique was developed first for the *E. coli* phage M13 (Parmley & Smith 1988), but has since been extended to other phages such as T4 and λ (Ren & Black 1998, Santini *et al.* 1998).

The M13 phage particle consists of a single-stranded DNA molecule surrounded by a coat consisting of several thousand copies of the major coat protein,

P8. At one end of the particle are five copies each of the two minor coat proteins P9 and P7 and at the other end five copies each of P3 and P6. In early examples of phage display, a random DNA cassette (see above) was inserted into either the P3 or the P8 gene at the junction between the signal sequence and the native peptide. *E. coli* transfected with the recombinant DNA molecules secreted phage particles that displayed on their surface the amino acids encoded by the foreign DNA. Particular phage displaying peptide motifs with, for example, antibody-binding properties were isolated by affinity chromatography (Fig. 7.13). Several rounds of affinity chromatography and phage propagation can be used to further enrich for phage with the desired binding characteristics. In this way, millions of random peptides have been screened for their ability to bind to an anti-peptide antibody or to streptavidin (Cwirla *et al.* 1990, Devlin *et al.* 1990, Scott & Smith 1990), and variants of human growth hormone with improved affinity and receptor specificity have been isolated (Lowman *et al.* 1991).

One disadvantage of the original method of phage display is that polypeptide inserts greater than 10 residues compromise coat-protein function and so cannot be efficiently displayed. This problem can be solved by the use of phagemid display (Bass *et al.* 1990). In this system, the starting-point is a plasmid

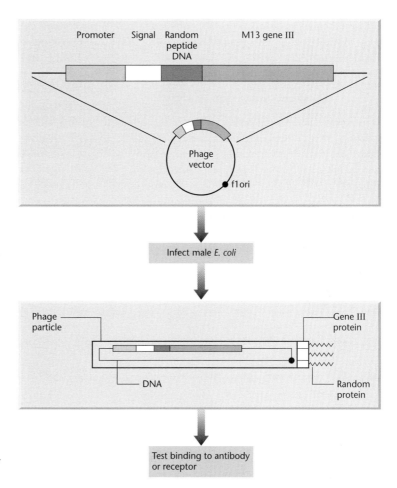

Fig. 7.13 The principle of phage display of random peptides.

carrying a single copy of the P3 or P8 gene from M13 plus the M13 *ori* sequence (i.e. a phagemid, see p. 70). As before, the random DNA sequence is inserted into the P3 or P8 gene downstream from the signal peptide-cleavage site and the construct transformed into *E. coli*. Phage particles displaying the amino acid sequences encoded by the DNA insert are obtained by superinfecting the transformed cells with helper phage. The resulting phage particles are phenotypically mixed and their surfaces are a mosaic of normal coat protein and fusion protein.

Specialized phagemid display vectors have been developed for particular purposes. For example, phagemids have been constructed that have an amber (chain-terminating) codon immediately downstream from the foreign DNA insert and upstream

from the body of P3 or P8. When the recombinant phagemid is transformed into non-suppressing strains of *E. coli*, the protein encoded by the foreign DNA terminates at the amber codon and is secreted into the medium. However, if the phagemid is transformed into cells carrying an amber suppressor, the entire fusion protein is synthesized and displayed on the surface of the secreted phage particles (Winter *et al.* 1994). Other studies (Jespers *et al.* 1995, Fuh & Sidhu 2000, Fuh *et al.* 2000) have shown that proteins can be displayed as fusions to the carboxy terminus of P3, P6 and P8. Although amino-terminal display formats are likely to dominate established applications, carboxy-terminal display permits constructs that are unsuited to amino-terminal display.

For a detailed review of phage and phagemid display, the reader should consult Sidhu (2000) and Sidhu *et al.* (2000).

Directed mutation *in vivo*

In vitro mutagenesis methods are very useful when working with small plasmids carrying cloned genes. However, if it is desired to modify a single residue on a very large vector, such as a yeast artificial chromosome (YAC), P1-derived artificial chromosome (PAC) or bacterial artificial chromosome (BAC), *in vitro* methods are not appropriate. With such large plasmids, directed mutagenesis has to be done *in vivo* and a number of different methods have been described (Muyrers *et al.* 2000, Lalioti & Heath 2001). The simplest is that of Yu *et al.* (2000), which involves recombination between linear DNA fragments carrying the desired mutation and the homologous region on the chromosome.

Unlike many organisms, *E. coli* is not readily transformed by linear DNA fragments. The main reason for this is the rapid degradation of the DNA by the intracellular RecBCD exonuclease. Mutant strains lacking the exonuclease do not degrade linear DNA very rapidly, but such strains grow poorly, are defective for recombination and do not support efficient replication of plasmid vectors. The solution devised by Yu *et al.* (2000) was to use an *E. coli* strain containing a λ prophage harbouring the genes *exo*, *bet* and *gam* under the control of a temperature-sensitive λcI repressor. The Gam gene product stops the RecBCD nuclease from attacking the linear DNA and the Exo and Bet gene products generate recombination activity to enable DNA exchange to occur.

CHAPTER 8

Cloning in bacteria other than *Escherichia coli*

Introduction

For many experiments it is convenient to use *E. coli* as a recipient for genes cloned from eukaryotes or other prokaryotes. Transformation is easy and there is available a wide range of easy-to-use vectors with specialist properties, e.g. regulatable high-level gene expression. However, use of *E. coli* is not always practicable because it lacks some auxiliary biochemical pathways that are essential for the phenotypic expression of certain functions, e.g. degradation of aromatic compounds, antibiotic synthesis, pathogenicity, sporulation, etc. In such circumstances, the genes have to be cloned back into species similar to those whence they were derived.

There are three prerequisites for cloning genes in a new host. First, there needs to be a method for introducing the DNA of interest into the potential recipient. The methods available include transformation, conjugation and electroporation, and these will be discussed in more detail later. Secondly, the introduced DNA needs to be maintained in the new host. Either it must function as a replicon in its new environment or it has to be integrated into the chromosome or a pre-existing plasmid. Finally, the uptake and maintenance of the cloned genes will only be detected if they are expressed. Thus the inability to detect a cloned gene in a new bacterial host could be due to failure to introduce the gene, to maintain it or to express it, or to a combination of these factors. Another cause of failure could be restriction. For example, the frequency of transformation of *Pseudomonas putida* with plasmid RSF1010 is 10^5 transformants/μg DNA, but only if the plasmid is prepared from another *P. putida* strain. Otherwise, no transformants are obtained (Bagdasarian *et al.* 1979). Wilkins *et al.* (1996) have noted that conjugative transfer of promiscuous IncP plasmids is unusually sensitive to restriction.

Introducing DNA into bacterial cells

DNA can be transferred between different strains of *E. coli* by the three classical methods of conjugation, transduction and transformation, as well as by the newer method of electroporation. For gene-manipulation work, transformation is nearly always used. The reasons for this are threefold. First, it is relatively simple to do, particularly now that competent cells are commercially available. Secondly, it can be very efficient. Efficiencies of 10^8–10^9 transformants/μg plasmid DNA are readily achievable and are more than adequate for most applications. Thirdly, self-transmissible cloning vectors are much larger than their non-transmissible counterparts because they have to carry all the genes required for conjugal transfer.

A large number of bacteria from different taxonomic groups, including archaebacteria, are known to be transformable. For example, Lorenz and Wackernagel (1994) listed over 40 species and the numbers are still growing. However, transformation in these organisms differs in a number of respects from that in *E. coli*. First, transformation in these organisms occurs naturally, whereas transformation in *E. coli* is artificially induced. With the exception of *Neisseria gonorrhoeae*, competence for transformation is a transient phenomenon. Secondly, transformation can be sequence-independent, as in *Bacillus subtilis* and *Acinetobacter calcoaceticus*, but in other species (*Haemophilus influenzae, N. gonorrhoeae*) is dependent on the presence of specific uptake sequences. Thirdly, the mechanism of natural transformation involves breakage of the DNA duplex and degradation of one of the two strands so that a linear single strand can enter the cell (for review, see Dubnau 1999). This mechanism is not compatible with efficient plasmid transformation (see Box 8.1). Geneticists working with *B. subtilis* and *Streptomyces*

Box 8.1 Transforming *Bacillus subtilis* with plasmid DNA

Although it is very easy to transform *B. subtilis* with fragments of chromosomal DNA, there are problems associated with transformation by plasmid molecules. Ehrlich (1977) first reported that competent cultures of *B. subtilis* can be transformed with covalently closed circular (CCC) plasmid DNA from *Staphylococcus aureus* and that this plasmid DNA is capable of autonomous replication and expression in its new host. The development of competence for transformation by plasmid and chromosomal DNA follows a similar time course and in both cases transformation is first-order with respect to DNA concentration, suggesting that a single DNA molecule is sufficient for successful transformation (Contente & Dubnau 1979). However, transformation of *B. subtilis* with plasmid DNA is very inefficient in comparison with chromosomal transformation, for only one transformant is obtained per 10^3–10^4 plasmid molecules.

An explanation for the poor transformability of plasmid DNA molecules was provided by Canosi *et al.* (1978). They found that the specific activity of plasmid DNA in transformation of *B. subtilis* was dependent on the degree of oligomerization of the plasmid genome. Purified monomeric CCC forms of plasmids transform *B. subtilis* several orders of magnitude less efficiently than do unfractionated plasmid preparations or multimers. Furthermore, the low residual transforming activity of monomeric CCC DNA molecules can be attributed to low-level contamination with multimers (Mottes *et al.* 1979). Using a recombinant plasmid capable of replication in both *E. coli* and *B. subtilis* (pHV14) (see p. 149), Mottes *et al.* (1979) were able to show that plasmid transformation of *E. coli* occurs regardless of the degree of oligomerization, in contrast to the situation with *B. subtilis*. Oligomerization of linearized plasmid DNA by DNA ligase resulted in a substantial increase of specific transforming activity when assayed with *B. subtilis* and caused a decrease when used to transform *E. coli*. An explanation of the molecular events in transformation which generate the requirement for oligomers has been presented by De Vos *et al.* (1981). Basically, the plasmids are cleaved into linear molecules upon contact with competent cells, just as

chromosomal DNA is cleaved during transformation of *Bacillus*. Once the linear single-stranded form of the plasmid enters the cell, it is not reproduced unless it can circularize; hence the need for multimers to provide regions of homology that can recombine. Michel *et al.* (1982) have shown that multimers, or even dimers, are not required, provided that part of the plasmid genome is duplicated. They constructed plasmids carrying direct internal repeats 260–2000 bp long and found that circular or linear monomers of such plasmids were active in transformation.

Canosi *et al.* (1981) have shown that plasmid monomers will transform recombination-proficient *B. subtilis* if they contain an insert of *B. subtilis* DNA. However, the transformation efficiency of such monomers is still considerably less than that of oligomers. One consequence of the requirement for plasmid oligomers for efficient transformation of *B. subtilis* is that there have been very few successes in obtaining large numbers of clones in *B. subtilis* recipients (Keggins *et al.* 1978, Michel *et al.* 1980). The potential for generating multimers during ligation of vector and foreign DNA is limited.

Transformation by plasmid rescue

An alternative strategy for transforming *B. subtilis* has been suggested by Gryczan *et al.* (1980). If plasmid DNA is linearized by restriction-endonuclease cleavage, no transformation of *B. subtilis* results. However, if the recipient carries a homologous plasmid and if the restriction cut occurs within a homologous marker, then this same marker transforms efficiently. Since this rescue of donor plasmid markers by a homologous resident plasmid requires the *B. subtilis recE* gene product, it must be due to recombination between the linear donor DNA and the resident plasmid. Since DNA linearized by restriction-endonuclease cleavage at a unique site is monomeric, this rescue system (*plasmid rescue*) bypasses the requirement for a multimeric vector. The model presented by De Vos *et al.* (1981) to explain the requirement for oligomers (see above) can be adapted to account for transformation by monomers by means of plasmid

continued

Box 8.1 *continued*

rescue. In practice, foreign DNA is ligated to monomeric vector DNA and the *in vitro* recombinants are used to transform *B. subtilis* cells carrying a homologous plasmid. Using such a 'plasmid-rescue' system, Gryczan *et al.* (1980) were able to clone various genes from *B. licheniformis* in *B. subtilis*.

One disadvantage of the plasmid-rescue method is that transformants contain both the recombinant molecule and the resident plasmid. Incompatibility will result in segregation of the two plasmids. This may require several subculture steps, although Haima *et al.* (1990) observed very rapid segregation. Alternatively, the recombinant plasmids can be transformed into plasmid-free cells.

Transformation of protoplasts

A third method for plasmid DNA transformation in *B. subtilis* involves polyethylene glycol (PEG) induction of DNA uptake in protoplasts and subsequent regeneration of the bacterial cell wall (Chang & Cohen 1979). The procedure is highly efficient and yields up to 80% transformants, making the method suitable for the introduction even of cryptic plasmids. In addition to its much higher yield of plasmid-containing transformants, the protoplast transformation system differs in two respects from the 'traditional' system using physiologically competent cells. First, linear plasmid DNA and non-supercoiled circular plasmid DNA molecules constructed by ligation *in vitro* can be introduced at high efficiency into *B. subtilis* by the protoplast transformation system, albeit at a frequency 10–1000 lower than the frequency observed for CCC plasmid DNA. However, the efficiency of shotgun cloning is much lower with protoplasts than with competent cells (Haima *et al.* 1988). Secondly, while competent cells can be transformed easily for genetic determinants located on the *B. subtilis* chromosome, no detectable transformation with chromosomal DNA is seen using the protoplast assay. Until recently, a disadvantage of the protoplast system was that the regeneration medium was nutritionally complex, necessitating a two-step selection procedure for auxotrophic markers. Details have been presented of a defined regeneration medium by Puyet *et al.* 1987.

Table B8.1 Comparison of the different methods of transforming *B. subtilis*.

System	Efficiency (transformants/µg DNA)		Advantages	Disadvantages
Competent cells	Unfractionated plasmid	2×10^4	Competent cells readily prepared	Requires plasmid oligomers or
	Linear	0	Transformants can be selected	internally duplicated plasmids,
	CCC monomer	4×10^4	readily on any medium	which makes shotgun experiments
	CCC dimer	8×10^3	Recipient can be Rec⁻	difficult unless high DNA
	CCC multimer	2.6×10^5		concentrations and high vector/
				donor DNA ratios are used
				Not possible to use phosphatase-treated vector
Plasmid rescue	Unfractionated plasmid	2×10^6	Oligomers not required	Transformants contain resident
			Can transform with linear DNA	plasmid and incoming plasmid and
			Transformants can be selected on	these have to be separated by
			any medium	segregation or retransformation
				Recipient must be Rec⁺
Protoplasts	Unfractionated plasmid	3.8×10^6	Most efficient system	Efficiency lower with molecules
	Linear	2×10^4	Gives up to 80% transformants	which have been cut and religated
	CCC monomer	3×10^6	Does not require competent cells	Efficiency also very size-dependent,
	CCC dimer	2×10^6	Can transform with linear DNA and	and declines steeply as size increases
	CCC multimer	2×10^6	can use phosphatase-treated vector	

sp. have developed specialized methods for over-coming these problems. For work with other species, electroporation (see p. 18) offers a much simpler alternative, although the efficiency may not be high e.g. in *Methanococcus voltae*, it was only 10^2 transformants/μg plasmid DNA (Tumbula & Whitman 1999).

Given that plasmid transformation is difficult in many non-enteric bacteria, conjugation represents an acceptable alternative. The term *promiscuous plasmids* was originally coined for those plasmids which are self-transmissible to a wide range of other Gram-negative bacteria (see p. 45), where they are stably maintained. The best examples are the IncP alpha plasmids RP4. RP1, RK2, etc., which are about 60 kb in size, but the IncW plasmid Sa (29.6 kb) has been used extensively in *Agrobacterium tumefaciens*. Self-transmissible plasmids carry *tra* (transfer) genes encoding the conjugative apparatus. IncP plasmids are able to promote the transfer of other compatible plasmids, as well as themselves. For example, they can mobilize IncQ plasmids, such as RSF1010. More important, transfer can be mediated between *E. coli* and Gram-positive bacteria (Trieu-Cuot *et al.* 1987, Gormley & Davies 1991), as well as to yeasts and fungi (Heinemann & Sprague 1989, Hayman & Bolen 1993, Bates *et al.* 1998). However, the transfer range of a plasmid may be greater than its replication maintenance or host range (Mazodier & Davies 1991).

Self-transmissible plasmids have been identified in many different Gram-positive genera. The transfer regions of these plasmids are much smaller than those for plasmids from Gram-negative bacteria. One reason for this difference may be the much simpler cell-wall structure in Gram-positive bacteria. As well as being self-transmissible, many of these Gram-positive plasmids are promiscuous. For example, plasmid pAMβ1 was originally isolated from *Enterococcus faecalis* but can transfer to staphylococci, streptococci, bacilli and lactic acid bacteria (De Vos *et al.* 1997), as well as to Gram-negative bacteria, such as *E. coli* (Trieu-Cuot *et al.* 1988). The self-transmissible plasmids of Gram-positive bacteria, like their counterparts in the Gram-negative bacteria, can also mobilize other plasmids between different genera (Projan & Archer 1989, Charpentier *et al.* 1999). Non-self-transmissible plasmids can also be mobilized within and between genera by conjugative transposons (Salyers *et al.* 1995, Charpentier *et al.* 1999).

It should be noted that conjugation is not a replacement for transformation or electroporation. If DNA is manipulated *in vitro*, then it has to be transferred into a host cell at some stage. In many cases, this will be *E. coli* and transformation is a suitable procedure. Once in *E. coli*, or any other organism for that matter, it may be moved to other bacteria directly by conjugation, as an alternative to purifying the DNA and moving it by transformation or electroporation.

Maintenance of recombinant DNA in new hosts

For recombinant DNA to be maintained in a new host cell, either it must be capable of replication or it must integrate into the chromosome or a plasmid. In most instances, the recombinant will be introduced as a covalently closed circle (CCC) plasmid and maintenance will depend on the host range of the plasmid. As noted in Chapter 4 (p. 45), the host range of a plasmid is dependent on the number of proteins required for its replication which it encodes. Some plasmids have a very narrow host range, whereas others can be maintained in a wide range of Gram-negative or Gram-positive genera. Some, such as the plasmids from *Staphylococcus aureus* and RSF1010, can replicate in both Gram-negative and Gram-positive species (Lacks *et al.* 1986, Gormley & Davies 1991, Leenhouts *et al.* 1991).

As noted in Chapter 5, there is a very wide range of specialist vectors for use in *E. coli*. However, most of these vectors have a very narrow host range and can be maintained only in enteric bacteria. A common way of extending the host range of these vectors is to form hybrids with plasmids from the target species. The first such *shuttle* vectors to be described were fusions between the *E. coli* vector pBR322 and the *S. aureus*/*B. subtilis* plasmids pC194 and pUB110 (Ehrlich 1978). The advantage of shuttle plasmids is that *E. coli* can be used as an efficient *intermediate* host for cloning. This is particularly important if transformation or electroporation of the alternative host is very inefficient.

Integration of recombinant DNA

When a recombinant plasmid is transferred to an unrelated host cell, there are a number of possible outcomes:

• It may be stably maintained as a plasmid.
• It may be lost.
• It may integrate into another replicon, usually the chromosome.
• A gene on the plasmid may recombine with a homologous gene elsewhere in the cell.

Under normal circumstances, a plasmid will be maintained if it can replicate in the new host and will be lost if it cannot. Plasmids which will be lost in their new host are particularly useful for delivering transposons (Saint *et al.* 1995, Maguin *et al.* 1996). If the plasmid carries a cloned insert with homology to a region of the chromosome, then the outcome is quite different. The homologous region may be excised and incorporated into the chromosome by the normal recombination process, i.e. substitution occurs via a double crossover event. If only a single crossover occurs, the entire recombinant plasmid is integrated into the chromosome.

It is possible to favour integration by transferring a plasmid into a host in which it cannot replicate and selecting for a plasmid-borne marker. For example, Stoss *et al.* (1997) have constructed an integrative vector by cloning a neomycin resistance gene and part of the *amyE* (alpha amylase) gene of *B. subtilis* in plasmid pBR322. When this plasmid is transferred to *B. subtilis*, it is unable to replicate, but, if selection is made for neomycin resistance, then integration of the plasmid occurs at the *amyE* locus (Fig. 8.1).

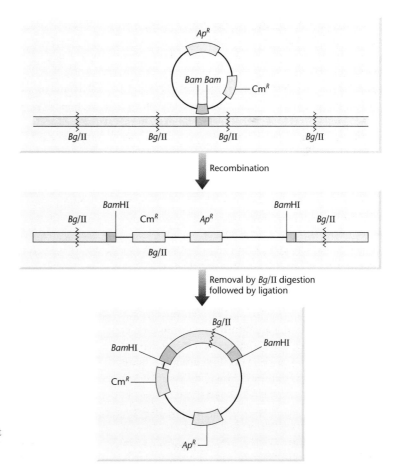

Fig. 8.1 Cloning DNA sequences flanking the site of insertion. The red bar on the plasmid represents a *Bam*HI fragment of *B. subtilis* chromosomal DNA carrying the *amyE* gene. Note that the plasmid has no *Bgl*II sites. (See text for details.)

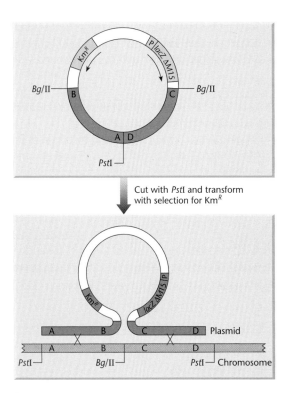

Cut with *Pst*I and transform
with selection for Km^R

Fig. 8.2 Insertion of plasmid DNA into the chromosome by
a double crossover event. The *B. subtilis* DNA is shown in
grey and the letters A to D represent different chromosomal
sequences. Vector DNA is shown in white and other
vector-borne genes in pink.

This technique is particularly useful if one wishes
to construct a recombinant carrying a single copy of
a foreign gene.

Once a recombinant plasmid has integrated into
the chromosome, it is relatively easy to clone adjacent sequences. Suppose, for example, that a vector
carrying *B. subtilis* DNA in the *Bam*HI site (Fig. 8.1)
has recombined into the chromosome. If the recombinant plasmid has no *Bgl*II sites, it can be recovered
by digesting the chromosomal DNA with *Bgl*II, ligating the resulting fragments and transforming *E. coli*
to Ap^R. However, the plasmid which is isolated will
be larger than the original one, because DNA flanking the site of insertion will also have been cloned.
In this way, Niaudet *et al.* (1982) used a plasmid
carrying a portion of the *B. subtilis ilvA* gene to clone
the adjacent *thyA* gene.

Genes cloned into a plasmid and flanked by regions
homologous to the chromosome can also integrate

without tandem duplication of the chromosomal
segments. In this case, the plasmid DNA is linearized
before transformation, as shown in Fig. 8.2. The
same technique can be used to generate deletions.
The gene of interest is cloned, a portion of the gene
replaced *in vitro* with a fragment bearing an antibiotic marker and the linearized plasmid transformed into *B. subtilis*, with selection made for
antibiotic resistance.

Cloning in Gram-negative bacteria other than *E. coli*

To clone DNA in non-enteric bacteria, a plasmid
cloning vehicle is required which can replicate in the
selected organism(s). Under normal circumstances,
E. coli will be used as an intermediate host for transformation of the ligation mix and screening for
recombinant plasmids. Therefore, the vector must
be able to replicate in *E. coli* as well. The options
which are available are to generate a shuttle vector
or to use a broad-host-range plasmid as a vector. If
a small plasmid can be isolated from the bacterium
of interest, then it is easy to splice it into an existing
E. coli vector to generate a shuttle vector. Recent
examples of this approach are the construction
of vectors for use in *Pasteurella* (Bills *et al.* 1993),
Desulfovibrio (Rousset *et al.* 1998) and *Thermus* (De
Grado *et al.* 1999). This approach is particularly
useful if the selectable markers used in *E. coli*
also function in the new host. Then one can take
advantage of the many different specialist vectors
(see Chapter 5) which already exist, e.g. expression vectors, secretion vectors, etc. If the selectable
markers are not expressed in the new host, then
extensive manipulations may be necessary just to
enable transformants to be detected.

With broad-host-range plasmids, there is a high
probability that the selectable markers will be expressed in the new host and confirming that this is
indeed the case is easy to do. However, the naturally
occurring plasmids do not fulfil all the criteria for an
ideal vector, which are:
• small size;
• having multiple selectable markers;
• having unique sites for a large number of restriction enzymes, preferably in genes with readily
scorable phenotypes.

Consequently the natural plasmids have been extensively modified, but few approach the degree of sophistication found in the standard *E. coli* vectors.

Vectors derived from the IncQ-group plasmid RSF1010

Plasmid RSF1010 is a multicopy replicon which specifies resistance to two antimicrobial agents, sulphonamide and streptomycin. The plasmid DNA, which is 8684 bp long, has been completely sequenced (Scholz *et al.* 1989). A detailed physical and functional map has been constructed (Bagdasarian *et al.* 1981, Scherzinger *et al.* 1984). The features mapped are the restriction-endonuclease recognition sites, RNA polymerase binding sites, resistance determinants, genes for plasmid mobilization (*mob*), three replication proteins (Rep A, B and C) and the origins of vegetative (*ori*) and transfer (*nic*) replication.

Plasmid RSF1010 has unique cleavage sites for *Eco*RI, *Bst*EII, *Hpa*I, *Dra*II, *Nsi*I and *Sac*I and, from the nucleotide sequence data, is predicted to have unique sites for *Afl*III, *Ban*II, *Not*I, *Sac*II, *Sfi*I and *Spl*I. There are two *Pst*I sites, about 750 bp apart, which flank the sulphonamide-resistance determinant (Fig. 8.3). None of the unique cleavage sites is

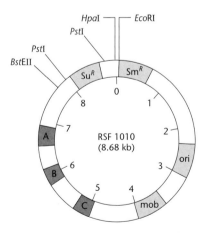

Fig. 8.3 The structure of plasmid RSFI010. The pink tinted areas show the positions of the *Sm*R and *Su*R genes. The region marked *ori* indicates the location of the origin of replication. The *mob* function is required for conjugal mobilization by a compatible self-transmissible plasmid. A, B and C are the regions encoding the three replication proteins.

located within the antibiotic-resistance determinants and none is particularly useful for cloning. Before the *Bst*, *Eco* and *Pst* sites can be used, another selective marker must be introduced into the RSF1010 genome. This need arises because the *Sm*R and *Su*R genes are transcribed from the same promoter (Bagdasarian *et al.* 1981). Insertion of a DNA fragment between the *Pst* sites inactivates both resistance determinants. Although the *Eco* and *Bst* sites lie outside the coding regions of the *Sm*R gene, streptomycin resistance is lost if a DNA fragment is inserted at these sites unless the fragment provides a new promoter. Furthermore, the *Su*R determinant which remains is a poor selective marker.

A whole series of improved vectors has been derived from RSF1010 but only a few are mentioned here. The earliest vectors contained additional unique cleavage sites and more useful antibiotic-resistance determinants. For example, plasmids KT230 and KT231 encode *Km*R and *Sm*R and have unique sites for *Hind*III, *Xma*I, *Xho*RI and *Sst*I which can be used for insertional inactivation. These two vectors have been used to clone in *P. putida* genes involved in the catabolism of aromatic compounds (Franklin *et al.* 1981). Vectors for the regulated expression of cloned genes have also been described. Some of these make use of the *tac* promoter (Bagdasarian *et al.* 1983, Deretic *et al.* 1987) or the phage T7 promoter (Davison *et al.* 1989), which will function in *P. putida* as well as *E. coli*. Another makes use of positively activated twin promoters from a plasmid specifying catabolism of toluene and xylenes (Mermod *et al.* 1986). Expression of cloned genes can be obtained in a wide range of Gram-negative bacteria following induction with micromolar quantities of benzoate, and the product of the cloned gene can account for 5% of total cell protein.

Vectors derived from the IncP-group plasmids

Both the IncP alpha plasmids (R18, R68, RK2, RP1 and RP4), which are 60 kb in size, and the smaller (52 kb) IncP beta plasmid R751 have been completely sequenced (Pansegrau *et al.* 1994, Thorsted *et al.* 1998). As a result, much is known about the genes carried, the location of restriction sites, etc. Despite this, the P-group plasmids are not widely

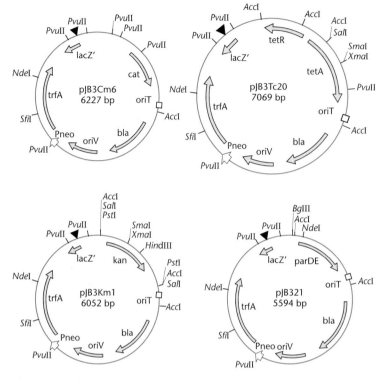

Fig. 8.4 Map and construction of general-purpose broad-host-range cloning vectors derived from plasmid RP4. The restriction sites in the polylinker downstream of the *lacZ* promoter are marked (▼), and the sites are, in the counterclockwise direction, *Hind*III, *Sph*I, *Pst*I, *Sal*I/*Hinc*II/*Acc*I, *Xba*I, *Bam*HI, *Xma*I/*Sma*I, *Kpn*I, *Sac*I and *Eco*RI. Sites in the polylinker that are not unique are indicated elsewhere on each vector. Note that the sites for *Nde*I and *Sfi*I are unique for all of the vectors except pJB321. *Pneo*, promoter from the neomycin resistance gene; *bla*, *kan*, *tet* and *cat*, genes encoding ampicillin, kanamycin, tetracycline and chloramphenicol resistance, respectively. (Figure modified from Blatny *et al.* 1997.)

used as vectors because their large size makes manipulations difficult.

A number of groups have developed mini-IncP plasmids as vectors and good examples are those of Blatny *et al.* (1997). Their vectors are only 4.8–7.1 kb in size but can still be maintained in a wide range of Gram-negative bacteria. All the vectors share a common polylinker and *lacZ'* region, thereby simplifying cloning procedures and identification of inserts by blue/white screening (see p. 35) and most carry two antibiotic-resistance determinants. All the vectors retain the *oriT* (origin of transfer) locus, enabling them to be conjugally transferred in those cases where the recipient cannot be successfully transformed or electroporated. Two other features of these vectors deserve mention. First, the *parDE* region from the parent plasmid has been included in some of the vectors, since this greatly enhances their segregative stability in certain hosts. Secondly, the *trfA* locus on the vectors contains unique sites for the restriction enzymes *Nde*I and *Sfi*I. Removal of the *Nde*I–*Sfi*I fragment results in an increased copy number. Expression vectors have also been devel-

oped by the inclusion of controllable promoters. Representative examples of these vectors are shown in Fig. 8.4.

An alternative way of using P-group plasmids as cloning vectors has been described by Kok *et al.* (1994). Their method combines the advantages of high-copy-number pBR322 vectors with the convenience of conjugative plasmids. This is achieved by converting the pBR322 vector into a transposable element. Most pBR322 derivatives contain the β-lactamase gene and one of two 38 bp inverted repeats of transposon Tn2. By adding a second inverted repeat, a transposable element is created (Fig. 8.5). All that is missing is transposase activity and this is provided by another plasmid, which is a pSC101 derivative carrying the *tnpA* gene. To use this system, the desired DNA sequence is cloned into the transposition vector. The recombinant molecules are transformed into an *E. coli* strain carrying the P-group plasmid (e.g. R751) and the pSC101 *tnpA* derivative and selection is made for the desired characteristics. Once a suitable transformant has been selected, it is conjugated with other Gram-negative

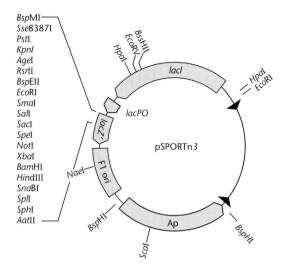

Fig. 8.5 A transposable vector derived from pBR322. The solid arrowheads indicate the inverted repeats required for transposition. Insertion of a DNA fragment in the multiple-cloning site results in inactivation of the *lacZα* gene and regulated gene expression from the wild-type *lac* promoter.

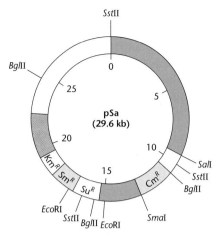

Fig. 8.6 The structure of plasmid Sa. The grey area encodes the functions essential for plasmid replication. The dark red areas represent the regions containing functions essential for self-transmission, the one between the *Sst* and *Sal* sites being responsible for suppression of tumour induction by *Agrobacterium tumefaciens*.

bacteria and selection made for the ampicillin-resistance marker carried on the transposon.

Vectors derived from the IncW plasmid Sa

Although a group-W plasmid, such as plasmid pSa (Fig. 8.6) can infect a wide range of Gram-negative bacteria, it has been developed mainly as a vector for use with the oncogenic bacterium *A. tumefaciens* (see p. 224). Two regions of the plasmid have been identified as involved in conjugal transfer of the plasmid and one of them has the unexpected property of suppressing oncogenicity by *A. tumefaciens* (Tait *et al.* 1982). Information encoding the replication of the plasmid in *E. coli* and *A. tumefaciens* is contained within a 4 kb DNA fragment. Leemans *et al.* (1982b) have described four small (5.6–7.2 MDa), multiply marked derivatives of pSa. The derivatives contain single-target sites for a number of the common restriction endonucleases and at least one marker in each is subject to insertional inactivation. Although these Sa derivatives are non-conjugative, they can be mobilized by other conjugative plasmids. Tait *et al.* (1983) have also constructed a set of broad-host-range vectors from pSa. The properties of their derivatives are similar to those of Leemans *et al.* (1982b), but one of them also contains the bacterio-phage λ *cos* sequence and hence functions as a cosmid. Specialist vectors for use in *Agrobacterium* and which are derived from a natural *Agrobacterium* plasmid have been described by Gallie *et al.* (1988).

Vectors derived from pBBR1

Plasmid BBR1 is a broad-host-range plasmid originally isolated from *Bordatella bronchiseptica* that is compatible with IncP, IncQ and IncW plasmids and replicates in a wide range of bacteria. Kovach *et al.* (1995) have developed a series of vectors from pBBR1 that are relatively small (< 5.3 kb), possess multiple cloning sites, allow direct selection of recombinants in *E. coli* by blue/white screening and are mobilizable by IncP plasmids. Newman and Fuqua (1999) have developed an expression vector from one of these pBBR1 derivatives by incorporating the *araBAD/araC* cassette from *E. coli* (see p. 76) and shown that the promoter is controllable in *Agrobacterium*. Sukchawalit *et al.* (1999), using a similar vector, have shown that the promoter is also controllable in *Xanthomonas*.

Cloning in Gram-positive bacteria

In Gram-positive bacteria, the base composition of the different genomes ranges from < 30% GC to > 70% GC. Given this disparity in GC content, the preferred codons and regulatory signals used by organisms at one end of the % GC spectrum will not be recognized by organisms at the other end. This in turn means that there are no universal cloning vehicles for use with all Gram-positive bacteria. Rather, one set of systems has been developed for high-GC organisms (e.g. streptomycetes) and another for low-GC organisms. This latter group comprises bacteria from the unrelated genera *Bacillus, Clostridium* and *Staphylococcus* and the lactic acid bacteria *Streptococcus, Lactococcus* and *Lactobacillus*.

Vectors for cloning in *Bacillus subtilis* and other low-GC organisms

The development of *B. subtilis* vectors began with the observation (Ehrlich 1977) that plasmids from *S. aureus* (Table 8.1) can be transformed into *B. subtilis*, where they replicate and express antibiotic resistance normally.

As can be seen from Table 8.1, none of the natural *S. aureus* plasmids carries more than one selectable marker and so improved vectors have been constructed by gene manipulation, e.g. pHV11 is pC194 carrying the Tc^R gene of pT127 (Ehrlich 1978). In general, these plasmids are stable in *B. subtilis*, but segregative stability is greatly reduced following insertion of exogenous DNA (Bron & Luxen 1985). Reasoning that stable host–vector systems in *B.*

subtilis are more likely if endogenous plasmids are used, Bron and colleagues have developed the cryptic *Bacillus* plasmid pTA1060 as a vector (Haima *et al.* 1987, Bron *et al.* 1989).

Because of the difficulties experienced in direct cloning in *B. subtilis*, hybrid plasmids were constructed which can replicate in both *E. coli* and *B. subtilis*. Originally most of these were constructed as fusions between pBR322 and pC194 or pUB110. With such plasmids, *E. coli* can be used as an efficient intermediate host for cloning. Plasmid preparations extracted from *E. coli* clones are subsequently used to transform competent *B. subtilis* cells. Such preparations contain sufficient amounts of multimeric plasmid molecules to be efficient in *B. subtilis*-competent cell transformation (see p. 140).

Table 8.2 lists some of the commonly used shuttle plasmids. Note that some of them carry some of the features described earlier for *E. coli* plasmids, e.g. the *E. coli lacZα*-complementation fragment, multiple cloning sites (MCS) (see p. 53) and the phage f1 origin for subsequent production of single-stranded DNA in a suitable *E. coli* host (see p. 70).

The influence of mode of replication: vectors derived from pAMβ1

Early in the development of *B. subtilis* cloning vectors, it was noted that only short DNA fragments could be efficiently cloned (Michel *et al.* 1980) and that longer DNA segments often undergo rearrangements (Ehrlich *et al.* 1986). This structural instability is independent of the host recombination systems, for it still occurs in Rec⁻ strains (Peijnenburg *et al.* 1987).

Table 8.1 Properties of some *S. aureus* plasmids used as vectors in *B. subtilis*.

Plasmid	Phenotype conferred on host cell	Size	Copy no.	Other comments
pC194	Chloramphenicol resistance	2906 bp	15	Generates large amount of high-molecular-weight DNA when carrying heterologous inserts
pE194	Erythromycin resistance	3728 bp	10	*cop*-6 derivative has copy number of 100. Plasmid is naturally temperature-sensitive for replication
pUB110	Kanamycin resistance	4548 bp	50	Virtually the complete sequence is involved in replication maintenance, site-specific plasmid recombination or conjugal transfer

Table 8.2 *B. subtilis–E. coli* shuttle plasmids.

Plasmid	Size (kbp)	Replicon		Markers		Comments
		E. coli	*B. subtilis*	*E. coli*	*B. subtilis*	
pHV14	4.6	pBR322	pC194	Ap, Cm	Cm	pBR322/pC194 fusion. Sites: *Pst*I, *Bam*HI, *Sal*I, *Nco*I (Ehrlich 1978)
pHV15	4.6	pBR322	pC194	Ap, Cm	Cm	pHV14, reversed orientation of pC194 relative to pBR322
pHV33	4.6	pBR322	pC194	Ap, Tc, Cm	Cm	Revertant of pHV14 (Primrose & Ehrlich 1981)
pEB10	8.9	pBR322	pUB110	Ap, Km	Km	pBR322/pUB110 fusion (Bron *et al.* 1988)
pLB5	5.8	pBR322	pUB110	Ap, Cm, Km	Cm, Km	Deletion of pBR322/pUB110 fusion, *Cm*^{R} gene of pCl94 Segregationally unstable (Bron & Luxen 1985). Sites: *Bam*HI, *Eco*RI, *Bgl*III (in *Km*^{R} gene), *Nco*I (in *Cm*^{R} gene)
pHP3	4.8	pBR322	pTA1060	Em, Cm	Em, Cm	Segregationally stable pTA1060 replicon (Peeters *et al.* 1988). Copy number *c.* 5. Sites: *Nco*I (*Cm*^{R} gene), *Bcl*I and *Hpa*I (both *Em*^{R} gene)
pHP3Ff	5.3	pBR322	pTA1060	Em, Cm	Em, Cm	Like pHP3; phage f1 replication origin and packaging signal
pGPA14	5.8	pBR322	pTA1060	Em	Em	Stable pTA1060 replicon. Copy number *c.* 5. α-Amylase-based selection vector for protein export functions (Smith *et al.* 1987). MCS of M13*mp*11 in *lacZ*α
pGPB14	5.7	pBR322	pTA1060	Em	Em	As pGPA14, probe gene TEM-β-lactamase
pHP13	4.9	pBR322	pTA1060	Em, Cm	Em, Cm	Stable pTA1060 replicon. Copy number *c.* 5. Efficient (shotgun) cloning vector (Haima *et al.* 1987). MCS of M13*mp*9 in *lacZ*α *LacZ*α not expressed in *B. subtilis*. Additional sites: *Bcl*I and *Hpa*I (both *Em*^{R} gene)
pHV1431	10.9	pBR322	pAMβ1	Ap, Tc, Cm	Cm	Efficient cloning vector based on segregationally stable pAMβ1 (Jannière *et al.* 1990). Copy number *c.* 200. Sites: *Bgm*HI, *Sal*I, *Pst*I, *Nco*I. Structurally unstable in *E. coli*
pHV1432	8.8	pBR322	pAMβ1	Ap, Tc, Cm	Cm	pHV1431 lacking stability fragment orfH. Structurally stable in *E. coli*
pHV1436	8.2	pBR322	pTB19	Ap, Tc, Cm	Cm	Low-copy-number cloning vector (Jannière *et al.* 1990) Structurally stable

A major contributing factor to structural instability of recombinant DNA in *B. subtilis* appears to be the mode of replication of the plasmid vector (Gruss & Ehrlich 1989, Jannière *et al.* 1990). All the *B. subtilis* vectors described above replicate by a rolling-circle mechanism (see Box 8.2). Nearly every step in the process digresses or could digress from its usual function, thus effecting rearrangements. Also, single-stranded DNA is known to be a reactive intermediate

in every recombination process, and single-stranded DNA is generated during rolling-circle replication.

If structural instability is a consequence of rolling-circle replication, then vectors which replicate by the alternative theta mechanism could be more stable. Jannière *et al.* (1990) have studied two potentially useful plasmids, pAMβ1 and pTB19, which are large (26.5 kb) natural plasmids derived from *Streptococcus (Enterococcus) faecalis* and *B. subtilis*,

Box 8.2 The two modes of replication of circular DNA molecules

There are two modes of replication of circular DNA molecules: via theta-like structures or by a rolling-circle type of mechanism. Visualization by electron microscopy of the replicating intermediates of many circular DNA molecules reveals that they retain a ring structure throughout replication. They always possess a theta-like shape that comes into existence by the initiation of a replicating bubble at the origin of replication (Fig. B8.1). Replication can proceed either uni- or bidirectionally. As long as each chain remains intact, even minor untwisting of a section of the circular double helix results in the creation of positive supercoils in the other direction. This supercoiling is relaxed by the action of topoisomerases (see Fig. 4.1), which create single-stranded breaks (relaxed molecules) and then reseal them.

An alternative way to replicate circular DNA is the rolling-circle mechanism (Fig. B8.2). DNA synthesis starts with a cut in one strand at the origin of replication. The 5′ end of the cut strand is displaced from the duplex, permitting the addition of deoxyribonucleotides at the free 3′ end. As replication proceeds, the 5′ end of the cut strand is rolled out as a free tail of increasing length. When a full-length tail is produced, the replicating machinery cuts it off and ligates the two ends together. The double-stranded progeny can reinitiate replication, whereas the single-stranded progeny must first be converted to a double-stranded form. Gruss and Ehrlich (1989) have suggested how deletants and defective molecules can be produced at each step in the rolling-circle process.

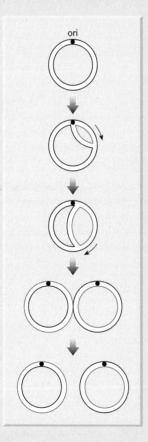

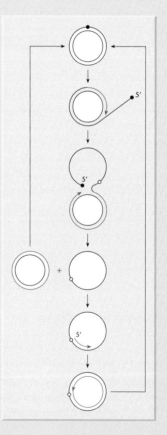

Fig. B8.1 Theta replication of a circular DNA molecule. The original DNA is shown in black and the newly-synthesized DNA in red. ● represents the origin of replication and the arrow shows the direction of replication.

Fig. B8.2 Rolling-circle replication of a circular DNA molecule. The original DNA is shown in black and the newly synthesized DNA in red. The solid and open circles represent the positions of the replication origins of the outer (+) and inner (−) circles, respectively.

respectively. Replication of these plasmids does not lead to accumulation of detectable amounts of single-stranded DNA, whereas the rolling-circle mode of replication does. Also, the replication regions of these two large plasmids share no sequence homology with the corresponding highly conserved regions of the rolling-circle-type plasmids. It is worth noting that the classical *E. coli* vectors, which are derived from plasmid ColE1, all replicate via theta-like structures.

Renault *et al.* (1996) have developed a series of cloning vectors from pAMβ1. All the vectors carry a gene essential for replication, *repE*, and its regulator, *copF*. The latter gene can be inactivated by inserting a linker into its unique *Kpn*I site. Since *copF* down-regulates the expression of *repE*, its inactivation leads to an increase in the plasmid copy number per cell. The original low-copy-number state can be restored by removal of the linker by cleavage and religation. This new replicon has been used to build vectors for making transcriptional and translational fusions and for expression of native proteins. Poyart and Trieu-Cuot (1997) have constructed a shuttle vector based on pAMβ1 for the construction of transcriptional fusions; it can be conjugally transferred between *E. coli* and a wide range of Gram-positive bacteria.

Transcription and translation

The composition of the core RNA polymerase in *B. subtilis* and other low-GC hosts resembles that of *E. coli*. The number of sigma factors is different in each of the various genera but the principal sigma factor is sigma A. Analysis of many sigma A-dependent *Bacillus* promoters shows that they contain the canonical –35 and –10 sequences found in *E. coli* promoters. In *B. subtilis*, at least, many promoters contain an essential TGTG motif (–16 region) upstream of the –10 region. Mutations of this region significantly reduce promoter strength (Helmann 1995, Voskuil & Chambliss 1998). The promoters also have conserved polyA and polyT tracts upstream of the –35 region. Although the –16 region is found in some *E. coli* promoters, such promoters often lack the –35 region, whereas this never occurs in *B. subtilis*.

The translation apparatus of *B. subtilis* differs significantly from that of *E. coli* (for review, see

Vellanoweth 1993). This is demonstrated by the observation that *E. coli* ribosomes can support protein synthesis when directed by mRNA from a range of Gram-positive and Gram-negative organisms, whereas ribosomes from *B. subtilis* recognize only homologous mRNA (Stallcup *et al.* 1974). The explanation for the selectivity of *B. subtilis* ribosomes is that they lack a counterpart of the largest *E. coli* ribosomal protein, S1 (Higo *et al.* 1982, Roberts & Rabinowitz 1989). Other Gram-positive bacteria, such as *Staphylococcus*, *Streptococcus*, *Clostridium* and *Lactobacillus*, also lack an S1-equivalent protein and they too exhibit mRNA selectivity. The role of S1 is believed to be to bind RNA non-specifically and bring it to the decoding site of the 30S subunit, where proper positioning of the Shine–Dalgarno (S-D) sequence and initiation codon signals can take place. This is reflected in a more extensive complementarity between the S-D sequences and the 3′ end of the 16S ribosomal RNA (rRNA) than found in bacteria which have ribosomal protein S1.

The additional sequence requirements for efficient transcription and translation in *B. subtilis* and other low-GC organisms probably explain why many *E. coli* genes are not expressed in these hosts.

Controlled expression in *B. subtilis* and other low-GC hosts

The first controlled expression system to be used in *B. subtilis* was the *Spac* system (Yansura & Henner 1984). This consists of the *E. coli* lacI gene and the promoter of phage SPO-1 coupled to the *lac* operator. More recently, the *E. coli* T7 system (see p. 74) has been successfully implemented in *B. subtilis* (Conrad *et al.* 1996). This was achieved by inserting the T7 RNA polymerase gene (*rpoT7*) into the chromosome under the control of a xylose-inducible promoter and cloning the gene of interest, coupled to a T7 promoter, on a *B. subtilis* vector. Of course, expression of the heterologous gene can be made simpler by putting it directly under the control of the xylose-inducible promoter (Kim *et al.* 1996). A similar xylose-inducible system has been developed in staphylococci (Sizemore *et al.* 1991, Peschel *et al.* 1996) and *Lactobacillus* (Lokman *et al.* 1997). Many different controllable promoters are available in *Lactococcus lactis* (for reviews see Kuipers *et al.* 1995,

Table 8.3 Some inducible systems in *L. lactis*.

Promoter	Inducer
lacA or *lacR*	Lactose
dnaJ	High temperature
sodA	Aeration
PA170	Low pH, low temperature
trpE	Absence of tryptophan
φ31 and *ori*	φ31 infection
nisA or *nisF*	Nisin

De Vos *et al.* 1997) and some representative examples are shown in Table 8.3.

In the φ31 system of *L. lactis*, the gene of interest is placed under the control of a phage middle promoter inserted in a low-copy-number vector carrying the phage *ori* region. Following infection of the host cell with φ31, the plasmid copy number rapidly increases and this is followed by expression from the phage promoter. Following induction in this manner, the level of expression of the cloned gene can increase over 1000-fold (O'Sullivan *et al.* 1996). Similar levels of expression can be achieved by using the *nisA* and *nisF* systems but with the added advantage that the exact level of expression depends on the amount of nisin added to the medium (Kuipers *et al.* 1995, De Ruyter *et al.* 1996).

Secretion vectors for low-GC bacteria

The export mechanism in *Bacillus* and other low-GC bacteria resembles that of *E. coli* (for review, see Tjalsma *et al.* 2000). However, there are differences in the signal peptides compared with those found in *E. coli* and eukaryotes. The NH$_2$ termini are more positively charged. The signal peptides are also larger and the extra length is distributed among all three regions of the signal peptide. Hols *et al.* (1992) developed two probe vectors for the identification of Gram-positive secretion signals. These vectors made use of a silent reporter gene encoding the mature α-amylase from *Bacillus licheniformis*. The disadvantage of this system is that detection of secreted amylase involves flooding starch-containing media with iodine and this kills the bacteria in the colonies.

Consequently, replica plates must be made before iodine addition.

Poquet *et al.* (1998) have developed an alternative probe system which uses the *S. aureus*-secreted nuclease as a reporter. This nuclease is a small (168 amino acid), stable, monomeric enzyme that is devoid of cysteine residues and the enzymatic test is nontoxic to bacterial colonies. The probe vectors have the nuclease gene, lacking its signal sequence, located downstream from an MCS. Cloning DNA in the vectors results in the synthesis of fusion proteins and those containing signal sequences are detected by nuclease activity in the growth medium. Le Loir *et al.* (1998) have noted that inclusion of a nine-residue synthetic propeptide immediately downstream of the signal-peptide cleavage site significantly enhances secretion.

Vectors for systematic gene inactivation

With the advent of mass sequencing of genomes (see p. 2), many genes have been discovered whose function is unknown. One way of determining function is to inactivate the gene and then monitor the effect of this on cell fitness under different growth conditions. To study the functions of uncharacterized open reading frames in *B. subtilis*, Vagner *et al.* (1998) constructed a series of vectors to perform directed insertional mutagenesis in the chromosome. These vectors, which have been given the designation pMUTIN, have the following properties:
• an inability to replicate in *B. subtilis*, which allows insertional mutagenesis;
• a reporter *lacZ* gene to facilitate the measurement of expression of the target gene;
• the inducible Pspac promoter to allow controlled expression of genes downstream of and found in the same operon as the target gene.

A typical pMUTIN vector is shown in Fig. 8.7 and their mode of use is as follows. An internal fragment of the target gene is amplified by polymerase chain reaction (PCR) and cloned in a pMUTIN vector and the resulting plasmid is used to transform *B. subtilis*. Upon integration, the target gene is interrupted and a transcriptional fusion is generated between its promoter and the reporter *lacZ* gene (Fig. 8.8). If the targeted gene is part of an operon, then any genes downstream of it are placed under the control of the

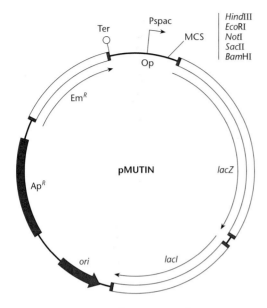

Fig. 8.7 A typical pMUTIN vector. The Em^R, *lacI* and *lacZ* genes are expressed in *B. subtilis* and the Ap^R gene is expressed in *E. coli*. 'Ter' indicates the presence of a terminator to prevent run-through transcription from the Em^R gene. 'Op' represents the LacI operator.

Pspac promoter. It should be noted that the procedure shown in Fig. 8.8 simultaneously generates two types of mutants: an absolute (null) mutation in orf2 through gene inactivation, and a conditional mutation in orf3, which can be relieved by induction with isopropyl-β-D-thiogalactoside (IPTG).

Cloning in streptomycetes

Cloning in *Streptomyces* has attracted a lot of interest because of the large number of antibiotics that are made by members of this genus. Although *Streptomyces coelicolor* is the model species for genetic studies (Hopwood 1999), many other species are the subject of intensive study and methods developed for one species may not work particularly well in another.

Streptomycete DNA has a G+C content of 70–75% and this affects the frequency of restriction sites. As might be expected, AT-rich recognition sites are rare and this can be useful if large-sized fragments of DNA are wanted. For the construction of gene libraries, the most commonly used enzymes are ones with a high GC content in their recognition sequence, e.g. *Bam*H1 (G′GATCC), *Bgl*II (A′GATCT), and *Bcl*I (T′GATCA).

In *Streptomyces*, promoters may be several hundred base pairs upstream of the start of the gene and so can be lost during gene cloning. Also, many *Streptomyces* promoters are complex and may include tandem sites for recognition by different sigma factors. Streptomycetes are good at expressing genes (promoters, ribosome binding sites, etc.) from low-G+C organisms, but *Streptomyces* genes are usually difficult to express in *E. coli* because most promoters do not function, and translation may be inefficient unless the initial amino acid codons are changed to lower-G+C alternatives.

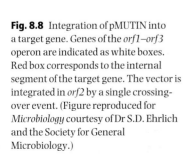

Fig. 8.8 Integration of pMUTIN into a target gene. Genes of the *orf1–orf3* operon are indicated as white boxes. Red box corresponds to the internal segment of the target gene. The vector is integrated in *orf2* by a single crossing-over event. (Figure reproduced for *Microbiology* courtesy of Dr S.D. Ehrlich and the Society for General Microbiology.)

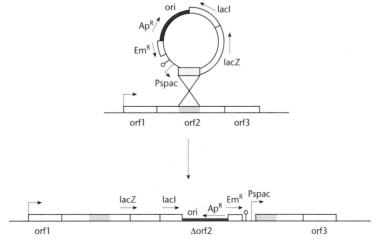

There are several ways in which DNA can be introduced into streptomycetes, including transformation, transfection and conjugation. Transformation is achieved by using protoplasts, rather than competent cells, and high frequencies of plasmid DNA uptake can be achieved in the presence of polyethylene glycol (Bibb *et al.* 1978). Plasmid monomers that are covalently closed will yield 10^6–10^7 transformants/µg of DNA, even with plasmids up to 60 kb in size. Open circular and linearized molecules with sticky ends transform with 10–100-fold lower efficiency (Bibb *et al.* 1980). The number of transformants obtained with non-replicating plasmids that integrate by homologous recombination into the recipient chromosome is greatly stimulated by simple denaturation of the donor DNA (Oh & Chater 1997). This stimulation reflects an increased frequency of recombination rather than an increased frequency of DNA uptake. Electroporation has been used to transform streptomycetes, since it bypasses the need to develop protoplast regeneration procedures (Pigac & Schrempf 1995, Tyurin *et al.* 1995). For electroporation, limited non-protoplasting lysozyme treatment is used to weaken the cell wall and improve DNA uptake. Intergeneric conjugation of mobilizable plasmids from *E. coli* into streptomycetes (see p. 142) is increasingly being used, because the required constructs can be made easily in *E. coli* and the conjugation protocols are simple. For intergeneric conjugation to occur, the vectors have to carry the *oriT* locus from RP4 and the *E. coli* strain needs to supply the transfer functions *in trans* (Mazodier *et al.* 1989).

Transformants are generally identified by the selection of appropriate phenotypes. However, antibiotic resistance has much less utility than in other organisms, because many streptomycetes produce antibiotics and hence have innate resistance to them. One particularly useful phenomenon is that clones harbouring conjugative plasmids can be detected by the visualization of pocks. The property of pock formation, also known as *lethal zygosis*, is exhibited if a strain containing a conjugative plasmid is replica-plated on to a lawn of the corresponding plasmid-free strain. Under these conditions, clones containing plasmids are surrounded by a narrow zone in which the growth of the plasmid-free strain is retarded (Chater & Hopwood 1983).

Table 8.4 *Streptomyces* plasmids that have been used in the development of vectors.

Plasmid	Size	Mode of replication	Copy number	Host range
pIJ101	8.8 kb	Rolling circle	300	
pJV1	11.1 kb	Rolling circle		Broad
pSG5	12.2 kb	Rolling circle	20–50	Broad
SCP2*	31 kb	Theta	1–4	
SLP1	17.2 kb	Rolling circle	Integrating	Limited
pSAM2	10.9 kb	Rolling circle	Integrating	Broad

Vectors for streptomycetes

With the exception of RSF1010 (see p. 145), no plasmid from any other organism has been found to replicate in *Streptomyces*. For this reason, all the cloning vectors used in streptomycetes are derived from plasmids and phages that occur naturally in them. The different replicons that have been subjugated as vectors are listed in Table 8.4. Nearly all *Streptomyces* plasmids carry transfer functions that permit conjugative plasmid transfer and provide different levels of chromosome-mobilizing activity. These transfer functions are very simple, consisting of a single transfer gene and a repressor of gene function.

Plasmid SCP2* is a deriviative (Lydiate *et al.* 1985) of the sex plasmid SCP2. Both plasmids have a size of 31.4 kb and are physically indistinguishable, although SCP2* exhibits a much more pronounced lethal zygosis reaction. SCP2* is important because it is the progenitor of many very low-copy-number, stable vectors. High-copy-number derivatives have also been isolated with the exact copy number (10 or 1000) being dependent on the sequences from the replication region that are present. SLP1 and pSAM2 are examples of *Streptomyces* plasmids that normally reside integrated into a specific highly conserved chromosomal transfer RNA (tRNA) sequence (Kieser & Hopwood 1991). Many different specialist vectors have been derived from these plasmids, including cosmids, expression vectors, vectors with promoterless reporter genes, positive-selection vectors, temperature-sensitive vectors, etc., and full details can be found in Kieser *et al.* (2000).

The temperate phage ϕC31 is the streptomycete equivalent of phage λ and has been subjugated as a vector. ϕC31-derived vectors have upper and lower size limits for clonable fragments with an average insert size of 8 kb. In contrast, there are no such size constraints on plasmid cloning, although recombinant plasmids of a size greater than 35 kb are rare with the usual vectors. However, phage vectors do have one important advantage: plaques can be obtained overnight, whereas plasmid transformants can take up to 1 week to sporulate. Plasmid-integrating vectors can be generated by incorporating the integration functions of ϕC31.

As noted earlier, a major reason for cloning in streptomycetes is to analyse the genetics and regulation of antibiotic synthesis. Although all the genes for a few complete biosynthetic pathways have been cloned (Malpartida & Hopwood 1985, Kao *et al.* 1994, Schwecke *et al.* 1995), some gene clusters may be too large to be cloned in the standard vectors. For this reason, Sosio *et al.* (2000) generated bacterial artificial chromosomes (BACs) that can accommodate up to 100 kb of streptomycete DNA. These vectors can be shuttled between *E. coli*, where they replicate autonomously, and *Streptomyces*, where they integrate site-specifically into the chromosome.

Homoeologous recombination

Homoeologous recombination is the recombination between DNA sequences that are only partially homologous. In most bacteria, homoeologous recombination fails to occur because of mismatch repair, but it does occur in streptomycetes, although the frequency of recombination is about 10^5-fold lower than for homologous recombination. The significance of homoeologous recombination is that it permits the formation of hybrid genes, gene clusters or even species and can lead to the formation of new antibiotics (Baltz 1998).

CHAPTER 9

Cloning in *Saccharomyces cerevisiae* and other fungi

Introduction

The analysis of eukaryotic DNA sequences has been facilitated by the ease with which DNA from eukaryotes can be cloned in prokaryotes, using the vectors described in previous chapters. Such cloned sequences can be obtained easily in large amounts and can be altered *in vivo* by bacterial genetic techniques and *in vitro* by specific enzyme modifications. To determine the effects of these experimentally induced changes on the function and expression of eukaryotic genes, the rearranged sequences must be taken out of the bacteria in which they were cloned and reintroduced into a eukaryotic organism. Despite the overall unity of biochemistry, there are many functions common to eukaryotic cells which are absent from prokaryotes, e.g. localization of ATP-generating systems to mitochondria, association of DNA with histones, mitosis and meiosis, and obligate differentiation of cells. The genetic control of such functions must be assessed in a eukaryotic environment.

Ideally these eukaryotic genes should be reintroduced into the organism from which they were obtained. In this chapter we shall discuss the potential for cloning these genes in *Saccharomyces cerevisiae* and other fungi and in later chapters we shall consider methods for cloning in animal and plant cells. It should be borne in mind that yeast cells are much easier to grow and manipulate than plant and animal cells. Fortunately, the cellular biochemistry and regulation of yeast are very like those of higher eukaryotes. For example, signal transduction and transcription regulation by mammalian steroid receptors can be mimicked in strains of *S. cerevisiae* expressing receptor sequences (Metzger *et al.* 1988, Schena & Yamamoto 1988). There are many yeast homologues of human genes, e.g. those involved in cell division. Thus yeast can be a very good surrogate host for studying the structure and function of eukaryotic gene products.

Introducing DNA into fungi

Like *Escherichia coli*, fungi are not naturally transformable and artificial means have to be used for introducing foreign DNA. One method involves the use of spheroplasts (i.e. wall-less cells) and was first developed for *S. cerevisiae* (Hinnen *et al.* 1978). In this method, the cell wall is removed enzymatically and the resulting spheroplasts are fused with ethylene glycol in the presence of DNA and $CaCl_2$. The spheroplasts are then allowed to generate new cell walls in a stabilizing medium containing 3% agar. This latter step makes subsequent retrieval of cells inconvenient. Electroporation provides a simpler and more convenient alternative to the use of spheroplasts. Cells transformed by electroporation can be selected on the surface of solid media, thus facilitating subsequent manipulation. Both the spheroplast technique and electroporation have been applied to a wide range of yeasts and filamentous fungi.

DNA can also be introduced into yeasts and filamentous fungi by conjugation. Heinemann and Sprague (1989) and Sikorski *et al.* (1990) found that enterobacterial plasmids, such as R751 (IncPβ) and F (IncF), could facilitate plasmid transfer from *E. coli* to *S. cerevisiae* and *Schizosaccharomyces pombe*. The bacterial plant pathogen *Agrobacterium tumefaciens* contains a large plasmid, the Ti plasmid, and part of this plasmid (the transfered DNA (T-DNA)) can be conjugally transferred to protoplasts of *S. cerevisiae* (Bundock *et al.* 1995) and a range of filamentous fungi (De Groot *et al.* 1998). T-DNA can also be transferred to hyphae and conidia.

The fate of DNA introduced into fungi

In the original experiments on transformation of *S. cerevisiae*, Hinnen *et al.* (1978) transformed a leucine auxotroph with the plasmid pYeLeu 10. This plasmid is a hybrid composed of the enterobacterial plasmid

ColE1 and a segment of yeast DNA containing the LEU2$^+$ gene and is unable to replicate in yeast. Analysis of the transformants showed that in some of them there had been reciprocal recombination between the incoming LEU2$^+$ and the recipient Leu2$^-$ alleles. In the majority of the transformants, ColE1 DNA was also present and genetic analysis showed that in some of them the LEU2$^+$ allele was closely linked to the original Leu2$^-$ allele, whereas in the remaining ones the LEU2$^+$ allele was on a different chromosome.

The results described above can be confirmed by restriction-endonuclease analysis, since pYeLeu 10 contains no cleavage sites for *Hind*III. When DNA from the Leu2$^-$ parent was digested with endonuclease *Hind*III and electrophoresed in agarose, multiple DNA fragments were observed but only one of these hybridized with DNA from pYeLeu 10. With the transformants in which the Leu2$^-$ and LEU2$^+$ alleles were linked, only a single fragment of DNA hybridized to pYeLeu 10, but this had an increased size, consistent with the insertion of a complete pYeLeu 10 molecule into the original fragment. These data are consistent with there being a tandem duplication of the Leu2 region of the chromosome (Fig. 9.1). With the remaining transformants, two DNA fragments that hybridized to pYeLeu 10 could be found on electrophoresis. One fragment corresponded to the fragment seen with DNA from the recipient cells, the other to the plasmid genome which had been inserted in another chromosome (see Fig. 10.1). These results represented the first unambiguous demonstration that foreign DNA, in this case cloned ColE1 DNA, can integrate into the genome of a eukaryote. A plasmid such as pYeLeu 10 which can do this is known as a yeast integrating plasmid (YIp).

During transformation, the integration of exogenous DNA can occur by recombination with a homologous or an unrelated sequence. In most cases, non-homologous integration is more common than homologous recombination (Fincham 1989), but this is not so in *S. cerevisiae* (Schiestl & Petes 1991). In the experiments of Hinnen *et al.* (1978) described above, sequences of the yeast retrotransposon Ty2 were probably responsible for the integration of the plasmid in novel locations of the genome, i.e. the 'illegitimate' recombinants were the result of homologous crossovers within a repeated element (Kudla & Nicolas 1992). Based on a similar principle, a novel vector has been constructed by Kudla and Nicolas (1992) which allows integration of a cloned DNA sequence at different sites in the genome. This feature is provided by the inclusion in the vector of a repeated yeast *sigma* sequence present in approximately 20–30 copies per genome and spread over most or all of the 16 chromosomes.

When T-DNA from the Ti plasmid of *Agrobacterium* is transferred to yeast, it too will insert in different parts of the genome by illegitimate recombination (Bundock & Hooykaas 1996).

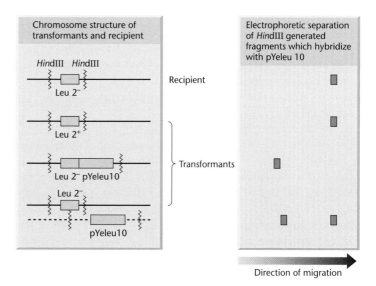

Fig. 9.1 Analysis of yeast transformants. (See text for details.)

Schiestl and Petes (1991) developed a method for forcing illegitimate recombination by transforming yeast with *Bam*H1-generated fragments in the presence of the *Bam*H1 enzyme. Not only did this increase the frequency of transformants but the transformants which were obtained had the exogenous DNA integrated into genomic *Bam*H1 sites. This technique, which is sometimes referred to as restriction-enzyme-mediated integration (REMI), has been extended to other fungi, such as *Cochliobolus* (Lu *et al.* 1994), *Ustilago* (Bolker *et al.* 1995) and *Aspergillus* (Sanchez *et al.* 1998).

Plasmid vectors for use in fungi

If the heterologous DNA introduced into fungi is to be maintained in an extrachromosomal state then plasmid vectors are required which are capable of replicating in the fungal host. Four types of plasmid vector have been developed: yeast episomal plasmids (YEps), yeast replicating plasmids (YRps), yeast centromere plasmids (YCps) and yeast artificial chromosomes (YACs). All of them have features in common. First, they all contain unique target sites for a number of restriction endonucleases. Secondly, they can all replicate in *E. coli*, often at high copy number. This is important, because for many experiments it is necessary to amplify the vector DNA in *E. coli* before transformation of the ultimate yeast recipient. Finally, they all employ markers that can be selected readily in yeast and which will often complement the corresponding mutations in *E. coli* as well. The four most widely used markers are His3, Leu2, Trp1 and Ura3. Mutations in the cognate chromosomal markers are recessive, and non-reverting mutants are available. Two yeast selectable markers, Ura3 and Lys2, have the advantage of offering both positive and negative selection. Positive selection is for complementation of auxotrophy. Negative selection is for ability to grow on medium containing a compound that inhibits the growth of cells expressing the wild-type function. In the case of Ura3, it is 5-fluoro-orotic acid (Boeke *et al.* 1984) and for Lys2 it is α-aminoadipate (Chatoo *et al.* 1979). These inhibitors permit the ready selection of those rare cells which have undergone a recombination or loss event to remove the plasmid DNA sequences. The Lys2 gene is not utilized frequently because it is

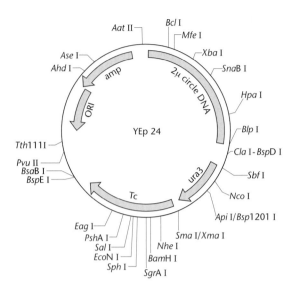

Fig. 9.2 Schematic representation of a typical yeast episomal plasmid (YEp 24). The plasmid can replicate both in *E. coli* (due to the presence of the pBR 3 22 origin of replication) and in *S. cerevisiae* (due to the presence of the yeast 2 μm origin of replication). The ampicillin and tetracycline determinants are derived from pBR 3 22 and the URA3 gene from yeast.

large and contains sites within the coding sequence for many of the commonly used restriction sites.

Yeast episomal plasmids

YEps were first constructed by Beggs (1978) by recombining an *E. coli* cloning vector with the naturally occurring yeast 2 μm plasmid. This plasmid is 6.3 kb in size, has a copy number of 50–100 per haploid cell and has no known function. A representative YEp is shown in Fig. 9.2.

Yeast replicating plasmids

YRps were initially constructed by Struhl *et al.* (1979). They isolated chromosomal fragments of DNA which carry sequences that enable *E. coli* vectors to replicate in yeast cells. Such sequences are known as *ars* (autonomously replicating sequence). An *ars* is quite different from a centromere: the former acts as an origin of replication (Palzkill & Newlon 1988, Huang & Kowalski 1993), whereas the latter is involved in chromosome segregation.

Although plasmids containing an *ars* transform yeast very efficiently, the resulting transformants are exceedingly unstable. For unknown reasons, YRps tend to remain associated with the mother cell and are not efficiently distributed to the daughter cell. (Note: *S. cerevisiae* does not undergo binary fission but buds off daughter cells instead.) Occasional stable transformants are found and these appear to be cases in which the entire YRp has integrated into a homologous region on a chromosome in a manner identical to that of YIps (Stinchcomb *et al.* 1979, Nasmyth & Reed 1980).

Yeast centromere plasmids

Using a YRp vector, Clarke and Carbon (1980) isolated a number of hybrid plasmids containing DNA segments from around the centromere-linked *leu2*, *cdc*10 and *pgk* loci on chromosome III of yeast. As expected for plasmids carrying an *ars*, most of the recombinants were unstable in yeast. However, one of them was maintained stably through mitosis and meiosis. The stability segment was confined to a 1.6 kb region lying between the *leu2* and *cdc*10 loci and its presence on plasmids carrying either of two *ars* tested resulted in those plasmids behaving like minichromosomes (Clarke & Carbon 1980, Hsiao & Carbon 1981). Genetic markers on the minichromosomes acted as linked markers segregating in the first meiotic division as centromere-linked genes and were unlinked to genes on other chromosomes.

Structurally, plasmid-borne centromere sequences have the same distinctive chromatin structure that occurs in the centromere region of yeast chromosomes (Bloom & Carbon 1982). Functionally YCps exhibit three characteristics of chromosomes in yeast cells. First, they are mitotically stable in the absence of selective pressure. Secondly, they segregate during meiosis in a Mendelian manner. Finally, they are found at low copy number in the host cell.

Yeast artificial chromosomes

All three autonomous plasmid vectors described above are maintained in yeast as circular DNA molecules – even the YCp vectors, which possess yeast centromeres. Thus, none of these vectors resembles the normal yeast chromosomes which have a linear structure. The ends of all yeast chromosomes, like those of all other linear eukaryotic chromosomes, have unique structures that are called *telomeres*. Telomere structure has evolved as a device to preserve the integrity of the ends of DNA molecules, which often cannot be finished by the conventional mechanisms of DNA replication (for detailed discussion see Watson 1972). Szostak and Blackburn (1982) developed the first vector which could be maintained as a linear molecule, thereby mimicking a chromosome, by cloning yeast telomeres into a YRp. Such vectors are known as yeast artificial chromosomes (YACs).

One advantage of YACs is that, unlike the other plasmid vectors, their stability increases as the size of the insert increases. Thus, there is no practical limitation to the size of a YAC and they are essential tools in any genome-sequencing project. The method for cloning large DNA sequences in YACs developed by Burke *et al.* (1987) is shown in Fig. 9.3.

Retrovirus-like vectors

The genome of *S. cerevisiae* contains 30–40 copies of a 5.9 kb mobile genetic element called Ty (for review see Fulton *et al.* 1987). This transposable element shares many structural and functional features with retroviruses (see p. 193) and the copia element of *Drosophila*. Ty consists of a central region containing two long open reading frames (ORFs) flanked by two identical terminal 334 bp repeats called *delta* (Fig. 9.4). Each delta element contains a promoter as well as sequences recognized by the transposing enzyme. New copies of the transposon arise by a replicative process, in which the Ty transcript is converted to a progeny DNA molecule by a Ty-encoded reverse transcriptase. The complementary DNA can transpose to many sites in the host DNA.

The Ty element has been modified *in vitro* by replacing its delta promoter sequence with promoters derived from the phosphoglycerate kinase or galactose-utilization genes (Garfinkel *et al.* 1985, Mellor *et al.* 1985). When such constraints are introduced into yeast on high-copy-number vectors, the Ty element is overexpressed. This results in the formation of large numbers of virus-like particles

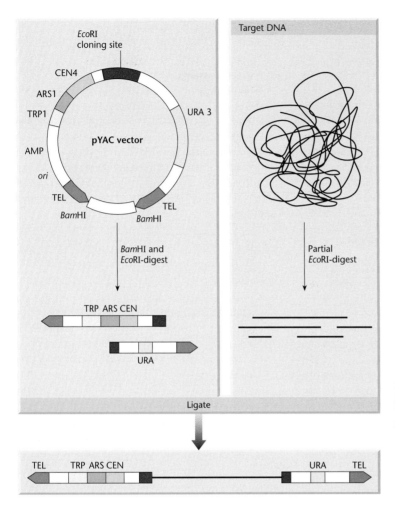

Fig. 9.3 Construction of a yeast artificial chromosome containing large pieces of cloned DNA. Key regions of the pYAC vector are as follows: TEL, yeast telomeres; ARS 1, autonomously replicating sequence; CEN 4, centromere from chromosome 4; *URA3* and *TRP1*, yeast marker genes; Amp, ampicillin-resistance determinant of pBR322; *ori*, origin of replication of pBR322.

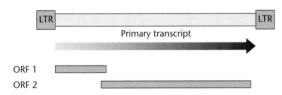

Fig. 9.4 Structure of a typical Ty element. ORF 1 and ORF 2 represent the two open reading frames. The delta sequences are indicated by LTR (long terminal repeats).

(VLPs), which accumulate in the cytoplasm (Fig. 9.5). The particles, which have a diameter of 60–80 nm, have reverse-transcriptase activity. The major structural components of VLPs are proteins produced by proteolysis of the primary translation product of

ORF 1. Adams *et al.* (1987) have shown that fusion proteins can be produced in cells by inserting part of a gene from human immunodeficiency virus (HIV) into ORF 1. Such fusion proteins formed hybrid HIV:Ty-VLPs.

The Ty element can also be subjugated as a vector for transposing genes to new sites in the genome. The gene to be transposed is placed between the 3′ end of ORF 2 and the 3′ delta sequence (Fig. 9.6). Providing the inserted gene lacks transcription-termination signals, transcription of the 3′ delta sequence will occur, which is a prerequisite for transposition. Such constructs act as amplification cassettes, for, once introduced into yeast, transposition of the new gene occurs to multiple sites in the genome (Boeke *et al.* 1988, Jacobs *et al.* 1988).

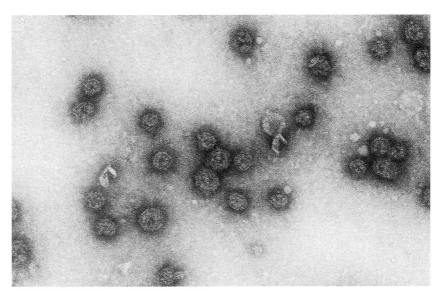

Fig. 9.5 Ty virus-like particles (magnification 80 000) carrying the entire HIV1 TAT coding region. (Photograph courtesy of Dr S. Kingsman.)

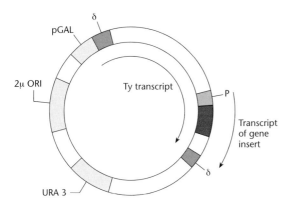

Fig. 9.6 Structure of the multicopy plasmid used for inserting a modified Ty element, carrying a cloned gene, into the yeast chromosome. pGAL and P are yeast promoters, δ represents the long terminal repeats (delta sequences) and the red region represents the cloned gene. (See text for details.)

Choice of vector for cloning

There are three reasons for cloning genes in yeast. The first of these relates to the potential use of yeast as a cloning host for the overproduction of proteins of commercial value. Yeast offers a number of advantages, such as the ability to glycosylate proteins during secretion and the absence of pyrogenic toxins. Commercial production demands overproduction and

the factors affecting expression of genes in yeast are discussed in a later section (see p. 165). Yeast is also used in the production of food and beverages. The ability to clone in yeast without the introduction of bacterial sequences by using vectors like those of Chinery and Hinchliffe (1989) is particularly beneficial.

A second reason for cloning genes in yeast is the ability to clone large pieces of DNA. Although there is no theoretical limit to the size of DNA which can be cloned in a bacterial plasmid, large recombinant plasmids exhibit structural and segregative instability. In the case of bacteriophage-λ vectors, the size of the insert is governed by packaging constraints. Many DNA sequences of interest are much larger than this. For example, the gene for blood Factor VIII covers about 190 kbp, or about 0.1% of the human X chromosome, and the Duchenne muscular dystrophy gene spans more than a megabase. Long sequences of cloned DNA have greatly facilitated efforts to sequence the human genome. YACs offer a convenient way to clone large DNA fragments but are being replaced with BACs and PACs (see p. 67) because these have greater stability. Nevertheless, the availability of YACs with large inserts was an essential prerequisite for the early genome-sequencing projects and they were used in the sequencing of the entire *S. cerevisiae* chromosome III (Oliver *et al.* 1992). The method for

Table 9.1 Properties of the different yeast vectors.

Vector	Transformation frequency	Copy no./cell	Loss in non-selective medium	Disadvantages	Advantages
YIp	10^2 transformants per μg DNA	1	Much less than 1% per generation	1 Low transformation frequency 2 Can only be recovered from yeast by cutting chromosomal DNA with restriction endonuclease which does not cleave original vector containing cloned gene	1 Of all vectors, this kind give most stable maintenance of cloned genes 2 An integrated YIp plasmid behaves as an ordinary genetic marker, e.g. a diploid heterozygous for an integrated plasmid segregates the plasmid in a Mendelian fashion 3 Most useful for surrogate genetics of yeast, e.g. can be used to introduce deletions, inversions and transpositions (see Botstein & Davis 1982)
YEp	10^3–10^5 transformants per μg DNA	25–200	1% per generation	Novel recombinants generated *in vivo* by recombination with endogenous 2 μm plasmid	1 Readily recovered from yeast 2 High copy number 3 High transformation frequency 4 Very useful for complementation studies
YRp	10^4 transformants per μg DNA	1–20	Much greater than 1% per generation but can get chromosomal integration	Instability of transformants	1 Readily recovered from yeast 2 High copy number. Note that the copy number is usually less than that of YEp vectors but this may be useful if cloning gene whose product is deleterious to the cell if produced in excess 3 High transformation frequency 4 Very useful for complementation studies 5 Can integrate into the chromosome
YCp	10^4 transformants per μg DNA	1–2	Less than 1% per generation	Low copy number makes recovery from yeast more difficult than that with YEp or YRp vectors	1 Low copy number is useful if product of cloned gene is deleterious to cell 2 High transformation frequency 3 Very useful for complementation studies 4 At meiosis generally shows Mendelian segregation
YAC		1–2	Depends on length: the longer the YAC the more stable it is	Difficult to map by standard techniques	1 High-capacity cloning system permitting DNA molecules greater than 40 kb to be cloned 2 Can amplify large DNA molecules in a simple genetic background
Ty	Depends on vector used to introduce Ty into cell	~20	Stable, since integrated into chromosome	Needs to be introduced into cell in another vector	Can get amplification following chromosomal integration

cloning large DNA sequences developed by Burke *et al.* (1987) is shown in Fig. 9.3).

For many biologists the primary purpose of cloning is to understand what particular genes do *in vivo*. Thus most of the applications of yeast vectors have been in the surrogate genetics of yeast. One advantage of cloned genes is that they can be analysed easily, particularly with the advent of DNA-sequencing methods. Thus nucleotide-sequencing analysis can reveal many of the elements that control expression of a gene, as well as identifying the sequence of the gene product. In the case of the yeast actin gene (Gallwitz & Sures 1980, Ng & Abelson 1980) and some yeast transfer RNA (tRNA) genes (Peebles *et al.* 1979, Olson 1981), this kind of analysis revealed the presence within these genes of non-coding sequences which are copied into primary transcripts. These *introns* are subsequently eliminated by a process known as *splicing*. Nucleotide-sequence analysis can also reveal the direction of transcription of a gene, although this can be determined *in vivo* by other methods. For example, if the yeast gene is expressed in *E. coli* using bacterial transcription signals, the direction of reading can be deduced by observing the orientation of a cloned fragment required to permit expression. Finally, if a single transcribed yeast gene is present on a vector, the chimera can be used as a probe for quantitative solution hybridization analysis of transcription of the gene.

The availability of different kinds of vectors with different properties (see Table 9.1) enables yeast geneticists to perform manipulations in yeast like those long available to *E. coli* geneticists with their sex factors and transducing phages. Thus cloned genes can be used in conventional genetic analysis by means of recombination using YIp vectors or linearized YRp vectors (Orr-Weaver *et al.* 1981). Complementation can be carried out using YEp, YRp, YCp or YAC vectors, but there are a number of factors which make YCps the vectors of choice (Rose *et al.* 1987). For example, YEps and YRps exist at high copy number in yeast and this can prevent the isolation of genes whose products are toxic when overexpressed, e.g. the genes for actin and tubulin. In other cases, the overexpression of genes other than the gene of interest can suppress the mutation used for selection (Kuo & Campbell 1983). All the yeast vectors can be used to create partial diploids or partial polyploids and the extra gene sequences can be integrated or extrachromosomal. Deletions, point mutations and frame-shift mutations can be introduced *in vitro* into cloned genes and the altered genes returned to yeast and used to replace the wild-type allele. Excellent reviews of these techniques have been presented by Botstein and Davis (1982), Hicks *et al.* (1982), Struhl (1983) and Stearns *et al.* (1990).

Plasmid construction by homologous recombination in yeast

During the process of analysing a particular cloned gene it is often necessary to change the plasmid's selective marker. Alternatively, it may be desired to move the cloned gene to a different plasmid, e.g. from a YCp to a YEp. Again, genetic analysis may require many different alleles of a cloned gene to be introduced to a particular plasmid for subsequent functional studies. All these objectives can be achieved by standard *in vitro* techniques, but Ma *et al.* (1987) have shown that methods based on recombination *in vivo* are much quicker. The underlying principle is that linearized plasmids are efficiently repaired during yeast transformation by recombination with a homologous DNA restriction fragment.

Suppose we wish to move the HIS3 gene from pBR328, which cannot replicate in yeast, to YEp420 (see Fig. 9.7). Plasmid pRB328 is cut with *Pvu*I and *Pvu*II and the HIS3 fragment selected. The HIS3 fragment is mixed with YEp420 which has been linearized with *Eco*RI and the mixture transformed into yeast. Two crossover events occurring between homologous regions flanking the *Eco*RI site of YEp420 will result in the generation of a recombinant YEp containing both the HIS3 and URA3 genes. The HIS3 gene can be selected directly. If this were not possible, selection could be made for the URA3 gene, for a very high proportion of the clones will also carry the HIS3 gene.

Many other variations of the above method have been described by Ma *et al.* (1987), to whom the interested reader is referred for details.

Expression of cloned genes

When gene manipulation in fungi first became possible, there were many unsuccessful attempts to

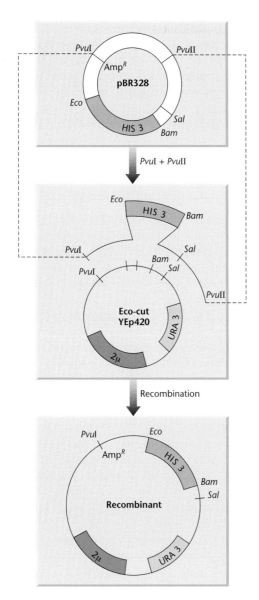

Fig. 9.7 Plasmid construction by homologous recombination in yeast. pRB 328 is digested with *Pvu*I and *Pvu*II and the HIS 3-containing fragment is transformed into yeast along with the *Eco*RI-cut YEp420. Homologous recombination occurs between pBR 322 sequences, shown as thin lines, to generate a new plasmid carrying both HIS 3 and URA 3.

express heterologous genes from bacteria or higher eukaryotes. This suggested that fungal promoters have a unique structure, a feature first shown for *S. cerevisiae* (Guarente 1987). Four structural elements can be recognized in the average yeast promoter

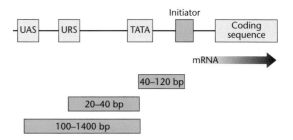

Fig. 9.8 Structure of typical yeast promoters. (See text for details.)

(Fig. 9.8). First, several consensus sequences are found at the transcription-initiation site. Two of these sequences, TC(G/A)A and PuPuPyPuPu, account for more than half of the known yeast initiation sites (Hahn *et al.* 1985, Rudolph & Hinnen 1987). These sequences are not found at transcription-initiation sites in higher eukaryotes, which implies a mechanistic difference in their transcription machinery compared with yeast.

The second motif in the yeast promoter is the TATA box (Dobson *et al.* 1982). This is an AT-rich region with the canonical sequence TATAT/AAT/A, located 60–120 nucleotides before the initiation site. Functionally, it can be considered equivalent to the Pribnow box of *E. coli* promoters (see p. 51).

The third and fourth structural elements are upstream activating sequences (UASs) and upstream repressing sequences (URSs). These are found in genes whose transcription is regulated. Binding of positive-control proteins to UASs increases the rate of transcription and deletion of the UASs abolishes transcription. An important structural feature of UASs is the presence of one or more regions of dyad symmetry (Rudolph & Hinnen 1987). Binding of negative-control proteins to URSs reduces the transcription rate of those genes that need to be negatively regulated.

The level of transcription can be affected by sequences located within the gene itself and which are referred to as downstream activating sequences (DASs). Chen *et al.* (1984) noted that, when using the phosphoglycerate kinase (*PGK*) promoter, several heterologous proteins accumulate to 1–2% of total cell protein, whereas phosphoglycerate kinase itself accumulates to over 50%. These disappointing amounts of heterologous protein reflect the levels of mRNA which were due to a lower level of initiation rather than a reduced mRNA half-life (Mellor *et al.* 1987). Addition of downstream PGK sequences restored the rate of mRNA transcription, indicating the presence of a DAS. Evidence for these DASs has been found in a number of other genes.

Overexpression of proteins in fungi

The first overexpression systems developed were for *S. cerevisiae* and used promoters from genes encoding abundant glycolytic enzymes, e.g. alcohol dehydrogenase (ADH1), PGK or glyceraldehyde-3-phosphate dehydrogenase (GAP). These are strong promoters and mRNA transcribed from them can accumulate up to 5% of total. They were at first thought to be constitutive but later were shown to be induced by glucose (Tuite *et al.* 1982). Now there is a large variety of native and engineered promoters available (Table 9.2), differing in strength, regulation and induction ratio. These have been reviewed in detail by Romanos *et al.* (1992).

Table 9.2 Common fungal promoters used for manipulation of gene expression.

Species	Promoter	Gene	Regulation
General			
S. cerevisiae	PGK	Phosphoglycerate kinase	Glucose-induced
	GAL1	Galactokinase	Galactose-induced
	PHO5	Acid phosphatase	Phosphate-repressed
	ADH2	Alcohol dehydrogenase II	Glucose-repressed
	CUP1	Copper metallothionein	Copper-induced
	MFα1	Mating factor α1	Constitutive but temperature-induced variant available
Candida albicans	MET3	ATP sulphur lyase	Repressed by methionine and cysteine
Methanol utilizers			
Candida boidnii	AOD1	Alcohol oxidase	Methanol-induced
Hansenula polymorpha	MOX	Alcohol oxidase	Methanol-induced
Pichia methanolica	AUG1	Alcohol oxidase	Methanol-induced
Pichia pastoris	AOX1	Alcohol oxidase	Methanol-induced
	GAP	Glyceraldehyde-3-phosphate dehydrogenase	Strong constitutive
	FLD1	Formaldehyde dehydrogenase	Methanol- or methylamine-induced
	PEX8	Peroxin	Methanol-induced
	YPT1	Secretory GTPase	Medium constitutive
Lactose utilizer			
Kluyveromyces lactis	LAC4	β-Galactosidase	Lactose-induced
	PGK	Phosphoglycerate kinase	Strong constitutive
	ADH4	Alcohol dehydrogenase	Ethanol-induced
Starch utilizer			
Schwanniomyces occidentalis	AMY1	α-Amylase	Maltose- or starch-induced
	GAM1	Glucoamylase	Maltose- or starch-induced
Xylose utilizer			
Pichia stipitis	XYL1		Xylose-induced
Alkane utilizer			
Yarrowia lipolytica	XPR2	Extracellular protease	Peptone-induced
	TEF	Translation elongation factor	Strong constitutive
	RPS7	Ribosomal protein S7	Strong constitutive

The ideal promoter is one that is tightly regulated so that the growth phase can be separated from the induction phase. This minimizes the selection of non-expressing cells and can permit the expression of proteins normally toxic to the cell. The ideal promoter will also have a high induction ratio. One promoter which has these characteristics and which is now the most widely used is that from the *GAL1* gene. Galactose regulation in yeast is now extremely well studied and has become a model system for eukaryotic transcriptional regulation (see Box 9.1).

Following addition of galactose, GAL1 mRNA is rapidly induced over 1000-fold and can reach 1% of total mRNA. However, the promoter is strongly repressed by glucose and so in glucose-grown cultures this induction only occurs following depletion of glucose. To facilitate galactose induction in the presence of glucose, mutants have been isolated which are insensitive to glucose repression (Matsumoto *et al.* 1983, Horland *et al.* 1989). The *trans*-activator GAL4 protein is present in only one or two molecules per cell and so GAL1 transcription is limited. With multicopy expression vectors, GAL4 limitation is exacerbated. However, GAL4 expression can be made autocatalytic by fusing the GAL4 gene to a GAL10 promoter (Schultz *et al.* 1987), i.e. GAL4 expression is now regulated (induced) by galactose.

In recent years, methylotrophic yeasts, such as *Pichia pastoris*, have proved extremely popular as hosts for the overexpression of heterologous proteins. There are a number of reasons for this. First, the alcohol oxidase (AOX1) promoter is one of the strongest and most regulatable promoters known. Second, it is possible to stably integrate expression plasmids at specific sites in the genome in either single or multiple copies. Third, the strains can be cultivated to very high density. To date, over 300 foreign proteins have been produced in *P. pastoris* (Cereghino & Cregg 1999, 2000). Promoters for use in other yeasts are shown in Table 9.2.

Specialist vectors

Many different specialist yeast vectors have been developed which incorporate the useful features found in the corresponding *E. coli* vectors (see p. 70), e.g. an f1 origin to permit sequencing of inserts production of the cloned gene product as a purification fusion, etc. Some representative examples are shown in Fig. 9.9. Two aspects of these vectors warrant further discussion: secretion and surface display.

In yeast, proteins destined for the cell surface or for export from the cell are synthesized on and translocated into the endoplasmic reticulum. From there they are transported to the Golgi body for processing and packaging into secretory vesicles. Fusion of the secretory vesicles with the plasma membrane then occurs constitutively or in response to an external signal (reviewed by Rothman & Orci 1992). Of the proteins naturally synthesized and secreted by yeast, only a few end up in the growth medium, e.g. the mating pheromone α factor and the killer toxin. The remainder, such as invertase and acid phosphatase, cross the plasma membrane but remain within the periplasmic space or become associated with the cell wall.

Polypeptides destined for secretion have a hydrophobic amino-terminal extension, which is responsible for translocation to the endoplasmic reticulum (Blobel & Dobberstein 1975). The extension is usually composed of about 20 amino acids and is cleaved from the mature protein within the endoplasmic reticulum. Such signal sequences precede the mature yeast invertase and acid phosphatase sequences. Rather longer leader sequences precede the mature forms of the α mating factor and the killer toxin (Kurjan & Herskowitz 1982, Bostian *et al.* 1984). The initial 20 amino acids or so are similar to the conventional hydrophobic signal sequences, but cleavage does not occur in the endoplasmic reticulum. In the case of α factor, which has an 89 amino acid leader sequence, the first cleavage occurs after a Lys–Arg sequence at positions 84 and 85 and happens in the Golgi body (Julius *et al.* 1983, 1984).

To date, a large number of non-yeast polypeptides have been secreted from yeast cells containing the appropriate recombinant plasmid and in almost all cases the α-factor signal sequence has been used. There is a perception that *S. cerevisiae* has a lower secretory capacity than *P. pastoris* and other yeasts (Muller *et al.* 1998), but the real issue may be the type of vector used. For example, Parekh *et al.* (1996) found that *S. cerevisiae* strains containing one stably integrated copy of an expression cassette

Box 9.1 Galactose metabolism and its control in *Saccharomyces cerevisiae*

Galactose is metabolized to glucose-6-phosphate in yeast by an identical pathway to that operating in other organisms (Fig. B9.1). The key anzymes and their corresponding genes are a kinase (*GAL1*), a transferase (*GAL7*), an epimerase (*GAL10*) and a mutase (*GAL5*). Melibiose (galactosyl-glucose) is metabolized by the same enzymes after cleavage by an α-galactosidase encoded by the *MEL1* gene. Galactose uptake by yeast cells is via a permease encoded by the *GAL2* gene. The *GAL5* gene is constitutively expressed. All the others are induced by growth on galactose and repressed during growth on glucose.

The *GAL1*, *GAL7* and *GAL10* genes are clustered on chromosome II but transcribed separately from individual promoters. The *GAL2* and *MEL1* genes are on other chromosomes. The *GAL4* gene encodes a protein that activates transcription of the catabolic genes by binding UAS 5′ to each gene. The *GAL80* gene encodes a repressor that binds directly to *GAL4* gene product, thus preventing it from activating transcription. The *GAL3* gene product catalyses the conversion of galactose to an inducer, which combines with the *GAL80* gene product, preventing it from inhibiting the *GAL4* protein from binding to DNA (Fig. B9.2).

The expression of the *GAL* genes is repressed during growth on glucose. The regulatory circuit responsible for this phenomenon, termed catabolite repression, is superimposed upon the circuit responsible for induction of *GAL* gene expression. Very little is known about its mechanism.

For a review of galactose metabolism in *S. cerevisiae*, the reader should consult Johnston (1987).

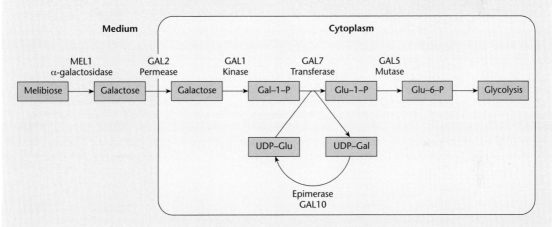

Fig. B9.1 The genes and enzymes associated with the metabolism of galactose by yeast.

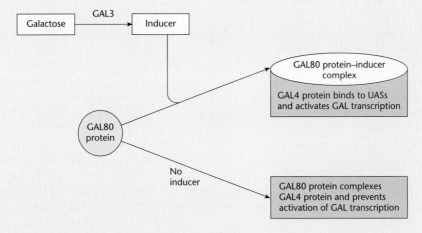

Fig. B9.2 The regulation of transcription of the yeast galactose genes.

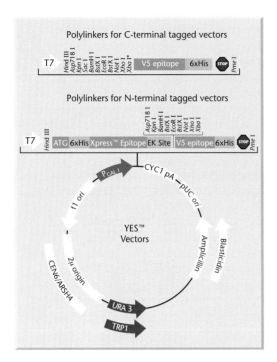

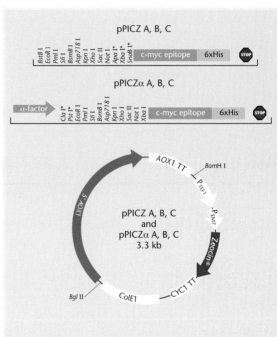

Fig. 9.9 Two specialized vectors for use in *Saccharomyces* (YES vectors) and *Pichia* (pPICZ). V5, Express, 6XHis and c-myc encode epitopes which can be readily detected and purified by affinity chromatography. The YES vectors offer a choice of 2 μm origin for high copy or CEN6/ARSH4 origin for low copy in yeast, a choice of URA 3 or TRP 1 genes for auxotrophic selection on minimal medium, blasticidin resistance for dominant selection in any strain and an f1 origin to facilitate DNA sequencing. In the pPICZ vectors, the zeocin-resistance gene is driven by the EM-7 promoter for selection in *E. coli* and the TEF 1 promoter for selection in *Pichia*. More details of these features can be found in Chapter 5. (Figure reproduced courtesy of Invitrogen Corporation.)

secreted more bovine pancreatic trypsin inhibitor than strains with the same expression cassette on a 2 μm multicopy vector. Optimal expression was obtained with 10 integrated copies. With those proteins which tend to accumulate in the endoplasmic reticulum as denatured aggregates, secretion may be enhanced by simultaneously overexpressing chaperons (Shusta *et al.* 1998). Secretion may also be enhanced by minor amino acid changes. Katakura *et al.* (1999) noted a sixfold increase in lactoglobulin secretion by conversion of a tryptophan residue to tyrosine.

Yeast surface display

S. cerevisiae can be used to elucidate and dissect the function of a protein in a manner similar to phage-display systems. Either can be used to detect protein–ligand interactions and to select mutant proteins with altered binding capacity (Shusta *et al.* 1999). However, phage-display systems often cannot display secreted eukaryotic proteins in their native functional conformation, whereas yeast surface display can.

Yeast surface display makes use of the cell surface receptor α-agglutinin (Aga), which is a two-subunit glycoprotein. The 725-residue Aga1p subunit anchors the assembly to the cell wall via a covalent linkage. The 69-residue binding subunit (Aga2p) is linked to Aga1p by two disulphide bonds. To achieve surface display, the appropriate gene is inserted at the C terminus of a vector-borne Aga2p gene under the control of the GAL1 promoter. The construct is then transformed into a yeast strain carrying a chromosomal copy of the Aga1p gene, also under the control of the GAL1 promoter. If the cloned gene has

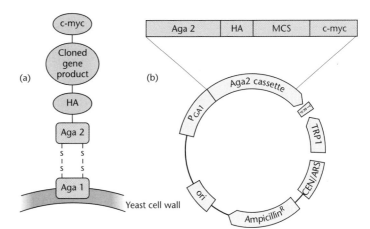

Fig. 9.10 Yeast surface display of heterologous proteins. (a) Schematic representation of surface display. (b) A vector used for facilitating surface display. MCS, multiple cloning site; HA, haemagglutinin epitope; c-myc, c-myc epitope. See text for details.

been inserted in the correct translational reading frame, its gene product will be synthesized as a fusion with the Aga2p subunit. The fusion product will associate with the Aga1p subunit within the secretory pathway and be exported to the cell surface (Boder & Wittrup 1997).

In practice, the gene fusion is somewhat more complicated. Usually the cloned gene product is sandwiched between two simple epitopes to permit quantitation of the number of fusion proteins per cell and to determine the accessibility of different domains of the fusion protein (Fig. 9.10).

Detecting protein–protein interactions

Chien *et al.* (1991) have made use of the properties of the GAL4 protein to develop a method for detecting protein–protein interactions. The GAL4 protein has separate domains for the binding to UAS DNA and for transcriptional activation. Plasmids were constructed which encode two hybrid proteins. The first consisted of the DNA-binding domain (residues 1–147) of the GAL4 protein fused to a test protein. The second consisted of residues 768–881 of the GAL4 protein, representing the activation domain, fused to protein sequences encoded by a library of yeast genomic DNA fragments. Interaction between the test protein (bait) and a protein encoded by one of the library plasmids (prey) led to transcriptional activation of a reporter gene (Fig. 9.11). This method is known as the two-hybrid system. The two-hybrid system has become a major tool in the study of protein–protein and protein–ligand interactions (see

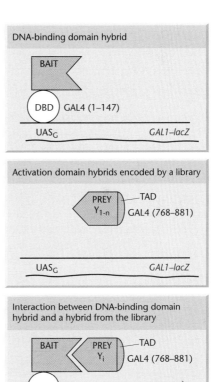

Fig. 9.11 Strategy to detect interacting proteins using the two-hybrid system. UAS$_G$ is the upstream activating sequence for the yeast *GAL* genes, which binds the GAL4 protein, DBD is a DNA-binding domain and TAD is a transcription-activation domain. Interaction is detected by expression of a *GAL1–lacZ* gene fusion. (Reproduced courtesy of Dr S. Fields and the National Academy of Sciences.)

Box 9.2 Variations of the two-hybrid system

Post-translational modifications

A limitation of using the two-hybrid system is that certain protein–protein interactions are not possible in *S. cerevisiae* because of the absence of the appropriate post-translational modifying enzymes. This is a particular problem with baits relevant to signal transduction in higher eukaryotes, because of the prevalence of site-specific phosphorylation by tyrosine kinase as a means of regulating protein–protein interactions. Osborne *et al.* (1996) were able to solve this problem by using a strain of yeast expressing the mammalian tyrosine kinase Lck.

Protein–peptide interactions

The utility of the two-hybrid system can be extended to interactions between proteins and very small (< 16 residues) peptides (Yang *et al.* 1995, Colas *et al.* 1996). Small peptides can be used not only as the prey, as in the studies just cited, but also as the bait (Geyer *et al.* 1999, Norman *et al.* 1999). In another variation, Stagljar *et al.* (1996) have described a rapid method to identify small interaction-specific sequences within larger proteins. In their method, complementary DNA (cDNA) encoding a known interacting protein is sheared by sonication and shotgun-cloned into a prey vector. The resulting targeted library is cotransformed with the bait of interest and direct selection made for fragments capable of conferring the desired interaction.

Three-component systems

In some cases, a bait and prey may not directly interact, or else form a transient contact, because a third component is missing. However, co-expression of the bait with a previously identified partner protein may permit expression of the reporter gene (Tomashek *et al.* 1996). The ternary partner in such three-component interactions need not be a protein but can be a small molecule, such as a hormone (Lee *et al.* 1995) or a drug (Chiu *et al.* 1994).

Bait and hook

The two-hybrid system can be modified to study the interaction between proteins and RNA (SenGupta *et al.* 1999) or proteins and drugs (Licitra & Liu 1996). To do this a 'hook' is used to display an RNA molecule or a drug to the incoming prey, thereby forming a non-protein bridge to the reporter gene (Fig. B9.3).

Reverse two-hybrid system

The original two-hybrid system relies on transcriptional activation of reporter genes to detect protein–protein interactions (Fig. 9.11). However, the two-hybrid system can be 'turned on its head' to monitor *loss* of protein–protein interactions. For example, Vidal *et al.* (1996a,b), have described a screen for mutations that disrupted the ability of the transcription factor E2F1 to associate with its partner DP1. In the absence of interaction between the two partner proteins, transcription of the URA3 reporter gene was abolished. This in turn allowed the host yeasts to grow in the presence of 5-fluoro-orotic acid, a compound that is toxic to cells capable of synthesizing uracil. This type of reverse selection can be used to screen for peptides or other small molecules that disrupt protein–protein interactions and that could have utility as drugs (Huang & Schreiber 1997).

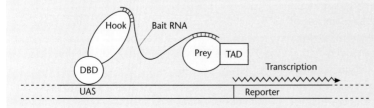

Fig. B9.3 The bait-and-hook variation of the two-hybrid system.

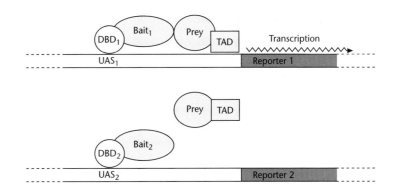

Fig. 9.12 The dual-bait two-hybrid system. The first DNA-binding domain fusion protein (DBD$_1$–Bait$_1$) drives the expression of reporter 1 through UAS$_1$. A second separate DNA-binding domain fused to a distinct bait (DBD$_2$–Bait$_2$) directs expression of reporter 2 through UAS$_2$. In this example, the prey can interact with Bait 1 but not Bait 2.

Box 9.2 and the reviews by Colas & Brent 1998, and Uetz & Hughes 2000).

A major operational problem with the two-hybrid system has been the high frequency of false positives, i.e. cells in which the reporter gene is active even though the bait and prey do not interact in nature and/or in yeast. There are two general solutions to this problem. In the first of these (James *et al.* 1996), a host strain is used that contains three reporter genes under the control of different GAL4-inducible promoters. An alternative approach is to use the two-bait system (Fig. 9.12). In this system, the yeast expresses a single prey and two independent baits, each targeted to distinct reporter genes (Xu *et al.* 1997). In weeding out false positives, one looks for preys that interact with only one bait. The rationale here is that the more related the two baits are, the more likely it is that a prey interacting with only one of them is a natural interaction.

Identifying genes encoding particular cellular activities

As a result of a major genome-sequencing project, the sequences of genes encoding every biochemical activity of *S. cerevisiae* are known. The next task is to connect specific biochemical activities with particular genes. Martzen *et al.* (1999) have developed a rapid and sensitive method for doing this which is applicable to almost any activity. The starting-point was a yeast vector containing the glutathione-S-transferase (GST) gene under the control of the CUP1 promoter. A large number of different DNA fragments bearing yeast ORFs were cloned in this vector to generate GST–ORF fusions. In all, 6144

constructs were made and these were transformed back into *S. cerevisiae*.

To correlate genes with activity, the different transformants were dispensed into 64 96-well microtitre plates ($64 \times 96 = 6144$). Initially, all the clones from each plate were pooled and the 64 pools analysed for a specific biochemical activity. Analysis is facilitated since the GST–ORF fusions are readily purified from other cellular activities by affinity chromatography using immobilized glutathione. Once a plate was identified as having the desired activity, pools of clones were made from each of the eight rows and 12 columns of wells and these 20 pools reassayed. If only one clone expresses the desired activity, then one row and one column should be identified (Fig. 9.13) and the point of intersection identifies the wanted clone. Sequencing of the ORF then allows identification of the gene corresponding to the activity.

Determining functions associated with particular genes

Determining the function of a particular gene product usually involves determining the expression profile of the gene, the subcellular location of the protein, and the phenotype of a null strain lacking the protein. Conditional alleles of the gene are often created as an additional tool. These procedures usually require multiple independent manipulations. Ross-Macdonald *et al.* (1997) have developed a multifunctional, transposon-based system that simultaneously generates constructs for all of the above analyses and is suitable for mutagenesis of any given *S. cerevisiae* gene. The transposons used by them are shown in Fig. 9.14. Each carries a reporter gene (β-galactosidase

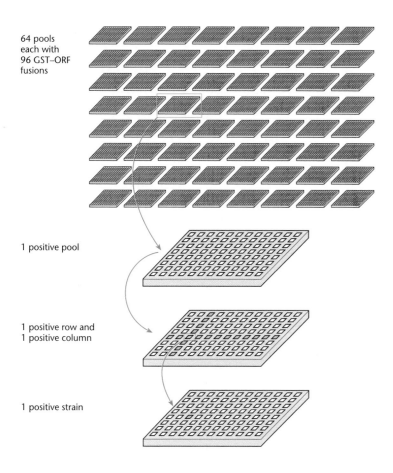

64 pools
each with
96 GST–ORF
fusions

1 positive pool

1 positive row and
1 positive column

1 positive strain

Fig. 9.13 Method for selecting glutathione-*S*-transferase open-reading-frame fusions. (Figure reproduced from *Trends in Genetics* courtesy of Elsevier Press.)

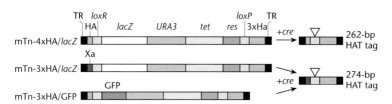

Fig. 9.14 Schematic representation of the transposons, and the derived HAT tag elements, for use in yeast. Each transposon carries the tetracycline-resistance determinant, a functional yeast URA3 gene, and the *res* element from transposon Tn3. The function of the *res* element is to resolve transposition cointegrates. (Figure reproduced courtesy of Dr Michael Snyder and the National Academy of Sciences.)

or green fluorescent protein) lacking a promoter. In-frame fusions between a yeast coding region and the transposon can be detected by β-galactosidase activity or fluorescence. The transposon insertion creates a truncation of the gene, thereby generating a null phenotype. The method is also adaptable to genome-wide analyses, using a high-throughput screening methodology (Ross-Macdonald *et al.* 1999).

To use this transposon system, yeast genes are cloned in an *E. coli* strain that overexpresses the Tn3 transposase. The transposon, carried on a derivative of the sex factor F, is introduced by mating and transposition ensues. Yeast DNA containing the transposon is excised from the *E. coli* vector and transformed back into yeast, with selection made for URA3 activity. Two types of transformants are obtained. If homologous

(a) PCR deletion cassette

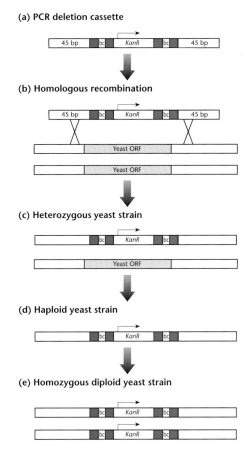

(b) Homologous recombination

(c) Heterozygous yeast strain

(d) Haploid yeast strain

(e) Homozygous diploid yeast strain

Fig. 9.15 'Knockout' strain collection. (a) The bar-coded deletion cassette is generated in two rounds of PCR. It contains a kanamycin-resistance gene for selection in yeast and two molecular bar-codes 20 bp long that are unique to each deletion. The molecular bar-codes are flanked by common sequence elements that allow for the PCR amplification of all the bar-codes in a population of yeast strains. At the end of the cassette are 45 bp sequences that are homologous to sequences flanking the gene of interest. (b) The deletion cassette is transformed into diploid yeast strains. The 45 bp of yeast sequences allows one copy of the gene of interest to be replaced with the deletion cassette via homologous recombination. The resulting heterozygous strains (c) can then be sporulated and dissected to generate a collection of haploid mutant strains, if viable (d). Homozygous mutant strains (e) are made by the selective matings of haploid segregants. (Figure reproduced courtesy of Elsevier Science Ltd and the authors.)

recombination occurs at the URA3 locus, then no reporter activity is seen. If integration occurs at the locus of the gene carrying the transposon, reporter activity is retained but gene function is lost (Fig. 9.14). This reporter activity can be used to study the expression of the cloned gene under different growth conditions. If the transformants also carry a phage P1 *cre* gene (see p. 68), under the control of the GAL1 promoter, excision of most of the transposon sequences can be induced by growth of the host strain in the presence of galactose. In most instances, this results in restoration of gene activity and the gene product can be purified by affinity purification, using antibodies to the haemagglutinin determinants.

An alternative approach to the large-scale mutational analysis of the yeast genome is to individually delete each gene using a polymerase chain reaction (PCR)-based methodology (Winzeler *et al.* 1999). For each gene, a deletion cassette is constructed (Fig. 9.15) that contains a kanamycin resistance gene, two 'molecular bar-codes' and yeast sequences homologous to the upstream and downstream flanking sequences. This construct is amplified by PCR and transformed into a diploid yeast strain, where the deletion cassette replaces one of the two chromosomal gene sequences. The diploid strains are then allowed to sporulate and those haploid segregants that are viable are recovered.

The molecular bar-codes are particularly important since they are unique to each deletion strain within a pool of many different strains (Shoemaker *et al.* 1996). The bar-code can be amplified by PCR, using common primer sequences that flank the bar-code. A pool of yeast mutant strains can thus be analysed in a single experiment by amplifying the bar-codes and probing a DNA microarray (p. 116) containing sequences complementary to the bar-codes. Variations on this technique have been used in other organisms to identify key functions associated with particular activities. For example, *signature-tagged mutagenesis* has been used to identify genes essential for virulence (for review, see Lehoux & Levesque 2000).

The application of genome-wide mutagenesis techniques to the understanding of the ways in which the different components of the yeast cell interact has been reviewed by Delneri *et al.* (2001) and Vidan and Snyder (2001).

CHAPTER 10

Gene transfer to animal cells

Introduction

Gene transfer to animal cells has been practised now for over 40 years. Techniques are available for the introduction of DNA into many different cell types in culture, either to study gene function and regulation or to produce large amounts of recombinant protein. Animal cells are advantageous for the production of recombinant animal proteins because they perform authentic post-translational modifications that are not carried out by bacterial cells and fungi. Cell cultures have therefore been used on a commercial scale to synthesize products such as antibodies, hormones, growth factors, cytokines and viral coat proteins for immunization. There has been intense research into the development of efficient vector systems and transformation methods for animal cells. Although this research has focused mainly on the use of mammalian cell lines, other systems have also become popular, such as the baculovirus expression system, which is used in insects. More recently, research has focused on the introduction of DNA into animal cells *in vivo*. The most important application of this technology is *in vivo* gene therapy, i.e. the introduction of DNA into the cells of live animals in order to treat disease. Viral gene-delivery vectors are favoured for therapeutic applications because of their efficiency, but safety concerns have prompted research into alternative DNA-mediated transfer procedures.

This chapter focuses on the introduction of DNA into somatic cells. Unlike the situation in plants, most animal cells are restricted in terms of their developmental potential and cannot be used to generate transgenic animals. Mouse embryonic stem cells (ES cells) are exceptional in this respect because they are derived from the early embryo prior to the formation of the germ line. Gene transfer to ES cells is therefore discussed in the next chapter, along with the introduction of DNA into oocytes, eggs and cells of the early embryo.

Overview of gene-transfer strategies

Gene transfer to animal cells can be achieved essentially via three routes. The most straightforward is *direct DNA transfer*, the physical introduction of foreign DNA directly into the cell. For example, in cultured cells this can be done by microinjection, whereas for cells *in vivo* direct transfer is often achieved by bombardment with tiny DNA-coated metal particles. The second route is termed *transfection*, and this encompasses a number of techniques, some chemical and some physical, which can be used to persuade cells to take up DNA from their surroundings. The third is to package the DNA inside an animal virus, since viruses have evolved mechanisms to naturally infect cells and introduce their own nucleic acid. The transfer of foreign DNA into a cell by this route is termed *transduction*. Whichever route is chosen, the result is *transformation*, i.e. a change of the recipient cell's genotype caused by the acquired foreign DNA, the *transgene*. Transformation can be transient or stable, depending on how long the foreign DNA persists in the cell.

DNA-mediated transformation

Transformation techniques

DNA/calcium phosphate coprecipitate method

The ability of mammalian cells to take up exogenously supplied DNA from their culture medium was first reported by Szybalska and Szybalski (1962). They used total uncloned genomic DNA to transfect human cells deficient for the enzyme hypoxanthine-guanine phosphoribosyltransferase (HPRT). Rare HPRT-positive cells, which had presumably taken up fragments of DNA containing the functional gene, were identified by selection on HAT medium (p. 177). At this time, the actual mechanism of DNA uptake was not understood. Much later, it was appreciated that successful DNA transfer in such experiments was dependent on the formation of a fine DNA/calcium phosphate coprecipitate, which first settles on to the cells and is then internalized. The precipitate must be formed freshly at the time of transfection. It is thought that small granules of calcium phosphate associated with DNA are taken up by endocytosis and transported to the nucleus, where some DNA escapes and can be expressed (Orrantia & Chang, 1990). The technique became generally accepted after its application, by Graham and Van der Erb (1973), to the analysis of the infectivity of adenoviral DNA. It is now established as a general method for the introduction of DNA into a wide range of cell types in culture. However, since the precipitate must coat the cells, this method is suitable only for cells growing in monolayers, not those growing in suspension or as clumps.

As originally described, calcium phosphate transfection was limited by the variable and rather low proportion of cells that took up DNA (1–2%). Only a small number of these would be stably transformed. Improvements to the method have increased the transfection frequency to 20% for some cell lines (Chu & Sharp 1981). A variant of the technique, using a different buffer system, allows the precipitate to form slowly over a number of hours, and this can increase stable transformation efficiency by up to 100-fold when using high-quality plasmid DNA (Chen & Okayama 1987, 1988).

Other chemical transfection methods

The calcium phosphate method is applicable to many cell types, but some cell lines are adversely affected by the coprecipitate due to its toxicity and are hence difficult to transfect. Alternative chemical transfection methods have been developed to address this problem. One such method utilizes diethylaminoethyl dextran (DEAE-dextran), a soluble polycationic carbohydrate that promotes interactions between DNA and the endocytotic machinery of the cell. This technique was initially devised to introduce viral DNA into cells (McCutchan & Pango 1968) but was later adapted as a method for plasmid DNA transfer (Milman & Herzberg 1981). The efficiency of the original procedure was improved by Lopata *et al.*

(1984) and Sussman and Milman (1984) by adding after-treatments, such as osmotic shock or exposure to chloroquine, the latter having been shown to inhibit the acidification of endosomal vesicles (Luthmann & Magnusson 1983). Although efficient for the transient transfection of many cell types, DEAE-dextran cannot be used to generate stably transformed cell lines. Another polycationic chemical, the detergent Polybrene, has been used for the transfection of Chinese hamster ovary (CHO) cells, which are not amenable to calcium phosphate transfection (Chaney *et al.* 1986).

Phospholipids as gene-delivery vehicles

An alternative to chemical transfection procedures is to package DNA inside a phosopholipid vesicle, which interacts with the target cell membrane and facilitates DNA uptake. The first example of this approach was provided by Schaffner (1980), who used bacterial protoplasts containing plasmids to transfer DNA into mammalian cells. Briefly, bacterial cells were transformed with a suitable plasmid vector and then treated with chloramphenicol to amplify the plasmid copy number. Lysozyme was used to remove the cell walls, and the resulting protoplasts were gently centrifuged on to a monolayer of mammalian cells and induced to fuse with them, using polyethylene glycol. A similar strategy was employed by Wiberg *et al.* (1987), who used the haemoglobin-free ghosts of erythrocytes as delivery vehicles. The procedures are very efficient in terms of the number of transformants obtained, but they are also labour-intensive and so have not been widely adopted as a general transfection method. However, an important advantage is that they are gentle, allowing the transfer of large DNA fragments without shearing. Yeast cells with the cell wall removed (sphaeroplasts) have therefore been used to introduce yeast artificial chromosome (YAC) DNA into mouse ES cells by this method, for the production of YAC transgenic mice (see Chapter 11).

More widespread use has been made of artificial phospholipid vesicles, which are called *liposomes* (Schaefer-Ridder *et al.* 1982). Initial liposome-based procedures were hampered by the difficulty encountered in encapsulating the DNA, and provided a transfection efficiency no better than the calcium phosphate method. However, a breakthrough came with the discovery that cationic/neutral lipid mixtures can spontaneously form stable complexes with DNA that interact productively with the cell membrane, resulting in DNA uptake by endocytosis (Felgner *et al.* 1987, 1994). This low-toxicity transfection method, commonly known as *lipofection*, is one of the simplest to perform and is applicable to many cell types that are difficult to transfect by other means, including cells growing in suspension (e.g. Ruysscharet *et al.* 1994). This technique is suitable for transient and stable transformation, and is sufficiently gentle to be used with YACs and other large DNA fragments. The efficiency is also much higher than that of chemical transfection methods – up to 90% of cells in a culture dish can be transfected. A large number of different lipid mixtures is available, varying in efficiency for different cell lines. A unique benefit of liposome gene-delivery vehicles is their ability to transform cells in live animals following injection into target tissues or even the bloodstream. Transfection efficiency has been improved and targeting to specific cell types achieved by combining liposomes with viral proteins that promote cell fusion, nuclear targeting signals and various molecular conjugates that recognize specific cell surface molecules. The development of liposomes for gene therapy has been comprehensively reviewed (Scheule & Cheng 1996, Tseng & Huang 1998).

Electroporation

Electroporation involves the generation of transient, nanometre-sized pores in the cell membrane, by exposing cells to a brief pulse of electricity. DNA enters the cell through these pores and is transported to the nucleus. This technique was first applied to animal cells by Wong and Neumann (1982), who successfully introduced plasmid DNA into mouse fibroblasts. The electroporation technique has been adapted to many other cell types (Potter *et al.* 1984). The most critical parameters are the intensity and duration of the electric pulse, and these must be determined empirically for different cell types. However, once optimal electroporation parameters have been established, the method is simple to carry out and highly reproducible. The technique has high input costs, because a specialized

capacitor discharge machine is required that can accurately control pulse length and amplitude (Potter 1988). Additionally, larger numbers of cells may be required than for other methods because, in many cases, the most efficient electroporation occurs when there is up to 50% cell death.

In an alternative method, pores are created using a finely focused laser beam (Kurata *et al.* 1986). Although very efficient (up to 0.5% stable transformation), this technique is applicable to only small numbers of cells and has not gained widespread use.

Direct transfer methods

A final group of methods considered in this section encompasses those in which the DNA is transferred directly into the cell nucleus. One such procedure is microinjection, a technique that is guaranteed to generate successful hits on target cells but that can only be applied to a few cells in any one experiment. This technique has been applied to cultured cells that are recalcitrant to other transfection methods (Capecchi 1980), but its principal use is to introduce DNA and other molecules into large cells, such as oocytes, eggs and the cells of early embryos, as discussed in Chapter 11. Particle bombardment is another direct delivery method, initially developed for the transformation of plants (Chapter 12). This involves coating small metal particles with DNA and then accelerating them into target tissues using a powerful force, such as a blast of high-pressure gas or an electric discharge through a water droplet. In animals, this technique is most often used to transfect multiple cells in tissue slices rather than cultured cells (e.g. Arnold *et al.* 1994, Lo *et al.* 1994). It has also been used to transfer DNA into skin cells *in vivo* (Haynes *et al.* 1996).

Transformation with non-replicating DNA

Transient and stable transformation

DNA-mediated transformation of animal cells occurs in two stages, the first involving the introduction of DNA into the cell (the transfection stage) and the second involving its incorporation into the genome (the integration stage). Transfection is much more efficient than integration; hence a large proportion of transfected cells never integrate the foreign DNA they contain. The DNA is maintained in the nucleus in an extrachromosomal state and, assuming it does not contain an origin of replication that functions in the host cell, it persists for just a short time before it is diluted and degraded. This is known as *transient transformation* (the term *transient transfection* is also used), reflecting the fact that the properties of the cell are changed by the introduced transgene, but only for a short duration. In a small proportion of transfected cells, the DNA will integrate into the genome, forming a new genetic locus that will be inherited by all clonal descendants. This is known as *stable transformation*, and results in the formation of a 'cell line' carrying and expressing the transgene. Since integration is such an inefficient process, the rare stably transformed cells must be isolated from the large background of non-transformed and transiently transformed cells by selection, as discussed below. While stable transformation is required for long-term analytical experiments and the production of large amounts of recombinant protein over a prolonged time, transient transformation is sufficient for many types of short-term experiments, such as determining the efficiency of promoter sequences attached to a reporter gene (Box 10.1).

Cotransformation and selection of stable transformants

Following the general acceptance of the calcium phosphate transfection method, it was shown that mouse cells deficient for the enzyme thymidine kinase (TK) could be stably transformed to a wild-type phenotype by transfecting them with the herpes simplex virus (HSV) *Tk* gene (Wigler *et al.* 1977). As for the HPRT⁺ transformants discussed earlier in the chapter, cells positive for TK can be selected on HAT medium. This is because both enzymes are required for nucleotide biosynthesis via the *salvage pathway* (Fig. 10.1). In mammals, nucleotides are produced via two alternative routes, the *de novo* and salvage pathways. In the *de novo* pathway, nucleotides are synthesized from basic precursors, such as sugars and amino acids, while the salvage pathway recycles nucleotides from DNA and RNA. If the *de novo* pathway is blocked, nucleotide synthesis becomes

dependent on the salvage pathway, and this can be exploited for the selection of cells carrying functional *Hprt* and *Tk* genes. The drug aminopterin blocks the *de novo* synthesis of both inosine monophosphate (IMP) and thymidine monophosphate (TMP) by inhibiting key enzymes in the *de novo* pathway. Cells exposed to animopterin can thus survive only if they have functional *Hprt* and *Tk* genes and a source of hypoxanthine and thymidine. *Hprt*⁺ and *Tk*⁺ transformants can therefore both be selected using HAT medium, which contains hypoxanthine, aminopterin and thymidine.

In these early experiments, the transgene of interest conferred a selectable phenotype on the cell. However, most genes do not generate a conveniently selectable phenotype, and the isolation of transformants in such experiments was initially problematic. A breakthrough was made when it was discovered that transfection with two physically unlinked DNAs resulted in *cotransformation*, i.e. the integration of both transgenes into the genome (Wigler *et al.* 1979). To obtain cotransformants, cultured cells were transfected with the HSV *Tk* gene and a vast excess of well-defined DNA, such as the plasmid pBR322. Cells selected on HAT medium were then tested by Southern blot hybridization for the presence of the non-selected plasmid DNA. Wigler and colleagues found evidence for the presence of non-selected DNA in nearly 90% of the TK⁺ cells, indicating that the HSV *Tk* gene could be used as a *selectable marker*. In subsequent experiments, it was shown that the initially unlinked donor DNA fragments were incorporated into large concatemeric structures up to 2 Mbp in length prior to integration (Perucho *et al.* 1980b).

Other selectable markers for animal cells

The phenomenon of cotransformation allows the stable introduction of any foreign DNA sequence into mammalian cells as long as a selectable marker is introduced at the same time. The HSV *Tk* gene is representative of a class of genes known as *endogenous markers*, because they confer a property that is already present in wild-type cells. A number of such markers have been used, all of which act in redundant metabolic pathways (Table 10.1). The major disadvantage of endogenous markers is that they

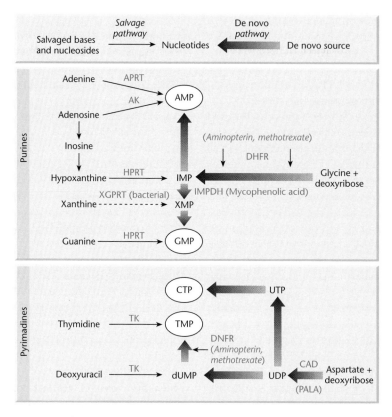

Fig. 10.1 Simplified representation of the *de novo* and salvage nucleotide synthesis pathways. Top panel: purine synthesis. *De novo* purine nucleotide synthesis (shown on the right) initially involves the formation of inosine monophosphate (IMP) which is then converted into either adenosine monophosphate (AMP) or, via xanthine monophosphate (XMP), guanosine monophosphate (GMP). The *de novo* synthesis of IMP requires the enzyme dihydrofolate reductase (DHFR), whose activity can be blocked by aminopterin or methotrexate. In the presence of such inhibitors, cell survival depends on nucleotide salvage, as shown on the left. Cells lacking one of the essential salvage enzymes, such as HPRT or APRT, therefore cannot survive in the presence of aminopterin or methotrexate unless they are transformed with a functional copy of the corresponding gene. Thus, the genes encoding salvage-pathway enzymes can be used as selectable markers. Note that the enzyme XGPTR, which converts xanthine to XMP, is found only in bacterial cells and not in animals. Bottom panel: pyrimidine synthesis. *De novo* pyrimidine nucleotide synthesis (shown on the right) initially involves the formation of uridine diphosphate (UDP). This step requires a multifunctional enzyme, CAD, whose activity can be blocked by *N*-phosphonacetyl-L-aspartate (PALA). UDP is then converted into either thymidine monophosphate (TMP) or cytidine triphosphate (CTP), the latter via uridine triphospate (UTP). *De novo* TMP synthesis requires DHFR, so the reaction can be blocked in the same way as in *de novo* purine synthesis, making cell survival dependent on the salvage enzyme thymidine kinase (TK). Thus, the *Tk* gene can be used as a selectable marker. There is no salvage pathway for CTP.

can only be used with mutant cell lines in which the corresponding host gene is non-functional. This restricts the range of cells that can be transfected. Endogenous markers have therefore been largely superseded by so-called *dominant selectable markers*, which confer a phenotype that is entirely novel to the cell and hence can be used in any cell type. Such markers are usually drug-resistance genes of bacterial origin, and transformed cells are selected on

a medium that contains the drug at an appropriate concentration. For example, *Escherichia coli* transposons Tn 5 and Tn*601* contain distinct genes encoding neomycin phosphotransferase, whose expression confers resistance to aminoglycoside antibiotics (kanamycin, neomycin and G418). These are protein-synthesis inhibitors, active against bacterial and eukaryotic cells, and can therefore be used for selection in either bacteria or animals. By attaching the

Table 10.1 Commonly used endogenous selectable marker genes. Most of these markers are involved in the redundant endogenous nucleotide biosynthetic pathways (Fig. 10.1). They can also be used as counterselectable markers. For example, negative selection for TK activity is achieved using toxic thymidine analogs (e.g. **5-bromodeoxyuridine**, **ganciclovir**), which are incorporated into DNA only if there is TK activity in the cell.

Marker	Product	Selection	References
Ada	Adenosine deaminase	**Xyl-A** (9-β-D-xylofuranosyl adenosine) and 2′-deoxycoformycin	Kaufman *et al.* 1986
Aprt	Adenine phosphoribosyltransferase	Adenine plus **azaserine**, to block *de novo* dATP synthesis	Lester *et al.* 1980
Cad	Multifunctional enzyme	**PALA** (*N*-phosphonacetyl-L-aspartate) inhibits the aspartate transcarbamylase activity of CAD*	De Saint-Vincent *et al.* 1981
Hprt	Hypoxanthine-guanine phosphoribosyltransferase	Hypoxanthine and **aminopterin**, to block *de novo* IMP synthesis. Selected on HAT medium	Lester *et al.* 1980
Tk	Thymidine kinase	Thymidine and **aminopterin** to block *de novo* dTTP synthesis. Selected on HAT medium	Wigler *et al.* 1977

*CAD: carbamyl phosphate synthetase/aspartate transcarbamylase/dihydroorotase.

selectable marker to the simian virus 40 (SV40) early promoter (Berg 1981) or the HSV *Tk* promoter (Colbère-Garapin *et al.* 1981), both of which function in many cell types, neomycin phosphotransferase was shown to confer antibiotic resistance in a variety of non-mutant cell lines. This marker continues to be used in many contemporary expression vectors. The power of aminoglycoside antibiotic resistance as a selective system in eukaryotes is now very evident. It also has application in yeast (Chapter 9) and plants (Chapter 12). Other commonly used dominant selectable markers are listed in Table 10.2.

Selectable markers and transgene amplification

If animal cells are exposed to toxic concentrations of certain drugs, rare individual cells can survive because they have spontaneously undergone a mutation that confers resistance to that drug. The first such compound to be investigated in this manner was the folic acid analog methotrexate, which is a competitive inhibitor of the enzyme dihydrofolate reductase (DHFR). The analysis of surviving cells showed that some had undergone point mutations in the *Dhfr* locus, producing an enzyme with resistance to the inhibitor. Others had undergone muta-

tions in other loci, for example preventing the uptake of the drug. The most interesting group comprised those cells that had survived by amplifying the *Dhfr* locus, therefore providing enough enzyme to outcompete the inhibitor (Schimke *et al.* 1978). This type of amplification mutation is useful because the drug dose can be progressively increased, resulting in the stepwise selection of cells with massively amplified *Dhfr* gene arrays. Such cells can survive in media containing methotrexate concentrations up to 10 000 times higher than the nominal dose lethal to wild-type cells. The amplified loci are often maintained within the chromosome as extended *homogeneously staining regions* or alternatively as small extra chromosomes called *double minutes*.

Cells with high copy numbers of the *Dhfr* locus are not generated in response to methotrexate exposure, but arise randomly in the population and are selected on the basis of their resistance to the drug. The random nature of the amplification is confirmed by the fact that, as well as the *Dhfr* gene itself, extensive regions of flanking DNA are also amplified, even though they confer no advantage on the cell. This phenomenon can be exploited to co-amplify transgenes introduced along with a *Dhfr* marker gene, resulting in high-level expression. Wigler *et al.*

Table 10.2 Commonly used dominant selectable marker genes in animals (see Box 12.2 for selectable markers used in plants).

Marker	Product (and source)	Principles of selection	References
as	Asparagine synthase (*Escherichia coli*)	Toxic glutamine analog **albizziin**	Andrulis & Siminovitch 1981
ble	Glycopeptide-binding protein (*Streptoalloteichus hindustantus*)	Confers resistance to glycopeptide antibiotics **bleomycin, pheomycin, Zeocin**™	Genilloud *et al.* 1984
bsd	Blasticidin deaminase (*Aspergillus terreus*)	Confers resistance to basticidin S	Izumi *et al.* 1991
gpt	Guanine-xanthine phosphoribosyltransferase (*E. coli*)	Analogous to *Hprt* in mammals, but possesses additional xanthine phosphoribosyltransferase activity, allowing survival in medium containing **aminopterin** and **mycophenolic acid** (Fig 10.1)	Mulligan & Berg 1981b
hisD	Histidinol dehydrogenase (*Salmonella typhimurium*)	Confers resistance to **histidinol**	Mantei *et al.* 1979
hpt	Hygromycin phosphotransferase (*E. coli*)	Confers resistance to **hygromycin-B**	Blochlinger & Diggelmann 1984
neo (nptII)	Neomycin phosphotransferase (*E. coli*)	Confers resistance to **aminoglycoside antibiotics** (e.g. neomycin, kanamycin, G418)	Colbère-Garapin *et al.* 1981
pac	Puromycin N-acetyltransferase (*Streptomyces alboniger*)	Confers resistance to **puromycin**	Vara *et al.* 1986
trpB	Tryptophan synthesis (*E. coli*)	Confers resistance to **indole**	Hartman & Mulligan 1988

(1980) demonstrated this principle by transfecting methotrexate-sensitive mouse cells with genomic DNA from the methotrexate-resistant cell line A29, which contains multiple copies of an altered *Dhfr* gene. They linearized the A29 genomic DNA with the restriction enzyme *Sal*I, and ligated it to *Sal*I-linearized pBR322 DNA prior to transfection. Following stepwise drug selection of the transformed cells, Southern-blot hybridization showed that the amount of pBR322 DNA had increased more than 50-fold.

Methotrexate selection has been used for the large-scale expression of many recombinant proteins, including tissue plasminogen activator (Kaufman *et al.* 1985), hepatitis B surface antigen (HBSAg) (Patzer *et al.* 1986) and human factor VIII (Kaufman *et al.* 1988). CHO cells are preferred as hosts for this expression system because of the availability of a number of *dhfr⁻* mutants (Urlaub *et al.* 1983). In non-mutant cell lines, non-transformed cells can survive selection by amplifying the endogenous *Dhfr* genes, generating a background of 'false positives'. Alternative strategies have been developed that allow DHFR selection to be used in wild-type cells. Expressing the *Dhfr* marker using a strong constitutive promoter (Murray *et al.* 1983) or using a methotrexate-resistant allele of the mouse gene (Simonsen & Levinson 1983) allows selection at methotrexate concentrations much higher than wild-type cells can tolerate. The *E. coli dhfr* gene is also naturally resistant to methotrexate, although for this reason it cannot be used for amplifiable selection (O'Hare *et al.* 1981). Another useful strategy is to employ a second marker gene, such as *neo*, allowing non-transformed cells to be eliminated using G418 (Kim & Wold 1985). Although *dhfr* is the most widely used amplifiable marker, many others have been evaluated, as shown in Table 10.3.

Table 10.3 Common markers used for *in situ* gene amplification. Many amplifiable markers can also be used as endogenous or dominant selectable markers, but, in some cases, the drug used for amplification may not be the same as that used for standard selection.

Marker	Product	Amplifying selective drug	References
Ada	Adenosine deaminase	Deoxycoformycin	Kaufman *et al.* 1986
as	Asparagine synthase	β-Aspartylhydroxamate	Cartier *et al.* 1987
Cad	Aspartate transcarbamylase	*N*-Phosphonacetyl-L-aspartate	Wahl *et al.* 1984
Dhfr	Dihydrofolate reductase	Methotrexate	Kaufman *et al.* 1985
gpt	Xanthine-guanine phosphoribosyltransferase	Mycophenolic acid	Chapman *et al.* 1983
GS	Glutamine synthase	Methionine sulphoxamine	Cockett *et al.* 1990
Hprt	Hypoxanthine-guanine phosphoribosyltransferase	Aminopterin	Kanalas & Suttle 1984
Impdh	Inosine monophosphate dehydrogenase	Mycophenolic acid	Collart & Huberman 1987
Mt-1	Metallothionein 1	Cd^{2+}	Beach & Palmiter 1981
M^{res}	Multidrug resistance: P-glycoprotein 170 gene	Adriamycin, colchicine, others	Kane *et al.* 1988
Odc	Ornithine decarboxylase	Difluoromethylornithine	Chiang & McConlogue 1988
Tk	Thymidine kinase	Aminopterin	Roberts & Axel 1982
Umps	Uridine monophosphate synthases	Pyrazofurin	Kanalas & Suttle 1984

Plasmid vectors for DNA-mediated gene transfer

Stable transformation by integration can be achieved using any source of DNA. The early gene-transfer experiments discussed above were carried out using complex DNA mixtures, e.g. genomic DNA, bacterial plasmids and phage. Calcium phosphate transfection was used in most of these experiments, and the specific donor DNA was often bulked up with a non-specific carrier, such as cleaved salmon-sperm DNA. However, it is generally more beneficial to use a purified source of the donor transgene. This principle was originally demonstrated by Wigler *et al.* (1977), who transfected cultured mouse cells with a homogeneous preparation of the HSV *Tk* gene. Later, this gene was cloned in *E. coli* plasmids to provide a more convenient source. The use of plasmid vectors for transfection provides numerous other advantages, depending on the modular elements included on the plasmid backbone.

1 The convenience of bacterial plasmid vectors can be extended to animal cells, in terms of the ease of subcloning, *in vitro* manipulation and purification of recombinant proteins (p. 76).

2 More importantly, modular elements can be included to drive transgene expression, and these can be used with any transgene of interest. The pSV and pRSV plasmids are examples of early expression vectors for use in animal cells. As discussed in Box 10.2, transcriptional control sequences from SV40 and Rous sarcoma virus are functional in a wide range of cell types. The incorporation of these sequences into pBR322 generated convenient expression vectors in which any transgene could be controlled by these promoters when integrated into the genome of a transfected cell (Fig. 10.2).

3 The inclusion of a selectable marker gene obviates the need for cotransformation, since the transgene and marker remain linked when they cointegrate into the recipient cell's genome. A range of pSV and pRSV vectors were developed containing alternative selectable marker genes, e.g. pSV2-neo (Southern & Berg 1982), pSV2-gpt (Mulligan & Berg 1980) and pSV2-dhfr (Subramani *et al.* 1981).

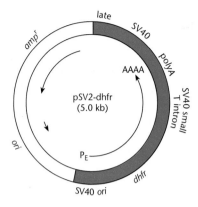

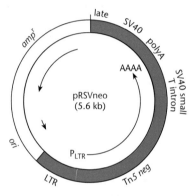

Fig. 10.2 The mammalian plasmid expression vectors pSV2-dhfr and pRSV-neo.

4 As discussed below, some plasmid vectors for gene transfer to animal cells are designed to be shuttle vectors, i.e. they contain origins of replication functional in animal cells, allowing the vector to be maintained as an episomal replicon.

Non-replicating plasmid vectors for transient transformation

One application in which the use of plasmid vectors is critical is transient transformation. Here, the goal is to exploit the short-term persistence of extrachromosomal DNA. Such experiments have a variety of uses, including transient assays of gene expression and the recovery of moderate amounts of recombinant protein. Generally, transient transformation is used as a test system, e.g. to assay regulatory elements using reporter genes (Box 10.1), to check the correct function of an expression construct before going to the expense of generating stable cell lines or to recover moderate amounts of recombinant protein for verification purposes. Transient transformation is particularly useful for testing large numbers of alternative constructs in parallel. No regime of selection is required because stable cell lines are not recovered – the cells are generally transfected, assayed after 1 or 2 days and then discarded. The simplest way to achieve the transient transformation of animal cells is to use a plasmid

---

Box 10.1 Reporter genes and promoter analysis

Reporter genes are also known as screenable marker genes. These differ from selectable marker genes in that they do not confer a property that allows transformed cells to survive under selective conditions. Instead, they encode a product that can be detected using a simple and inexpensive assay.

When controlled by a strong constitutive promoter, reporter genes are often used as markers to confirm transient or stable transformation, since only cells containing the reporter-gene construct can express the corresponding protein. Importantly, the assays used to detect reporter-gene activity are *quantitative*,

so they can also be used to *measure* transformation efficiency. If attached to a cloned promoter, reporter genes can therefore be used to determine transcriptional activity in different cell types and under different conditions. Transient reporter assays have been widely used to characterize and dissect the regulatory elements driving eukaryotic genes, as shown in the example below. The use of reporters is advantageous, because it circumvents the necessity to derive different assays for individual genes and also allows the activities of transgenes and homologous endogenous genes to be distinguished in the same cell.

continued

Box 10.1 *continued*

An example of *in vitro* promoter analysis using chloramphenicol acetyltransferase

The first reporter gene to be used in animal cells was *cat*, derived from *E. coli* transposon Tn*9* (Gorman *et al*. 1982b); it has also been used to a certain extent in plants (Herrera-Estrella *et al*. 1983a,b). This gene encodes the enzyme chloramphenicol acetyltransferase (CAT), which confers resistance to the antibiotic chloramphenicol by transferring acetyl groups on to the chloramphenicol molecule from acetyl-CoA. If ^{14}C-labelled chloramphenicol is used as the substrate, CAT activity produces a mixture of labelled monoacetylated and diacetylated forms, which can be separated by thin-layer chromatography and detected by autoradiography. The higher the CAT activity, the more acetylated forms of chloramphenicol are produced. These can be quantified in a scintillation counter or using a phosphorimager. Gorman *et al*. (1982a,b) placed the *cat* gene downstream of the SV40 and Rous sarcoma virus (RSV) promoters in the expression vectors pSV2 and pRSV2, to create the pSV-CAT and pRSV-CAT constructs, respectively. These vectors, and derivatives thereof, have been widely used to analyse transient-transfection efficiency, because the promoters are active in many animal cells and CAT activity can be assayed rapidly in cell homogenates.

The *cat* gene has also been used to test regulatory elements by attaching it to a 'minimal promoter', typically a simple TATA box. This basic construct generates only low-level background transcription in transiently transfected cells. The activity of other regulatory elements, such as promoters and enhancers, and *response elements*, which activate transcription in response to external signals, can be tested by subcloning them upstream of the minimal promoter and testing their activity in appropriate cell types. In an early example of this type of experiment, Walker *et al*. (1986) attached the promoter and 5'-flanking sequences of human and rat insulin and rat chymotrypsin genes, which are expressed at a high level only in the pancreas, to the *cat* gene. Each gene is expressed in clearly distinct cell types: insulin is synthesized in endocrine β cells and chymotrypsin in exocrine cells. Plasmid DNA was introduced into either pancreatic endocrine or pancreatic exocrine cell lines in culture and, after a subsequent 44 h incubation, cell extracts were assayed for CAT activity. It was found that the constructs retained their preferential expression in the appropriate cell type. The insulin 5'-flanking DNA conferred a high level of CAT expression in the endocrine but not the exocrine cell line, with the converse being the case for the chymotrypsin 5'-flanking DNA. The analysis was extended by creating deletions in the 5'-flanking sequences and testing their effects on expression. From such experiments it could be concluded that sequences located 150–300 bp upstream of the transcription start are essential for appropriate cell-specific transcription.

Other reporter genes

The *cat* reporter gene has been widely used for *in vitro* assays but has a generally low sensitivity and is dependent on a rather cumbersome isotopic assay format. An alternative reporter gene, *SeAP* (secreted alkaline phosphatase), has been useful in many cases because various sensitive colourimetric, fluorometric and chemiluminescent assay formats are available. Also, since the reporter protein is secreted, it can be assayed in the growth medium, so there is no need to kill the cells (Berger *et al*. 1988, Cullen & Malim 1992). The bacterial genes *lacZ* and *gusA* have been used as reporters for *in vitro* assays using colourimetric and fluorometric substrates. An important advantage of these markers is that they can also be used for *in situ* assays, since histological assay formats are available. More recently, bioluminescent markers, such as luciferase and green fluorescent protein, have become popular because these can be assayed in live cells and whole animals and plants. The *lacZ*, *gusA*, luciferase and green fluorescent protein reporter genes are discussed in more detail in Box 13.1.

vector lacking an origin of replication functional in the host. Although the vector cannot replicate, gene expression from a mammalian transcription unit is possible for as long as the plasmid remains stable, which depends on the host cell's propensity to break down extrachromosomal DNA. Linear DNA is degraded very quickly in mammalian cells, so high-quality supercoiled plasmid vectors are used. Even covalently closed circular DNA tends to remain stable for only 1 or 2 days in most animal cells, but this is sufficient for the various transient-expression assays. Some cell types, however, are renowned for their ability to maintain exogenous DNA for longer periods. In the human embryonic kidney cell line 293, for example, supercoiled plasmid DNA can remain stable for up to 80 h (Gorman *et al.* 1990).

Transformation with replicon vectors

Runaway polyomavirus replicons

Transient transformation can also be achieved using replicon vectors that contain origins of replication derived from certain viruses of the polyomavirus family, such as SV40 and the murine polyomavirus. These viruses cause lytic infections, i.e. the viral genome replicates to a very high copy number, resulting in cell lysis and the release of thousands of progeny virions. During the infection cycle, viral gene products accumulate at high levels, so there has been considerable interest in exploiting this strategy to produce recombinant proteins. SV40 was the first animal virus to be characterized in detail at the molecular level, and for this reason it was also the first to be developed as a vector. The productive host range of the virus is limited to certain simian cells. The development of SV40 replicon vectors is discussed in some detail below, but first it is necessary to understand a little of the molecular biology of the virus itself.

SV40 has a small icosahedral capsid and a circular double-stranded DNA genome of approximately 5 kb. The genome has two transcription units, known as the early and late regions, which face in opposite directions. Both transcripts produce multiple products by alternative splicing (Fig. 10.3). The early region produces regulatory proteins, while the late region produces components of the viral capsid.

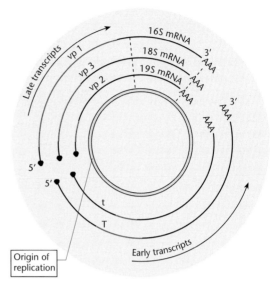

Fig. 10.3 Transcripts and transcript processing of SV40. Intron sequences which are spliced out of the transcripts are shown by red lines.

Transcription is controlled by a complex regulatory element located between the early and late regions, and this includes early and late promoters, an enhancer and the origin of replication. During the first stage of the SV40 infection cycle, the early transcript produces two proteins, known as the large T and small t tumour antigens. The function of the T antigen is particularly important, as this protein binds to the viral origin of replication and is absolutely required for genome replication. All vectors based on SV40 must therefore be supplied with functional T antigen, or they cannot replicate, even in permissive cells. The T antigen also acts as an oncoprotein, interacting with the host's cell-cycle machinery and causing uncontrolled cell proliferation.

The first SV40 vectors were viral vectors and were used to introduce foreign genes into animal cells by transduction. The small size of the viral genome made *in vitro* manipulation straightforward. Either the late region (Goff & Berg 1976) or the early region (Gething & Sambrook 1981) could be replaced with foreign DNA. However, since both these regions are essential for the infection cycle, their functions had to be provided *in trans* initially by a co-introduced helper virus. The use of early replacement vectors was considerably simplified by the development of the COS cell line, a derivative of the African green-monkey cell line CV-1 containing an integrated partial copy of the SV40 genome. The integrated fragment included the entire T-antigen coding sequence and provided this protein *in trans* to any SV40 recombinant in which the early region had been replaced with foreign DNA (Gluzman 1981). For example, using this system, Gething and Sambrook (1981) made recombinant viruses that expressed influenza-virus haemagglutinin in COS cells.

The major problem with these initial SV40 vectors was that the capacity of the viral capsid allowed a maximum of only about 2.5 kb of foreign DNA to be incorporated. The discovery that plasmids carrying the SV40 origin of replication behaved in the same manner as the virus itself, i.e. replicating to a high copy number in permissive monkey cells, was a significant breakthrough. Since these SV40 replicons were not packaged into viral capsids, there were no size constraints on the foreign DNA. Many laboratories developed plasmid vectors on this principle (Myers & Tjian 1980; Lusky & Botchan 1981; Mellon *et al.* 1981). In general, these vectors consisted of a small SV40 DNA fragment (containing the origin of replication) cloned in an *E. coli* plasmid vector. Some vectors also contained a T-antigen coding region and could be used in any permissive cell line, while others contained the origin alone and could only replicate in COS cells. Permanent cell lines are not established when SV40 replicons are transfected into COS cells, because the massive vector replication eventually causes cell death. However, even though only a low proportion of cells are transfected, the high copy number (10^5 genomes per cell) is compensatory, allowing the transient expression of cloned genes and the harvesting of large amounts of recombinant protein.

Similar vectors have been designed that incorporate the murine-polyomavirus origin, which is functional in mouse cells. These vectors allow transient, high-level recombinant-protein expression in permissive mouse cells, such as MOP-8 (which contains an integrated polyomavirus genome and supplies T antigen *in trans* (Muller *et al.* 1984)) and WOP (which is latently infected with the virus). A series of vectors currently marketed by Invitrogen are particularly versatile, because they contain both the SV40 and the murine-polyomavirus origins,

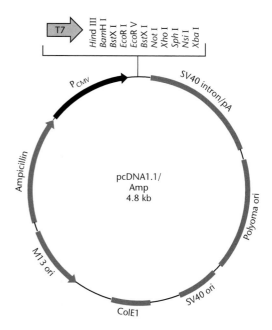

Fig. 10.4 The transient expression vector pcDNA1.1/Amp, which has origins of replication for both SV40 and murine polyomavirus. (Reproduced with permission from Invitrogen.)

allowing high-level protein expression in cells that are permissive for either virus. For example, the vector pcDNA1.1/Amp contains the SV40 and polyoma origins, a transcription unit comprising the human cytomegalovirus promoter and SV40 intron/polyadenylaton site, an interstitial polylinker to insert the transgene and the ampicillin-resistance marker for selection in *E. coli* (Fig. 10.4).

Stable transformation with BK and BPV replicons

As an alternative to integration, stable transformation can be achieved using a recombinant vector that is maintained as an episomal replicon. While viruses such as SV40 use a lytic infection strategy, others, such as human BK virus and Epstein–Barr virus (EBV), cause latent infections, where the viral genome is maintained as a low- to moderate-copy-number replicon that does not interfere with host cell growth. Plasmids that contain such latent origins behave in a similar manner to the parental virus, except they are not packaged in a viral capsid. Such vectors are advantageous, because the DNA does not need to integrate in order to be stably maintained, thus stable transformation occurs with an efficiency equal to that normally achieved with transient transformation. Furthermore, while the expression of integrated transgenes is often affected by the surrounding DNA, a problem known as the position effect (see Box 11.1), these episomal vectors are not subject to such influences.

The human BK polyomavirus infects many cell types and is maintained with a copy number of about 500 genomes per cell. Plasmid vectors containing the BK origin replicate in the same manner as the virus when the BK T antigen is provided *in trans*. However, by incorporating a selectable marker such as *neo* into the vector, it is possible to progressively increase the concentration of antibiotic in the medium and select for cells with higher vector copy number. Cell lines have been propagated for over a year under such conditions (De Benedetti & Rhoads 1991) and stable lines with up to 9000 copies of the genome have been produced. Any transgene incorporated into the vector is expressed at very high levels, leading to the recovery of large amounts of recombinant protein (Sabbioni *et al.* 1995). SV40 itself can also be used as a stable episomal vector if the availability of T antigen is rationed to prevent runaway replication. This has been achieved using a conditional promoter and by mutating the coding region to render the protein temperature-sensitive (Gerard & Gluzman 1985, Rio *et al.* 1985).

The first virus to be developed as an episomal replicon was bovine papillomavirus (BPV). The papillomaviruses are distantly related to the polyomaviruses and cause papillomas (warts) in a range of mammals. BPV has been exploited as a stable expression vector because it can infect mouse cells without yielding progeny virions. Instead, the viral genome is maintained as an episomal replicon, with a copy number of about 100. The molecular biology of BPV is considerably more complex that that of SV40, but the early part of the infection cycle is similar, involving the production of a T antigen that: (i) is required for viral replication; and (ii) causes oncogenic transformation of the host cell. The early functions of BPV are carried on a 5.5 kb section of the genome, which is called the 69% transforming fragment (BPV_{69T}). This was cloned in the *E. coli* plasmid pBR322 and was shown to be sufficient to

establish and maintain episomal replication, as well as inducing cell proliferation (Sarver *et al.* 1981a). Initially, the ability of the virus to cause uncontrolled cell proliferation was used to identify transformed cells, but this limited the range of cell types that could be used. The incorporation of selectable markers, such as *neo*, allows transformants to be selected for resistance to G418, permitting the use of a wider range of cell types (Law *et al.* 1983). BPV_{69T} replicons have been used to express numerous proteins, including rat preproinsulin (Sarver *et al.* 1981b) and human β-globin (Di Maio *et al.* 1982). Generally, the plasmids are maintained episomally, but the copy number varies from 10 to over 200 vector molecules per cell. Some investigators have reported the long-term maintenance of such episomal transformants (Fukunaga *et al.* 1984), while others found a tendency for the construct to integrate into the genome (Ostrowski *et al.* 1983, Sambrook *et al.* 1985). Recombination within the vector or between vectors is also a fairly common observation, resulting in unpredictable spontaneous deletions and plasmid oligomerization. Recombination events and copy number appear to be affected by multiple factors, including the host cell type, the incorporated transgene and the structure of the vector itself (see Mecsas & Sugden 1987).

Stable transformation with EBV replicons

Unlike BPV replicons, vectors based on EBV replicate very stably in mammalian cells. EBV is a herpesvirus (see also discussion of herpes simplex virus later in the chapter), with a large double-stranded DNA genome (approximately 170 kb), which predominantly infects primate and canine cells. It is also naturally lymphotrophic, infecting B cells in humans and causing infectious mononucleosis. In cultured lymphocytes, the virus becomes established as an episomal replicon with about 1000 copies per cell (Miller 1985). Although the virus itself only infects lymphocytes, the genome is maintained in a wide range of primate cells if introduced by transfection. Only two relatively small regions of the genome are required for episomal maintenance – the latent origin (*oriP*) and a gene encoding a *trans*-acting regulator called Epstein–Barr nuclear antigen 1 (EBNA1) (Yates *et al.* 1984, Reisman *et al.* 1985).

These sequences have formed the basis of a series of latent EBV-based plasmid expression vectors, which are maintained at a copy number of two to 50 copies per cell. The first EBV vectors comprised the *oriP* element cloned in a bacterial plasmid and could replicate only if EBNA1 was supplied *in trans*. Yates *et al.* (1984) described the construction of a shuttle vector, pBamC, comprising *oriP*, a bacterial origin and ampicillin-resistance gene derived from pBR322, and the neomycin phosphotransferase marker for selection in animal cells. A derivative vector, pHEBO, contained a hygromycin-resistance marker, instead of *neo* (Sugden *et al.* 1985). The presence of a selectable marker in such vectors is important, because, if selection pressure is not applied, EBV replicons are lost passively from the cell population at about 5% per cell generation. Yates *et al.* (1985) added the EBNA1 gene to pHEBO, producing a construct called p201 that was capable of replicating independently. Similar constructs have been developed in other laboratories (Lupton & Levine 1985). EBV replicons have been used to express a wide range of proteins in mammalian cell lines, including the epidermal growth-factor receptor (Young *et al.* 1988), the tumour-necrosis-factor receptor (Heller *et al.* 1990) and an Na$^+$K$^+$ ATPase (Canfield *et al.* 1990). Generally, such studies have resulted in high-level and long-term gene expression (reviewed by Margolskee 1992, Sclimenti & Calos 1998). It has been suggested that the stability of EBV replicons may reflect the fact that replication is limited to once per cell cycle, unlike BPV replicons, which replicate continuously throughout the S phase (Gilbert & Cohen 1987, DuBridge & Calos 1988). This, together with the large genome size, allows EBV-derived vectors to carry large DNA fragments, including mammalian complementary DNAs (cDNAs) and genes. EBV replicons have therefore been used for the construction of episomal cDNA libraries for expression cloning (Margolskee *et al.* 1988) and, more recently, for the preparation of genomic libraries (Sun *et al.* 1994). The *oriP* element has also been incorporated into YAC vectors carrying large human genomic DNA fragments. These linear vectors were then circularized *in vitro* and introduced into human cells expressing EBNA1, whereupon they were maintained as episomal replicons (Simpson *et al.* 1996). EBV can also replicate lytic-ally in primate B cells, and this requires a separate origin (*oriLyt*) and a distinct *trans*-acting regulator called ZEBRA. Vectors have been developed containing both origins, and these are maintained at a moderate copy number in cells latently infected with the virus, but can be amplified up to 400-fold if ZEBRA is supplied by transfecting the cells with the corresponding viral gene under the control of a constitutive promoter (Hammerschmidt & Sugden 1988). Interestingly, while the latent vector remains as a circular replicon, vector DNA isolated from induced cells is present as multicopy concatemers, reflecting the rolling-circle mechanism of lytic replication.

Gene transfer by viral transduction

General principles of viral vectors

This section describes the use of viruses as vectors for transduction, i.e. the introduction of genes into animal cells by exploiting the natural ability of the virus particle, within which the transgene is packaged, to adsorb to the surface of the cell and gain entry. Due to the efficiency with which viruses can deliver their nucleic acid into cells and the high levels of replication and gene expression it is possible to achieve, viruses have been used as vectors not only for gene expression in cultured cells but also for gene transfer to living animals. Four classes of viral vector have been developed for use in human gene therapy and have reached phase 1 clinical trials. These are the retrovirus, adenovirus, herpesvirus and adeno-associated virus (AAV) vectors (Robbins *et al.* 1998).

Before introducing the individual vector systems, we discuss some general properties of viral transduction vectors. Transgenes may be incorporated into viral vectors either by addition to the whole genome or by replacing one or more viral genes. This is generally achieved either by ligation (many viruses have been modified to incorporate unique restriction sites) or homologous recombination. If the transgene is added to the genome or if it replaces one or more genes that are non-essential for the infection cycle in the expression host being used, the vector is described as *replication-competent* or *helper-independent*, because it can propagate independently. However, if the transgene replaces an essential viral gene, this renders the vector *replication-defective* or *helper-dependent*, so that missing functions must be supplied *in trans*. This can be accomplished by co-introducing a helper virus or transfecting the cells with a helper plasmid, each of which must carry the missing genes. Usually steps are taken to prevent the helper virus completing its own infection cycle, so that only the recombinant vector is packaged. It is also desirable to try and prevent recombination occurring between the helper and the vector, as this can generate wild-type replication-competent viruses as contaminants. An alternative to the co-introduction of helpers is to use a *complementary cell line*, which is transformed with the appropriate genes. These are sometimes termed 'packaging lines'. For many applications, it is favourable to use vectors from which all viral coding sequences have been deleted. These *amplicons* (also described as '*gutless vectors*') contain just the *cis*-acting elements required for packaging and genome replication. The advantage of such vectors is their high capacity for foreign DNA and the fact that, since no viral gene products are made, the vector has no intrinsic cytotoxic effects. The choice of vector depends on the particular properties of the virus and the intended host, whether transient or stable expression is required and how much DNA needs to be packaged. For example, icosahedral viruses such as adenoviruses and retroviruses package their genomes into preformed capsids, whose volume defines the maximum amount of foreign DNA that can be accommodated. Conversely, rod-shaped viruses such as the baculoviruses form the capsid around the genome, so there are no such size constraints. There is no ideal virus for gene transfer – each has its own advantages and disadvantages. However, in recent years, a number

of hybrid viral vectors have been developed incorporating the beneficial features of two or more viruses. The interested reader can consult recent reviews on this subject (Robbins *et al.* 1998, Reynolds *et al.* 1999). Also note that most of the principles discussed above also apply to the use of plant viruses as vectors (Chapter 12).

Adenovirus

Adenoviruses are DNA viruses with a linear, double-stranded genome of approximately 36 kb. The genome of serotype Ad5, from which many adenovirus vectors are derived, is shown in Fig. 10.5. There are six early-transcription units, most of which are essential for viral replication, and a major late transcript that encodes components of the capsid. Adenoviruses have been widely used as gene transfer and expression vectors, because they have many advantageous features, including stability, a high capacity for foreign DNA, a wide host range that includes non-dividing cells, and the ability to produce high-titre stocks (up to 10^{11} plaque-forming units (pfu)/ml) (reviewed by Berkner 1992). They are suitable for transient expression in dividing cells because they do not integrate efficiently into the genome, but prolonged expression can be achieved in post-mitotic cells, such as neurons (e.g. see Davidson *et al.* 1993, LaSalle *et al.* 1993). Adenoviruses are particularly attractive as gene therapy vectors, because the virions are taken up efficiently by cells *in vivo* and adenovirus-derived vaccines have been used in humans with no reported side-effects. However, the recent death of a patient following an extreme inflammatory response to adenoviral

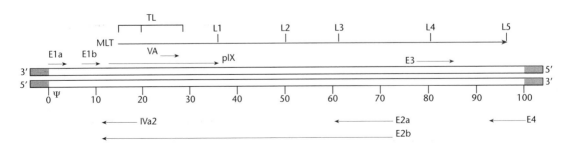

Fig. 10.5 Map of the adenovirus genome, showing the positions of the early transcription units (E), the major late transcript (MLT), the tripartite leader (TL) and other genes (VA, pIX, IVa2). Terminal repeats are shown in pink, ψ is the packaging site.

gene-therapy treatment underlines the necessity for rigorous safety testing (Marshall 1999). A number of strategies are being developed to control the activity of the immune system when the vectors are first introduced into the host (reviewed by Benihoud *et al.* 1999).

Most early adenoviral vectors were replication-deficient, lacking the essential E1a and E1b genes and often the non-essential gene E3. These first-generation 'E1 replacement vectors' had a maximum capacity of about 7 kb and were propagated in the human embryonic kidney line 293. This is transformed with the leftmost 11% of the adenoviral genome, comprising the E1 transcription unit, and hence supplies these functions *in trans* (Graham *et al.* 1977). Although these vectors have been used with great success, they suffer from two particular problems: cytotoxic effects, resulting from low-level expression of the viral gene products, and the tendency for recombination to occur between the vector and the integrated portion of the genome, resulting in the recovery of replication-competent viruses. Higher-capacity vectors have been developed, which lack the E2 or E4 regions in addition to E1 and E3, providing a maximum cloning capacity of about 10 kb. These must be propagated on complementary cell lines providing multiple functions, and such cell lines have been developed in several laboratories (e.g. Brough *et al.* 1996, Gao *et al.* 1996, Gorziglia *et al.* 1996, Zhou *et al.* 1996). The use of E1/E4 deletions is particularly attractive, as the E4 gene is responsible for many of the immunological effects of the virus (Gao *et al.* 1996, Dedieu *et al.* 1997). Unwanted recombination has been addressed through the use of a refined complementary cell line transformed with a specific DNA fragment corresponding exactly to the E1 genes (Imler *et al.* 1996). An alternative strategy is to insert a large fragment of 'stuffer DNA' into the non-essential E3 gene, so that recombination yields a genome too large to be packaged (Bett *et al.* 1993). Gutless adenoviral vectors are favoured for *in vivo* gene transfer, because they have a large capacity (up to 37 kb) and minimal cytotoxic effects (reviewed by Morsey & Caskey 1999). Therefore, transgene expression persists for longer than can be achieved using first-generation vectors (e.g. see Schneider *et al.* 1998). Complementary cell lines supplying all

adenoviral functions are not available at present, so gutless vectors must be packaged in the presence of a helper virus, which presents a risk of contamination.

Adeno-associated virus

AAV is not related to adenovirus, but is so called because it was first discovered as a contaminant in an adenoviral isolate (Atchison *et al.* 1965). AAV is a single-stranded DNA virus, a member of the parvovirus family, and is naturally replication-defective, such that it requires the presence of another virus (usually adenovirus or herpesvirus) to complete its infection cycle. In adenovirus- or herpesvirus-infected cells, AAV replicates lytically and produces thousands of progeny virions (Buller *et al.* 1981). However, in the absence of these helpers, the AAV DNA integrates into the host cell's genome, where it remains as a latent provirus (Berns *et al.* 1975). In human cells, the provirus integrates predominantly into the same genetic locus on chromosome 19 (Kotin *et al.* 1990). Subsequent infection by adenovirus or herpesvirus can 'rescue' the provirus and induce lytic infection.

The dependence of AAV on a heterologous helper virus provides an unusual degree of control over vector replication, making AAV theoretically one of the safest vectors to use for gene therapy. Proviral integration is considered advantageous for increasing the persistence of transgene expression, while at the same time the site specificity of this process theoretically limits the chances of insertional mutagenesis. Other advantages include the wide host range, which encompasses non-dividing cells (reviewed by Muzyczka 1992, Rabinowitz & Samulski 1998).

The AAV genome is small (about 5 kb) and comprises a central region containing *rep* (replicase) and *cap* (capsid) genes flanked by 145-base inverted terminal repeats (Fig. 10.6). In the first AAV vectors, foreign DNA replaced the *cap* region and was expressed from an endogenous AAV promoter (Hermonat & Muzyczka 1984). Heterologous promoters were also used, although in many cases transgene expression was inefficient because the Rep protein inhibited their activity (reviewed by Muzyczka, 1992). Rep interference with endogenous promoters is also responsible for many of the cytotoxic effects of the virus. Several groups therefore

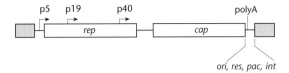

Fig. 10.6 Map of the adeno-associated virus genome, showing the three promoters, the *rep* and *cap* genes, and the *cis*-acting sites required for replication (*ori*), excision (rescue), (*res*), packaging (*pac*) and integration (*int*). Terminal repeats are shown as pink blocks.

developed vectors in which both genes were deleted and the transgene was expressed from either an endogenous or a heterologous promoter (McLaughlin *et al.* 1988, Samulski *et al.* 1989). From such experiments, it was demonstrated that the repeats are the only elements required for replication, transcription, proviral integration and rescue. All current AAV vectors are based on this principle. *In vitro* manipulation of AAV is facilitated by cloning the inverted terminal repeats in a plasmid vector and inserting the transgene between them. Traditionally, recombinant viral stocks are produced by transfecting this construct into cells along with a helper plasmid to supply AAV products, and then infecting the cells with adenovirus to stimulate lytic replication and packaging. This has generally yielded recombinant AAV titres too low to use for human gene therapy and contaminated with helper AAV and adenovirus. The recent development of transfection-based adenoviral helper plasmids, packaging lines and the use of affinity chromatography to isolate AAV virions has helped to alleviate such problems (reviewed by Monahan & Samulski 2000).

AAV vectors have been used to introduce genes efficiently into many cell types, including liver (Snyder *et al.* 1997), muscle (Pruchnic *et al.* 2000) and neurons (Davidson *et al.* 2000). However, deletion of the *rep* region abolishes the site specificity of proviral integration, so the vector integrates at essentially random positions, which may increase the risk of insertional gene inactivation (Weitzman *et al.* 1994, Yang *et al.* 1997, Young *et al.* 2000). It is also unclear whether the persistence of the vector and prolonged transgene expression are due primarily to vector integration or to episomal maintenance of concatemeric double-stranded DNA (dsDNA) copies of the genome (for discussion see Monahan &

Samulski 2000). The fact that AAV uses concatemeric replication intermediates has been used to circumvent perhaps the most serious disadvantage of AAV vectors, which is the limited capacity for foreign DNA. This strategy involves cloning a large cDNA as two segments in two separate vectors, which are co-introduced into the same cell. The 5′ portion of the cDNA is cloned in one vector, downstream of a promoter and upstream of a splice donor site. The 3′ portion is cloned in another vector, downstream of a splice acceptor. Concatemerization results in the formation of heterodimers and transcription across the junction yields an mRNA that can be processed to splice out the terminal repeats of the vector. In this way, cDNAs of up to 10 kb can be expressed (Nakai *et al.* 2000, Sun *et al.* 2000).

Baculovirus

Baculoviruses have rod-shaped capsids and large, dsDNA genomes. They productively infect arthropods, particularly insects. One group of baculoviruses, known as the nuclear polyhedrosis viruses, have an unusual infection cycle that involves the production of *nuclear occlusion bodies*. These are proteinaceous particles in which the virions are embedded, allowing the virus to survive harsh environmental conditions, such as desiccation (reviewed by Fraser 1992). Baculovirus vectors are used mainly for high-level transient protein expression in insects and insect cells (King & Possee 1992, O'Reilley *et al.* 1992). The occlusion bodies are relevant to vector development because they consist predominantly of a single protein, called polyhedrin, which is expressed at very high levels. The nuclear-occlusion stage of the infection cycle is non-essential for the productive infection of cell lines; thus the polyhedrin gene can be replaced with foreign DNA, which can be expressed at high levels under the control of the endogenous polyhedrin promoter. Two baculoviruses have been extensively developed as vectors, namely the *Autographa californica* multiple nuclear polyhedrosis virus (AcMNPV) and the *Bombyx mori* nuclear polyhedrosis virus (BmNPV). The former is used for protein expression in insect cell lines, particularly those derived from *Spodoptera frugiperda* (e.g. Sf9, Sf21). The latter infects the silkworm and has been used for the production of recombinant protein in live silkworm larvae. One limitation of this expression system is that the glycosylation pathway in insects differs from that in mammals, so recombinant mammalian proteins may be incorrectly glycosylated and hence immunogenic (reviewed by Fraser 1992). This has been addressed by using insect cell lines chosen specifically for their ability to carry out mammalian-type post-translational modifications, e.g. those derived from *Estigmene acrea* (Ogonah *et al.* 1996). An innovative approach to this problem is to exploit the indefinite capacity of baculovirus vectors to coexpress multiple transgenes and thus modify the glycosylation process in the host cell line by expressing appropriate glycosylation enzymes along with the transgene of interest. Wagner *et al.* (1996) used this strategy to coexpress fowl-plague haemagglutinin and β-1,2-N-acetylglucosaminyltransferase, resulting in the synthesis of large amounts of haemagglutinin correctly modified with *N*-acetylglucosamine residues. Hollister *et al.* (1998) have developed an Sf9 cell line that expresses 1,4-galactosyltransferase under the control of a baculovirus immediate early promoter, such that gene expression is induced by baculovirus infection. These cells were infected with a recombinant baculovirus vector carrying a tissue plasminogen-activator transgene, resulting in the production of recombinant protein that was correctly galactosylated.

Polyhedrin gene-replacement vectors are the most popular due to the high level of recombinant protein that can be expressed (up to 1 mg/10^6 cells). The polyhedrin upstream promoter and 5′ untranslated region are important for high-level foreigngene expression and these are included in all polyhedrin replacement vectors (Miller *et al.* 1983, Maeda *et al.* 1985). The highest levels of recombinant-protein expression were initially achieved if the transgene was expressed as a fusion, incorporating at least the first 30 amino acids of the polyhedrin protein (Lucklow & Summers 1988). However, this was shown to be due not to the stabilizing effects of the leader sequence on the recombinant protein, but to the presence of regulatory elements that overlapped the translational start site. Mutation of the polyhedrin start codon has allowed these sequences to be incorporated as part of the 5′ untranslated region of the foreign gene cassette, so that native proteins can be expressed (Landford 1988). Replacement of the

polyhedrin gene also provides a convenient method to detect recombinant viruses. The occlusion bodies produced by wild-type viruses cause the microscopic viral plaques to appear opalescent if viewed under an oblique light source (OB⁺), while recombinant plaques appear clear (OB⁻) (Smith *et al.* 1983). Insertion of the *E. coli lacZ* gene in frame into the polyhedrin coding region allows blue–white screening of recombinants in addition to the OB assay (Pennock *et al.* 1984), and many current baculovirus expression systems employ *lacZ* as a screenable marker to identify recombinants. Substitution of *lacZ* with the gene for green fluorescent protein allows the rapid identification of recombinants by exposing the plaques to UV light (Wilson *et al.* 1997).

The construction of baculovirus expression vectors involves inserting the transgene downstream of the polyhedrin promoter. Since the genome is large, this is usually achieved by homologous recombination using a plasmid vector carrying a baculovirus homology region. A problem with homology-based strategies for introducing foreign DNA into large viral genomes is that recombinants are generated at a low efficiency. For baculovirus, recombinant vectors are recovered at a frequency of 0.5–5% of total virus produced. The proportion of recombinants has been increased by using linear derivatives of the wild-type baculovirus genome containing large deletions, which can be repaired only by homologous recombination with the targeting vector. Compatible targeting vectors span the deletion and provide enough flanking homologous DNA to sponsor recombination between the two elements and to generate a viable, recombinant genome. Such approaches result in the production of up to 90% recombinant plaques (Kitts & Possee 1993; Fig. 10.7). Recently, alternative systems have been described in which the baculovirus genome is maintained as a low-copy-number replicon in bacteria or yeast, allowing the powerful genetics of these microbial systems to be exploited. In the bacterial system marketed by Gibco-BRL under the name 'Bac-to-Bac', the baculovirus genome is engineered to contain an origin of replication from the *E. coli* F plasmid. This hybrid replicon, called a *bacmid*, also contains the target site for the transposon Tn*7*, inserted in frame within the *lacZ* gene, which is itself

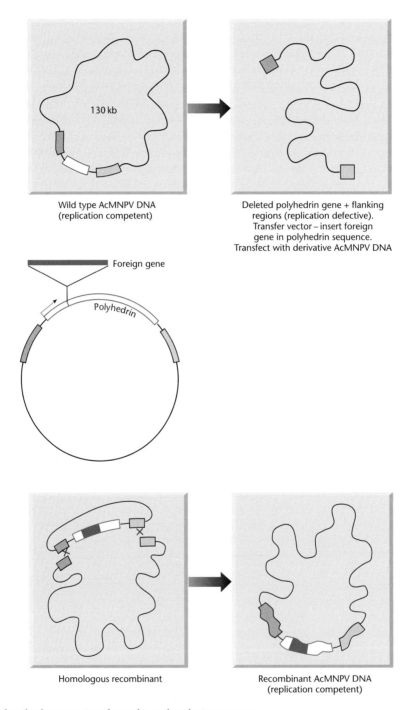

Wild type AcMNPV DNA
(replication competent)

Deleted polyhedrin gene + flanking
regions (replication defective).
Transfer vector – insert foreign
gene in polyhedrin sequence.
Transfect with derivative AcMNPV DNA

Foreign gene

Polyhedrin

Homologous recombinant

Recombinant AcMNPV DNA
(replication competent)

Fig. 10.7 Procedure for the generation of recombinant baculovirus vectors.

downstream of the polyhedrin promoter. The foreign gene is cloned in another plasmid between two Tn7 repeats and introduced into the bacmid-containing bacteria, which also contains a third plasmid expressing Tn7 transposase. Induction of transposase synthesis results in the site-specific transposition of the transgene into the bacmid, generating a recombinant baculovirus genome that can be isolated for transfection into insect cells. Transposition of the transgene into the bacmid interrupts the *lacZ* gene, allowing recombinant bacterial colonies to be identified by blue–white screening.

Although baculoviruses productively infect insect cells, they can also be taken up by mammalian cells, although without producing progeny virions. The number of mammalian cell lines reported to be transduced by baculoviruses is growing (reviewed by Kost & Condreay 1999) and this suggests that recombinant baculoviruses could be developed as vectors for gene therapy. A number of baculovirus-borne transgenes have been expressed using constitutive promoters, such as the cytomegalovirus immediate early promoter (Hoffman *et al.* 1995) and the Rous sarcoma virus promoter (Boyce & Bucher 1996). More recently, cell lines have been generated that are stably transformed with baculovirus vectors, although it is not clear whether the viral genome has integrated in these cells (Condreay *et al.* 1999).

Herpesvirus vectors

The herpesviruses are large dsDNA viruses that include EBV (discussed above) and the HSVs (e.g. HSV-I, varicella zoster). Most HSVs are transmitted without symptoms (varicella zoster virus is exceptional) and cause prolonged infections. Unlike EBV, which is used as a replicon vector, HSV-I has been developed as a transduction vector. Viral replication can occur in many cell types in a wide range of species if the genome is introduced by transfection, but HSV vectors are particularly suitable for gene therapy in the nervous system, because the virus is remarkably neurotropic. As with other large viruses, recombinants can be generated in transfected cells by homologous recombination, and such vectors may be replication-competent or helper-dependent (Marconi *et al.* 1996). Alternatively, plasmid-based

amplicon vectors can be constructed, which carry only those *cis*-acting elements required for replication and packaging. These require packaging systems to provide the missing functions *in trans* (e.g. Stavropoulos & Strathdee 1998). Therapeutic use of herpesvirus vectors has been limited, but a number of genes have been successfully transferred to neurons *in vivo* (e.g. Boviatsis *et al.* 1994, Lawrence *et al.* 1995). Generally, transgene expression is transient, although prolonged expression has been observed in some neuronal populations (see reviews by Vos *et al.* 1996, Simonato *et al.* 2000). Note that HSV is also transmitted across neuronal synapses during lytic infections, a phenomenon that can be exploited to trace axon pathways (Norgren & Lehman 1998).

Retrovirus vectors

Retroviruses are RNA viruses that replicate via a dsDNA intermediate. The infection cycle involves the precise integration of this intermediate into the genome of the host cell, where it is transcribed to yield daughter genomes that are packaged into virions.

Retroviruses have been developed as vectors for a number of reasons (reviewed by Miller 1992). First, certain retroviruses are acutely oncogenic because they carry particular genes that promote host cell division. Investigation of such viruses has shown that these *viral oncogenes* are in fact gain-of-function derivatives of host genes, *proto-oncogenes*, which are normally involved in the regulation of cell growth. In most cases, the viral oncogenes are found to be expressed as fusions with essential viral genes, rendering the virus replication-defective. These *acute transforming retroviruses* therefore demonstrate the natural ability of retroviruses to act as replication-defective gene-transfer vectors. Secondly, most retroviruses do not kill the host, but produce progeny virons over an indefinite period. Retroviral vectors can therefore be used to make stably transformed cell lines. Thirdly, viral gene expression is driven by strong promoters, which can be subverted to control the expression of transgenes. In the case of murine mammary-tumour virus, transcription from the viral promoter is inducible by glucocorticoids, allowing transgenes controlled by this promoter to be switched on and off (Lee *et al.* 1981, Scheidereit *et al.* 1983). Fourthly, some retroviruses, such as amphotropic strains of murine leukaemia virus (MLV), have a broad host range, allowing the transduction of many cell types. Finally, retroviruses make efficient and convenient vectors for gene transfer because the genome is small enough for DNA copies to be manipulated *in vitro* in plasmid cloning vectors, the vectors can be propagated to high titres (up to 10^8 pfu/ml) and the efficiency of infection *in vitro* can approach 100%.

The major disadvantage of oncoretroviral vectors is that they only productively infect dividing cells, which limits their use for gene-therapy applications (Miller *et al.* 1990, Roe *et al.* 1993). However, lentiviruses, such as human immunodeficiency virus (HIV), are more complex retroviruses that have the ability to infect non-dividing cells (Lewis & Emmerman 1994). These were initially developed as vectors for the stable transduction of cells displaying CD4 (Poznansky *et al.* 1991, Shimada *et al.* 1991, Buchschacher & Panganiban 1992, Parolin *et al.* 1994), but recent advances in lentiviral vector design provide improved safety and allow the transduction of multiple cell types (Naldini *et al.* 1996; reviewed by Naldini 1998).

Before discussing the development of retroviral vectors, it is necessary to briefly describe the genome structure and the molecular biology of the infection cycle (for a comprehensive account, see Weiss *et al.* 1985). A typical retroviral genome map is shown in Fig. 10.8. The infection cycle begins when the viral envelope interacts with the host cell's plasma membrane, delivering the particle into the cell. The capsid contains two copies of the RNA genome, as

well as reverse transcriptase/integrase. Thus, immediately after infection, the RNA genome is reverse-transcribed to produce a cDNA copy. This is a complex process involving two template jumps, with the result that the terminal regions of the RNA genome are duplicated in the DNA as long terminal repeats (LTRs). The DNA intermediate then integrates into the genome at an essentially random site (there may be some preference for actively transcribed regions). The integrated provirus has three genes (*gag*, *pol* and *env*). The *gag* gene encodes a viral structural protein, *pol* encodes the reverse transcriptase and integrase and the *env* gene encodes viral envelope proteins. Viral genomic RNA is synthesized by transcription from a single promoter located in the left LTR and ends at a polyadenylation site in the right LTR. Thus, the full-length genomic RNA is shorter than the integrated DNA copy and lacks the duplicated LTR structure. The genomic RNA is capped and polyadenylated, allowing the *gag* gene to be translated (the *pol* gene is also translated by read-through, producing a Gag–Pol fusion protein, which is later processed into several distinct polypeptides). Some of the full-length RNA also undergoes splicing, eliminating the *gag* and *pol* genes and allowing the downstream *env* gene to be translated. Two copies of the full-length RNA genome are incorporated into each capsid, which requires a specific *cis*-acting packaging site termed ψ. The reverse transcriptase/integrase is also packaged.

Strategies for vector construction

Most retroviral vectors are replication-defective, because removal of the viral genes provides the maximum capacity for foreign DNA (about 8 kb).

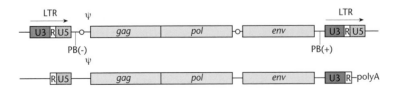

Fig. 10.8 Generic map of an oncoretrovirus genome. Upper figure shows the structure of an integrated provirus, with long terminal repeats (LTRs) comprising three regions U3, R and U5, enclosing the three open reading frames *gag*, *pol* and *env*. Lower figure shows the structure of a packaged RNA genome, which lacks the LTR structure and possesses a poly(A) tail. PB represents primer binding sites in the viral replication cycle, and ψ is the packaging signal. The small circles represent splice sites.

Only the *cis*-acting sites required for replication and packaging are left behind. These include the LTRs (necessary for transcription and polyadenylation of the RNA genome as well as integration), the packaging site ψ, which is upstream of the *gag* gene, and 'primer-binding sites', which are used during the complex replication process. The inclusion of a small portion of the *gag* coding region improves packaging efficiency by up to 10-fold (Bender *et al.* 1987). Deleted vectors can be propagated only in the presence of a replication-competent helper virus or a packaging cell line. The former strategy leads to the contamination of the recombinant vector stock with non-defective helper virus. Conversely, packaging lines can be developed where an integrated provirus provides the helper functions but lacks the *cis*-acting sequences required for packaging (Mann *et al.* 1983). Many different retroviruses have been used to develop packaging lines and, since these determine the type of envelope protein inserted into the virion envelope, they govern the host range of the vector (they are said to *pseudotype* the vector). Packaging lines based on amphotropic MLVs allow retroviral gene transfer to a wide range of species and cell types, including human cells (e.g. see Cone & Mulligan 1984, Danos & Mulligan 1988). It is still possible for recombination to occur between the vector and the integrated helper provirus, resulting in the production of wild-type contaminants. The most advanced 'third-generation' packaging lines limit the extent of homologous sequence between the helper virus and the vector and split up the coding regions so that up to three independent crossover events are required to form a replication-competent virus (e.g. see Markowitz *et al.* 1988).

The simplest strategy for the high-level constitutive expression of single genes in retroviral vectors is to delete all coding sequences and place the foreign gene between the LTR promoter and the viral polyadenylation site. Alternatively, an internal heterologous promoter can be used to drive transgene expression. However, many investigators have reported interference between the heterologous promoter and the LTR promoter (e.g. see Emmerman & Temin 1984, Wu *et al.* 1996). Yu *et al.* (1986) addressed this problem by devising *self-inactivating vectors*, containing deletions in the 3′ LTR, which are copied to the 5′ LTR during vector replication,

thus inactivating the LTR promoter while leaving internal promoters intact. This strategy also helps to alleviate additional problems associated with the LTR promoter: (i) that adjacent endogenous genes may be activated following integration; and (ii) that the entire expression cassette may be inactivated by DNA methylation after a variable period of expression in the target cell (Naviaux & Verma 1992).

Since retroviral vectors are used for the production of stably transformed cell lines, it is necessary to co-introduce a selectable marker gene along with the transgene of interest. The expression of two genes can be achieved by arranging the transgene and marker gene in tandem, each under the control of a separate promoter, one of which may be the LTR promoter. This leads to the production of full-length and subgenomic RNAs from the integrated provirus. Alternatively, if the first gene is flanked by splice sites, only a single promoter is necessary, because the RNA is spliced in a manner reminiscent of the typical retroviral life cycle, allowing translation of the downstream gene (Cepko *et al.* 1984). Vectors in which the downstream gene is controlled by an internal ribosome entry site have also been used (e.g. Dirks *et al.* 1993, Sugimoto *et al.* 1994; see Box 13.3). Since the viral replication cycle involves transcription and splicing, an important consideration for vector design is that the foreign DNA must not contain sequences that interfere with these processes. For example, polyadenylation sites downstream of the transgene should be avoided, as these will cause truncation of the RNA, blocking the replication cycle (Miller *et al.* 1983). Retroviruses also remove any introns contained within the transgene during replication (Shimotohno & Temin 1982).

Sindbis virus and Semliki forest virus (alphaviruses)

The alphaviruses are a family of enveloped viruses with a single-strand positive-sense RNA genome. One of the advantages of using such RNA viruses for gene transfer is that integration into the host genome is guaranteed never to occur. Alphavirus replication takes place in the cytoplasm, and produces a large number of daughter genomes, allowing very high-level expression of any transgene (reviewed by Berglung *et al.* 1996). To date, these are the only animal viruses with a replication cycle based solely on RNA to be extensively developed as expression vectors. Sindbis virus (Xiong *et al.* 1989) and Semliki Forest virus (SFV) (Liljestrom & Garoff 1993) have been the focus of much of this research. These display a broad host range and cell tropism, and mutants have been isolated with reduced cytopathic effects (Agapov *et al.* 1998, Frolov *et al.* 1999; reviewed by Schlesinger & Dubensky 1999).

The wild-type alphavirus comprises two genes: a 5′ gene encoding viral replicase and a 3′ gene encoding a polyprotein from which the capsid structural proteins are autocatalytically derived. Since the genome is made of RNA, it can act immediately as a substrate for protein synthesis. However, because protein synthesis in eukaryotes is dependent on mRNAs possessing a specialized 5′ cap structure, only the replicase gene is initially translated. The replicase protein produces a negative-sense complementary strand, which in turn acts as a template for the production of full-length daughter genomes. However, the negative strand also contains an internal promoter, which allows the synthesis of a subgenomic positive-sense RNA containing the capsid polyprotein gene. This subgenomic RNA is subsequently capped and translated.

A number of different strategies have been used to express recombinant proteins using alphavirus vectors (Berglung *et al.* 1996, Lundstrom 1997). For example, replication-competent vectors have been constructed in which an additional subgenomic promoter is placed either upstream or downstream of the capsid polyprotein gene. If foreign DNA is introduced downstream of this promoter, the replicase protein produces two distinct subgenomic RNAs, one corresponding to the transgene. Such insertion vectors tend to be unstable and have been largely superseded by replacement vectors in which the capsid polyprotein gene is replaced by the transgene. The first 120 bases of both the Sindbis and the SFV structural polyprotein genes include a strong enhancer of protein synthesis, which significantly increases the yield of recombinant protein (Frolov & Schlesinger 1994, Sjoberg *et al.* 1994). This is downstream of the translational initiation site, so in many vectors this enhancer region is included so that the foreign gene is expressed as an N-terminal fusion protein. This can result in extremely high levels of recombinant protein synthesis – up to 50% of the total cellular protein.

Both plasmid replicon and viral transduction vectors have been developed from the alphavirus genome. A versatile Sindbis replicon vector, pSinRep5, is currently marketed by Invitrogen (Fig. 10.9). The vector is a plasmid containing bacterial backbone elements, the Sindbis replicase genes and packaging

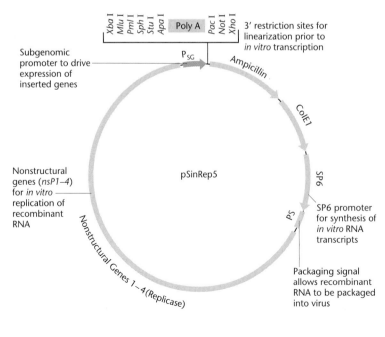

Fig. 10.9 The Sindbis virus-based vector pSinRep5. (Reproduced with permission from Invitrogen.)

site, and an expression cassette featuring a Sindbis subgenomic promoter, a multiple cloning site and a polyadenylation site. There is an SP6 promoter upstream of the replicase genes and expression cassette for generating full-length *in vitro* transcripts. There is a second set of restriction sites downstream from the polylinker, allowing the vector to be linearized prior to *in vitro* transcription. Foreign DNA is cloned in the expression cassette, the vector is linearized and transcribed and the infectious recombinant Sindbis RNA thus produced is transfected into cells, from which recombinant protein can be recovered. An alternative strategy is to place the entire alphavirus genome under the control of a standard eukaryotic promoter, such as SV40, and to transfect cells with the DNA. In this case, the DNA is expressed in the nucleus and the recombinant vector RNA is exported into the cytoplasm. Such DNA vectors have been described for Sindbis and SFV (Herweijer *et al.* 1995, Johanning *et al.* 1995, Berglund *et al.* 1996). Transduction is a more suitable delivery procedure for gene therapy applications, and in this case a replicon vector, such as pSinRep5, is co-introduced with a defective helper plasmid supplying the missing structural proteins. This facilitates one round of replication and packaging and the production of recombinant viral particles, which can be isolated from the extracellular fluid. In the original packaging system, the replicon and helper vectors could undergo recombination to produce moderate amounts of contaminating wild-type virus. This problem has been addressed by supplying the structural-protein genes on multiple plasmids (Smerdou & Liljestrom 1999) and by the development of complementary cell lines producing the structural proteins (Polo *et al.* 1999).

Vaccinia and other poxvirus vectors

Vaccinia virus is closely related to variola virus, the agent responsible for smallpox. A worldwide vaccination programme using vaccinia virus resulted in the elimination of smallpox as an infectious disease. The success of the programme raised hopes that recombinant vaccinia viruses, carrying genes from other pathogens, could be used as live vaccines for other infectious diseases. This expectation appears to have been borne out by recent successes, as discussed below.

The poxviruses have a complex structure and a large double-stranded linear DNA genome (up to 300 kb). Unusually for a DNA virus, the poxviruses replicate in the cytoplasm of the infected cell rather than its nucleus. Part of the reason for the large genome and structural complexity is that the virus must encode and package all its own DNA replication and transcription machinery, which most DNA viruses 'borrow' from the host cell nucleus. The unusual replication strategy and large size of the vaccinia genome make the design and construction of expression vectors more complex than for other viruses. Since the virus normally packages its own replication and transcription enzymes, recombinant genomes introduced into cells by transfection are non-infectious. Recombinant viruses are therefore generated by homologous recombination, using a targeting plasmid transfected into virus-infected cells. More recently, direct ligation vectors have been developed, and these are transfected into cells containing a helper virus to supply replication and transcription enzymes *in trans* (Merchlinsky *et al.* 1997). Recombinant vectors can be identified by hybridization to the large viral plaques that form on permissive cells. However, the efficiency of this process can be improved by various selection regimes. In one strategy, the transgene is inserted into the viral *Tk* gene and negative selection, using the thymidine analogue 5-bromodeoxyuridine, is carried out to enrich for potential recombinants (Mackett *et al.* 1982). In another, the transgene is inserted into the viral haemagglutinin locus. If chicken erythrocytes are added to the plate of infected cells, wild-type plaques turn red whereas the recombinant plaques remain clear (Shida 1986). Since vaccinia vectors have a high capacity for foreign DNA, selectable markers, such as *neo* (Franke *et al.* 1985), or screenable markers, such as *lacZ* (Chakrabarti *et al.* 1985) or *gusA* (Carroll & Moss 1995), can be co-introduced with the experimental transgene to identify recombinants.

Transgene expression usually needs to be driven by an endogenous vaccinia promoter, since transcription relies on proteins supplied by the virus. The highest expression levels are provided by late promoters such as P11, allowing the production of up to 1 μg of protein per 10^6 cells, but other promoters such as P7.5 and 4b are used, especially where expression early in the infection cycle is desired (Bertholet *et al.* 1985, Cochran *et al.* 1985). A synthetic late promoter, whose use allows up to 2 µg of protein to be produced per 10^6 cells, has also been developed (Lundstrom 1997). Since the cytoplasm lacks not only host transcription factors but also the nuclear splicing apparatus, vaccinia vectors cannot be used to express genes with introns. Furthermore, the sequence TTTTTNT must be removed from all foreign DNA sequences expressed in vaccinia vectors, since the virus uses this motif as a transcriptional terminator. A useful binary expression system has been developed, in which the transgene is driven by the bacteriophage T7 promoter and the T7 polymerase itself is expressed in a vaccinia vector under the control of a vaccinia promoter (Fuerst *et al.* 1986). Initially, the transgene was placed on a plasmid vector and transfected into cells infected with the recombinant vaccinia virus, but higher expression levels (up to 10% total cellular protein) can be achieved using two vaccinia vectors, one carrying the T7-driven transgene and the other expressing the polymerase (Fuerst *et al.* 1987).

In an early demonstration that vaccinia virus could be used to express antigens from other infectious agents, Smith, G.L. *et al.* (1983) replaced the vaccinia *Tk* locus with a transgene encoding hepititis B surface antigen (HBSAg). The transgene was cloned in a plasmid containing the vaccinia *Tk* gene, interrupted by one of the vaccinia early promoters. This plasmid was transfected into vaccinia-infected monkey cells, and recombinant vectors carrying the transgene were selected using 5-bromodeoxyuridine. Cells infected with this virus secreted large amounts of HBSAg into the culture medium, and vaccinated rabbits rapidly produced high-titre antibodies to HBSAg. Similarly, vaccinia viruses expressing the influenza haemagglutinin gene were used to immunize hamsters and induce resistance to influenza (Smith, G.E. *et al.* 1983). Recombinant vaccinia viruses have been constructed expressing a range of important antigens, including HIV and HTLV-III envelope proteins

Table 10.4 Summary of major expression systems used in animal cells.

System	Host	Major applications
Non-replicating plasmid vectors		
No selection	Many cell lines	Transient assays
Dominant selectable markers	Many cell lines	Stable transformation, long-term expression
DHFR/methotrexate	CHO cells	Stable transformation, high-level expression
Plasmids with viral replicons		
SV40 replicons	COS cells	High-level transient expression
BPV replicons	Various murine	Stable transformation (episomal)
EBV replicons	Various human	Stable transformation (episomal), library construction
Viral transduction vectors		
Adenovirus E1 replacement	293 cells	Transient expression
Adenovirus amplicons	Various mammalian	*In vivo* transfer
Adeno-associated virus	Various mammalian	*In vivo* transfer
Baculovirus	Insects	High-level transient expression
	Various mammalian	*In vivo* transfer
Oncoretrovirus	Various mammalian and avian	Stable transformation
	ES cells	Transgenic mice (Chapter 11)
Lentivirus	Non-dividing cells, mammalian	*In vivo* transfer
Sindbis, Semliki Forest virus	Various mammalian	High-level transient expression
Vaccinia virus	Various mammalian	High-level transient expression

(Chakrabarti *et al.* 1986, Hu *et al.* 1986). In many cases, recombinant vectors have been shown to provide immunity when administered to animals (reviewed by Moss 1996). For example, monkeys infected with recombinant vaccinia and canarypox vectors have shown resistance to simian immunodeficiency virus (SIV) and HIV-2 (Hirsch *et al.* 1996, Myagkikh *et al.* 1996). Several recombinant poxviruses have progressed to phase 1 clinical trials, including a vaccine against the Epstein–Barr major membrane antigen (Gu *et al.* 1996). The immunization of wild foxes, using 'spiked' meat, with a recombinant vaccinia virus expressing the rabies-virus glycoprotein appears to have eliminated the disease from Belgium (Brochier *et al.* 1991).

Summary of expression systems for animal cells

A variety of gene transfer and expression systems have been discussed in this chapter, and Table 10.4 provides a summary of these systems and their major applications. Many factors influence the expression of foreign genes in animal cells, and an understanding of these factors allows transgene expression to be controlled. For the production of recombinant proteins, it is often appropriate to maximize transgene expression, and principles for achieving this are shown in Box 10.2. In other cases, it may be desirable to switch the transgene on and off, using inducible expression systems. We explore these considerations in more detail in Chapter 13.

Box 10.2 Construct design for high-level transgene expression in animal cells

Many of the expression systems discussed in this chapter are used in experiments where the production of recombinant protein is the ultimate aim. In such cases, it is appropriate for transgene expression to be maximized. Although the different expression systems vary in their total potential yield, in terms of vector design and experimental methodology, the following considerations should apply when high-level expression is required.

The use of a strong and constitutive promoter

Very active promoters provide the highest levels of transgene expression. In viral vectors, transgenes are often expressed under the control of the strongest endogenous promoters, e.g. the baculovirus polyhedrin promoter, the adenoviral E1 promoter and the vaccinia virus p7.5 promoter. Certain viruses contain strong promoters and enhancers that function in a wide range of cell types, and several of these have been subverted for use in plasmid vectors. The elements most commonly used in mammalian cells are the SV40 early promoter and enhancer (Mulligan & Berg 1981b), the Rous sarcoma virus long-terminal-repeat promoter and enhancer (Gorman *et al.* 1982a) and the human cytomegalovirus immediate early promoter (Boshart *et al.* 1985). Although these function widely, they are not necessarily active in all mammalian cells, e.g. the SV40 promoter functions poorly in the human embryonic kidney line 293 (Gorman *et al.* 1989).

The inclusion of an intron

The presence of an intron in a eukaryotic expression unit usually enhances expression. Evidence for the positive effect of introns accumulated in the early years of cDNA expression in animal cells, although there are also many studies in which efficient gene expression was obtained in the absence of an intron (see Kaufman 1990). Nevertheless, most mammalian expression vectors in current use incorporate a heterologous intron, such as the SV40 small t-antigen intron or the human growth-hormone intron, or modified hybrid introns that match the consensus splice donor and acceptor-site sequences. Note that introns may not be used in some expression systems, such as vaccinia virus. The presence of an intron is very important in constructs that are to be expressed in transgenic animals (see Box 11.1 for more discussion).

continued

Box 10.2 *continued*

The inclusion of a polyadenylation signal

Polyadenylation signals (terminators) are required in eukaryotic genes to generate a defined 3′ end to the mRNA. In most cases, this defined end is extended by the addition of several hundred adenosine residues to generate a poly(A) (polyadenylate) tail. This tail is required for the export of mRNA into the cytoplasm, and also increases its stability. In the absence of such a site, the level of recombinant protein produced in transformed cells can fall by as much as 90% (Kaufman 1990). Poly(A) sites from the SV40 early transcription unit or mouse β-globin gene are often incorporated into mammalian expression vectors.

The removal of unnecessary untranslated sequence

Eukaryotic mRNAs comprise a coding region (which actually encodes the gene product) bracketed by untranslated regions (UTRs) of variable lengths. Both the 5′ and 3′ UTRs can influence gene expression in a number of ways (Kozak 1999). For example, the 5′ UTR may contain one or more AUG codons upstream of the authentic translational start site, and these are often detrimental to translational initiation. The 3′ UTR may contain regulatory elements that control mRNA stability (e.g. AU-rich sequences that reduce stability have been identified (see Shaw & Kamen 1986)). Furthermore, both the 5′ and 3′ UTRs may be rich in secondary structure, which prevents efficient translation. In animal systems, UTR sequences are generally removed from transgene constructs to maximize expression.

Optimization of the transgene for translational efficiency

The sequence around the translational initiation site should conform to Kozak's consensus, which is defined as 5′-CCRCCAUGG-3′ (Kozak 1986, 1999). Of greatest importance is the purine at the −3 position (identified as R) and the guanidine at position +4. The adenosine of the AUG initiation codon (underlined) is defined as position +1, and the immediately preceding base is defined as

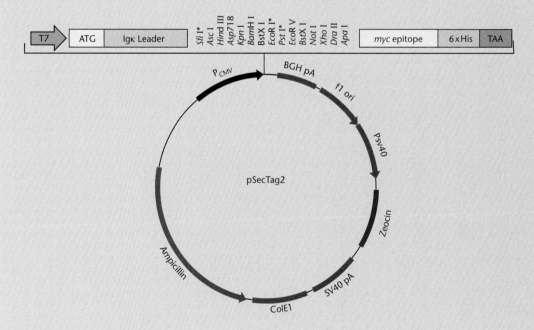

continued

Box 10.2 *continued*

position −1. The expression of foreign genes in animals can also be inefficient in some cases due to suboptimal codon choice, which reflects the fact that different organisms prefer to use different codons to specify the same amino acid. If a transgene contains a codon that is commonly used in the source organism but rarely used in the host, translation may pause at that codon due to the scarcity of the corresponding transfer RNA (tRNA). This will reduce the rate of protein synthesis and may even lead to truncation of the protein or frame-shifting. It may therefore be beneficial to 'codon-optimize' transgenes for the expression host.

The incorporation of a targeting signal

If the goal of an expression study is to recover large amounts of a functional *eukaryotic* protein, it is necessary to consider whether that protein needs to be post-translationally modified in order to function correctly. For example, many proteins intended for therapeutic use require authentic glycosylation patterns not only for correct function, but also to prevent an immune response in the patient. Since specific types of modification occur in particular cell compartments, it is necessary to consider strategies for targeting the recombinant protein to the correct compartment to ensure that it is appropriately modified. Proteins that need to be glycosylated, for example, must be targeted to the secretory pathway using a *signal peptide*. Many mammalian expression vectors are available for this purpose, and they incorporate heterologous signal peptides. The figure opposite shows the Invitrogen vector pSecTag2, which incorporates a sequence encoding the murine immunoglobulin light-chain signal peptide for high-efficiency targeting to the secretory pathway. Note also that the C terminus of the recombinant protein is expressed as a fusion to two different epitope tags to facilitate protein purification (see p. 76).

CHAPTER 11

Genetic manipulation of animals

Introduction

The genetic manipulation of animals has revolutionized our understanding of biology, by making it possible to test gene expression and function at the whole-animal level. Gene-transfer techniques can be used to produce *transgenic animals*, in which every cell carries new genetic information, as well as designer mutants with specific preselected modifications to the genome. The whole animal is the ultimate assay system in which to investigate gene function, particularly for complex biological processes, such as development.

In the case of plants, gene transfer to tissues or cultured cells is often the first step in the production of a transgenic organism, since whole fertile plants can regenerate from such cells and explants under the appropriate culture conditions. Two fundamental differences between plants and animals make this strategy impossible in animals. First, animal cells become progressively restricted in terms of developmental potency as development proceeds, which means that differentiated animal cells are normally unable to fully dedifferentiate and recapitulate the developmental programme.* Secondly, in most animals, the somatic cells and germ cells (the cells that give rise to gametes) separate at an early developmental stage. Therefore, the only way to achieve germ-line transformation in animals is to introduce

DNA into totipotent cells prior to the developmental stage at which the germ line forms.

In most cases, this involves introducing DNA directly into the developing oocyte, egg or early embryo. In mice, it is also possible to use cultured embryonic stem cells (ES cells), which are derived from the preimplantation embryo and can contribute to all the tissues of the developing animal (including the germ line) if introduced into a host embryo at the correct developmental stage. These cells are also remarkably amenable to homologous recombination, which allows them to be used for gene targeting, the accurate replacement of a segment of the endogenous genome with a homologous segment of exogenous DNA. Gene targeting can be used to replace endogenous genes with a completely non-functional copy (a null allele) or to make subtle changes, both allowing the function of the endogenous gene to be tested. The same technology can be used for the opposite purpose, i.e. to replace a mutant allele with a functional copy.

While rapid progress has been made with a range of model organisms, especially the fruit fly *Drosophila melanogaster*, the mouse and, more recently, the African clawed frog, *Xenopus*, great potential exists in the production of transgenic farm animals with improved or novel traits. The technology for introducing DNA into animals such as chickens, pigs, cattle and sheep is still in its infancy and is much less efficient than that of mice. Furthermore, it has proved impossible to isolate amenable ES cells from any species except mice and chickens. The desire to generate transgenic livestock has driven research in a different direction, that of nuclear transfer. Although differentiated animal cells are developmentally restricted, their nuclei still contain all the genetic information required to recapitulate the whole of development. Dolly, the first sheep produced following nuclear transfer from a differentiated somatic cell to an enucleated egg, has opened

---

* Recently, adult haemopoietic stem cells have been shown to contribute to fetal blood-cell development when injected into the blastocysts of mouse embryos (Geiger *et al.* 1998). Under similar circumstances, bone marrow and neural stem cells have been shown to contribute to specific developing tissues (Petersen *et al.* 1999, Clarke *et al.* 2000). This suggests that mammalian development may be more plastic than previously assumed, and could open the way for novel routes to animal transgenesis.

the way for animal cloning and the rapid production of élite transgenic herds.

Genetic manipulation of mammals

Methods for producing transgenic mice

The ability to introduce DNA into the germ line of mice is one of the greatest achievements of the twentieth century and has paved the way for the transformation of other mammals. Genetically modified mammals have been used not only to study gene function and regulation, but also as bioreactors producing valuable recombinant proteins, e.g. in their milk. Several methods for germ-line transformation have been developed, all of which require the removal of fertilized eggs or early embryos from donor mothers, brief culture *in vitro* and then their return to foster-mothers, where development continues to term. These methods are discussed below and summarized in Fig. 11.1.

Pronuclear microinjection

Direct microinjection of DNA was the first strategy used to generate transgenic mice. Simian virus 40 (SV40) DNA was injected into the blastocoele cavities of preimplantation embryos by Jaenisch and Mintz (1974). The embryos were then implanted into the uteri of foster-mothers and allowed to develop. The DNA was taken up by some of the embryonic cells and occasionally contributed to the germ line, resulting in transgenic mice containing integrated SV40 DNA in the following generation. Transgenic mice have also been recovered following the injection of viral DNA into the cytoplasm of the fertilized egg (Harbers *et al.* 1981).

The technique that has become established is the injection of DNA into one of the pronuclei of the egg (reviewed by Palmiter & Brinster 1986). The technique is shown in Fig. 11.2. Just after fertilizaton, the small egg nucleus (female pronucleus) and the large sperm nucleus (male pronucleus) are discrete.

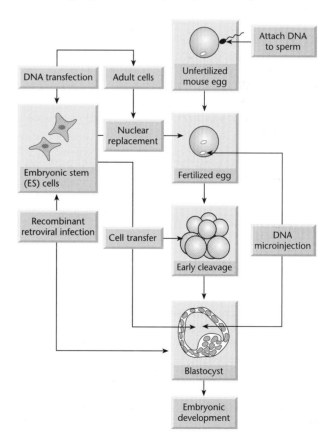

Fig. 11.1 Summary of methods for producing transgenic mammals.

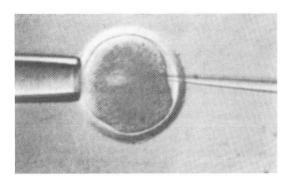

Fig. 11.2 Pronuclear microinjection of a fertilized mouse egg. The two pronuclei are visible, and the egg is held using a suction pipette. The DNA is introduced through a fine glass needle. (Photograph courtesy of Roberta Wallace, Roslin Institute.)

Since the male pronucleus is larger, this is usually chosen as the target for injection. About 2 pl of DNA solution is transferred into the nucleus through a fine needle, while the egg is held in position with a suction pipette. The injected embryos are cultured *in vitro* to the morula stage and then transferred to pseudopregnant foster-mothers (Gordon & Ruddle, 1981). The procedure requires specialized microinjection equipment and considerable dexterity from the handler. The exogenous DNA may integrate immediately or, less commonly, may remain extrachromosomal for one or more cell divisions. Thus the resulting animal may be transgenic or may be chimeric for transgene insertion. The technique is reliable, although the efficiency varies, so that 5–40% of mice developing from manipulated eggs contain the transgene (Lacy *et al.* 1983). However, once the transgene is transmitted through the germ line, it tends to be stably inherited over many generations. The exogenous DNA tends to form head-to-tail arrays prior to integration, and the copy number varies from a few copies to hundreds. The site of integration appears random and may depend on the occurrence of natural chromosome breaks. Extensive deletions and rearrangements of the genomic DNA often accompany transgene integration (Bishop & Smith 1989).

Recombinant retroviruses

As discussed in Chapter 10, recombinant retroviruses provide a natural mechanism for stably introducing DNA into the genome of animal cells. Retroviruses are able to infect early embryos and ES cells (see below), so recombinant retroviral vectors can be used for germ-line transformation (Robertson *et al.* 1986). An advantage over the microinjection technique is that only a single copy of the retroviral provirus is integrated, and the genomic DNA surrounding the transgenic locus generally remains intact. The infection of preimplantation embryos with a recombinant retrovirus is technically straightforward and, once the infected embryos are implanted in the uterus of a foster-mother, can lead to germline transmission of the transgene. However, there are also considerable disadvantages to this method, including the limited amount of foreign DNA that can be carried by the virus, the possible interference of viral regulatory elements with the expression of surrounding genes and the susceptibility of the virus to *de novo* methylation, resulting in transgene silencing (see Box 13.2). The founder embryos are always chimeric with respect to transgene integration (reviewed by Jaenisch 1988). Retroviral transduction is therefore not favoured as a method for generating fully transgenic animals, but it is useful for generating transgenic sectors of embryos. For example, the analysis of chicken-limb buds infected with recombinant retroviruses has allowed many of the genes involved in limb development to be functionally characterized (see review by Tickle & Eichele 1994).

Transfection of ES cells

ES cells are derived from the inner cell mass of the mouse blastocyst and thus have the potential to contribute to all tissues of the developing embryo (Evans & Kaufman 1981, Martin 1981). The ability of ES cells to contribute to the germ line was first demonstrated by Bradley *et al.* (1984) and requires culture conditions that maintain the cells in an undifferentiated state (Joyner 1998). Since these cells can be serially cultured, like any other established cell line, DNA can be introduced by transfection or viral transduction and the transformed cells can be selected using standard markers, as discussed in

Chapter 10. This is an important advantage, since there is no convenient way to select for eggs or embryos that have taken up foreign DNA, so, instead, each potential transgenic mouse must be tested by Southern-blot hybridization or the polymerase chain reaction (PCR) to confirm transgene integration. ES cells are also particularly efficient at carrying out homologous recombination (see below), so, depending on the design of the vector, DNA introduced into ES cells may integrate randomly or may target and replace a specific locus.

Whichever strategy is chosen, the recombinant ES cells are then introduced into the blastocoele of a host embryo at the blastocyst stage, where they mix with the inner cell mass. This creates a true chimeric embryo (i.e. an embryo comprising cells from different sources). The contribution of ES cells to the germ line can thus be confirmed using visible markers. Most ES cell lines in common use are derived from mouse strain 129, which has the dominant coat colour agouti. A popular strategy is to use host embryos from a mouse strain such as C57BL/6J, which has a recessive black coat colour. Colonization of the embryo by vigorous ES cells can be substantial, generating chimeras with patchwork coats of black and agouti cell clones. If the ES cells have contributed to the germ line, mating chimeric males with black females will generate heterozygous transgenic offspring with the agouti coat colour, confirming germ-line transmission of the foreign DNA. Most ES cells in use today are derived from male embryos, resulting in a large sex bias towards male chimeras (McMahon & Bradley, 1990). This is desirable because male chimeras sire many more offspring than females.

Gene targeting with ES cells

Pronuclear microinjection and retroviral transfer are useful for the addition of DNA to the mouse genome. However, in many cases it is more desirable to replace endogenous gene sequences with exogenous DNA, since this would allow the introduction of specific mutations into any preselected gene. In yeast, gene targeting by homologous recombination occurs with high efficiency (Chapter 9). In contrast, when the first gene-targeting experiments in animal cells were carried out in the 1980s, only a very low

frequency of targeted recombination was achieved. These experiments involved the correction of mutations in selectable markers such as *neo*, which had been introduced into cell lines as transgenes by standard methods (e.g. Thomas *et al.* 1986). Smithies *et al.* (1985) were the first to demonstrate targeting of an endogenous gene. They introduced a modified β-globin gene containing the bacterial marker *supF* into a human fibroblast × mouse erythroleukaemia cell line and screened large numbers of potential recombinants by reisolating the modified locus, using *supF* as a cloning tag. This experiment demonstrated that the frequency of homologous recombination was up to 1000-fold lower than that of random integration. Targeting occurs with significantly higher efficiency in certain cell lines, including mouse ES cells. The combination of pluripotency, amenability for *in vitro* manipulation and capacity for homologous recombination makes ES cells uniquely suitable for the generation of targeted mutant mice, i.e. mice carrying the same mutation in every cell and transmitting it through the germ line. Gene targeting in ES cells was first achieved by Thomas and Capecchi (1987), who disrupted the *hprt* gene with the *neo* marker and selected recombinant cells using either G418 or 6-thioguanine, a toxic guanine analogue that is only incorporated into DNA if the nucleotide salvage pathway is functional (see p. 178). Doetschman *et al.* (1987) also successfully targeted the *hprt* locus, although they used a mutant recipient cell line and repaired the locus with homologous DNA, subsequently selecting on HAT medium (see p. 177).

Design of targeting vectors

Targeting vectors are specialized plasmid vectors, which promote homologous recombination when introduced into ES cells. This is achieved by the inclusion of a homology region, i.e. a region that is homologous to the target gene, allowing the targeting vector to synapse with the endogenous DNA. Both the size of the homology region and the level of sequence identity have been shown to play an important role in the efficiency of gene targeting (Hasty *et al.* 1991b, Deng & Capecchi 1992, Te Riele *et al.* 1992). Recombination is also more efficient if the vector is linearized prior to transfection.

Most gene targeting experiments have been used to disrupt endogenous loci, resulting in targeted null alleles (this strategy is often termed 'gene knockout'). Two types of targeting vector have been developed for this purpose: insertion vectors and replacement (or transplacement) vectors (Thomas & Capecchi 1987; Fig. 11.3). Insertion vectors are linearized *within* the homology region, resulting in the insertion of the entire vector into the target locus. This type of vector disrupts the target gene but leads to a duplication of the sequences adjacent to the selectable marker. This is not always a desirable configuration, since duplication of the target sequences can lead to a subsequent homologous recombination event that restores the wild-type genotype (Fiering *et al.* 1995). Replacement vectors are designed so that the homology region is collinear with the target. The vector is linearized *outside* the homology region prior to transfection, resulting in crossover events in which the endogenous DNA is replaced by the incoming DNA. With this type of vector, only sequences within the homology region (not the vector backbone) are inserted, so, to achieve gene knockout, the homology region itself must be interrupted. Insertion and replacement vectors are thought to be equally efficient, but replacement vectors have been used in the majority of knockout experiments. In both cases, however, it is possible for transcription to occur through the targeted locus, producing low amounts of RNA. This may be spliced in such a way as to remove the targeted exon, resulting in a residual amount of functional protein (e.g. Dorin *et al.* 1994, Dahme *et al.* 1997).

Selection strategy

The first gene-targeting experiments involved the selectable *Hprt* locus, which is present on the X chromosome, allowing targeted events to be selected in male ES cells without the requirement for homozygosity. The first non-selectable genes to be targeted were *int-2* (also known as *fgf-3*) (Mansour *et al.* 1988) and the oncogene *c-abl* (Schwartzberg *et al.* 1989). In each case, it was necessary to include a selectable marker in the targeting vector to identify transformed cells. In the case of insertion vectors this is placed anywhere on the vector backbone,

while in replacement vectors the marker must interrupt the homology region. The *neo* marker has been most commonly used, allowing transformed ES cells to be selected using G418. However, other dominant markers are equally applicable (e.g. see Von Melchner *et al.* 1992), and *Hprt* can be used in combination with *hprt⁻* mutant ES cells (Matzuk *et al.* 1992).

The use of a single marker fails to discriminate between targeted cells and those where the construct has integrated randomly. This problem can be addressed by combined positive–negative selection using *neo* and the herpes simplex virus (HSV) *Tk* gene (Mansour *et al.* 1988). If the *neo* marker is used to interrupt the homology region in a replacement vector, transformed cells can be selected using G418. The HSV *Tk* gene is placed outside the homology region, such that it is inserted by random integration but not by homologous recombination. Therefore, cells that have undergone homologous recombination will survive in the presence of the toxic thymidine analogues ganciclovir or FIAU,* while in those cells containing randomly integrated copies of the *Tk* gene, the analogues will be incorporated into the DNA, resulting in cell death.

A different strategy is to make expression of the *neo* gene dependent on homologous recombination (e.g. see Schwartzberg *et al.* 1989, Mansour *et al.* 1993). Another alternative to positive–negative selection, which is used in many laboratories, is simply to screen large numbers of G418-resistant transfected cells by PCR to identify genuine recombinants (Hogan & Lyons 1988, Zimmer & Gruss 1989). Screening can be carried out relatively quickly without recourse to cloning.

Introducing subtle mutations

While gene targeting has often been used to disrupt and hence inactivate specific endogenous genes by introducing large insertions, more refined approaches can be used to generate subtle mutations. The precise effects of minor deletions or point mutations cannot be assessed using the simple targeting strategies discussed above, which necessarily leave the

---

* 1-(2-deoxy-2-fluoro-β-D-arabinofuranosyl)-5-iodouracil.

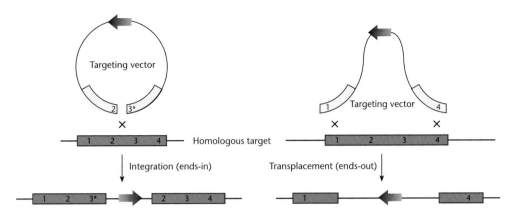

Fig. 11.3 Structure and integration mechanisms of the two major types of targeting vector, (a) insertion and (b) transplacement. Regions of homology between the vector and target are designated by the same number.

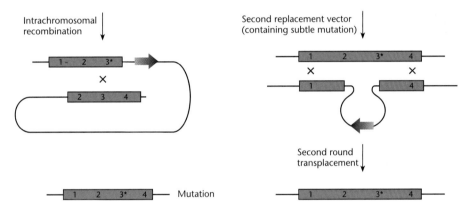

Fig. 11.4 Procedure for the introduction of subtle mutations using (a) an insertion vector and (b) two transplacement vectors. Regions of homology between the vector and target are designated by the same number. An asterisk indicates the position of a point mutation in the vector, which is to be introduced at the target locus.

selectable marker and, in some cases, the entire targeting vector integrated at the target site. Furthermore, many investigators have reported that the strong promoters used to drive marker-gene expression can affect the regulation of neighbouring genes, in some cases up to 100 kb away from the targeted locus (Pham *et al.* 1996). Two major strategies for the introduction of subtle mutations have been devised, each involving two rounds of homologous recombination. These are the 'hit-and-run' strategy, involving a single insertion vector, and the 'tag-and-exchange' strategy (Fig. 11.4), involving two replacement vectors. Strategies involving Cre recombinase have also been developed and are discussed on p. 257.

The 'hit-and-run' or 'in–out' strategy (Hasty *et al.* 1991a, Valancius & Smithies 1991) involves the use of an insertion vector carrying two selectable markers, such as *neo* and *Tk*. The insertion event is positively selected using G418. As discussed above, the use of an insertion vector results in duplication of the homology region in the targeted clone, although in this case the homology region derived from the vector is modified to contain the desired subtle mutation. The success of this strategy relies on a second intrachromosomal homologous recombination event, which replaces the endogenous allele with the mutant and deletes the markers, allowing the second-round recombinants to survive in the presence of ganciclovir. However, the second homologous recombination event occurs at a very low frequency and, in 50% of cases, restores the locus to its original wild-type configuration.

The 'tag-and-exchange' strategy also requires two homologous recombination events, but in this case two replacement-type vectors are used. For example, Moore *et al.* (1995) demonstrated the principle using *neo* for positive selection and HSV *Tk* for negative selection. The first 'tag' vector was designed to mutate the target gene by inserting a large cassette containing the selectable markers. This event was positively selected with G418. The second 'exchange' vector introduced the desired mutation and eliminated the selectable markers, allowing the second-round recombinants to be selected for the absence of *Tk*.

Applications of genetically modified mice

Applications of transgenic mice

Transgenic mice, i.e. mice containing additional transgenes, as opposed to those with targeted mutations, have been used to address many aspects of gene function and regulation. A vast literature has accumulated on this subject, and genes concerning every conceivable biological process have been investigated (e.g. see Houdebine 1997). As well as their use for basic scientific investigation, transgenic mice can be used for more applied purposes, such as models for human disease and the production of valuable pharmaceuticals. Many mouse models for human diseases have been generated by gene knockout (see below), but gain-of-function models have also been generated by adding transgenes. For example, much information concerning the pathology of prion diseases has arisen from the study of transgenic mice expressing mutant prion transgenes (reviewed by Gabizon & Taraboulos 1997). Transgenic mice expressing oncogenes have been extensively used to study cancer (reviewed by Macleod & Jacks 1999).

For illustrative purposes, we now consider some early experiments that demonstrated how transgenic mice can be used for the analysis of gene function and regulation, but also highlighted some limitations of the transgenic approach. Brinster *et al.* (1981) constructed plasmids in which the promoter of the mouse metallothionein-1 (*MMT*) gene was fused to the coding region of the HSV *Tk* gene. The thymidine kinase (TK) enzyme can be assayed readily and provides a convenient reporter of *MMT* promoter function. The endogenous *MMT* promoter is inducible by glucocorticoid hormones and heavy metals, such as cadmium and zinc, so it was envisaged that the hybrid transgene, *MK* (metallothionein-thymidine kinase), would be similarly regulated. The gene was injected into the male pronucleus of fertilized eggs, which were then incubated *in vitro* in the presence or absence of cadmium ions (Brinster *et al.* 1982). As expected, TK activity was found to be induced by the metal. By making a range of deletions of mouse sequences upstream of the *MMT* promoter sequences, the minimum region necessary for inducibility was localized to a stretch of DNA 40–180 nucleotides upstream of the transcription-initiation site. Additional sequences that potentiate both basal and induced activities extended to at least 600 bp upstream of the transcription-initiation site. The mouse egg was therefore being used in the same way as transfected cell lines, to dissect the activity of a functional promoter (Gorman *et al.* 1982b; see Box 10.1).

The same *MK* fusion gene was injected into embryos, which were raised to transgenic adults (Brinster *et al.* 1981). Most of these mice expressed the *MK* gene and in such mice there were from one to 150 copies of the gene. The reporter activity was inducible by cadmium ions and showed a tissue distribution very similar to that of metallothionein itself (Palmiter *et al.* 1982b). Therefore these experiments showed that DNA sequences necessary for heavy-metal induction and tissue-specific expression could be functionally dissected in both eggs and transgenic mice. For unknown reasons, there was no response to glucocorticoids in either the egg or the transgenic-mouse experiments.

In a dramatic series of experiments, Palmiter *et al.* (1982a) fused the *MMT* promoter to the rat growth-hormone gene. This hybrid gene (*MGH*) was constructed using the same principles as the *MK* fusion. Of 21 mice that developed from microinjected eggs, seven carried the *MGH* fusion gene and six of these grew significantly larger than their littermates. The mice were fed zinc to induce transcription of the *MGH* gene, but this did not appear to be absolutely necessary, since they showed an accelerated growth rate before being placed on the zinc diet. Mice containing

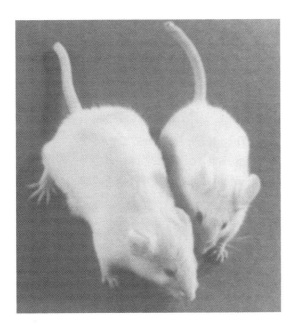

Fig. 11.5 Transgenic mouse containing the mouse metallothionein promoter fused to the rat growth-hormone gene. The photograph shows two male mice at about 10 weeks old. The mouse on the left contains the *MGH* gene and weighs 44 g; his sibling without the gene weighs 29 g. In general, mice that express the gene grow two to three times as fast as controls and reach a size up to twice the normal. (Photograph by courtesy of Dr R.L. Brinster.)

high copy numbers of the *MGH* gene (20–40 copies per cell) had very high concentrations of growth hormone in their serum, some 100–800 times above normal. Such mice grew to almost double the weight of littermates at 74 days old (Fig. 11.5).

The similarities between the tissue distribution of normal *MMT* expression and that of the hybrid transgenes encouraged the hope that transgenic mice would provide a general assay for functionally dissecting DNA sequences responsible for tissue-specific or developmental regulation of a variety of genes. However, there were also some unexpected findings. For example, independently derived transgenic mice carrying the *MK* transgene showed significant variations in the levels and patterns of transgene expression. Furthermore, while transgenic founders transmitted the construct to their progeny as expected, when reporter activity was assayed in these offspring the amount of expression could be very different from that in the parent.

Examples of increased, decreased or even totally extinguished expression were found. In some, but not all, cases, the changes in expression correlated with changes in methylation of the gene sequences (Palmiter *et al.* 1982b). These results provided the first examples of two complex phenomena, *position effects* (Box 11.1) and *de novo transgene silencing* (Box 13.2), which often affect integrated transgenes.

Yeast artificial chromosome (YAC) transgenic mice

Studies of the *MMT* promoter and others have demonstrated the principle that transgenes with minimal flanking sequences tend not to be expressed in the same manner as the corresponding endogenous gene. In many cases, it has also been shown that authentic patterns and levels of protein expression occur only when the intact gene is used, and this can span tens or hundreds of kilobase pairs of DNA (Box 11.1). The transfer of large DNA segments to the mouse genome has been achieved by transformation with yeast artificial chromosome (YAC) vectors. Jakobovits *et al.* (1993) were the first to report transformation of ES cells with a YAC vector, via fusion with yeast sphaeroplasts. The vector contained the entire human *HPRT* locus, nearly 700 kb in length. The disadvantage of this method is that the endogenous yeast chromosomes were co-introduced with the vector. Alternative strategies involve isolation of the vector DNA by pulsed-field gel electrophoresis (p. 10), followed by introduction of the purified YAC DNA into mouse eggs by pronuclear microinjection or transfection into ES cells. The latter technique is more suitable because microinjection involves shear forces that break the DNA into fragments. YAC transfer to ES cells has been achieved by lipofection, as discussed in Chapter 10. YAC transgenics have been used to study gene regulation, particularly by long-range regulatory elements, such as locus-control regions (reviewed by Lamb & Gerhart 1995). They have also been used to introduce the entire human immunoglobulin locus into mice, for the production of fully humanized antibodies (Mendez *et al.* 1997). It is also possible to introduce chromosomes and chromosome fragments into ES cells, using a technique called microcell-mediated fusion. This involves the prolonged mitotic

Box 11.1 Position effects

Independently derived transgenic animals and plants carrying the same expression construct often show variable levels and patterns of transgene expression. In many cases, such variation is dependent on the site of transgene integration, and this phenomenon has been termed the *position effect* (reviewed by Wilson *et al*. 1990). Position effects result from the influence of local regulatory elements on the transgene, as well as the architecture of the surrounding chromatin. For example, an integrated transgene may come under the influence of a local enhancer, resulting in the alteration of its expression profile to match that of the corresponding endogenous gene. The position dependence of the phenomenon has been demonstrated in mice by isolating the entire transgenic locus from such an anomalous line and microinjecting it into the pronuclei of wild-type eggs, resulting in 'secondary' transgenic lines with normal transgene expression profiles (Al-Shawi *et al*. 1990). Position effects are also revealed by enhancer-trap constructs, which contain a minimal promoter linked to a reporter gene (O'Kane & Gehring 1987; see Chapter 13).

Unlike the specific influences of nearby regulatory elements, chromatin-mediated position effects are generally non-specific and repressive. They reflect the integration of the transgene into a chromosomal region containing repressed chromatin (heterochromatin). The molecular features of heterochromatin, including its characteristic nucleosome structure, deacetylated histones and, in many cases, hypermethylated DNA, spread into the transgene, causing it to be inactivated (Huber *et al*. 1996, Pikaart *et al*. 1998). In some cases, variegated transgene expression has been reported due to cell-autonomous variations in the extent of this spreading process (reviewed by Heinkoff 1990). Negative chromosomal position effects can be troublesome in terms of achieving desirable transgene expression levels and patterns; thus a number of different strategies have been used to combat them.

Incorporating dominantly acting transcriptional control elements

Certain regulatory elements are thought to act as master-switches, regulating the expression of genes or gene clusters by helping to establish an open chromatin domain. The locus control region (LCR) of the human β-globin gene cluster is one example (Forrester *et al*. 1987). Transgenic mice carrying a human β-globin transgene driven by its own promoter show a low frequency of expression and, in those mice that do express the transgene, only a low level of the mRNA is produced (e.g. Magram *et al*. 1985, Townes *et al*. 1985). However, inclusion of the LCR in the expression construct confers high-level and position-independent expression (Grosveld *et al*. 1987). There is evidence that LCRs induce chromatin remodelling over large distances. For example, the murine immunoglobulin heavy-chain LCR has been shown to induce histone deacetylation in a linked *c-myc* gene (Madisen *et al*. 1998). This suggests that LCRs could protect against position effects by converting heterochromatin to open euchromatin at the site of transgene integration (Festenstein *et al*. 1996, Milot *et al*. 1996). The interested reader can consult several comprehensive reviews of LCR research (Bonifer 1999, Grosveld 1999, Li *et al*. 1999).

Using boundary elements/matrix attachment regions

Boundary elements (insulators) are sequences that can block the activity of enhancers when placed between the enhancer and a test transgene driven by a minimal promoter. For example, an 'A element' with insulator activity is found upstream of the chicken lysozyme gene. This inhibits the activity of a reporter gene when interposed between the promoter and an upstream enhancer, but not when placed elsewhere in the construct (Stief *et al*. 1989). However, by flanking the entire construct with a pair of

continued

Box 11.1 *continued*

A elements, the transgene is protected from chromosomal position effects (Stief *et al.* 1989). This protective effect works not only in cell lines, but also in transgenic animals (McKnight *et al.* 1992) and plants (Mlynarova *et al.* 1994). Many boundary elements are associated with matrix-attachment regions (MARs), sequences dispersed throughout the genome that attach to the nuclear matrix, dividing chromosomes into topologically independent loops (reviewed by Spiker & Thompson 1996). It is therefore possible that transgenes flanked by such elements are maintained in an isolated chromatin domain into which heterochromatin cannot spread. However, not all boundary elements are associated with MARs (e.g. see Mirkovitch *et al.* 1984). Similarly, some MARs do not function as boundary elements but as facilitators of gene expression (e.g. Van der Geest & Hall 1997).

Using large genomic transgenes

Conventional transgenes generally comprise complementary DNAs (cDNAs) or intronless 'minigenes' expressed under the control of viral promoters or cell-type-specific regulatory elements. Such transgenes are highly sensitive to position effects. Over the last few years, there has been an increasing appreciation that the regulation of eukaryotic gene expression is far more complex and involves much more upstream and downstream DNA than previously thought (reviewed by Bonifer 1999, 2000). The correct, high-level expression of transgenes is favoured by the use of genomic constructs that include introns and large amounts of flanking sequence from the source gene (e.g. see Bonifer *et al.* 1990, Lien *et al.* 1997, Nielsen *et al.* 1998). Such constructs are likely to include multiple enhancers, dominant regulatory elements, such as LCRs, and boundary elements, which all act together to protect the transgene from position effects.

Dominantly acting transgenes (transgene rescue)

Some conventional transgenes, including β-globin and α-fetoprotein (Chada *et al.* 1986, Kollias *et al.* 1986, Hammer *et al.* 1987) are very sensitive to position effects and *de novo* silencing. Other genes appear to be less sensitive to these phenomena, e.g. immunoglobulin and elastase (Storb *et al.* 1984, Swift *et al.* 1984, Davis & MacDonald 1988). Although the reason for this is not clear, the less sensitive transgenes are assumed to in some way induce or define an open chromatin domain. In some cases, such sequences have been used to protect more susceptible transgenes from negative position effects by introducing the two transgenes simultaneously (e.g. see Clark *et al.* 1992).

Site-specific integration

Site-specific recombination systems (see Chapter 13) can be used to introduce transgenes into a locus known to lack negative position effects, if a target site for the recombinase can be introduced at such a locus. The Cre-*loxP* system has been used to this effect in mammalian cells (see Fukushige & Sauer 1992).

---

arrest of cultured human cells, using an inhibitor such as colchicine. Eventually, the nucleus breaks up into vesicles containing individual chromosomes, which can be rescued as microcells comprising a nuclear vesicle surrounded by a small amount of cytoplasm and a plasma membrane (Fournier & Ruddle 1977). Transgenic mice have been generated using ES cells that were fused to human microcells, and evidence for germ-line transmission and ex-pression of the human chromosome was obtained (Tomizuka *et al.* 1997).

Applications of gene targeting

Since the first reports of gene targeting in ES cells, an ever-increasing number of targeted mutant mice have been produced. These have been discussed in several comprehensive reviews (Brandon *et al.*

1995a,b,c, Soriano 1995, Muller 1999) and a number of Internet databases have been established to keep track of the results (see Sikorski & Peters 1997). The phenotypes of homozygous, null mutant mice provide important clues to the normal function of the gene. Some gene knockouts have resulted in surprisingly little phenotypic effect, much less severe than might have been expected. For example, *myoD*, whose expression in transfected fibroblasts causes them to differentiate into muscle cells, and which was therefore a good candidate as a key regulator of myogenesis, is not necessary for development of a viable animal (Rudnicki *et al.* 1992). Similarly, the retinoic acid γ receptor is not necessary for viable mouse development in knockout mice (Lohnes *et al.* 1993), even though this receptor is a necessary component of the pathway for signalling by retinoids and has a pattern of expression quite distinct from other retinoic acid receptors in embryos. Such observations have prompted speculation that genetic redundancy may be common in development, and may include compensatory up-regulation of some members of a gene family when one member is inactivated. An example of this may be the up-regulation of *myf-5* in mice lacking *myoD* (Rudnicki *et al.* 1992). Gene knockouts have also been used as mouse models of human diseases, such as cystic fibrosis, β-thalassaemia and fragile X syndrome (reviewed by Bedell *et al.* 1997; see Chapter 14).

While most gene-targeting experiments in mice have been used to introduce mutations into genes (either disruptive insertional mutations or subtle changes), the scope of the technique is much wider. The early gene-targeting experiments demonstrated that this approach could also be used to correct mutated genes, with obvious applications in gene therapy. Homologous recombination has also been used to exchange the coding region of one gene for that of another, a strategy described as 'gene knock-in'. This has been used, for example, to test the ability of the transcription factors Engrailed-1 and Engrailed-2 to compensate for each other's functions. Hanks *et al.* (1995) replaced the coding region of the *engrailed-1* gene with that of *engrailed-2*, and showed that the *engrailed-1* mutant phenotype could be rescued. A more applied use of gene knock-in is the replacement of parts of the murine immunoglobulin genes with their human counterparts,

resulting in the production of humanized antibodies in transgenic mice (Moore *et al.* 1995). The Cre-*loxP* site-specific recombinase system has been used extensively in ES cells to generate mice in which conditional or inducible gene targeting is possible and to produce defined chromosome deletions and translocations as models for human disease. We shall discuss the many applications of Cre-*loxP* and other site-specific recombinase systems in Chapter 13.

Other mammals and birds

Traditional techniques

The three major routes for producing transgenic mice have also been used in other mammals and birds, particularly in farm animals. The efficiency of each procedure is much lower than in mice. Pronuclear microinjection in mammals such as sheep and cows, for example, typically results in less than 1% of the injected eggs giving rise to transgenic animals. Added to this, the recovery of eggs from donor animals and the reimplantation of transformed eggs into foster-mothers is a less efficient procedure and requires, at great expense, a large number of donors and recipients. The eggs themselves are also more difficult to manipulate – they are very delicate and tend to be opaque. It is often necessary to centrifuge the eggs in order to see the pronuclei. In chickens, it is possible to remove eggs just after fertilization and microinject DNA into the cytoplasm of the germinal disc, where the male and female pronuclei are to be found. However, it is not possible to return the manipulated eggs to a surrogate mother, so they must be cultured *in vitro*. Using this procedure, Love *et al.* (1994) obtained seven chicks, equivalent to about 5% of the eggs injected, that survived to sexual maturity. One cockerel transmitted the transgene to a small proportion of his offspring, indicating that he was chimeric for transgene integration.

The use of retroviruses to produce transgenic chickens has been reported by Bosselman *et al.* (1989). These investigators injected a replication-defective recombinant reticuloendotheliosis virus carrying the *neo* gene into laid eggs and found that approximately 8% of male birds carried vector sequences. The transgene was transmitted through the germ line in a proportion of these birds and was stably expressed in 20 transgenic lines. It is necessary for the nuclear envelope to break down for most retroviral infections (lentiviruses such as human immunodeficiency virus (HIV) are exceptional) and, since this only occurs during mitosis, most retroviruses are unable to infect non-dividing cells (Chapter 10). The nuclear envelope also breaks down during meiosis, and this was exploited by Chan *et al.* (1998) to produce transgenic cattle following the injection of replication-defective retroviral vectors into the perivitelline space of isolated bovine oocytes. Retroviral integration occurred during the second meiotic division, resulting in the production of a number of transgenic offspring. Remarkably, the same technique has recently been used to generate the first ever transgenic primate, a rhesus monkey named ANDi (Chan *et al.* 2001). In primates, the technique was inefficient. Two hundred and twenty-four oocytes were injected to produce one live transgenic monkey; a number of further transgenic fetuses failed to develop to term.

After more than a decade of research, it has also proved impossible to derive reliable ES cell lines from any domestic species other than mice and, more recently, chickens (Pain *et al.* 1999, Prelle *et al.* 1999)*. However, there have been great advances in the isolation and transfection of primordial germ cells (PGCs), the embryonic cells that give rise to gametes. These can be transfected directly, or cultured as embryonic germ cells (EG cells), which are morphologically very similar to ES cells and could provide a route for the direct transformation of the germ line (Resnick *et al.* 1992). Chicken PGCs have been isolated from the germinal crescent, infected with a recombinant retrovirus and replaced in the embryo, leading to the development of chimeric birds producing transgenic offspring (Vick *et al.* 1993). Mammalian PGCs have also been transformed, although it has been difficult to persuade the cells to contribute to the germ line once introduced into the host animal (Labosky *et al.* 1994).

Intracytoplasmic sperm injection

The injection of sperm heads directly into the cytoplasm of the egg (intracytoplasmic sperm injection

(ICSI)) can overcome infertility in humans. It has been shown that sperm heads bind spontaneously to naked plasmid DNA *in vitro*, suggesting that sperm injections could be used to achieve transformation. This was demonstrated by Perry *et al.* (1999), who mixed mouse sperm with plasmid DNA carrying the gene for green fluorescent protein (GFP). These sperm were injected into unfertilized oocytes, and a remarkable 94% of the resulting embryos showed GFP activity. Random transfer of these embryos to pseudopregnant females resulted in development to term, and in about 20% of cases the mice were transgenic. This method could be adaptable to other animals. Rhesus monkey oocytes fertilized in the same manner gave rise to a number of embryos with GFP activity, but this only lasted until the blastula stage, suggesting that there was no stable integration. However, several monkeys developed to term, showing that the procedure was compatible with normal development (Chan *et al.* 2000).

Nuclear transfer technology

The failure of traditional transgenesis techniques to yield routine procedures for the genetic modification of mammals other than mice has driven researchers in search of other methods. Fifty years ago, Briggs and King (1952) established the principle of nuclear transfer in amphibians by transplanting nuclei from the blastula of the frog *Rana pipens* to an enucleated egg, obtaining a number of normal embryos in the process. In *Xenopus laevis*, nuclei from various types of cell in the swimming tadpole can be transplanted to an egg that has been UV-irradiated to destroy the peripheral chromosomes, and similar results are obtained (reviewed by Gurdon 1986, 1991). The important principle here is that, while animal cells become irreversibly committed to their fate as development proceeds, the nuclei of most cells still retain all the genetic information required for the entire developmental programme and can, under appropriate circumstances, be reprogrammed by the cytoplasm of the egg to recapitulate development. In all species, it appears that the earlier the developmental stage at which nuclei are isolated, the greater their potential to be reprogrammed. Nuclear transplantation can be used to generate clones of animals with the same genotype by transplanting many

---

* Human ES cells and Human ES cell lines are available.

Fig. 11.6 Megan and Morag, the first sheep produced by nuclear transfer from cultured cells. Reproduced by kind permission of the Roslin Institute, Edinburgh.

Fig. 11.7 Dolly and her lamb Bonnie. Dolly was the first mammal to be generated by nuclear transfer from an adult cell. Reproduced by kind permission of the Roslin Institute, Edinburgh.

somatic nuclei from the same individual into a series of enucleated eggs (King & Briggs 1956). This allows animals with specific and desirable traits to be propagated. If possible in mammals, this would have obvious applications in farming.

Nuclear transfer in mammals has been practised with success for the last decade, although rabbits and farm animals, such as sheep, pigs and cows, are far more amenable to the process than mice. In each case, donor nuclei were obtained from the morula or blastocyst-stage embryo and transferred to an egg or oocyte from which the nucleus had been removed with a pipette (Smith & Wilmut 1989, Willadsen 1989, Collas & Robl 1990, McLaughlin *et al.* 1990). The donor nucleus can be introduced by promoting fusion between the egg and a somatic cell. A brief electric pulse is often used to achieve this, as it also activates embryonic development by stimulating the mobilization of calcium ions.

A major advance was made in 1995, when two live lambs, Megan and Morag (Fig. 11.6), were produced by nuclear transfer from cultured embryonic cells (Campbell *et al.* 1996). This demonstrated the principle that mammalian nuclear transfer was possible using a cultured cell line. The same group later reported the birth of Dolly (Fig. 11.7), following nuclear transfer from an adult mammary epithelial cell line (Wilmut *et al.* 1997). This was the first mammal to be produced by nuclear transfer from a differentiated adult cell, and aroused much debate

among both scientists and the public concerning the possibility of human cloning (see Johnson 1998). It was suggested that a critical factor in the success of the experiment was the quiescent state of the cells in culture, allowing synchronization between the donor and recipient cell cycles. For the production of Dolly, this was achieved by lowering the level of serum in the culture medium, causing the cells to withdraw from the cell cycle due to lack of growth factors. However, the success rate was very low: only one of 250 transfer experiments produced a viable lamb. Similar transfer experiments have since been carried out in mice, cows, pigs and goats (Cibelli *et al.* 1998, Wakayama *et al.* 1998, Baguisi *et al.* 1999, Polejaeva *et al.* 2000).

The success of nuclear transfer in domestic mammals provides a new route for the production of transgenic animals. This involves the introduction of DNA into cultured cells, which are then used as a source of donor nuclei for nuclear transfer. Such a cell-based strategy has many advantages over traditional techniques, such as microinjection, including the ability to screen transformed cells for high-level transgene expression prior to the nuclear-transfer step. The production of a transgenic mammal by nuclear transfer from a transfected cell line was first achieved by Schnieke *et al.* (1997), who introduced the gene for human factor IX into fetal sheep fibroblasts and transferred the nuclei to enucleated eggs. The resulting sheep, Polly, produces the recombinant protein in her milk and can therefore be used as a bioreactor (Chapter 14). More recently, McCreath *et al.* (2000) succeeded in producing a transgenic sheep by nuclear transfer from a somatic cell whose genome had been specifically modified by gene targeting. A foreign gene was introduced into the *COL1A1* locus and was expressed at high levels in the lamb.

DNA transfer to other vertebrates

Gene transfer to *Xenopus*

Xenopus *oocytes as a heterologous expression system*

Since Gurdon *et al.* (1971) first showed that oocytes synthesized large amounts of globin after they had been microinjected with rabbit globin mRNA, the *Xenopus* oocyte expression system has been a valuable tool for expressing a very wide range of proteins from plants and animals (Colman 1984). *X. laevis* is an African clawed frog. Oocytes can be obtained in large numbers by removal of the ovary of an adult female. Each fully grown oocyte is a large cell (0.8–1.2 mm diameter) arrested at first meiotic prophase. This large cell has a correspondingly large nucleus (called the *germinal vesicle*), which is located in the darkly pigmented hemisphere of the oocyte.

Due to the large size of the oocytes, mRNA – either natural or synthesized by transcription *in vitro*, using phage-T7 RNA polymerase (Melton 1987) – can be readily introduced into the cytoplasm or nucleus by microinjection. This is achieved using a finely drawn glass capillary as the injection needle, held in a simple micromanipulator. DNA can also be injected. The oocyte nucleus contains a store of the three eukaryotic RNA polymerases, enough to furnish the needs of the developing embryo at least until the 60 000-cell stage. The RNA polymerases are available for the transcription of injected exogenous DNA. Using this system, it has therefore been possible to express complementary DNAs (cDNAs) linked to a heat-shock promoter or to mammalian

virus promoters (Ballivet *et al.* 1988, Ymer *et al.* 1989, Swick *et al.* 1992). In addition, vaccinia virus vectors (Chapter 10) can be used for gene expression in the cytoplasm (Yang *et al.* 1991).

An important aspect of the oocyte expression system is that recombinant proteins are usually correctly post-translationally modified and directed to the correct cellular compartment. For example, oocytes translate a wide variety of mRNAs encoding secretory proteins, modify them and correctly secrete them (Lane *et al.* 1980, Colman *et al.* 1981). Foreign plasma-membrane proteins are generally targeted to the plasma membrane of the oocyte, where they can be shown to be functional. The first plasma-membrane protein to be expressed in this system was the acetylcholine receptor from the electric organ of the ray, *Torpedo marmorata* (Sumikawa *et al.* 1981). Injected oocytes translated mRNA extracted from the electric organ and assembled functional multi-subunit receptor molecules in the plasma membrane (Barnard *et al.* 1982). Following this work, the oocyte has become a standard heterologous expression system for plasma-membrane proteins, including ion channels, carriers and receptors. The variety of successfully expressed plasma-membrane proteins is very impressive. However, there are examples of foreign channels and receptors being non-functional in oocytes, either due to lack of coupling to second-messenger systems in the oocyte, incorrect post-translational modification, or other reasons (reviewed in Goldin 1991).

Xenopus *oocytes for functional expression cloning*

Functional expression cloning using oocytes was first developed by Noma *et al.* (1986), using a strategy outlined in Fig. 11.8. The following example, the cloning of the substance-K receptor, is illustrative. It has been found that oocytes can be made responsive to the mammalian tachykinin neuropeptide, substance K, by injecting an mRNA preparation from bovine stomach into the oocyte cytoplasm. The preparation contains mRNA encoding the substance-K receptor protein, which is evidently expressed as a functional protein and inserted into the oocyte membrane. Masu *et al.* (1987) exploited this property to isolate a cDNA clone encoding the receptor. The principle was to make a cDNA library from

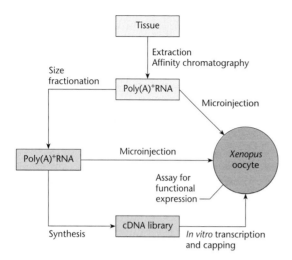

Fig. 11.8 Strategy for functional expression cloning, using *Xenopus* oocytes as a heterologous expression system.

stomach mRNA, using a vector in which the cDNA was flanked by a promoter for the SP6 or T7 RNA polymerase. This allowed *in vitro* synthesis of mRNA from the mixture of cloned cDNAs in the library.

The receptor clone was identified by testing for receptor expression following injection of synthetic mRNA into the oocyte cytoplasm. Repeated subdivision of the mixture of cDNAs in the library led to the isolation of a single cloned cDNA. The strategy described above can only be applied to cloning single-subunit proteins, not proteins composed of different subunits or proteins whose function in oocytes requires more than one foreign polypeptide. This limitation was overcome by Lubbert *et al.* (1987), who used a hybrid depletion procedure to clone a serotonin-receptor cDNA.

A prerequisite for using the oocyte in functional expression cloning is a knowledge of the oocyte's own ion channels, carriers and receptors. Endogenous activity may mask or interfere with the sought-after function (for a review, see Goldin 1991).

Transient gene expression in Xenopus *embryos*

Messenger RNA, synthesized and capped *in vitro*, can be microinjected into dejellied *Xenopus* embryos at the one- or two-cell stage. The mRNA is distributed among the descendants of the injected cells and is expressed during early development. This

approach has been exploited very widely for examining the developmental effects resulting from the overexpression of normal or altered gene products (reviewed by Vize & Melton 1991).

DNA can be introduced into *Xenopus* embryos in the same manner. However, unlike the situation in mammals, where the injected DNA integrates rapidly into the genome, exogenous DNA in *Xenopus* persists episomally and undergoes extensive replication (Endean & Smithies 1989). Bendig and Williams (1983) provide a typical example of this process. They injected a recombinant plasmid carrying *Xenopus* globin genes into the egg and showed that the amount of plasmid DNA increased 50- to 100-fold by the gastrula stage. In later development, the amount of DNA per embryo decreased, and most of the persisting DNA co-migrated with high-molecular-weight chromosomal DNA. This difference between mammals and amphibians probably reflects their distinct modes of early development. In mammals, cleavage divisions are slow and asynchronous. Gene expression occurs throughout early development and supplies the embryo with the proteins it requires at a steady rate. Conversely, there is no transcription in the early *Xenopus* embryo and yet the cleavage divisions are rapid and synchronous. DNA replication relies on stored maternal gene products, so there is a stockpile of chromatin assembly proteins and replication enzymes. Exogenous DNA injected into *Xenopus* eggs is therefore assembled immediately into chromatin and undergoes replication in tune with the rapid DNA synthesis already occurring in the nucleus (Leno & Laskey 1991). Etkin *et al.* (1987) have analysed the replication of a variety of DNAs injected into *Xenopus* embryos. It was found that various plasmids increase to different extents. This was not simply related to the size of the plasmid, but also reflected the presence of specific sequences that inhibited replication. Replication has also been found to depend upon the conformation and number of molecules injected (Marini *et al.* 1989).

Transgenic Xenopus

DNA injected into early *Xenopus* embryos is expressed in a mosaic fashion during development, regardless of the promoter used, which limits the use of this system for the analysis of gene expression and function. Some of the DNA does become incorporated into the genome and may be transmitted through the germ line (Rusconi and Schaffner 1981). However, integration occurs at a very low frequency and, given the long generation interval of *Xenopus laevis* (12–18 months from egg to adult), this is not an efficient way to generate transgenic frogs.

A simple and efficient process for large-scale transgenesis in *Xenopus* has become available only in the last few years (Kroll & Amaya 1996). In this technique, known as restriction-enzyme-mediated integration (REMI), linearized plasmids containing the transgene of interest are mixed with decondensed sperm nuclei and treated with limiting amounts of a restriction enzyme to introduce nicks in the DNA. The nuclei are then transplanted into unfertilized *Xenopus* eggs, where the DNA is repaired, resulting in the integration of plasmid DNA into the genome. This technique allows the production of up to 700 transgenic embryos per person per day, most of which survive at least to the swimming-tadpole stage. The decondensed nuclei are extremely fragile, so careful handling and transplantation within about 30 min are required for a good yield of normal transgenic embryos. In some cases, viable transgenic adults have been derived from the tadpoles and transgenic *X. laevis* lines have been established (Bronchain *et al.* 1999, Marsh-Armstrong *et al.* 1999). A disadvantage of *X. laevis* is that the species is tetraploid. Offield *et al.* (2000) have therefore established transgenic lines of the closely related but diploid species *Xenopus tropicalis*, which also has a shorter generation interval than its tetraploid cousin.

Since *Xenopus* is used worldwide as a developmental model organism, transgenic *Xenopus* technology has been rapidly adopted in many laboratories and is being used to examine (or in many cases re-examine) the roles of developmental genes. Thus far, the sophisticated tools used in transgenic mice have not been applied to *Xenopus*, but this is only a matter of time. Recently, an inducible expression system based on the use of a *Xenopus* heat-shock promoter was described, allowing inducible control of the GFP gene. This system has been used to investigate Wnt signalling in early *Xenopus* development (Wheeler *et al.* 2000). As discussed above, one of the

early successes in transgenic mouse methodology was the expression of rat growth hormone, resulting in transgenic mice up to twice the size of their non-transgenic siblings. The role of growth hormone in amphibian metamorphosis has now been examined by expressing *Xenopus* growth hormone in transgenic frogs. The transgenic tadpoles developed at the same rate as control tadpoles, but typically grew to twice the normal size (Huang & Brown 2000). After metamorphosis, the transgenic frogs also grew much more quickly than controls and showed skeletal defects.

Gene transfer to fish

Fish transgenesis can be used to study gene function and regulation, e.g. in model species, such as the zebrafish (*Danio rerio*) and medaka (*Oryzias latipes*), and to improve the traits of commercially important species, such as salmon and trout. Gene-transfer technology in fish has lagged behind that of mammals, predominantly due to the lack of suitable regulatory elements to control transgene expression. The first transgenic fish carried transgenes driven by mammalian or viral regulatory elements, and their performance varied considerably. For example, attempts to express growth-hormone genes in trout initially met with little success, and this may have been due to the inability of fish cells to correctly process mammalian introns (Betancourt *et al.* 1993). However, fish are advantageous assay systems for several reasons, including their fecundity, the fact that fertilization and development are external and the ease with which haploid and uniparental diploid embryos can be produced (Ihssen *et al.* 1990).

Like frogs, the injection of DNA into fish eggs and early embryos leads to extensive replication and expression from unintegrated transgenes, so that fish, like frogs, can be used for transient expression assays (Vielkind 1992). Some of the DNA integrates into the genome, leading to germ-line transmission and the production of transgenic fish lines (reviewed by Iyengar *et al.* 1996). There has been recent progress in the development of transgenic fish with enhanced growth characteristics, particularly through the use of expression constructs that are derived from the same species (e.g. Rahman *et al.* 1998; reviewed by Dunham 1999). It is likely that transgenic fish will be the first genetically modified animals to enter the food-chain.

DNA transfer to invertebrates

Transgenic flies

Drosophila *P elements*

P elements are transposable DNA elements that, under certain circumstances, can be highly mobile in the germ line of *D. melanogaster*. The subjugation of these sequences as specialized vector molecules in *Drosophila* was a landmark in *Drosophila* genetics. Through the use of P-element vectors, any DNA sequence can be introduced into the genome of the fly.

P elements cause a syndrome of related genetic phenomena called *P–M hybrid dysgenesis* (Bingham *et al.* 1982, Rubin *et al.* 1982). Dysgenesis occurs when males of a P (paternally contributing) strain are mated with females of an M (maternally contributing) strain, but not when the reciprocal cross is made. The syndrome predominantly affects the germ line and induces a high rate of mutation and frequent chromosomal aberrations, resulting in abnormal (dysgenic) hybrid offspring. In extreme cases, there is failure to produce any gametes at all.

Hybrid dysgenesis occurs because P strains contain transposable genetic elements, P elements, which are mobilized in the eggs of M-strain females (eggs that are permissive for P-element transposition are described as 'M-cytotype'). The P elements do not cause dysgenesis in crosses within P strains, because they are not mobilized in P-cytotype eggs. This is because the P element encodes a repressor of its own transposase, which prevents transposition. When a sperm from a P-cytotype male fertilizes the egg of an M-cytotype female, the absence of repressor in the egg results in temporary derepression of the transposase, such that P-element transposition occurs at a high frequency. The high rate of mutation characteristic of the dysgenesis syndrome reflects the insertion of P elements into multiple genetic loci.

Several members of the P-element family have been cloned and characterized (O'Hare & Rubin 1983). The prototype is a 2.9-kb element, while other members of the family appear to have arisen by internal deletion events. The elements are characterized by perfect 31-bp inverted terminal repeats, which are recognized by the transposase. The prototype element contains a single gene, comprising four exons, encoding the transposase (a truncated version of the transposase may act as the repressor). The transposase primary transcript is differentially spliced in germ cells and somatic cells, such that functional transposase is produced only in germ cells. Laski *et al.* (1986) showed this clearly by making a P-element construct in which the differentially spliced third intron was precisely removed. This element showed a high level of somatic transposition activity. Naturally occurring short P elements are generally defective, because they do not encode functional transposase. However, they do possess the inverted terminal repeats and can be activated *in trans* by transposase supplied by a non-defective P element in the same nucleus.

Spradling and Rubin (1982) devised an approach for introducing P-element DNA into *Drosophila* chromosomes. Essentially, a recombinant plasmid comprising a 2.9 kb P element together with some flanking *Drosophila* DNA sequences, cloned in the pBR322 vector, was microinjected into the posterior pole of M-cytotype embryos. The embryos were injected at the syncytial blastoderm stage, when the cytoplasm has not yet become partitioned into individual cells (Fig. 11.9). The posterior pole was chosen because this is where the germ line originates, and P-element DNA in this region was expected to be incorporated into the genome in a proportion of the germ cells.

A screen of progeny lines showed that P elements had indeed integrated at a variety of sites in each of the five major chromosomal arms, as revealed by *in situ* hybridization to polytene chromosomes. P-element integration occurred by transposition, not by random integration. This was proved by probing Southern blots of restricted DNA and showing that the integrated P element was not accompanied by the flanking *Drosophila* or pBR322 DNA sequences present in the recombinant plasmid (Spradling & Rubin 1982). The injected plasmid DNA must therefore have been expressed at some level *before* integration, so as to provide transposase.

These experiments showed that P elements could transpose with a high efficiency from injected plasmids

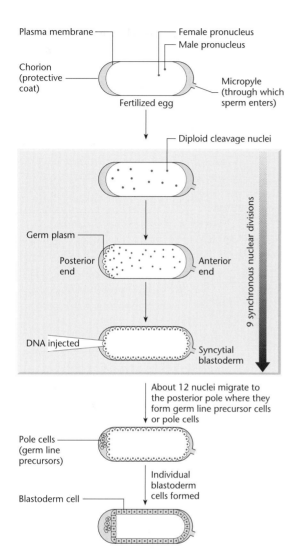

Fig. 11.9 Early embryogenesis of *Drosophila*. DNA injected at the posterior end of the embryo just prior to pole-cell formation is incorporated into germ-line cells.

into diverse sites in the chromosomes of germ cells. At least one of the integrated P elements in each progeny line remained functional, as evidenced by the hypermutability it caused in subsequent crosses to M-cytotype eggs.

Development of P-element vectors for gene transfer

Rubin and Spradling (1982) exploited their finding that P elements can be artificially introduced in the *Drosophila* genome. A possible strategy for using the

P element as a vector would be to attempt to identify a suitable site in the 2.9 kb P-element sequence where insertion of foreign DNA could be made without disrupting genes essential for transposition. However, an alternative strategy was favoured. A recombinant plasmid was isolated which comprised a short (1.2 kb), internally deleted member of the P-element family together with flanking *Drosophila* sequences, cloned in pBR322. This naturally defective P element did not encode transposase (O'Hare & Rubin 1983). Target DNA was ligated into the defective P element. The aim was to integrate this recombinant P element into the germ line of injected embryos by providing transposase function *in trans*. Two approaches for doing this were initially tested. In one approach a plasmid carrying the recombinant P element was injected into embryos derived from a P–M dysgenic cross, in which transposase activity was therefore expected to be high. A disadvantage of this approach was that frequent mutations and chromosomal aberrations would also be expected. In the other approach, the plasmid carrying the recombinant P element was co-injected with a plasmid carrying the non-defective 2.9 kb element.

In the first experiments of this kind, embryos homozygous for the *rosy* mutation were microinjected with a P-element vector containing a wild-type *rosy* gene. Both methods for providing complementing transposase were effective. Rosy⁺ progeny, recognized by their wild-type eye colour, were obtained from 20–50% of injected embryos. The chromosomes of these flies contained one or two copies of the integrated *rosy* transgene. The *rosy* gene is a particularly useful genetic marker. It produces a clearly visible phenotype: Rosy⁻ flies have brown eyes instead of the characteristic red colour of Rosy⁺ flies. The *rosy* gene encodes the enzyme xanthine dehydrogenase, which is involved in the production of a precursor of eye pigments. The *rosy* gene is not cell-autonomous: expression of *rosy* anywhere in the fly, for example in a genetically mosaic fly developing from an injected larva, results in a wild-type eye colour. Selectable markers have been used instead of visible markers to identify transformed flies. These include the alcohol dehydrogenase gene *adh* (Goldberg *et al.* 1983) and *neo* (Steller & Pirrotta 1985). Other eye-colour markers have been used, including *white* and

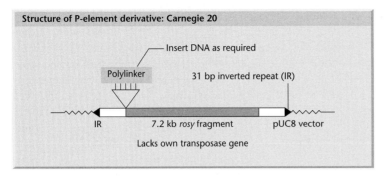

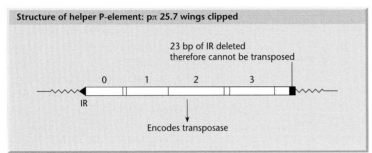

Fig. 11.10 P-element derivatives as a vector system. (See text for details.)

vermilion (Ashburner 1989, Fridell & Searles 1991), as well as alternative visible markers, such as *rough* (which restores normal eye morphology) and *yellow* (which restores normal body pigmentation and is particularly useful for scoring larvae) (Locket *et al.* 1992; Patton *et al.* 1992).

A simple P-element vector is shown in Fig. 11.10 (Rubin & Spradling 1983). It consists of a P element cloned in the bacterial vector pUC8. Most of the P element has been replaced by the *rosy* gene, but the terminal repeats essential for transposition have been retained. The vector includes a polylinker site for inserting foreign sequences. Transposition of the recombinant vector into the genome of injected larvae is brought about by co-injecting a helper P element, which provides transposase *in trans* but which cannot transpose itself because of a deletion in one of its terminal inverted repeats. Such an element is referred to as a *wings-clipped* element

(Karess & Rubin 1984). An alternative strategy is to inject purified transposase protein (Kaufman & Rio 1991). The capacity of P-element vectors is large, although increasing the size of the recombinant element appears to reduce the transposition frequency. Inserts of over 40 kb have been successfully introduced into flies (Haenlin *et al.* 1985) and this has allowed the construction of cosmid libraries using P-element vectors (Speek *et al.* 1988).

As well as their use for germ-line transformation, P elements have been exploited for insertional mutagenesis, as cloning tags and as entrapment vectors to detect genes and regulatory elements. Similar applications have been applied to other transposable elements, such as the *Ac–Ds* transposons of maize, and to other gene-transfer systems, such as retroviruses and the T-DNA of *Agrobacterium tumefaciens*. These diverse uses of gene-transfer vectors are discussed in Chapter 13.

CHAPTER 12

Gene transfer to plants

Introduction

The expression of foreign genes introduced into plants was first achieved in the early 1980s. In the 20 years following these initial successes, we have witnessed a revolution in plant genetic engineering, with the transformation of well over 100 different plant species now a routine procedure. Our ability to manipulate the plant genome has come about through intensive research into vector systems based on the soil bacterium *Agrobacterium tumefaciens* and alternative strategies involving direct DNA transfer. In addition, plant viruses have been developed as versatile episomal vectors for high-level transient gene expression. This research has immense biotechnological implications in the creation of plants with useful genetically engineered characteristics, such as insect resistance, herbicide tolerance and improved nutritional value (Chapter 14).

A fundamental difference between animals and plants is that organized, differentiated plant tissue shows a high degree of developmental plasticity. An isolated stem segment, for example, can regenerate into an entire new plant under appropriate culture conditions. For most plant species, some form of tissue-culture step is therefore a prerequisite for the successful production of *transgenic plants*, i.e. plants carrying the same foreign DNA sequence (the *transgene*) in every cell. Before considering the various gene-transfer systems available for plants, we shall therefore begin with a discussion of the basic principles of plant callus and cell culture. Notably, there is now increasing interest in the use of whole-plant (*in planta*) transformation strategies, in which the need for tissue culture is minimized or eliminated.

Plant callus and cell culture

Callus culture

Tissue culture is the process whereby small pieces of living tissue (*explants*) are isolated from an organism and grown aseptically for indefinite periods on a nutrient medium. For successful plant tissue culture, it is best to start with an explant rich in undetermined cells, e.g. those from the cortex or meristem, because such cells are capable of rapid proliferation. The usual explants are buds, root tips, nodal stem segments or germinating seeds, and these are placed on suitable culture media where they grow into an undifferentiated mass known as a callus (Fig. 12.1). Since the nutrient media used for plants can also support the growth of microorganisms, the explant is first washed in a disinfectant such as sodium hypochlorite or hydrogen peroxide. Once established, the callus can be propagated indefinitely by subdivision.

For plant cells to develop into a callus, it is essential that the nutrient medium contains the correct balance of plant hormones (phytohormones), i.e. auxins, cytokinins and gibberellins, to maintain the cells in an undifferentiated state (Fig. 12.2). The absolute amounts required vary for different tissue explants from different parts of the same plant, and

Fig. 12.1 Close-up view of a callus culture.

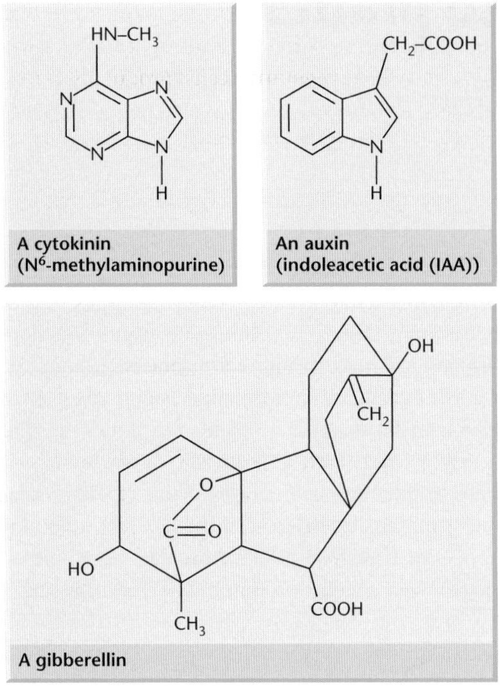

Fig. 12.2 The structures of some chemicals which are plant growth regulators, phytohormones.

for the same explant from different species. Thus there is no ideal medium. Most of the media in common use consist of inorganic salts, trace metals, vitamins, organic nitrogen sources (e.g. glycine), inositol, sucrose and growth regulators. More complex organic nutrients, such as casein hydrolysate, coconut water or yeast extract, and a gelling agent are optional extras.

Cell-suspension culture

When callus is transferred into liquid medium and agitated, the cell mass breaks up to give a suspension of isolated cells, small clusters of cells and much larger aggregates. Such suspensions can be maintained indefinitely by subculture but, by virtue of the presence of aggregates, are extremely heterogeneous. Genetic instability adds to this heterogeneity, so that long-term culture results in the accumulation of mutations (*somaclonal variation*), which can adversely affect the vitality and fertility of regenerated plants. Some species, such as *Nicotiana tabacum* (tobacco) and *Glycine max* (soybean), yield very

friable calli, and cell lines obtained from these species are much more homogeneous, allowing either continuous or batchwise cultivation.

If placed in a suitable medium, isolated single cells from suspension cultures are capable of division. As with animal cells, conditioned medium may be necessary for proliferation to occur. Conditioned medium is prepared by culturing high densities of cells in fresh medium for a few days and then removing the cells by filter sterilization. Medium conditioned in this way contains essential amino acids, such as glutamine and serine, as well as growth regulators, such as cytokinins. Provided conditioned medium is used, single cells can be plated on to solid media in exactly the same way as microorganisms, but, instead of forming a colony, plant cells proliferate and form a callus.

Protoplasts

Protoplasts are cells from which the cellulose walls have been removed. They are very useful for genetic manipulation because, under certain conditions, protoplasts from similar or contrasting cell types can be fused to yield somatic hybrids, a process known as protoplast fusion. Protoplasts can be produced from suspension cultures, callus tissue or intact tissues, e.g. leaves, by mechanical disruption or, preferably, by treatment with cellulolytic and pectinolytic enzymes. Pectinase is necessary to break up cell aggregates into individual cells and the cellulase digests away the cell wall. After enzyme treatment, protoplast suspensions are collected by centrifugation, washed in medium without the enzyme, and separated from intact cells and cell debris by flotation on a cushion of sucrose. When plated on to nutrient medium, protoplasts will synthesize new cell walls within 5–10 days and then initiate cell division (Fig. 12.3).

Regeneration of fertile plants

The developmental plasticity of plant cells means that whole fertile plants can often be regenerated from tissue explants, callus, cell suspensions or protoplasts by placing them on appropriate media. As discussed above, the maintenance of cells in an undifferentiated state requires the correct balance

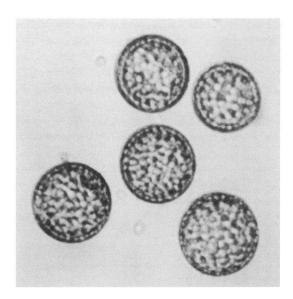

Fig. 12.3 Photomicrograph of tobacco protoplasts.

of phytohormones. However, only cytokinin is required for shoot culture and only auxin for root culture, therefore increasing the level of cytokinins available to the callus induces shoot formation and increasing the auxin level promotes root formation. Ultimately, plantlets arise through the development of adventitious roots on shoot buds or through the development of shoot buds from tissues formed by proliferation at the base of rootlets. The formation of roots and shoots on callus tissue is known as *organogenesis*. The culture conditions required to achieve organogenesis vary from species to species, and have not been determined for every type of callus. The adventitious organogenesis of shoots and roots can also occur directly from organized plant tissues, such as stem segments, without first passing through a callus stage.

Under certain conditions, cell suspensions or callus tissue of some plant species can be induced to undergo a different development process known as *somatic embryogenesis*. In this process, the cells undergo a pattern of differentiation similar to that seen in zygotes after fertilization, to produce *embryoids*. These structures are embryo-like but differ from normal embryos in being produced from somatic cells and not from the fusion of two germ cells. The embryoids can develop into fertile plants without the need to induce root and shoot formation on artificial media. Immature pollen or microspores can also be induced to form vegetative cells, producing haploid callus or embryoids. Such cells can be persuaded to undergo diploidization by treatment with mitotic inhibitors.

The ease with which plant material is manipulated and interconverted in culture provides many opportunities for the development of techniques for gene transfer and the recovery of transgenic plants (Fig. 12.4). DNA can be introduced into most types of plant material – protoplasts, cell suspensions, callus, tissue explants, gametes, seeds, zygotes, embryos, organs and whole plants – so the ability to recover fertile plants from such material is often the limiting step in plant genetic engineering rather than the DNA transfer process itself. It is also possible to maintain transformed plant cell lines or tissues (e.g. root cultures) producing recombinant proteins or metabolites, in the same way that cultured animal cells can be used as bioreactors for valuable products.

Overview of gene-transfer strategies

Gene transfer to plants is achieved using three different methods. The first exploits the natural ability of certain bacteria of the genus *Agrobacterium* to naturally transfer DNA to the genomes of infected plant cells. This generally results in the stable transformation of the infected cell, and the transferred DNA behaves as a new genetic locus. Initial limitations with respect to the host range of *Agrobacterium* prompted research into alternative methods based on direct DNA transfer. These include the chemically assisted transformation of protoplasts and the bombardment of plant material with DNA-coated microprojectiles. Such strategies can be used for both transient and stable transformation. Finally, plant viruses can been used as vectors for gene delivery. The viruses of plants never integrate into the genome and are not transmitted through seeds, so stable transformation cannot be achieved. However, plant viruses often cause systemic infections, resulting in the rapid production of high levels of recombinant protein throughout the plant, and they can be transmitted through normal infection routes or by grafting infected scions on to virus-free hosts.

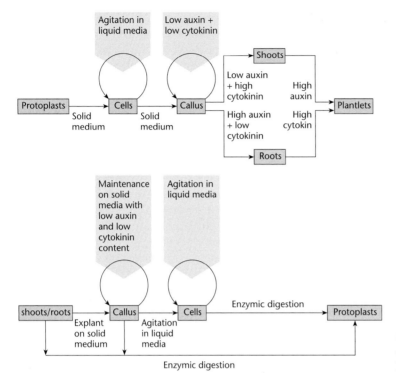

Fig. 12.4 Summary of the different cultural manipulations possible with plant cells, tissues and organs.

Agrobacterium-mediated transformation

Crown-gall disease

Crown-gall is a plant tumour that can be induced in a wide variety of gymnosperms and dicotyledonous angiosperms (dicots) by inoculation of wound sites with the Gram-negative soil bacterium *A. tumefaciens* (Fig. 12.5). The involvement of bacteria in this disease was established by Smith and Townsend (1907). It was subsequently shown that the crown-gall tissue represents true oncogenic transformation, since the undifferentiated callus can be cultivated *in vitro* even if the bacteria are killed with antibiotics, and retains its tumorous properties (Fig. 12.6). These properties include the ability to form a tumour when grafted on to a healthy plant, the capacity for unlimited growth as a callus in tissue culture even in the absence of phytohormones necessary for the *in vitro* growth of normal cells, and the synthesis of opines, such as octopine and

nopaline, which are unusual amino acid derivatives not found in normal plant tissue (Fig. 12.7).

The metabolism of opines is a central feature of crown-gall disease. Opine synthesis is a property conferred upon the plant cell when it is colonized by *A. tumefaciens*. The type of opine produced is determined not by the host plant but by the bacterial strain. In general, the bacterium induces the synthesis of an opine that it can catabolize and use as its sole carbon and nitrogen source. Thus, bacteria that utilize octopine induce tumours that synthesize octopine, and those that utilize nopaline induce tumours that synthesize nopaline (Bomhoff *et al.* 1976, Montaya *et al.* 1977).

Tumour-inducing (Ti) plasmids

Since the continued presence of *Agrobacterium* is not required to maintain plant cells in their transformed state, it is clear that some 'tumour-inducing principle' is transferred from the bacterium to the plant at the wound site. Zaenen *et al.* (1974) first noted

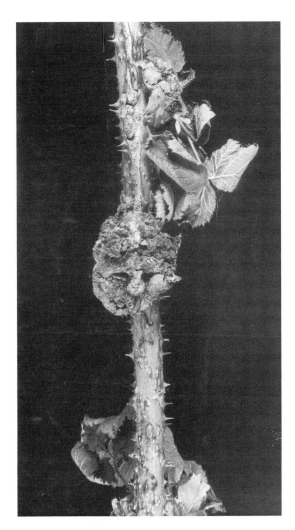

Fig. 12.5 Crown gall on blackberry cane. (Photograph courtesy of Dr C.M.E. Garrett, East Malling Research Station.)

that virulent strains of *A. tumefaciens* harbour large plasmids (140–235 kbp), and experiments involving the transfer of such plasmids between various octopine- and nopaline-utilizing strains soon established that virulence and the ability to use and induce the synthesis of opines are plasmid-borne traits. These properties are lost when the bacteria are cured of their resident plasmid (Van Larbeke *et al.* 1974, Watson *et al.* 1975) but are acquired by avirulent strains when a virulence plasmid is reintroduced by conjugation (Bomhoff *et al.* 1976, Gordon *et al.* 1979). The plasmids therefore became known as tumour-inducing plasmids (Ti plasmids).

Ti plasmids specify the type of opine that is synthesized in the transformed plant tissue and the opine utilized by the bacteria. Plasmids in the octopine group are closely related to each other, while those in the nopaline group are considerably more diverse. Between the groups, there are four regions of homology, including the genes directly responsible for tumour formation (Drummond & Chilton 1978, Engler *et al.* 1981; Fig. 12.8). It should be noted that the presence of a plasmid in *A. tumefaciens* does not mean that the strain is virulent. Many strains contain very large cryptic plasmids that do not confer virulence and, in some natural isolates, a cryptic plasmid is present together with a Ti plasmid.

T-DNA transfer

T-DNA sequences required (in cis) for transfer

Complete Ti-plasmid DNA is not found in plant tumour cells but a small, specific segment of the plasmid, about 23 kbp in size, is found integrated in the plant nuclear DNA at an apparently random site. This DNA segment is called T-DNA (transferred DNA) and carries genes that confer both unregulated growth and the ability to synthesize opines upon the transformed plant tissue. However, these genes are non-essential for transfer and can be replaced with foreign DNA (see below). The structure and organization of nopaline plasmid T-DNA sequences is usually simple, i.e. there is a single integrated segment. Conversely, octopine T-DNA comprises two segments: T_L (which carries the genes required for tumour formation) and T_R (which carries the genes for opine synthesis). The two segments are transferred to the plant genome independently and may be present as multiple copies. The significance of this additional complexity is not clear.

In the Ti plasmid itself, the T-DNA is flanked by 25 bp imperfect direct repeats known as *border sequences*, which are conserved between octopine and nopaline plasmids. The border sequences are not transferred intact to the plant genome, but they are involved in the transfer process. The analysis of junction regions isolated from plant genomic DNA has shown that the integrated T-DNA end-points lie internal to the border sequences. The right junction is rather precise, but the left junction can vary by

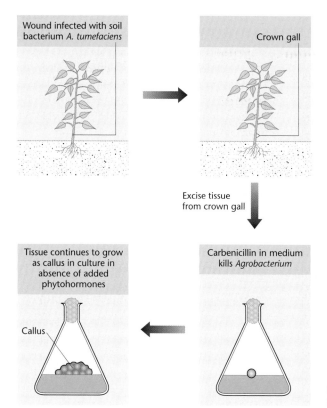

Fig. 12.6 *Agrobacterium tumefaciens* induces plant tumours.

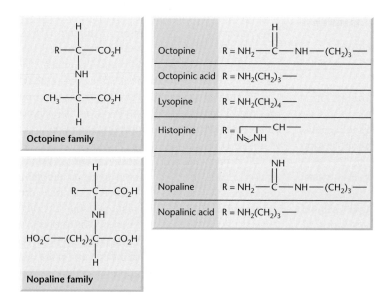

Fig. 12.7 Structures of some opines.

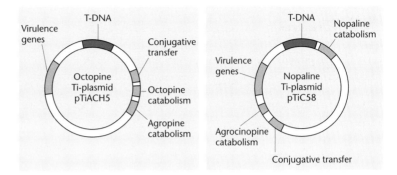

Fig. 12.8 Ti-plasmid gene maps.

about 100 nucleotides (Yadav *et al.* 1982, Zambryski *et al.* 1982). Deletion of the right border repeat abolishes T-DNA transfer, but the left-hand border surprisingly appears to be non-essential. Experiments in which the right border repeat alone has been used have shown that an enhancer, sometimes called the *overdrive sequence*, located external to the repeat is also required for high-efficiency transfer (Shaw *et al.* 1984, Peralta *et al.* 1986). The left border repeat has little transfer activity alone (Jen & Chilton 1986).

Genes required (in trans) for transfer

The genes responsible for T-DNA transfer are located in a separate part of the Ti plasmid called the *vir* (virulence) region. Two of these genes, *virA* and *virG*, are constitutively expressed at a low level and control the plant-induced activation of the other *vir* genes. VirA is a kinase that spans the inner bacterial membrane and acts as the receptor for certain phenolic molecules that are released by wounded plant cells. A large number of such compounds have been characterized, but one in particular, acetosyringone, has been the most widely used in the laboratory to induce *vir* gene expression (Stachel *et al.* 1985; Fig. 12.9). Notably, phenolic compounds such as acetosyringone do not attract bacteria to wounded plant cells. Rather, the bacteria appear to respond to simple molecules, such as sugars and amino acids, and the *vir* genes are induced after attachment (Parke *et al.* 1987, Loake *et al.* 1988). Many sugars also synergize the action of the phenolic signals to enhance *vir* gene expression (Shimada *et al.* 1990). Activated VirA transphosphorylates the VirG protein, which is a transcriptional activator of the other *vir* genes. The VirA and VirG proteins show similar-

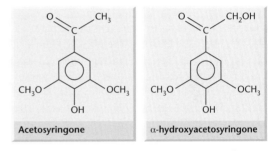

Fig. 12.9 Structures of signal molecules, produced by wounded plant tissue, which activate T-DNA transfer by *Agrobacterium tumefaciens*.

ities to other two-component regulatory systems common in bacteria (Winans 1992). In addition to *virG*, further genes on the bacterial chromosome also encode transcription factors that regulate *vir* gene expression (reviewed by Kado 1998).

The induction of *vir* gene expression results in the synthesis of proteins that form a conjugative pilus through which the T-DNA is transferred to the plant cell. The components of the pilus are encoded by genes in the *virB* operon (reviewed by Lai & Kado 2000). DNA transfer itself is initiated by an endonuclease formed by the products of the *virD1* and *virD2* genes. This introduces either single-strand nicks or a double-strand break at the 25 bp borders of the T-DNA, a process enhanced by the VirC12 and VirC2 proteins, which recognize and bind to the overdrive enhancer element. The VirD2 protein remains covalently attached to the processed T-DNA. Recent studies have suggested that the type of T-DNA intermediate produced (single- or double-stranded) depends on the type of Ti plasmid, with double-stranded T-DNA favoured by nopaline plasmids (where the T-DNA is a single element) and single 'T strands' favoured by octopine and succinopine plasmids, where the T-DNA is split into non-contiguous sections (Steck 1997). T strands are coated with VirE2, a single-stranded DNA-binding protein. The whole complex, sometimes dubbed the *firecracker complex* because of its proposed shape, is then transferred through the pilus and into the plant cell. It has been proposed that the VirD2 protein protects the T-DNA against nucleases, targets the DNA to the plant-cell nucleus and integrates it into the plant genome. The protein has two distinct nuclear localization signals, with the C-terminal signal thought to play the major role in targeting the T-DNA (Tinland *et al.* 1992). It has been observed that the nucleus of wounded plant cells often becomes associated with the cytosolic membrane close to the wound site, suggesting that the T-DNA could be transferred directly to the nucleus without extensive exposure to the cytosol (Kahl & Schell 1982). Once in the nucleus, the T-DNA is thought to integrate through a process of illegitimate recombination, perhaps exploiting naturally occurring chromosome breaks (Tinland 1996).

The *Agrobacterium* gene-transfer system appears to be a highly adapted form of bacterial conjugation. Many broad-host-range plasmids can transfer from *Agrobacterium* to the plant genome using their own mobilization functions (Buchanan-Wollaston *et al.* 1987) and the *vir* genes encode many components that are common with broad-host-range plasmid conjugation systems (reviewed by Kado 1998). In addition to plants, *Agrobacterium* can transfer DNA to other bacteria and to yeast. Recently, a novel insight into the scope of this gene-transfer mechanism was provided by Citovsky and colleagues (Kunik

et al. 2001) by demonstrating that gene transfer from *Agrobacterium* to cultured human cells was also possible! For the interested reader, T-DNA transfer has been discussed in several comprehensive reviews (Hooykaas & Schilperoort 1992, Hooykaas & Beijersbergen 1994, Sheng & Citovsky 1996, Tzfira *et al.* 2000, Zupan *et al.* 2000).

Disarmed Ti-plasmid derivatives as plant vectors

Function of T-DNA genes

Genetic maps of T-DNA have been obtained by studying spontaneous and transposon-induced mutants that affect tumour morphology, generating tumours that are larger than normal or that show 'shooty' or 'rooty' phenotypes. Although normal tumours can grow on medium lacking auxins and cytokinins, the tumour cells actually contain high levels of these hormones. Ooms *et al.* (1981) therefore proposed that the oncogenes carried on the T-DNA encoded products involved in phytohormone synthesis and that the abnormal morphologies of T-DNA mutants were due to a disturbance in the balance of plant hormones in the callus. The cloning and functional analysis of T-DNA genes has confirmed that those with 'shooty' mutant phenotypes encode enzymes for auxin biosynthesis and those with 'rooty' phenotypes are involved in cytokine production (Weiler & Schroder 1987). Other genes have been identified as encoding enzymes for opine synthesis, while the function of some genes remains unknown (Table 12.1). The transcript maps of

Table 12.1 Functions of some T-DNA genes in *Agrobacterium tumefaciens* Ti plasmids.

Gene	Product	Function
ocs	Octopine synthase	Opine synthesis
nos	Nopaline synthase	Opine synthesis
tms1 (*iaaH, auxA*)	Tryptophan-2-mono-oxygenase	Auxin synthesis
tms2 (*iaaM, auxB*)	Indoleacetamide hydrolase	Auxin synthesis
tmr (*ipt, cyt*)	Isopentyl transferase	Cytokinin synthesis
tml	Unknown	Unknown, mutations affect tumour size
frs	Fructopine synthase	Opine synthesis
mas	Mannopine synthase	Opine synthesis
ags	Agropine synthase	Opine synthesis

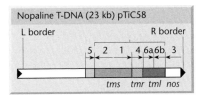

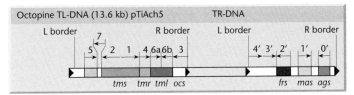

Fig. 12.10 Structure and transcription of T-DNA. The T-regions of nopaline and octopine Ti-plasmids have been aligned to indicate the common DNA sequences. The size and orientation of each transcript (numbered) is indicated by arrows. Genetic loci, defined by deletion and transposon mutagenesis, are shown as follows: *nos*, nopaline synthase; *ocs*, octopine synthase; *tms*, shooty tumour; *tmr*, rooty tumour.

T-DNAs from a nopaline plasmid (pTiC58) and an octopine plasmid (pTiAch5) are shown in Fig. 12.10 (Willmitzer *et al.* 1982, 1983, Winter *et al.* 1984).

Interestingly, nucleotide sequencing has revealed that the T-DNA genes have promoter elements and polyadenylation sites that are eukaryotic in nature (De Greve *et al.* 1982b, Depicker *et al.* 1982, Bevan *et al.* 1983a). This explains how genes from a bacterial plasmid come to be expressed when transferred to the plant nucleus. It is possible that the sequences may have been captured from plants during the evolution of the Ti plasmid. The ability of *Agrobacterium* to induce tumours in a wide variety of plants suggested that T-DNA promoters, such as those of the *ocs* (octopine synthase) and *nos* (nopaline synthase) genes, could be useful to drive transgene expression. These and other promoters used for transgene expression in plants are discussed in Box 12.1.

Prototype disarmed Ti vectors

We have seen that the Ti plasmid is a natural vector for genetically engineering plant cells because it can transfer its T-DNA from the bacterium to the plant genome. However, wild-type Ti plasmids are not suitable as general gene vectors because the T-DNA contains oncogenes that cause disorganized growth of the recipient plant cells. To be able to regenerate plants efficiently, we must use vectors in which the T-DNA has been *disarmed* by making it non-oncogenic. This is most effectively achieved simply by deleting all of its oncogenes. For example, Zambryski *et al.* (1983) substituted pBR322 sequences for almost all of the T-DNA of pTiC58, leaving only the left and right border regions and the *nos* gene. The

resulting construct was called pGV3850 (Fig. 12.11). *Agrobacterium* carrying this plasmid transferred the modified T-DNA to plant cells. As expected, no tumour cells were produced, but the fact that transfer had taken place was evident when the cells were screened for nopaline production and found to be positive. Callus tissue could be cultured from these nopaline-positive cells if suitable phytohormones were provided, and fertile adult plants were regenerated by hormone induction of plantlets.

The creation of disarmed T-DNA was an important step forward, but the absence of tumour formation made it necessary to use an alternative method to identify transformed plant cells. In the experiment described above, opine production was exploited as a screenable phenotype, and the *ocs* and *nos* genes have been widely used as screenable markers

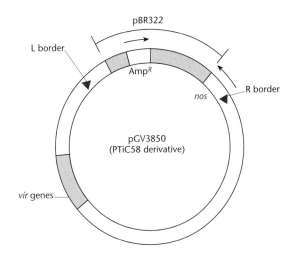

Fig. 12.11 Structure of the Ti-plasmid pGV3850, in which the T-DNA has been disarmed.

Box 12.1 Control of transgene expression in plants

Promoters

To achieve high-level and constitutive transgene expression in plants, a very active promoter is required. In dicots, promoters from the *Agrobacterium* nopaline synthase (*nos*), octopine synthase (*ocs*) and mannopine synthase (*mas*) genes have been widely used. These are constitutive and also moderately induced by wounding (An *et al.* 1990, Langridge *et al.* 1989). The most popular promoter for transgene expression in dicots is the 35S RNA promoter from cauliflower mosaic virus (CaMV 35S). This is very active, but can be improved still further by duplicating the enhancer region (Rathus *et al.* 1993). These promoters have a much lower activity in monocots, and duplicating the CaMV 35S enhancer has little effect. Alternative promoters have therefore been sought for transgene expression in cereals (reviewed by McElroy & Brettel 1994). The rice *actin-1* and maize *ubiquitin-1* promoters have been widely used for this purpose (McElroy *et al.* 1995, Christensen & Quail 1996). As well as constitutive promoters, a large number of promoters have been used to direct transgene expression in particular tissues. In monocots, promoters from seed storage-protein genes, such as maize zein, wheat glutenin and rice glutelins, have been used to target transgene expression to the seeds, which is beneficial for the accumulation of recombinant proteins (Wu *et al.* 1998; reviewed by Bilan *et al.* 1999). Promoters targeting transgene expression to green tissue are also useful (e.g. Graham *et al.* 1997, Datta *et al.* 1998). We discuss inducible expression systems for animals and plants in Chapter 13.

Other components of the expression vector

As discussed for animal cells (Box 10.2), other sequences in the expression vector also influence transgene expression. Generally, the presence of an intron in a plant expression cassette increases the activity of the promoter (Bilan *et al.* 1999). The insertion of a heterologous intron enhances the activity of the CaMV 35S promoter in monocots (e.g. see Mascarenhas *et al.* 1990, Vain *et al.* 1996) and constructs containing the actin or ubiquitin promoters generally include the first intron of the gene (McElroy *et al.* 1991). All transgenes must include a polyadenylation site, which in most cases is derived from the *Agrobacterium nos* gene or the CaMV 35S RNA. Whereas in animals it is conventional to remove untranslated regions from the expression construct, a number of such sequences in plants have been identified as translational enhancers. For example, the 5′ leader sequence of the tobacco mosaic virus RNA, known as the omega sequence, can increase transgene expression up to 80-fold (reviewed by Futterer & Hohn 1996, Gallie 1996). As in animals, the translational start site should conform to Kozak's consensus (Kozak 1999; see Box 10.2 for details) and the transgene should be codon-optimized for the expression host. A good example of the latter is the use of codon-optimized insecticidal toxin genes from *Bacillus thuringiensis* for expression in transgenic crops, leading to dramatically increased expression levels compared with the unmodified genes (Koziel *et al.* 1996). Also, the inclusion of targeting information in the expression cassette may be beneficial for the accumulation of recombinant proteins. For example, recombinant antibodies expressed in plants are much more stable if targeted to the endoplasmic reticulum (ER), since this provides a favourable molecular environment for folding and assembly. Targeting is achieved using an N-terminal signal sequence to direct the ribosome to the ER and a C-terminal tetrapeptide retrieval signal, KDEL, which causes accumulation in the ER lumen (Horvath *et al.* 2000).

(reviewed by Dessaux & Petit 1994). However, there are several drawbacks associated with this system, particularly the necessity to carry out enzymatic assays on all potential transformants. To provide a more convenient way to identify transformed plant cells, dominant selectable markers have been inserted into the T-DNA, so that transformed plant cells can be selected on the basis of drug or herbicide resistance. The use of selectable markers in plants is discussed in more detail in Box 12.2.

Box 12.2 Selectable markers for plants

Until recently, almost all selectable markers used in plants were dominant selectable markers, providing resistance to either antibiotics or herbicides (see table). Some plants, particularly monocots, are naturally tolerant of kanamycin, and this antibiotic may also interfere with regeneration. In these species, alternative systems, such as hygromycin or phosphinothricin selection, are preferred.

The introduction of markers such as *nptII*, *hpt* and *dhfr* into the T-DNA of disarmed Ti plasmids provided the first convenient methods to identify transformed plant tissue, and hence opened the way for *Agrobacterium* to be used as a general plant transformation system. The marker and experimental transgene can be cloned in tandem on the same T-DNA. In such cases, it is better for the selectable marker to be placed adjacent to the left border repeat, since this is transferred to the plant last (Sheng & Citovsky 1996). This strategy reduces the likelihood of obtaining plants under selection containing the marker alone and not the transgene of interest (see Hellens *et al.* 2000a). Alternatively, cotransformation can be achieved using *Agrobacterium* strains containing two plasmids or by co-inoculating plants with different *Agrobacterium* strains, each containing a single plasmid. Although there is some controversy surrounding the fate of co-introduced T-DNA sequences, it appears that nopaline-type plasmids favour the co-integration of multiple T-DNAs at the same locus, often in an inverted repeat pattern, while octopine-type plasmids favour independent integration sites, which can segregate in progeny plants (Depicker *et al.* 1985, Jones *et al.* 1987, Jorgensen *et al.* 1987, De Block & Debrouwer 1991). In direct transformation methods, the selectable marker can be included either on the same vector as the experimental transgene or on a separate vector, since cotransformation occurs at a high frequency (Schocher *et al.* 1986, Christou & Swain 1990).

Marker	Selection	References
Drug resistance markers		
aad (preferred for chloroplast transformation (see p. 240)	Trimethoprim, streptomycin, spectinomycin, sulphonamides	Svab *et al.* 1990a
ble	Bleomycin	Hille *et al.* 1986
dhfr	Methotrexate	Eichholtz *et al.* 1987
hpt	Hygromycin	Van den Elzen *et al.* 1985
nptII and *aphII*	Kanamycin, neomycin, G418	Pridmore 1987
gat	Gentamicin	Hayford *et al.* 1988
Herbicide resistance markers		
bar and *pat*	Phosphinothricin (bialaphos, glufosinate ammonium, Basta)	De Block *et al.* 1987
csr1-1	Chlorsulphuron	Haughn *et al.* 1988
dhps (*sul*)	Sulphonamides (Asualam)	Guerineau & Mullineaux 1989
epsp	Glyphosate	Shah *et al.* 1986

continued

Box 12.2 *continued*

Recently, public concern that antibiotic- and herbicide-resistance markers could pose a threat to health or the environment has prompted research into alternative innocuous marker systems. One example is the *E. coli manA* gene, which encodes mannose phosphate isomerase and confers upon transformed cells the ability to use mannose as a sole carbon source (Negrotto *et al.* 2000). Another is the *A. tumefaciens ipt* (isopentyl transferase) gene, located on the T-DNA, which induces cytokinin synthesis and can be used to select plants on the basis of their ability to produce shoots from callus on medium lacking cytokinins (Kunkel *et al.* 1999). Other strategies have also been explored, such as eliminating markers by sexual crossing (Komari *et al.* 1996), transposition (Goldsbrough *et al.* 1993) or site-specific

recombination (Dale & Ow 1991, Russel *et al.* 1992, Zubko *et al.* 2000). We consider the use of Cre-*loxP* for marker excision in transgenic organisms in Chapter 13. To verify the successful elimination of particular genes, counterselectable markers are required (e.g. see the discussion of *Tk* as a couterselectable marker for gene targeting in mice, p. 206). In plants, the *A. tumefaciens* T-DNA gene *tms2* has been used as a negative marker. This encodes indoleacetamide hydrolase, an enzyme that converts naphthaleneacetamide (NAM) into the potent auxin naphthaleneacetic acid. In the presence of exogenously applied NAM, transformed callus is unable to produce shoots due to the excess levels of auxin, therefore only callus lacking the gene is able to regenerate into full transgenic plants (Sundaresan *et al.* 1995).

Cointegrate vectors

Although disarmed derivatives of wild-type Ti plasmids can be used for plant transformation, they are not particularly convenient as experimental gene vectors, because their large size makes them difficult to manipulate *in vitro* and there are no unique restriction sites in the T-DNA. Initially, this problem was addressed by the construction of cointegrate vectors. T-DNA isolated from a parent Ti plasmid was subcloned in a conventional *Escherichia coli* plasmid vector for easy manipulation, producing a so-called *intermediate vector* (Matzke & Chilton 1981). These vectors were incapable of replication in *A. tumefaciens* and also lacked conjugation functions. Transfer was achieved using a 'triparental mating' in which three bacterial strains were mixed together: (i) an *E. coli* strain carrying a helper plasmid able to mobilize the intermediate vector *in trans*; (ii) the *E. coli* strain carrying the recombinant intermediate vector; and (iii) *A. tumefaciens* carrying the Ti plasmid. Conjugation between the two *E. coli* strains transferred the helper plasmid to the carrier of the intermediate vector, which was in turn mobilized and transferred to the recipient *Agrobacterium*.

Homologous recombination between the T-DNA sequences of the Ti plasmid and intermediate vector then resulted in the formation of a large cointegrate plasmid, from which the recombinant T-DNA was transferred to the plant genome (Fig. 12.12). In the cointegrate vector system, maintenance of the recombinant T-DNA is dependent on recombination, which is enhanced if there is an extensive homology region shared by the two plasmids, as in Ti plasmid pGV3850, which carries a segment of the pBR322 backbone in its T-DNA.

Binary vectors

Although intermediate vectors have been widely used, the large cointegrates are not necessary for transformation. The *vir* genes of the Ti plasmid function *in trans* and can act on any T-DNA sequence present in the same cell. Therefore, the *vir* genes and the disarmed T-DNA containing the transgene can be supplied on separate plasmids, and this is the principle of binary vector systems (Hoekma *et al.* 1983, Bevan 1984). The T-DNA can be subcloned on a small *E. coli* plasmid for ease of manipulation. This plasmid, called mini-Ti or micro-Ti, can be

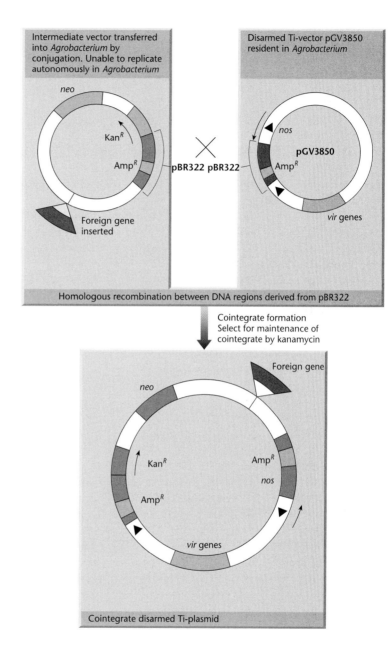

Intermediate vector transferred into *Agrobacterium* by conjugation. Unable to replicate autonomously in *Agrobacterium*

neo

Kan^R

Amp^R

Foreign gene inserted

pBR322 pBR322

Disarmed Ti-vector pGV3850 resident in *Agrobacterium*

nos

pGV3850

Amp^R

vir genes

Homologous recombination between DNA regions derived from pBR322

Cointegrate formation
Select for maintenance of cointegrate by kanamycin

Foreign gene

neo

Kan^R

Amp^R

Amp^R

nos

vir genes

Cointegrate disarmed Ti-plasmid

Fig. 12.12 Production of recombinant disarmed Ti plasmid by cointegrate formation.

introduced into an *Agrobacterium* strain carrying a Ti plasmid from which the T-DNA has been removed. The *vir* functions are supplied *in trans*, causing transfer of the recombinant T-DNA to the plant genome. The T-DNA plasmid can be introduced into *Agrobacterium* by triparental matings or by a more simple transformation procedure, such as electroporation (Cangelosi *et al.* 1991).

Most contemporary Ti-plasmid transformation systems are based on a binary principle, in which the T-DNA is maintained on a shuttle vector with a broad-host-range origin of replication, such as RK2 (which functions in both *A. tumefaciens* and *E. coli*), or separate origins for each species. An independently replicating vector is advantageous because maintenance of the T-DNA is not reliant

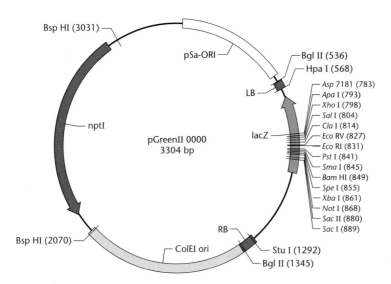

Fig. 12.13 The small and versatile binary vector pGreen, reproduced with permission of Roger Hellens and Phil Mullineaux.

on recombination, and the binary vector's copy number is not determined by the Ti plasmid, making the identification of transformants much easier. All the conveniences of bacterial cloning plasmids have been incorporated into binary vectors, such as multiple unique restriction sites in the T-DNA region to facilitate subcloning, the *lacZ* gene for blue–white screening (McBride & Summerfelt 1990) and a λ cos site for preparing cosmid libraries (Lazo *et al.* 1991, Ma *et al.* 1992). A current binary vector, pGreen, is shown in Fig. 12.13 (Hellens *et al.* 2000b). This plasmid is less than 5 kbp in size and has 18 unique restriction sites in the T-DNA, because the T-DNA is entirely synthetic. It has a *lacZ* gene for blue–white selection of recombinants, and a selectable marker that can be used both in bacteria and in the transformed plants. The progressive reduction in size has been made possible by removing essential genes required for replication in *Agrobacterium* and transferring these genes to the bacterium's genome or on to a helper plasmid. The pGreen plasmid, for example, contains the Sa origin of replication, which is much smaller than the more traditional Ri and RK2 regions. Furthermore, an essential replicase gene is housed on a second plasmid, called pSoup, resident within the bacterium. All conjugation functions have also been removed, so this plasmid can only be introduced into *Agrobacterium* by transformation (Hellens *et al.* 2000b).

A simple experimental procedure for *Agrobacterium*-mediated transformation

Once the principle of selectable, disarmed T-DNA vectors was established, there followed an explosion in the number of experiments involving DNA transfer to plants. Variations on the simple general protocol of Horsch *et al.* (1985) have been widely used for dicot plants (Fig. 12.14). In the original report, small discs (a few millimetres in diameter) were punched from leaves, surface-sterilized and inoculated in a medium containing *A. tumefaciens* transformed with the recombinant disarmed T-DNA (as a cointegrate or binary vector). The foreign DNA contained a chimeric *neo* gene conferring resistance to the antibiotic kanamycin. The discs were cultured for 2 days and transferred to medium containing kanamycin, to select for the transferred *neo* gene, and carbenicillin, to kill the *Agrobacterium*. After 2–4 weeks, developing shoots were excised from the callus and transplanted to root-induction medium. Rooted plantlets were subsequently transplanted to soil, about 4–7 weeks after the inoculation step.

This method has the advantage of being simple and relatively rapid. It is superior to previous methods, in which transformed plants were regenerated from protoplast-derived callus, the protoplasts having been transformed by cocultivation with the

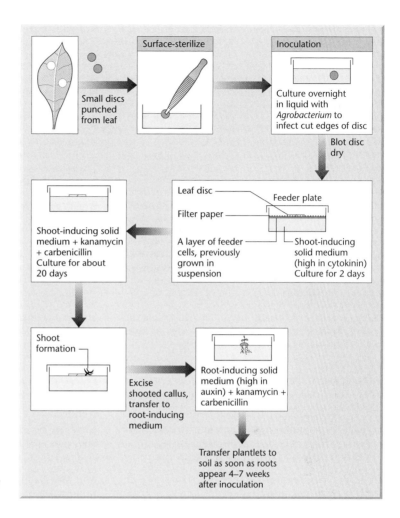

Fig. 12.14 Leaf-disc transformation by
Agrobacterium tumefaciens.

Agrobacterium (De Block *et al.* 1984, Horsch *et al.* 1984). Contemporary protocols for the *Agrobacterium*-mediated transformation of many solanaceous plants are variations on the theme of the leaf-disc protocol, although the optimal explant must be determined for each species. Alternative procedures are required for the transformation of monocots, as discussed below.

Agrobacterium and monocots

Until recently, most monocotyledonous plants (monocots) were thought to be outside the host range of *Agrobacterium*, prompting research into alternative transformation methods, as discussed below. During the 1980s, limited evidence accumulated showing

that some monocots might be susceptible to *Agrobacterium* infection (see, for example, the discussion of agroinfection with maize streak virus DNA on p. 243). However, in most cases there was no convincing evidence for T-DNA integration into the plant genome. In the laboratory, it proved possible to induce tumours in certain monocot species, such as asparagus (Hernalsteens *et al.* 1984) and yam (Schafer *et al.* 1987). In the latter case, an important factor in the success of the experiment was pretreatment of the *Agrobacterium* suspension with wound exudate from potato tubers. It has been argued that *Agrobacterium* infection of monocots is inefficient because wounded monocot tissues do not produce phenolics, such as acetosyringone, at sufficient levels to induce *vir* gene expression.

In the last 10 years, amazing progress has been made in the transformation of cereals using *Agrobacterium*. The first species to be transformed was rice. The *nptII* gene was used as a selectable marker, and successful transformation was demonstrated both by the resistance of transgenic callus to kanamycin or G418 and by the presence of T-DNA in the genome (Raineri *et al.* 1990, Chan *et al.* 1992, 1993). However, these antibiotics interfere with the regeneration of rice plants, so only four transgenic plants were produced in these early studies. The use of an alternative marker conferring resistance to hygromycin allowed the regeneration of large numbers of transgenic japonica rice plants (Hiei *et al.* 1994), and the same selection strategy has been used to produce transgenic rice plants representing the other important subspecies, indica and javanica (Dong *et al.* 1996, Rashid *et al.* 1996). More recently, efficient *Agrobacterium*-mediated transformation has become possible for other important cereals, including maize (Ishida *et al.* 1996), wheat (Cheng *et al.* 1997), barley (Tingay *et al.* 1997) and sugar cane (Arencibia *et al.* 1998).

The breakthrough in cereal transformation using *Agrobacterium* reflected the recognition of a number of key factors required for efficient infection and gene transfer to monocots. The use of explants containing a high proportion of actively dividing cells, such as embryos or apical meristems, was found to greatly increase transformation efficiency, probably because DNA synthesis and cell division favour the integration of exogenous DNA. In dicots, cell division is often induced by wounding, whereas wound sites in monocots tend to become lignified. This probably explains why traditional procedures, such as the leaf-disc method, are inefficient in monocots. Hiei *et al.* (1994) showed that the cocultivation of *Agrobacterium* and rice embryos in the presence of 100 mmol/l acetosyringone was a critical factor for successful transformation. Transformation efficiency is increased further by the use of vectors with enhanced virulence functions. The modification of *Agrobacterium* for increased virulence has been achieved by increasing the expression of *virG* (which in turn boosts the expression of the other *vir* genes) and/or the expression of *virE1*, which is a major limiting factor in T-DNA transfer (reviewed by Sheng & Citovsky 1996), resulting in so-called *supervirulent*

bacterial strains, such as AGL-1. Komari *et al.* (1996) used a different strategy, in which a portion of the virulence region from the Ti plasmid of supervirulent strain A281 was transferred to the T-DNA-carrying plasmid to generate a so-called '*superbinary vector*'. The advantage of the latter technique is that the superbinary vector can be used in any *Agrobacterium* strain.

High-capacity binary vectors

A precise upper limit for T-DNA transfer has not been established. It is greater than 50 kb (Herrera-Estrella *et al.* 1983a), but using standard vectors it is difficult to routinely transfer inserts larger than 30 kb due to instability in the bacterial host. The analysis of very large genes or the transfer of multiple genes (such as those encoding sequentially acting enzymes of a metabolic pathway) can now be achieved thanks to the development of high-capacity binary vectors based on the artificial-chromosome-type vectors used in *E. coli*. The first to be described was BIBAC2 (Hamilton 1997). This contains an F-plasmid origin of replication and is modelled on the bacterial artificial chromosome (BAC) (p. 67). The basic vector transforms tobacco with high efficiency, but the efficiency of transformation drops substantially when large inserts are used. This vector has been used to introduce 150 kbp of human DNA flanked by T-DNA borders into the tobacco genome, although virulence helper plasmids supplying high levels of VirG and VirE *in trans* were critical for successful DNA transfer (Hamilton *et al.* 1996). An alternative vector carrying a P1 origin of replication and modelled on the P1 artificial chromosome (PAC) (p. 68) was constructed by Liu *et al.* (1999). This transformation-competent bacterial artificial chromosome (TAC) vector was used to introduce up to 80 kbp of genomic DNA into *Arabidopsis* and, while there was some loss of efficiency with the larger inserts, it was still possible to produce many transgenic plants. Both vectors contain a kanamycin-resistance marker for selection in bacteria and *hpt* for hygromycin selection in transgenic plants. For the reasons discussed in Box 12.2, the *hpt* marker gene is placed adjacent to the right border T-DNA repeat. Both vectors also contain the Ri origin for maintenance in *Agrobacterium* and, within the T-DNA region, the *sacB* marker for negative selection in *E. coli*, interrupted by a poly-linker for cloning foreign DNA.

One of the most attractive uses of high-capacity binary vectors is for the positional cloning of genes identified by mutation. The ability to introduce large segments of DNA into the plant genome effect-ively bridges the gap between genetic mapping and sequencing, allowing the position of mutant genes to be narrowed down by complementation. Genomic libraries have been established for several plant species in BIBAC2 and TAC vectors (Hamilton *et al.* 1999, Shibata & Lui 2000) and a number of novel genes have been isolated (e.g. Sawa *et al.* 1999, Kubo & Kakimoto 2001).

Agrobacterium rhizogenes and Ri plasmids

Agrobacterium rhizogenes causes hairy-root disease in plants and this is induced by root-inducing (Ri) plasmids, which are analogous to the Ti plasmids of *A. tumefaciens*. The Ri T-DNA includes genes homo-logous to the *iaaM* (tryptophan 2-mono-oxygenase) and *iaaH* (indoleacetamide hydrolase) genes of *A. tumefaciens*. Four other genes present in the Ri T-DNA are named *rol* for *root locus*. Two of these, *rolB* and *rolC*, encode P-glucosidases able to hydrolyse indole- and cytokine-*N*-glucosides. *A. rhizogenes* therefore appears to alter plant physiology by releasing free hormones from inactive or less active conjugated forms (Estruch *et al.* 1991a,b).

Ri plasmids are of interest from the point of view of vector development, because opine-producing root tissue induced by Ri plasmids in a variety of dicots can be regenerated into whole plants by manipula-tion of phytohormones in the culture medium. Ri T-DNA is transmitted sexually by these plants and affects a variety of morphological and physiological traits, but does not in general appear deleterious. The Ri plasmids therefore appear to be already equivalent to disarmed Ti plasmids (Tepfer 1984). Transformed roots can also be maintained as hairy-root cultures, which have the potential to produce certain valuable secondary metabolites at higher levels than suspension cultures and are much more genetically stable (Hamil *et al.* 1987, Signs & Flores 1990). The major limitation for the commercial use

of hairy-root cultures is the difficulty involved in scale-up, since each culture comprises a heterogen-eous mass of interconnected tissue, with highly uneven distribution (reviewed by Giri & Narassu 2000).

Many of the principles explained in the context of disarmed Ti plasmids are applicable to Ri plasmids. A cointegrate vector system has been developed (Jensen *et al.* 1986) and applied to the study of nodulation in transgenic legumes. Van Sluys *et al.* (1987) have exploited the fact that *Agrobacterium* containing both an Ri plasmid and a disarmed Ti plasmid can frequently cotransfer both plasmids. The Ri plasmid induces hairy-root disease in recipient *Arabidopsis* and carrot cells, serving as a transforma-tion marker for the cotransferred recombinant T-DNA and allowing regeneration of intact plants. No drug-resistance marker on the T-DNA is necessary with this plasmid combination.

Direct DNA transfer to plants

Protoplast transformation

Until comparatively recently, the limited host range of *A. tumefaciens* precluded its use for the genetic manipulation of a large number of plant species, including most monocots. At first, the only alternat-ive to *Agrobacterium*-mediated transformation was the introduction of DNA into protoplasts. This pro-cess has much in common with the transfection of animal cells (Chapter 10). The protoplasts must initially be persuaded to take up DNA from their surroundings, after which the DNA integrates stably into the genome in a proportion of these transfected cells. Gene transfer across the protoplast membrane is promoted by a number of chemicals, of which polyethylene glycol has become the most widely used, due to the availability of simple transformation protocols (Negrutiu *et al.* 1987). Alternatively, DNA uptake may be induced by electroporation, which has also become a favoured technique (Shillito *et al.* 1985). As with animal cells, the introduction of a selectable marker gene along with the transgene of interest is required for the identification of stable transformants. This can be achieved using plasmid vectors carrying both the marker and the transgene of interest, but the use of separate vectors also results in a high frequency of cotransformation (Schocher *et al.* 1986). Putative transformants are transferred to selective medium, where surviving protoplasts regenerate their cell walls and commence cell division, producing a callus. Subsequent manipulation of the culture conditions then makes it possible to induce shoot and root development, culminating in the recovery of fertile transgenic plants. The major limitation of protoplast transformation is not the gene-transfer process itself, but the ability of the host species to regenerate from protoplasts. A general observation is that dicots are more amenable than monocots to this process. In species where regeneration is possible, an advantage of the technique is that protoplasts can be cryopreserved and retain their regenerative potential (DiMaio & Shillito 1989).

The first transformation experiments concentrated on species such as tobacco and petunia in which protoplast-to-plant regeneration is well documented. An early example is provided by Meyer *et al.* (1987), who constructed a plasmid vector containing the *nptII* marker gene, and a maize complementary DNA (cDNA) encoding the enzyme dihydroquercetin 4-reductase, which is involved in anthocyanin pigment biosynthesis. The transgene was driven by the strong and constitutive cauliflower mosaic virus (CaMV) 35S promoter. Protoplasts of a mutant, white-coloured petunia strain were transformed with the recombinant plasmid by electroporation and then selected on kanamycin-supplemented medium. After a few days, surviving protoplasts had given rise to microcalli, which could be induced to regenerate into whole plants. The flowers produced by these plants were brick-red instead of white, showing that the maize cDNA had integrated into the genome and was expressed.

After successful experiments using model dicots, protoplast transformation was attempted in monocots, for which no alternative gene-transfer system was then available. In the first such experiments, involving wheat (Lorz *et al.* 1985) and the Italian ryegrass *Lolium multiflorum* (Potrykus *et al.* 1985b), protoplast transformation was achieved and transgenic callus obtained, but it was not possible to recover transgenic plants. The inability of most monocots to regenerate from protoplasts may reflect the loss of competence to respond to tissue-culture conditions as the cells differentiate. In cereals and

grasses, this has been addressed to a certain extent by using embryogenic suspension cultures as a source of protoplasts. Additionally, since many monocot species are naturally tolerant to kanamycin, the *nptII* marker used in the initial experiments was replaced with alternative markers conferring resistance to hygromycin or phosphinothricin. With these modifications, it has been possible to regenerate transgenic plants representing certain varieties of rice and maize with reasonable efficiency (Shimamoto *et al.* 1988, Datta *et al.* 1990, Omirulleh *et al.* 1993). However, the extended tissue-culture step is unfavourable, often resulting in sterility and other phenotypic abnormalities in the regenerated plants. The transformation of protoplasts derived from stomatal guard cells has recently been identified as an efficient and genotype-independent method for the production of transgenic sugar beet (Hall *et al.* 1996).

Particle bombardment

An alternative procedure for plant transformation was introduced in 1987, involving the use of a modified shotgun to accelerate small (1–4 µm) metal particles into plant cells at a velocity sufficient to penetrate the cell wall (~250 m/s). In the initial test system, intact onion epidermis was bombarded with tungsten particles coated in tobacco mosaic virus (TMV) RNA. Three days after bombardment, approximately 40% of the onion cells that contained particles also showed evidence of TMV replication (Sanford *et al.* 1987). A plasmid containing the *cat* reporter gene driven by the CaMV 35S promoter was then tested to determine whether DNA could be delivered by the same method. Analysis of the epidermal tissue 3 days after bombardment revealed high levels of transient chloramphenicol transacetylase (CAT) activity (Klein *et al.* 1987).

The stable transformation of explants from several plant species was achieved soon after these initial experiments. Early reports included the transformation of soybean (Christou *et al.* 1988), tobacco (Klein *et al.* 1988b) and maize (Klein *et al.* 1988a). In each case, the *nptII* gene was used as a selectable marker and transformation was confirmed by the survival of callus tissue on kanamycin-supplemented medium. The ability to stably transform plant cells by this method offered the exciting possibility of generating transgenic plants representing species that were, at the time, intractable to other transformation procedures. In the first such report, transgenic soybean plants were produced from meristem tissue isolated from immature seeds (McCabe *et al.* 1988). In this experiment, the screenable marker gene *gusA* was introduced by particle bombardment and transgenic plants were recovered in the absence of selection by screening for β-glucuronidase (GUS) activity (Box 13.1). Other early successes included cotton, papaya, maize and tobacco (Finer & McMullen 1990, Fitch *et al.* 1990, Fromm *et al.* 1990, Gordon-Kamm *et al.* 1990, Tomes *et al.* 1990).

There appears to be no intrinsic limitation to the scope of this procedure, since DNA delivery is governed entirely by physical parameters. Many different types of plant material have been used as transformation targets, including callus, cell-suspension cultures and organized tissues, such as immature embryos, meristems and leaves. The number of species in which transgenic plants can be produced using variants of particle bombardment has therefore increased dramatically over the last 10 years. Notable successes include almost all of the commercially important cereals, i.e. rice (Christou *et al.* 1991), wheat (Vasil *et al.* 1992), oat (Somers *et al.* 1992, Torbert *et al.* 1995), sugar cane (Bower & Birch 1992) and barley (Wan & Lemaux 1994, Hagio *et al.* 1995).

The original gunpowder-driven device has been improved and modified, resulting in greater control over particle velocity and hence greater reproducibility of transformation conditions. An apparatus based on electric discharge (McCabe & Christou 1993) has been particularly useful for the development of variety-independent gene-transfer methods for the more recalcitrant cereals and legumes. Several instruments have been developed where particle acceleration is controlled by pressurized gas. These include a pneumatic apparatus (Iida *et al.* 1990), a 'particle inflow gun' using flowing helium (Finer *et al.* 1992, Takeuchi *et al.* 1992) and a device utilizing compressed helium (Sanford *et al.* 1991). Physical parameters, such as particle size and acceleration (which affect the depth of penetration and the amount of tissue damage), as well as the amount and conformation of the DNA used to coat the particles, must be optimized for each species and type of explant. However, the nature of the transformation target is probably the most important single variable in the success of gene transfer. The pretreatment of explants with an osmoticum has often been shown to improve transformation efficiency, probably by preventing the deflection of particles by films or droplets of water. Factors influencing the success of gene transfer by particle bombardment have been extensively reviewed (Sanford *et al.* 1993, Birch & Bower 1994).

Other direct DNA-transfer methods

There is a great diversity of approaches for gene transfer to animal cells and many of the same methods have been attempted in plants. Electroporation has been used to transform not only protoplasts (see above) but also walled plant cells, either growing in suspension or as part of intact tissues. In many cases, the target cells have been wounded or pretreated with enzymes in order to facilitate gene transfer (e.g. D'Halluin *et al.* 1992, Laursen *et al.* 1994). It has been shown, however, that immature rice, wheat and maize embryos can be transformed using electroporation without any form of pretreatment (Kloti *et al.* 1993, Xu & Li 1994). Other transformation methods also involve perforation of the cell, including the use of silicon carbide whiskers (Thompson *et al.* 1995, Nagatani *et al.* 1997), ultrasound (Zhang *et al.* 1991) or a finely focused laser beam (Hoffman 1996). In most of these cases, only transient expression of the introduced DNA has been achieved, although transgenic maize plants have been recovered following whisker-mediated transformation.

Finally, microinjection has been used to introduce DNA directly into the fertilized eggs of many animals (Chapter 10). In plants, microinjection of DNA into zygotes may also be the most direct way to produce transgenics, but so far the technique is inefficient and not widely used (Leduc *et al.* 1996, Holm *et al.* 2000).

In planta transformation

Until recently, gene transfer to plants involved the use of cells or explants as transformation targets and an obligatory tissue-culture step was needed for the regeneration of whole fertile plants. Experiments using the model dicot *Arabidopsis thaliana* have led the way in the development of so-called *in planta* transformation techniques, where the need for tissue culture is minimized or eliminated altogether. Such methods involve the introduction of DNA, either by *Agrobacterium* or by direct transfer, into intact plants. The procedure is carried out at an appropriate time in the plant's life cycle, so that the DNA becomes incorporated into cells that will contribute to the germ line, directly into the germ cells themselves (often at around the time of fertilization) or into the very early plant embryo. Generally, *in planta* transformation methods have a very low efficiency, so the small size of *Arabidopsis* and its ability to produce over 10 000 seeds per plant is advantageous. This limitation has so far prevented *in planta* techniques from being widely adopted for other plant species.

The first *in planta* transformation system involved imbibing *Arabidopsis* seeds overnight in an *Agrobacterium* culture, followed by germination (Feldmann & Marks 1987). A large number of transgenic plants containing T-DNA insertions were recovered, but in general this technique has a low reproducibility. Bechtold *et al.* (1993) has described a more reliable method, in which the bacteria are vacuum-infiltrated into *Arabidopsis* flowers. An even simpler technique called floral dip has become widely used (Clough & Bent 1998). This involves simply dipping *Arabidopsis* flowers into a bacterial suspension at the time of fertilization. In both these methods, the transformed plants are chimeric, but give rise to a small number of transgenic progeny (typically about 10 per plant). Similar approaches using direct DNA transfer have been tried in other species, but germ-line transformation has not been reproducible. For example, naked DNA has been injected into the floral tillers of rye plants (De La Pena *et al.* 1987) and post-fertilization cotton flowers (Zhou *et al.* 1983), resulting in the recovery of some transgenic plants. Transgenic tobacco has been produced following particle bombardment of pollen (Touraev *et al.* 1997).

An alternative to the direct transformation of germ-line tissue is the introduction of DNA into meristems *in planta*, followed by the growth of transgenic shoots. In *Arabidopsis*, this has been achieved simply by severing apical shoots at their bases and inoculating the cut tissue with *Agrobacterium* suspension (Chang *et al.* 1994). Using this procedure, transgenic plants were recovered from the transformed shoots at a frequency of about 5%. In rice, explanted meristem tissue has been transformed using *Agrobacterium* and particle bombardment, resulting in the proliferation of shoots that can be regenerated into transgenic plants (Park *et al.* 1996). Such procedures require only a limited amount of tissue culture.

Chloroplast transformation

So far, we have exclusively considered DNA transfer to the plant's nuclear genome. However, the chloroplast is also a useful target for genetic manipulation, because thousands of chloroplasts may be present in photosynthetic cells and this can result in levels of transgene expression up to 50 times higher than possible using nuclear transformation. Furthermore, transgenes integrated into chloroplast DNA do not appear to undergo silencing or suffer from position effects that can influence the expression levels of transgenes in the nuclear DNA (see Boxes 11.1 and 13.2). Chloroplast transformation also provides a natural containment method for transgenic plants, since the transgene cannot be transmitted through pollen (reviewed by Maliga 1993).

The first reports of chloroplast transformation were serendipitous, and the integration events were found to be unstable. For example, an early experiment in which tobacco protoplasts were cocultivated with *Agrobacterium* resulted in the recovery of one transgenic plant line, in which the transgene was transmitted maternally. Southern-blot analysis of chloroplast DNA showed directly that the foreign DNA had become integrated into the chloroplast genome (De Block *et al.* 1985). However, *Agrobacterium* does not appear to be an optimal system for chloroplast transformation, probably because the T-DNA complex is targeted to the nucleus. Therefore, direct DNA transfer has been explored as an alternative strategy. Stable chloroplast transformation was first achieved in the alga *Chlamydomonas reinhardtii*, which has a single large chloroplast occupying most of the volume of the cell (Boynton *et al.* 1988). Particle bombardment was used in this experiment. The principles established using this simple organism were extended to tobacco, allowing the recovery of stable *transplastomic* tobacco plants (Svab *et al.*

1990b). These principles included the use of vectors containing chloroplast homology regions, allowing targeted integration into the chloroplast genome, and the use of the selectable marker gene *aad* (aminoglycoside adenyltransferase), which confers resistance to streptomycin and spectinomycin (Zoubenko *et al.* 1994). Efficient chloroplast transformation has been achieved both through particle bombardment (e.g. Staub & Maliga 1992) and polyethylene glycol (PEG)-mediated transformation (Golds *et al.* 1993, Koop *et al.* 1996). The use of a combined selectable–screenable marker (*aad* linked to the gene for green fluorescent protein) allows the tracking of transplastomic sectors of plant tissue prior to chlorophyll synthesis, so that transformed plants can be rapidly identified (Khan & Maliga 1999).

Plant viruses as vectors

As an alternative to stable transformation using *Agrobacterium* or direct DNA transfer, plant viruses can be employed as gene-transfer and expression vectors. There are several advantages to the use of viruses. First, viruses are able to adsorb to and introduce their nucleic acid into intact plant cells. However, for many viruses, naked DNA or RNA is also infectious, allowing recombinant vectors to be introduced directly into plants by methods such as leaf rubbing. Secondly, infected cells yield large amounts of virus, so recombinant viral vectors have the potential for high-level transgene expression. Thirdly, viral infections are often systemic. The virus spreads throughout the plant, allowing transgene expression in all cells. Fourthly, viral infections are rapid, so large amounts of recombinant protein can be produced in a few weeks. Finally, all known plant viruses replicate episomally; therefore the transgenes they carry are not subject to the position effects that often influence the expression of integrated transgenes (Box 11.1). Since plant viruses neither integrate into nor pass through the germ line, plants cannot be stably transformed by viral infection and transgenic lines cannot be generated. However, this limitation can also be advantageous in terms of containment.

A complete copy of a viral genome can also be introduced into isolated plant cells or whole plants by *Agrobacterium* or direct DNA transfer. In this manner, it is possible to generate transiently transformed cell lines or transgenic plants carrying an integrated recombinant viral genome. In the case of RNA viruses, transcription of an integrated cDNA copy of the genome yields replication-competent viral RNA, which is amplified episomally, facilitating high-level transgene expression. Transgenic plants are persistently infected by the virus and can produce large amounts of recombinant protein. In the case of DNA viruses, *Agrobacterium*-mediated transient or stable transformation with T-DNA containing a partially duplicated viral genome can lead to the 'escape' of intact genomes, which then replicate episomally. The latter process, known as 'agroinfection' or 'agroinoculation', provides a very sensitive assay for gene transfer.

As well as their use for the expression of whole foreign proteins, certain plant viruses have recently been developed to present short peptides on their surfaces, similar to the phage-display technology discussed on p. 136. Epitope-display systems based on cowpea mosaic virus, potato virus X and tomato bushy stunt virus have been developed as a potential source of vaccines, particularly against animal viruses (reviewed by Johnson *et al.* 1997, Lomonossoff & Hamilton 1999).

DNA viruses as expression vectors

The vast majority of plant viruses have RNA genomes. However, the two groups of DNA viruses that are known to infect plants – the caulimoviruses and the geminiviruses – were the first to be developed as vectors, because of the ease with which their small, DNA genomes could be manipulated in plasmid vectors.

Cauliflower mosaic virus

The type member of the caulimoviruses is CaMV. The 8 kb double-stranded DNA (dsDNA) genome of several isolates has been completely sequenced, revealing an unusual structure characterized by the presence of three discontinuities in the duplex. A map of the CaMV genome is shown in Fig. 12.15. There are eight tightly packed genes, expressed as two major transcripts: the 35S RNA (which essentially represents the entire genome) and the 19S RNA (which contains the coding region for gene VI).

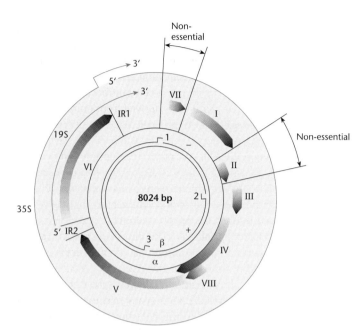

Fig. 12.15 Map of the cauliflower mosaic virus genome. The eight coding regions are shown by the thick grey arrows, and the different reading frames are indicated by the radial positions of the boxes. The thin lines in the centre indicate the (plus and minus) DNA strands with the three discontinuities. The major transcripts, 19S and 35S, are shown around the outside.

As discussed earlier in the chapter, the promoter and terminator sequences for both transcripts have been utilized in plant expression vectors, and the 35S promoter is particularly widely used (Box 12.1).

Only two of the genes in the CaMV genome are non-essential for replication (gene II and gene VII) and, since CaMV has an icosahedral capsid, the size of the genome cannot be increased greatly without affecting the efficiency of packaging. The maximum capacity of the CaMV capsid has been defined as 8.3 kb and, with the removal of all non-essential genes, this represents a maximum insert size of less than 1 kb (Daubert *et al.* 1983). This restriction in the capacity for foreign DNA represents a major limitation of CaMV vectors. Thus far it has been possible to express a number of very small transgenes, such as the 240 bp bacterial *dhfr* gene (Brisson *et al.* 1984), the 200 bp murine metallothionein cDNA (Lefebvre *et al.* 1987) and a 500 bp human interferon cDNA (De Zoeten *et al.* 1989). In Chapter 10, we describe how similar limitations were overcome for simian virus 40 (SV40), a virus that infects primate cells, through the development of replicon vectors and helper viruses or complementary cell lines supplying essential functions *in trans*. Unfortunately, such an approach is not possible with

CaMV, due to the high level of recombination that occurs, leading to rapid excision of the foreign DNA (Gronenborn *et al.* 1981).

Geminiviruses

Geminiviruses are characterized by their twin (geminate) virions, comprising two partially fused icosahedral capsids. The small single-stranded DNA genome is circular and, in some species, is divided into two segments called DNA A and DNA B. Interest in geminivirus vector development was stimulated by the discovery that such viruses use a DNA replicative intermediate, suggesting that they could be more stable than CaMV, whose RNA-dependent replication cycle is rather error-prone (Stenger *et al.* 1991). Of the three genera of geminiviruses, two have been developed as vectors. The begomoviruses have predominantly bipartite genomes; they are transmitted by the whitefly *Bemisia tabaci* and infect dicots. Species that have been developed as vectors include African cassava mosaic virus (ACMV) and tomato golden mosaic virus (TGMV). The mastreviruses have monopartite genomes; they are transmitted by leafhoppers and predominantly infect monocots. Species that have been developed as vectors include maize streak virus (MSV) and wheat dwarf virus (WDV).

An important additional distinction between these genera is that mastreviruses are not mechanically transmissible. MSV, for example, has never been introduced successfully into plants as native or cloned DNA. Grimsley *et al.* (1987) were able to overcome this problem using *Agrobacterium*, and were the first to demonstrate the principle of agroinfection. They constructed a plasmid containing a tandem dimer of the MSV genome. This dimer was inserted into a binary vector, and maize plants were infected with *A. tumefaciens* containing the recombinant T-DNA. Viral symptoms appeared within 2 weeks of inoculation. Agroinfection has been used to introduce the genomes of a number of different viruses into plants. It can be demonstrated that, if the T-DNA contains partially or completely duplicated genomes, single copies of the genome can escape and initiate infections. This may be mediated by homologous recombination or a replicative mechanism (Stenger *et al.* 1991). The study of Grimsley *et al.* (1987) incidentally provided the first evidence that *Agrobacterium* could transfer T-DNA to maize. Agroinfection is a very sensitive assay for transfer to the plant cell because of the amplification inherent in the virus infection and the resulting visible symptoms.

A number of geminiviruses have been developed as expression vectors, because of the possibility of achieving high-level recombinant-protein expression as a function of viral replication (reviewed by Stanley 1983, Timmermans *et al.* 1994, Palmer & Rybicki 1997). A generally useful strategy is the replacement of the coat-protein gene, since this is not required for replication and the strong promoter can be used to drive transgene expression. In the case of begomoviruses, which have bipartite genomes, the coat-protein gene is located on DNA A, along with all the functions required for DNA replication. Replicons based on DNA A are therefore capable of autonomous replication in protoplasts (e.g. Townsend *et al.* 1986). Geminivirus replicon vectors can facilitate the high-level transient expression of foreign genes in protoplasts. There appears to be no intrinsic limitation to the size of the insert, although larger transgenes tend to reduce the replicon copy number (e.g. Laufs *et al.* 1990, Matzeit *et al.* 1991). Generally,

it appears that mastrevirus replicons can achieve a much higher copy number in protoplasts than replicons based on begomiviruses. A WDV shuttle vector capable of replicating in both *E. coli* and plants was shown to achieve a copy number of greater than 3×10^4 in protoplasts derived from cultured maize endosperm cells (Timmermans *et al.* 1992), whereas the typical copy number achieved by TGMV replicons in tobacco protoplasts is less than 1000 (Kanevski *et al.* 1992). This may, however, reflect differences in the respective host cells, rather than the intrinsic efficiencies of the vectors themselves.

Geminiviruses are also valuable as expression vectors in whole plants. In the case of the mastreviruses, all viral genes appear to be essential for systemic infection, so coat-protein replacement vectors cannot be used in this manner. In contrast, the coat-protein genes of ACMV and TGMV are non-essential for systemic infection, but they are required for insect transmission (Briddon *et al.* 1990). Therefore, replicon vectors based on these viruses provide an *in planta* contained transient expression system. Note that viral movement functions are supplied by DNA B, so systemic infections occur only if DNA B is also present in the plant.

In an early study, Ward *et al.* (1988) replaced most of the ACMV AV1 gene with the *cat* reporter gene. In infected tobacco plants, high level CAT activity was detected for up to 4 weeks. Interestingly, they found that deletion of the coat-protein gene caused a loss of infectivity in plants, but this was restored upon replacement with *cat*, which is approximately the same size as the deleted gene. This and many subsequent reports indicated that, while there may be no intrinsic limit to the size of replicon vectors in protoplasts, systemic infection is dependent on preserving the size of the wild-type DNA-A component. A further limitation to this system is that the transmissibility of the recombinant genomes is poor, probably because they are not packaged. One way in which this can be addressed is to generate transgenic plants in which recombinant viral genomes are produced in every cell. This is achieved by transforming plants with DNA constructs containing a partially duplicated viral genome (Meyer *et al.* 1992). Intact replicons can excise from the delivered transgene in the same way as the MSV genome escapes during agroinfection, and autonomously replicating episomal copies can be detected. Transgenic tobacco plants have also been produced carrying an integrated copy of DNA B (Hayes *et al.* 1988, 1989). In the presence of replication functions supplied by DNA A, the DNA-B sequence is rescued from the transgene and can replicate episomally. DNA B can then provide movement functions for DNA A, facilitating the systemic spread of the vector.

RNA viruses as expression vectors

Most plant RNA viruses have a filamentous morphology, so the packaging constraints affecting the use of DNA viruses, such as CaMV, should not present a limitation in vector development. However, investigation into the use of RNA viruses as vectors lagged behind research on DNA viruses, awaiting the advent of robust techniques for the manipulation of RNA genomes.

A breakthrough was made in 1984, when a full-length clone corresponding to the genome of brome mosaic virus (BMV) was obtained. Infectious RNA could be produced from this cDNA by *in vitro* transcription (Ahlquist & Janda 1984, Ahlquist *et al.* 1984). The BMV genome comprises three segments: RNA1, RNA2 and RNA3. Only RNA1 and RNA2 are necessary for replication. RNA3, which is dicistronic, encodes the viral coat protein and movement protein. During BMV infection, a subgenomic RNA fragment is synthesized from RNA3, containing the coat-protein gene alone. It is therefore possible to replace the coat-protein gene with foreign DNA and still generate a productive infection (Ahlquist *et al.* 1987). This was demonstrated by French *et al.* (1986) in an experiment where the coat-protein gene was substituted with the *cat* reporter gene. Following the introduction of recombinant RNA3 into barley protoplasts, along with the essential RNA1 and RNA2 segments, high-level CAT activity was achieved.

This experiment showed that BMV was a potentially useful vector for foreign gene expression. However, to date, BMV has been used solely to study the function of genes from other plant viruses. Following the demonstration that infectious BMV RNA could be produced by *in vitro* transcription, the genomes of many other RNA viruses have prepared as cDNA

copies. Some of these viruses have been extensively developed as vectors for foreign gene expression (see comprehensive reviews by Scholthof *et al.* 1996, Porta & Lomonossoff 2001). Two examples are discussed below.

Tobacco mosaic virus

TMV is one of the most extensively studied plant viruses and was thus a natural choice for vector development. The virus has a monopartite RNA genome of 6.5 kb. At least four polypeptides are produced, including a movement protein and a coat protein, which are translated from subgenomic RNAs (Fig. 12.16). The first use of TMV as a vector was reported by Takamatsu *et al.* (1987). They replaced the coat-protein gene with *cat*, and obtained infected plants showing high-level CAT activity at the site of infection. However, the recombinant virus was unable to spread throughout the plant, because the coat protein is required for systemic infection.

Since there should be no packaging constraints with TMV, Dawson *et al.* (1989) addressed the deficiencies of the TMV replacement vector by generating a replication-competent *addition vector*, in which the entire wild-type genome was preserved. Dawson and colleagues added the bacterial *cat* gene, controlled by a duplicated coat-protein subgenomic promoter, between the authentic movement and coat-protein genes of the TMV genome. In this case, systemic infection occurred in concert with high-level CAT activity, but recombination events in infected plants resulted in deletion of the transgene and the production of wild-type TMV RNA. Homologous recombination can be prevented by replacing the TMV coat-protein gene with the equivalent sequence from the related *Odontoglossum* ringspot virus (Donson *et al.* 1991). This strategy has been used to produce a range of very stable expression vectors, which have been used to synthesize a variety of valuable proteins in plants, such as ribosome-inactivating protein (Kumagai *et al.* 1993) and single-chain Fv (scFv) antibodies (McCormick *et al.* 1999). It has also been possible to produce complete monoclonal antibodies by coinfecting plants with separate TMV vectors expressing the heavy and light immunoglobulin chains (Verch *et al.* 1998).

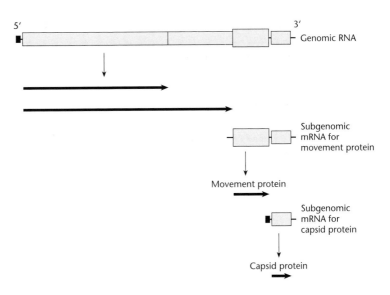

Fig. 12.16 Genome map and expression of tobacco mosaic virus.

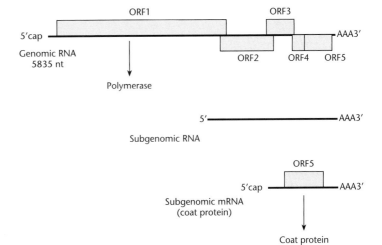

Fig. 12.17 Genome map and expression of potato virus X.

Potato virus X

Potato virus X (PVX) is the type member of the *Potexvirus* family. Like TMV, it has a monopartite RNA genome of approximately 6.5 kb, which is packaged in a filamentous particle. The genome map of PVX is shown in Fig. 12.17, and contains genes for replication, viral movement and the coat protein, the latter expressed from a subgenomic promoter. Reporter genes, such as *gusA* and green fluorescent protein, have been added to the PVX

genome under the control of a duplicated coat-protein subgenomic promoter and can be expressed at high levels in infected plants (Chapman *et al.* 1992, Baulcombe *et al.* 1995). As with the early TMV vectors, there is a tendency for the transgene to be lost by homologous recombination, but, in the case of PVX, no alternative virus has been identified whose coat-protein gene promoter can functionally substitute for the endogenous viral promoter. For this reason, PVX is generally not used for long-term expression, but has been widely employed as a transient expression vector. It has been used for the synthesis of valuable proteins, such as antibodies (e.g. Hendy *et al.* 1999, Franconi *et al.* 1999, Ziegler *et al.* 2000), and for the expression of genes that affect plant physiology (e.g. the fungal avirulence gene *avr9*) (Hammond-Kosack *et al.* 1995).

The stable transformation of plants with cDNA copies of the PVX genome potentially provides a strategy for extremely high-level transgene expression, because transcripts should be amplified to a high copy number during the viral replication cycle. However, instead of high-level expression, this strategy leads to potent and consistent transgene silencing, as well as resistance to viral infection (English *et al.* 1996, English & Baulcombe 1997). The basis of this phenomenon and its implications for transgene expression in plants are discussed in more detail in Chapter 13. In terms of vector development, however, it is notable that PVX-based vectors are probably most widely used to study virus-induced gene silencing and related phenomena (Dalmay *et al.* 2000) and to deliberately induce silencing of homologous plant genes (reviewed by Baulcombe 1999).

Advances in transgenic technology

Introduction

In the previous three chapters, we discussed the early development of transgenic technology, which focused on the evolution of reliable techniques for gene transfer and transgene expression in animals and plants. The standard use of this technology was to add genetic information to the genome, generating a *gain of function* in the resulting cells and transgenic organisms.

Now that such gene-transfer processes are routine, the focus of transgenic technology has shifted to the provision of more control over the behaviour of transgenes. New techniques have evolved in parallel in animals and plants. Advances have come about through the development of inducible expression systems that facilitate external regulation of transgenes and the exploitation of site-specific recombination systems to make precise modifications in target genomes. In mice, the combination of gene targeting, site-specific recombination and inducible transgene expression makes it possible to selectively switch on and off both transgenes and endogenous genes in a conditional manner. Other routes to gene silencing have also been explored, such as the expression of antisense RNA and the recently described phenomenon of RNA interference.

Transgenic animals and plants are also being used increasingly as tools for the analysis of genomes. We discuss the development of tagging systems for the rapid cloning of genes disrupted by insertional mutagenesis and the use of entrapment vectors, which allow the high-throughput analysis of gene expression and function.

Inducible expression systems

In many gene-transfer experiments, the production of a recombinant protein is the ultimate goal, in which case it is generally appropriate to maximize transgene expression using a strong constitutive promoter. However, there are also situations in which maximized transgene expression is undesirable. For example, if a recombinant protein is toxic, constitutive high-level expression would be lethal and would prevent the recovery of stably transformed cell lines. In other experiments, the timing of transgene expression is critical. This would be the case if we wished to study the effect of protein over-expression at a specific point in the cell cycle in cultured cells or at a particular developmental stage in transgenic animals and plants. These issues can be addressed through the use of *inducible expression systems*, in which transgene expression is controlled by an external stimulus.

Endogenous inducible promoters

A number of inducible expression systems have been developed based on promoters from endogenous cellular or viral genes. An early example is the *Drosophila* heat-shock promoter. Most cells respond to elevated temperature by synthesizing *heat-shock proteins*, which include molecular chaperons and other proteins with protective functions (Parsell & Lindquist 1983). The response is controlled at the level of transcription by a heat-labile transcription factor, which binds to heat-responsive promoters in the corresponding genes. The promoter of the *Drosophila hsp70* gene has been widely used to drive transgene expression, both in *Drosophila* itself (Lis *et al.* 1983) and in heterologous systems (e.g. Wurm *et al.* 1986). In transgenic flies, any gene linked to the *hsp70* promoter is more or less inactive at room temperature, but high-level expression in all cells can be induced by heating to 37°C for about 30 min.

The heat-shock promoter is unusual in that the stimulus that activates it is physical. Most inducible promoters respond to chemicals, which must be supplied to the transformed cells or transgenic organisms in order to activate the expression of a linked transgene. In mammals, several promoters are known to be activated by glucocorticoid hormones or synthetic analogues, such as dexamethasone. Two of these have been extensively used for inducible transgene expression – the mouse metallothionein promoter (Hager & Palmiter 1981) and the long-terminal-repeat (LTR) promoter of mouse mammary tumour virus (MMTV) (Lee *et al.* 1981). The metallothionein promoter is also induced by interferons and heavy metals, such as cadmium and zinc, allowing the transgene to be activated in transgenic animals, e.g. by including a source of heavy metals in the drinking-water. An example of zinc-induced activation of a human growth-hormone gene in transgenic mice is discussed on p. 208. A metal-inducible expression system has also been developed for plants, although the components of the system are derived from yeast (Mett *et al.* 1993). Endogenous inducible promoters used in plants include the *PR-1a* promoter, which is responsive to benzothiadiazole (Gorlach *et al.* 1996).

There are several limitations associated with the use of endogenous inducible promoters, and most systems based on such promoters suffer from one or more of the following problems. First, they tend to be somewhat leaky, i.e. there is a low to moderate level of background transcription in the absence of induction. Secondly, the level of stimulation achieved by induction (the *induction ratio*) is often quite low, typically less than 10-fold. Thirdly, there are often cytotoxic side-effects, caused by the activation of other endogenous genes that respond to the same inductive stimulus. The inducers of many of the commonly used endogenous promoters are also toxic if contact is prolonged. Fourthly, the kinetics of induction are often not ideal. For example, in transgenic animals and plants, there may be differential uptake of the inducer into different cell types and it may be eliminated slowly. There has therefore been great interest in the development of alternative inducible expression systems.

Recombinant inducible systems

Many of the disadvantages of endogenous inducible promoters can be addressed using recombinant systems. In the ideal system, induction is dependent on the supply of a non-toxic agent that specifically activates the transgene without interacting with any endogenous promoters. The agent is taken up rapidly and evenly, but has a short half-life, allowing rapid switching between induced and non-induced states. A range of inducers might be available with different properties. Steps towards this ideal system have been achieved using promoters and transcription factors that are either heterologous in the expression host or completely artificial.

The lac *and* tet *repressor systems*

The first heterologous systems were based on bacterial control circuits (Fig. 13.1). Hu and Davidson (1987) developed an inducible expression system for mammalian cells, incorporating the essential elements of the *lac* repressor control circuit. In *Escherichia coli*, transcription of the *lac* operon is switched off in the absence of lactose by a repressor protein encoded by the gene *lacI*. This protein binds to *operator* sites, the most important of which lies just downstream from the promoter, and thus inhibits transcriptional initiation. In the presence of lactose or a suitable analogue, such as isopropyl-β-D-thiogalactoside (IPTG), the *lac* repressor undergoes a conformational change that causes it to be released from the operator sites, allowing transcription to commence.

In order to use the *lac* circuit in eukaryotic cells, Hu and Davison (1987) modified the *lacI* gene by adding a eukaryotic initiation codon, and then made a hybrid construct in which this gene was driven by the Rous sarcoma virus (RSV) LTR promoter. The construct was introduced into mouse fibroblasts and a cell line was selected that constitutively expressed the *lac* repressor protein. This cell line was transiently transfected with a number of plasmids containing the *cat* reporter gene, driven by a modified simian virus 40 (SV40) promoter. Each of these plasmids also contained a *lac* operator site somewhere within the expression construct. The investigators found that, when the operator sites were placed in the promoter region, transcription from the reporter construct was blocked. However, transcription could be derepressed by supplying the cells with IPTG, resulting in strong chloramphenicol transacetylase (CAT) activity.

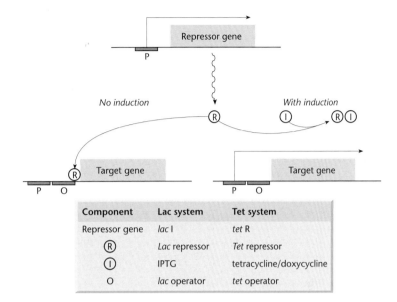

Fig. 13.1 Summary of repression-based inducible expression-control circuits based in the *lac* and *tet* operons of *E. coli*.

Component	Lac system	Tet system
Repressor gene	*lac* I	*tet* R
(R)	*Lac* repressor	*Tet* repressor
(I)	IPTG	tetracycline/doxycycline
O	*lac* operator	*tet* operator

A similar system, based on the *E. coli* tetracycline operon, was developed for tobacco plants (Gatz & Quail 1988). The *tet* operon is carried on a bacterial transposon that confers resistance to the antibiotic tetracycline. Similarly to the *lac* system, the *tet* operon is switched off by a repressor protein, encoded by the *tetR* gene, which binds to operator sites around the promoter and blocks transcriptional initiation. Tetracycline itself binds to this repressor protein and causes the conformational change that releases the *tet* operon from repression. Since tetracycline inhibits bacteria at very low concentrations, the *tet* repressor has a very high binding constant for the antibiotic, allowing derepression in the presence of just a few molecules. The *tet* repressor also has very high affinity for its operator sites. Therefore, cell cultures and transgenic tobacco plants expressing TetR were able to inhibit reporter gene expression from a cauliflower mosaic virus (CaMV) 35S promoter surrounded by three *tet* operator sites. This repressed state could be lifted rapidly by the application of tetracycline (Gatz *et al.* 1991).

The *lac* and *tet* repressor systems both show minimal background transcription in the presence of the appropriate repressor protein, and a high induction ratio is therefore possible. In the *lac* system the maximum induction ratio is approximately 50, whereas in the *tet* system up to 500-fold induction has been achieved (Figge *et al.* 1988, Gatz *et al.*

1992). Remarkably, the bacterial repressor proteins appear quite capable of interacting with the eukaryotic transcriptional apparatus and functioning as they do in bacteria, despite the many mechanistic differences in transcriptional control between prokaryotes and eukaryotes.

The tet *activator and reverse activator systems*

A disadvantage of repressor-based systems is that, in order to function effectively, high-level constitutive expression of repressor molecules is required to suppress background transgene activity. However, both the LacI and TetR proteins are cytotoxic at high levels.

To address these problems, TetR and LacI have been converted into activators by generating fusion proteins, in which the repressor is joined to the herpes simplex virus (HSV) VP16 transactivator (Labow *et al.* 1990, Gossen & Bujard 1992). In these systems, only the DNA-binding specificity of the repressor proteins is exploited. The binding of the modified bacterial proteins to operator sites within the transgene leads to transcriptional activation, because the VP16 protein acts positively on the transcriptional apparatus. The operator sites have effectively become enhancers and the inducers (IPTG and tetracycline) have effectively become repressors (Fig. 13.2). The *tet* transactivator (tTA) system has been more

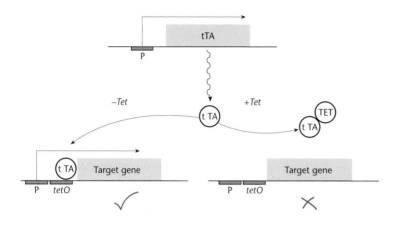

Fig. 13.2 The *tet* transactivator system (tTA).

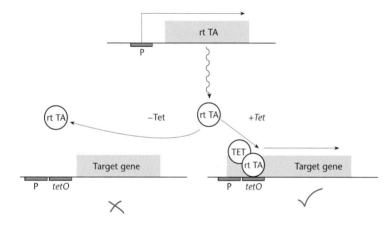

Fig. 13.3 The reverse *tet* transactivator system (rtTA).

widely used than the equivalent *lac* system, because very high levels of IPTG are required to inhibit LacI binding in mammalian cells and this is toxic (Figge *et al.* 1988). Many different proteins have been produced in mammalian cell lines using the tTA system, particularly cytotoxic proteins, whose constitutive expression would rapidly lead to cell death (Wu & Chiang 1996). In cells, a low background activity has been reported and an induction ratio of approximately 10^5 can be achieved. However, toxic effects of prolonged tetracycline exposure have been reported in transgenic animals, as well as unequal uptake of tetracycline into different organs, resulting in fluctuating basal transcription levels and cell-specific effects (reviewed by Saez *et al.* 1997).

A further modification to the *tet* system has led to marked improvements. A mutated form of the tTA protein has been generated whose DNA-binding

activity is *dependent on* rather than abolished by tetracycline (Gossen *et al.* 1995). This protein is called reverse tTA (rtTA) and becomes an activator in the *presence* of tetracycline. In this system, the antibiotic is once again an inducer, but there is no requirement for prolonged exposure (Fig. 13.3). An early example of the use of this system is described by Bohl *et al.* (1997). Myoblasts were transformed with the rtTA system using a retroviral vector in which the erythropoietin complementary DNA (cDNA) was placed under the control of a tetracycline-inducible promoter. These cells were implanted into mice, and erythropoietin secretion could be controlled by feeding the mice doxycycline, a derivative of tetracycline with a shorter half-life. An important finding was that long-term control of the secretion of this hormone was possible, with important implications for the use of inducible expression systems for gene therapy. A number of convenient plasmid- and virus-based vectors have been developed, allowing the components of the rtTA system to be transferred to the expression host in one experiment (for example, see Hofmann *et al.* 1996, Paulus *et al.* 1996, Schultze *et al.* 1996). A reverse-transactivator *lac* system has also been developed, with an induction ratio of approximately 10^4 (Baim *et al.* 1991).

Heterologous use of steroids: Drosophila *ecdysone in mammals and mammalian glucocorticoids in plants*

Steroid hormones are lipophilic molecules that penetrate cells rapidly and are eliminated within a few hours. The use of heterologous steroids for inducible transgene expression is therefore advantageous, because, in addition to their favourable kinetics, such molecules should not activate endogenous signalling pathways in the expression host and should therefore have limited toxicity.

Ecdysone is a steroid hormone found only in insects and is responsible for the extensive morphological changes that occur during moulting. As with other steroid-like signalling molecules, the hormone acts through a heterodimeric transcription factor of the nuclear receptor family. In *Drosophila*, this receptor comprises the products of the genes *ecdysone receptor* (*ecr*) and *ultraspiracle* (*usp*). The hormone and its signalling pathway are not found in mammalian cells. Therefore, transgenes including an ecdysone response element in the promoter can be induced by exogenously supplied ecdysone or its analogue, muristerone A, in cells or transgenic animals expressing the components of the *Drosophila* receptor. The unmodified *Drosophila* system has a poor induction ratio, but this can be improved using chimeric receptors and mammalian components (Yao *et al.* 1992, 1993). In a significant improvement, No and colleagues were able to achieve an induction ratio of 10^4 by generating a hybrid system in which the ecdysone receptor gene was expressed as a fusion with the HSV VP16 transactivator, and the ultraspiracle protein was replaced with a mammalian homologue, the retinoid X receptor. Background activity was reduced to near zero by altering the DNA sequence recognized by the hybrid receptor (No *et al.* 1996).

Although glucocorticoid induction has been used as an endogenous inducible expression system in mammalian cells, mammalian glucocorticoid receptors do not function properly in transgenic plants. However, the ligand-binding domain can be expressed as a fusion with other transcription factors and used to bring the activity of such transcription factors under inducible control. A potentially very powerful system has been described in which the glucocorticoid-receptor steroid-binding domain has been fused to the DNA-binding domain of the yeast transcription factor GAL4 and the VP16 transactivation domain. In this system, a CaMV 35S promoter modified to contain six GAL4-recognition sites is used to drive transgene expression. Genes placed under the control of this promoter can be induced 100-fold in the presence of dexamethasone (Aoyama & Chua 1997). An ecdysone-based system for plants has also been described, in which an agrochemical acts as the inducer, such that chemicals used in the field could function as inducers (Martinez *et al.* 1999).

Chemically induced dimerization

A novel strategy for inducible transgene regulation has been developed, exploiting essentially the same principles as the yeast two-hybrid system (p. 169). This technique, termed *chemically induced dimerization* (CID), involves the use of a synthetic divalent ligand to simultaneously bind and hence bring together separate DNA-binding and transactivation domains to generate a functional transcription factor. The initial system utilized the immunosuppressant drug FK-506. This binds with high specificity to an immunophilin protein called FKBP12, forming a complex that suppresses the immune system by inhibiting the maturation of T cells (reviewed by Schreiber 1991). For transgene induction, an artificial homodimer of FK-506 was created, which could simultaneously bind to two immunophilin domains. Therefore, by expressing fusion proteins in which the GAL4 DNA-binding domain and the VP16 transactivator were each joined to an immunophilin domain, the synthetic homodimer could recruit a functional hybrid transcription factor capable of activating any transgene carrying GAL4 recognition elements (Belshaw *et al.* 1996).

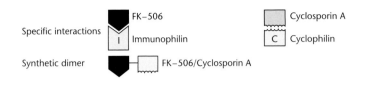

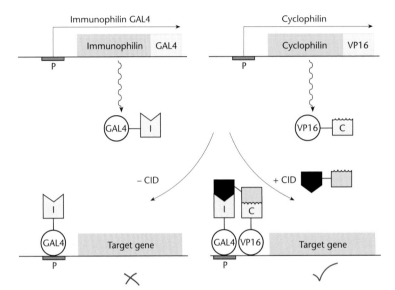

Fig. 13.4 Overview of chemically induced dimerization between the synthetic FK-506/cyclosporin A conjugate and fusion proteins containing immunophilin and cyclosporin domains. Dimerization assembles a functional transcription factor that can activate a promoter with GAL4 response elements.

Since this homodimer can also recruit non-productive combinations (e.g. two GAL4 fusions), an improved system has been developed using an artificial heterodimer specific for two different immunophilins (Belshaw *et al.* 1996). In this case, FK-506 was linked to cyclosporin A, a drug that binds specifically to a distinct target, cyclophilin. This heterodimer was shown to effectively assemble a transcription factor comprising an FKBP12–GAL4 fusion and a cyclophilin–VP16 fusion, resulting in strong and specific activation of a reporter gene in mammalian cells (Fig. 13.4). A more versatile system has been developed that exploits the ability of another immunosupressant drug, rapamycin, to mediate the heterodimerization of FKBP12 and a kinase known as FRAP (Rivera *et al.* 1996). In this system, FKBP12 and FRAP are each expressed as fusions with the components of a functional transcription factor. In the absence of rapamycin there is no interaction between these fusions, but when the drug is supplied they assemble into a hybrid transcription factor that can activate transgene expression. Transgenic mice containing a growth-hormone gene controlled by a CID-regulated promoter showed no expression in the absence of the inducer, but high levels of human growth hormone 24 h after induction with rapamycin (Magari *et al.* 1997). The advantage of this system is that rapamycin is rapidly taken up by cells *in vivo*, and it decays rapidly. The major disadvantage of immunosuppressant drugs as chemical inducers of dimerization is their pharmacological side-effects. An active area of current research is the design of modified CID systems that do not interact with endogenous targets in the immune system, and thus provide rapid and effective transgene induction with no side-effects (Liberles *et al.* 1997).

Inducible protein activity

The inducible expression systems discussed above are all regulated at the level of transcription, such

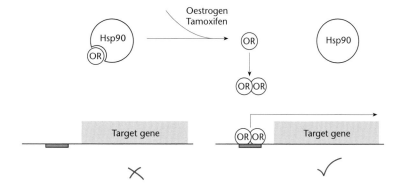

Fig. 13.5 Principle of the oestrogen-inducible expression system. The oestrogen receptor (OR) is normally sequestered by heat-shock protein 90 (Hsp90). However, in the presence of oestrogen or its analogue Tamoxifen, the receptor is released and can form dimers and activate target genes.

that there is often a significant delay between induction and response and between removal of induction and return to the basal state. Where a rapid response to induction is critical, inducible systems that operate at the post-translational level can be utilized. For example, the mammalian oestrogen receptor exists in an inert state in the absence of oestrogen, because the hormone-binding domain interacts with heat-shock protein 90 (Hsp90) to form an inactive complex. When oestrogen is present, it binds to its receptor, causing a conformational change that releases the receptor from Hsp90. The receptor is then free to dimerize and interact with DNA (Fig. 13.5). In principle, any protein expressed as a fusion with the oestrogen-binding domain will similarly interact with Hsp90 and form an inactive complex (Picard 1994). A recombinant protein can thus be expressed at high levels in an inactive state, but can be activated by feeding cells or transgenic animals with oestrogen or an analogue, such as Tamoxifen, which does not induce endogenous oestrogen-responsive genes (Littlewood *et al.* 1995). A similar system has been devised using a mutant-form progesterone receptor, which can no longer bind progesterone but can be induced with the antiprogestin RU486 (Garcia *et al.* 1992, Vegeto *et al.* 1992). An induction ratio of up to 3500 has been demonstrated in transgenic mice and, importantly, the inductive response occurs when the drug is supplied at a dose more than 100-fold below that required for it to function as an antiprogestin (Wang *et al.* 1997a,b).

Applications of site-specific recombination

Until recently, there was no generally applicable method for the precise *in vivo* manipulation of DNA sequences in animal and plant genomes. In mice, gene targeting by homologous recombination allows specific changes to be introduced into preselected genes, but it had proved impossible to extend the technique to other animals or to plants. Furthermore, even gene targeting is limited by the fact that the targeted gene is modified in the germ line; thus all cells in the mouse are similarly affected from the beginning of development and throughout its entire lifetime.

Over the last 10 years, general methods have become available that allow *in vivo* transgene manipulation in any animal or plant species. Importantly, by using such methods in concert with inducible or cell-type-specific expression systems, it is possible to generate transgenic organisms in which transgenes can be conditionally modified. In mice, the use of these methods in combination with gene targeting allows the production of conditional mutants (*conditional knockouts*), in which an endogenous gene is inactivated specifically in certain cell types or at a particular stage of development. These methods are based on a specialized genetic process, termed *site-specific recombination*.

Site-specific recombination

Site-specific recombination differs from homologous

Box 13.1 Visible marker genes

Reporter genes are widely used for *in vitro* assays of promoter activity (Box 10.1). However, reporters that can be used as cytological or histological markers are more versatile, because they allow gene expression profiles to be determined in intact cells and whole organisms.

β-galactosidase and β-glucuronidase

The *E. coli lacZ* gene encodes β-galactosidase, an enzyme that hydrolyses β-D-galactopyranosides, such as lactose, as well as various synthetic analogues. Like CAT, β-galactosidase activity can be assayed *in vitro*, although with the advantage that the assays are non-radioactive. For example, cell lysates can be assayed spectrophotometrically using the chromogenic substrate ONPG*, which yields a soluble yellow compound (Norton & Coffin 1985). Alternatively, a more sensitive fluorometric assay may be preferred, using the substrate MUG*. For histological staining, the substrate X-gal* yields an insoluble blue precipitate that marks cells brightly. The *lacZ* gene was first expressed in mammalian cells by Hall *et al.* (1983) to confirm transfection. For these experiments, the gene was linked to the SV40 early promoter and the mouse mammary tumour virus (MMTV) LTR promoter. Fusions between the *hsp70* promoter and *lacZ* were also constructed and shown to drive heat-shock-inducible β-galactosidase expression in *Drosophila* (Lis *et al.* 1983). One disadvantage of *lacZ* as a marker is that certain mammalian cells, and many plants, show a high level of endogenous β-galactosidase activity, which can obscure the analysis of chimeric genes (Helmer *et al.* 1984). The *E. coli gusA* gene, which encodes the enzyme β-glucuronidase (GUS), is an acceptable alternative (Jefferson *et al.* 1986). This marker is preferred in plants, due to the minimal background activity of the endogenous enzyme (Jefferson *et al.* 1987a), but has also been used successfully in animals (e.g. Jefferson *et al.* 1987b). Similar *in vitro* and histological assay formats to those described for β-galactosidase are also available for GUS, e.g. a histochemical substrate, X-gluc*, which yields an insoluble blue precipitate.

Luciferase

CAT, GUS and β-galactosidase are stable proteins, which persist in the cells that express them. One problem with stable reporter proteins is that, while they provide useful markers for gene activation, they are less useful for assaying transcriptional repression or rapid changes in gene activity. Luciferase was introduced as a novel reporter gene in 1986, for use in both plants (Ow *et al.* 1986) and animals (De Wet *et al.* 1987). The original marker gene, *luc*, was isolated from the North American firefly *Photinus pyralis* and encoded a single polypeptide of 550 amino acids. The enzyme catalyses the oxidation of luciferin, in a reaction requiring oxygen, ATP and the presence of magnesium ions. When excess substrate is supplied, a flash of light is emitted that is proportional to the amount of enzyme present. This can be detected using a luminometer, a scintillation counter as a luminometer or even photographic film (Wood & DeLuca 1987). Important advantages of the luciferase system include its very high sensitivity (more than 100-fold more sensitive than *lacZ*) and the rapid decay of light emission. Luciferase has therefore been used to analyse the activity of genes with oscillating expression profiles, such as the *Drosophila period* gene (Brandes *et al.* 1996). The amenability of the luciferase system has been expanded by the isolation of alternative luciferases from other organisms, which bioluminesce in different colours (e.g. see Thompson *et al.* 1990). A bacterial luciferase gene, *luxA*, has also been used as a marker in transgenic plants (Koncz *et al.* 1987).

Green fluorescent protein

The most recent addition to the growing family of reporters is green fluorescent protein (GFP), from the jellyfish *Aequoria victoria*. Over the last 5 years, this remarkable molecule has emerged as one of the most versatile tools in molecular and cellular biology and is being used to investigate an increasing variety of biological processes in bacteria, yeast, animals and plants (reviewed by Tsien 1998, Haseloff *et al.*

continued

Box 13.1 *continued*

1999, Ikawa *et al.* 1999, Naylor 1999). GFP is a bioluminescent marker that causes cells to emit bright green fluorescence when exposed to blue or ultraviolet light. However, unlike luciferase, GFP has no substrate requirements and can therefore be used as a vital marker to assay cellular processes in real time. Other advantages of the molecule include the fact that it is non-toxic, it does not interfere with normal cellular activity and it is stable even under harsh conditions (Ward & Bokman 1982).

GFP was first used as a heterologous marker in *Caenorhabditis elegans* (Chalfie *et al.* 1994). However, early experiments with GFP expression in a variety of other organisms, including *Drosophila* (Wang & Hazelrigg 1994), mammalian cell lines (Marshall *et al.* 1995) and plants (Haseloff & Amos 1995, Hu & Chen 1995, Sheen *et al.* 1995), identified a number of difficulties in the heterologous expression of the *gfp* gene. Modifications have been necessary for robust GFP expression in some plants (Chiu *et al.* 1996). In *Arabidopsis*, for example, the original *gfp* gene is expressed very poorly due to aberrant splicing. This problem was addressed by removing a cryptic splice site recognized in this plant (Haseloff *et al.* 1997). The original *gfp* gene has been extensively modified to alter various properties of the protein, such as the excitation and emission wavelengths, to increase the signal strength and to reduce photobleaching (e.g. Heim & Tsein 1996,

Zolotukhin *et al.* 1996, Cormack *et al.* 1997). As a result, a range of variant GFPs are available which can be used for dual labelling (e.g. Tsien & Miyawaki 1998; reviewed by Ellenberg *et al.* 1999). Fluorescent proteins of other colours are also available, including a red fluorescent protein from *Anthozoa* (Matz *et al.* 1999). A mutant form of this protein changes from green to red fluorescence over time, allowing it to be used to characterize temporal gene expression patterns (Terskikh *et al.* 2000).

GFP is particularly useful for generating fusion proteins, providing a tag to localize recombinant proteins in the cell. This facilitates the investigation of intracellular protein trafficking, and even the transport of proteins between cells. An early example of this application was the use of GFP to monitor the movement of ribonucleprotein particles during oogenesis in *Drosophila* (Wang & Hazelrigg 1994). Kohler *et al.* (1997) have used GFP to study the exchange of molecules between plant organelles, while Wacker *et al.* (1997) have investigated the transport of a GFP-tagged protein along the secretory pathway. The use of GFP to study the real-time dynamics of a systemic viral infection in plants was described by Padgett *et al.* (1996).

* Abbreviations: ONPG O-nitrophenyl-β-D-galactopyranoside; MUG 4-methylumbelliferyl-β-D-galactoside; X-gal: 5-bromo-4-chloro-3-indolyl-β-D-galactopyranoside; X-gluc: 5-bromo-4-chloro-3-indolyl-β-D-glucuronic acid.

---

recombination in several important respects. In terms of gene manipulation, the most important differences between these processes concern the availability of the recombinase and the size and specificity of its target sequence. Homologous recombination is a ubiquitous process that relies on endogenous recombinase enzymes present in every cell, whereas site-specific recombination systems are very specialized and different systems are found in different organisms. Homologous recombination occurs between DNA sequences with long regions of homology but no particular sequence specificity, whereas site-specific recombination occurs at short, specific recognition sites. This means that target

sites for site-specific recombination can be introduced easily and unobtrusively into transgenes, but recombination will only occur in a heterologous cell if a source of recombinase is also supplied. The power of site-specific recombination as a tool for genome manipulation thus relies on the ability of the experimenter to supply the recombinase enzyme on a conditional basis.

A number of different site-specific recombination systems have been identified and several have been studied in detail (reviewed by Craig 1988, Sadowski 1993). Some recombinases, such as bacteriophage λ integrase, require various accessory proteins for efficient recombination. However, the simplest systems require only the recombinase and its target sequence. Of these, the most extensively used are Cre recombinase from bacteriophage P1 (Lewanodski & Martin 1997) and FLP recombinase (flippase) from the 2 μm plasmid of the yeast *Saccharomyces cerevisiae* (Buchholz *et al.* 1998). These have been shown to function in many heterologous eukaryotic systems including mammalian cells and transgenic animals and plants (reviewed by Sauer 1994, Ow 1996, Metzger & Feil 1999). Both recombinases recognize 34 bp sites (termed *loxP* and *FRP*, respectively) comprising a pair of 13 bp inverted repeats surrounding an 8 bp central element. *FRP* possesses an additional copy of the 13 bp repeat sequence, although this has been shown to be non-essential for recombination. Cre recombinase has been used most extensively in mammals, because it works optimally at 37°C. The optimal temperature for FLP recombinase is 30°C (Buchholz *et al.* 1996).

Site-specific deletion of transgene sequences

Site-specific deletion of unwanted transgenes

The reaction catalysed by Cre recombinase is shown in Fig. 13.6. If two *loxP* sites are arranged as direct repeats, the recombinase will delete any intervening DNA, leaving a single *loxP* site remaining in the genome. If the *loxP* sites are arranged as inverted repeats, the intervening DNA segment is inverted. Both reactions are reversible. However, when the *loxP* sites are arranged in the same orientation, excision is favoured over reintegration, because the excised DNA fragment is rapidly degraded.

The ability of flanking *loxP* sites to delineate any sequence of interest for site-specific deletion has numerous applications. The most obvious of these is the deletion of unwanted sequences, such as marker genes. This approach has been used, for example, as a simplified strategy to generate targeted mutant mice containing point mutations. Recall from Chapter 11 that traditional strategies for generating subtle mutants in mice involve two rounds of homologous recombination in embryonic stem cells (ES cells) (p. 207). Such strategies are very inefficient, because homologous recombination is a rare event. However, in the Cre recombinase-based approach, a second round of homologous recombination is unnecessary (Kilby *et al.* 1993). As shown in Fig. 13.7, gene targeting is used to replace the wild-type allele of a given endogenous gene with an allele containing a point mutation, and simultaneously to introduce markers, such as *neo* and *Tk*, for positive and negative selection. The positive–negative selection markers within the homology region are flanked by *loxP* sites.

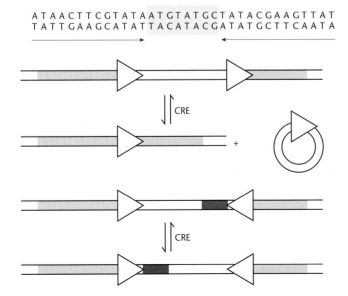

Fig. 13.6 Structure of the *loxP* site and reactions catalysed by Cre recombinase when paired *loxP* sites, shown as arrows, are arranged in different orientations.

Fig. 13.7 Gene targeting followed by marker excision, catalysed by Cre recombinase. Initially, positive and negative markers (*neo* and *Tk*), flanked by *loxP* sites, are introduced by homologous recombination (a second negative marker, in this case encoding diphtheria toxin, is included outside the homology region to eliminate random integration events). Following selection for *neo* on G418, Cre recombinase is used to excise both markers, leaving a single *loxP* site remaining in the genome. The excision event can be identified by selection for the absence of *Tk*, using ganciclovir or FIAU. Asterisk indicates mutation.

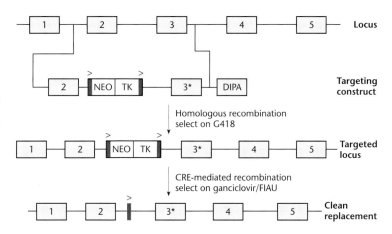

A second negative marker (e.g. the gene for diphtheria toxin) is included outside the homology region to select against random integration events. Cells that have lost the diphtheria-toxin gene and survive selection for *neo* are likely to represent authentic targeting events. Such cells are then transfected with a plasmid expressing Cre recombinase, which catalyses the excision of the remaining markers, leaving a clean point mutation and no evidence of tinkering except for a single *loxP* site remaining in one intron. Negative selection using ganciclovir or 1-(2-deoxy-2-fluoro-β-D-arabinofuranosyl)-5-iodouracil (FIAU) identifies cells that have lost the markers by selection against *Tk*.

Similar strategies can be used to remove marker genes from plants, as first demonstrated by Dale and Ow (1991). These investigators used *Agrobacterium* to transform tobacco-leaf explants with a CaMV 35S-luciferase reporter construct. The transfer DNA (T-DNA) also contained a selectable marker for hygromycin resistance, flanked by *loxP* sites. Transgenic plants were regenerated under hygromycin selection and leaf explants from these plants were then transformed with a second construct, in which Cre recombinase was driven by the CaMV 35S promoter. This construct also contained the *nptII* gene and the second-round transgenic plants were selected on kanamycin. Ten of the 11 plants tested were found to be hygromycin-sensitive, even though they continued to express luciferase, showing that the original marker had been excised. Since the *cre/nptII* construct was introduced separately, it was not linked to the original T-DNA and segregated in future generations, leaving 'clean' transgenic plants containing the luciferase transgene alone.

Site-specific transgene activation and switching

While commonly used as a method to inactivate transgenes by deletion, site-specific recombination can also activate transgenes or switch between the expression of two transgenes (Fig. 13.8). In one method, termed *recombinase-activated gene expression* (RAGE), a blocking sequence, such as a polyadenylation site, is placed between the transgene and its promoter, such that the transgene cannot be expressed. If this blocking sequence is flanked by *loxP* sites, Cre recombinase can be used to excise the sequence and activate the transgene.

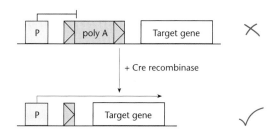

Fig. 13.8 Overview of the recombinase activation of gene expression (RAGE) strategy. A polyadenylation signal is inserted between the promoter and target gene, blocking its expression. However, if this signal is flanked by *loxP* sites, Cre recombinase can be used to excise the block, bringing the promoter and gene into juxtaposition and thus activating gene expression.

This strategy was first used in transgenic mice to study the effect of SV40 T-antigen expression in development (Pichel *et al.* 1993). In this case, Cre recombinase was expressed under the control of a developmentally regulated promoter. Essentially the same strategy was used in transgenic tobacco plants to activate a reporter gene in seeds (Odell *et al.* 1994). In this case, Cre recombinase was expressed under the control of a seed-specific promoter. An important feature of both these experiments was the use of two separate transgenic lines, one expressing Cre recombinase in a regulated manner and one containing the target gene. Crosses between these lines brought both transgenes together in the hybrid progeny, resulting in the conditional activation of the transgene based on the expression profile of Cre. This is an extremely versatile and widely used strategy, because it allows 'mix and match' between different Cre transgenic and 'responder' lines. We return to this subject below.

Site-specific transgene integration

Site-specific integration of transgenes can occur if the genome contains a recombinase recognition site. This may be introduced by random integration or (in mice) by gene targeting. Using an unmodified Cre–*loxP* system, transgene integration occurs at a low efficiency, because, as discussed above, the equilibrium of the reaction is shifted in favour of excision. Initial attempts to overcome this problem by providing transient Cre activity had limited success (see Sauer & Henderson 1990, Baubonis & Saur 1993). However, high-efficiency Cre-mediated integration has been achieved in plants (Albert *et al.* 1995) and mammalian cells (Feng *et al.* 1999) using mutated or inverted *loxP* sites. Site-specific transgene integration into mammalian cells has also been achieved using FLP recombinase (O'Gorman *et al.* 1991).

Transgene integration by site-specific recombination has many advantages over the random integration that is normally achieved by illegitimate recombination. For example, if a region of the genome can be identified that is not subject to negative position effects (Box 11.1), transgenic lines with a *loxP* site at this position can be used for the stable and high-level expression of any transgene (e.g. Fukushige & Sauer 1992). Also, due to the precise nature of site-specific recombination, transgenic loci generated by this method are likely to be less complex than loci generated by random integration.

Chromosome engineering

Site-specific recombination between widely separated target sites or target sites on different chromosomes can be used to generate large deletions, translocations and other types of chromosome mutation. Chromosome engineering by site-specific recombination was first reported by Golic (1991), using FLP recombinase in *Drosophila*, but similar experiments have now been carried out in plants and mice. Precise intrachromosomal deletions can be generated in mice by two rounds of gene targeting, introducing *loxP* sites at distant sites, followed by Cre-mediated recombination (Ramirez-Solis *et al.* 1995, Li *et al.* 1996). In plants, where gene targeting is very inefficient, an ingenious scheme has been developed where *loxP* sites are arranged in tandem on a transformation construct, one inside a *Ds* transposon and one outside (Fig. 13.9). The transposon is placed between a marker gene and its promoter. When this construct is introduced into tobacco plants containing the autonomous transposon *Ac* to provide a source of transposase, the *Ds* element can excise from the transgene, as revealed by marker-gene expression. In most heterologous plants, *Ac–Ds* elements reintegrate at a position that is linked to the original site. Although the site of reintegration cannot be controlled, this nevertheless defines a large chromosomal segment that can be excised by Cre recombinase (Medberry *et al.* 1995, Osbourne *et al.* 1995). Translocations are more difficult to engineer, because interchromosomal site-specific recombination is inefficient, and inventive selection strategies are required to identify the desired products (e.g. see Qin *et al.* 1994, Smith *et al.* 1995, Van Deursen *et al.* 1995).

Cre-mediated conditional mutants in mice

In mice, gene targeting and site-specific recombination can be used in a powerful combined approach to generate conditional knockout mutants. Essentially, targeting vectors are designed so that part of a selected endogenous gene becomes flanked by *loxP*

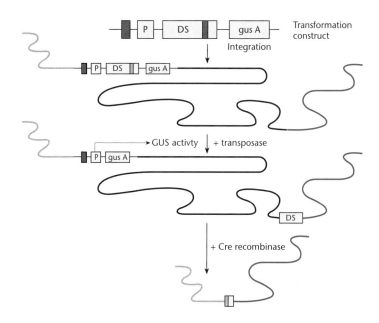

Fig. 13.9 Chromosome engineering in plants with Cre recombinase. A construct integrates into the plant genome. The construct contains a promoter separated from the reporter gene *gusA* by a *Ds* transposon containing a *loxP* site. Another *loxP* site is located just upstream of the promoter. The addition of transposase (e.g. by an *Ac* element elsewhere in the genome) causes the *Ds* element to excise, thereby bringing the *gusA* gene adjacent to the promoter. The excision event can thus be identified by the onset of GUS activity in the plant. The *Ds* element tends to insert a few tens of kilobases away on the same chromosome. Cre recombinase can then delete the large genomic region between the two *loxP* sites.

sites, or *floxed*. The usual strategy is to insert the *loxP* sites into introns flanking an essential exon, since this generally does not interfere with the normal expression of the gene. Cre recombinase is then supplied under the control of a cell-type-specific, developmentally regulated or inducible promoter, causing the gene segment defined by the *loxP* sites to be deleted in cells or at the developmental stage specified by the experimenter. This addresses a major limitation of traditional gene-knockout techniques, i.e. that, if the mutation has an embryonic lethal phenotype, only its earliest effects can be investigated.

The general methodology for such experiments, as we discussed earlier, is to cross two lines of transgenic mice, one carrying the floxed target gene and the other carrying the conditional *cre* transgene. As the number of reports of such experiments has increased, more and more transgenic mouse lines are becoming available, with Cre expressed under the control of different conditional promoters. For example, a mouse line with Cre expressed specifically in the lens of the eye was generated by Lasko *et al.* 1992. Lines are also available with Cre expressed specifically in the mammary gland (Wagner *et al.* 1997) and developing sperm (O'Gorman *et al.* 1997). Lines in which Cre is expressed in germ cells or in early development are known as 'deleter' lines and

are used to remove marker genes and generate Cre-mediated constitutive gene knockouts.

In the first examples of the conditional knockout approach, Gu *et al.* (1994) generated a Cre transgenic line expressing the recombinase under the control of the *lck* promoter, such that it was expressed only in T cells. This strain was crossed to targeted mice in which part of the DNA polymerase β gene was floxed, leading to T-cell-specific excision of an essential exon. Kuhn *et al.* (1995) mutated the same gene, but they used the metallothionein promoter to express Cre recombinase, allowing induction of site-specific recombination with interferon. Although successful, this experiment highlighted many of the inadequacies of inducible promoters. There was pronounced variation in the efficiency of excision in different tissues, probably reflecting differential uptake of the inducer. Furthermore, high-level background activity of Cre was observed in the spleen, resulting in excision of the gene segment in the absence of induction, probably caused by the presence of endogenous interferons. The tTA system has been used to bring Cre expression under the control of tetracycline administration, although a high level of background activity was also observed in this experiment, resulting in excision of the target gene prior to induction (St Ogne *et al.* 1996). Tighter control has been possible using post-translational induction. For example, Cre has been expressed as a fusion with the ligand-binding domain of the oestrogen receptor (Fiel *et al.* 1996). When this transgene was crossed into an appropriate responder strain, the background excision was minimal and Cre was strongly induced by Tamoxifen. Several strains of Cre mice are now available, in which Tamoxifen- or RU486-induced site-specific recombination has been shown to be highly efficient (e.g. Brocard *et al.* 1997, Wang *et al.* 1997a, Schwenk *et al.* 1998).

Further transgenic strategies for gene inhibition

Traditional gene-transfer strategies add new genetic information to the genome, resulting in a gain of function phenotype conferred by the transgene. Gene targeting and site-specific recombination now provide us with the ability to disrupt or delete specific parts of the mouse genome, allowing loss-of-function phenotypes to be studied, but this approach cannot be used in other animals or in plants. A range of alternative, more widely applicable transgenic strategies have therefore been developed for gene inhibition. These strategies involve the introduction of new genetic information into the genome, but, instead of conferring a gain of function, the transgene interferes with the expression of an endogenous gene, at either the RNA or the protein level. The actual target gene is not affected. The resulting loss of function effects are termed *functional knockouts* or *phenocopies*.

Gene inhibition at the RNA level

Antisense RNA transgenes

Antisense RNA has the opposite sense to mRNA. The presence of complementary sense and antisense RNA molecules in the same cell can lead to the formation of a stable duplex, which may interfere with gene expression at the level of transcription, RNA processing or possibly translation (Green *et al.* 1986). Antisense RNA is used as a natural mechanism to regulate gene expression in a number of prokaryote systems (Simons & Kleckner 1988) and, to a lesser extent, in eukaryotes (e.g. Kimelman & Kirchner 1989, Lee *et al.* 1993, Savage & Fallon 1995).

Transient inhibition of particular genes can be achieved by directly introducing antisense RNA or antisense oligonucleotides into cells. However, the transformation of cells with antisense transgenes (in which the transgene is inverted with respect to the promoter) allows the stable production of antisense RNA and thus the long-term inhibition of gene expression. This principle was established in transgenic animals and plants at about the same time. Katsuki *et al.* (1988) constructed an expression cassette in which the mouse myelin basic protein (MBP) cDNA was inverted with respect to the promoter, thus producing antisense RNA directed against the endogenous gene. In some of the transgenic mice, there was up to an 80% reduction in the levels of MBP, resulting in the absence of myelin from many axons and generating a phenocopy of the myelin-depleted 'shiverer' mutation. Smith *et al.* (1988) generated transgenic tomato plants carrying an antisense construct targeting the endogenous polygalacturonase (*pg*) gene. The product of this gene is an enzyme that causes softening and leads to overripening. The levels of *pg* mRNA in transgenic plants were reduced to 6% of the normal levels and the fruit had a longer shelf-life and showed resistance to bruising.

Antisense constructs have been widely used in transgenic animals and plants for gene inhibition. However, the efficiency of the technique varies widely and the effects can, in some cases, be non-specific. In some experiments, it has been possible to shut down endogenous gene activity almost completely, as demonstrated by Erickson *et al.* (1993), who used an inverted cDNA to generate antisense RNA against the mouse *wnt-1* gene and reduced endogenous mRNA levels to 2% of normal. Conversely, Munir *et al.* (1990) designed a construct to generate antisense RNA corresponding to the first exon and intron of the mouse *Hprt* gene, and observed no reduction in endogenous mRNA levels at all, even though the presence of antisense RNA was confirmed. The level of inhibition apparently does not depend on the size of the antisense RNA or the part of the endogenous gene to which it is complementary. For example, Moxham *et al.* (1993) achieved a 95% reduction in the level of $G_{\alpha i2}$ protein through the expression of antisense RNA corresponding to only 39 bp of the gene's 5′ untranslated region.

Conditional gene silencing can be achieved by placing antisense constructs under the control of an inducible promoter. The expression of antisense c-*myc* under the control of the MMTV LTR promoter resulted in the normal growth of transformed cells in the absence of induction, but almost complete growth inhibition in the presence of dexamethasone (Sklar *et al.* 1991). Experiments in which the tTA system was used to control antisense expression in plants have also been reported (e.g. Kumar *et al.* 1995).

Ribozyme constructs

Ribozymes are catalytic RNA molecules that carry out site-specific cleavage and (in some cases) ligation reactions on RNA substrates. The incorporation of ribozyme catalytic centres into antisense RNA allows the ribozyme to be targeted to particular mRNA molecules, which are then cleaved and degraded (reviewed by Rossi 1995, James & Gibson 1998). An important potential advantage of ribozymes over antisense RNA is that ribozymes are recycled after the cleavage reaction and can therefore inactivate many mRNA molecules. Conversely, antisense inhibition relies on stoichiometric binding between sense and antisense RNA molecules.

The use of ribozyme constructs for specific gene inhibition in higher eukaryotes was established in *Drosophila*. In the first such report, Heinrich *et al.* (1983) injected *Drosophila* eggs with a P-element vector containing a ribozyme construct targeted against the *white* gene. They recovered transgenic flies with reduced eye pigmentation, indicating that expression of the endogenous gene had been inhibited. A ribozyme construct has also been expressed under the control of a heat-shock promoter in *Drosophila* (Zhao & Pick 1983). In this case, the target was the developmental regulatory gene *fushi tarazu* (*ftz*). It was possible to generate a series of conditional mutants with *ftz* expression abolished at particular stages of development, simply by increasing the temperature to 37°C.

Ribozymes have also been used in mammalian cell lines, predominantly for the study of oncogenes and in attempts to confer resistance to viruses (reviewed by Welch *et al.* 1998). There has been intensive research into ribozyme-mediated inhibition of HIV, and remarkable success has been achieved using retroviral vectors, particularly vectors carrying multiple ribozymes (reviewed by Welch *et al.* 1998, Muotri *et al.* 1999). So far, there have been relatively few reports of ribozyme expression in transgenic mice. Larsson *et al.* (1994) produced mice expressing three different ribozymes targeted against β_2-macroglobulin mRNA, and succeeded in reducing endogenous RNA levels by 90%. Tissue-specific expression of ribozymes has also been reported. A ribozyme targeted against glucokinase mRNA was expressed in transgenic mice under the control of the insulin promoter, resulting in specific inhibition of the endogenous gene in the pancreas (Efrat *et al.* 1994). Recently, retroviral delivery of anti-neuregulin ribozyme constructs into chicken embryos has been reported (Zhao & Lemke 1998). Inhibition of neuregulin expression resulted in embryonic lethality, generating a very close phenocopy of the equivalent homozygous null mutation in mice.

Cosuppression

Cosuppression refers to the ability of a sense transgene to suppress the expression of a homologous endogenous gene. This surprising phenomenon was first demonstrated in plants, in a series of experiments designed to increase the levels of an endogenous protein by introducing extra copies of the corresponding gene. In an attempt to increase the amount of pigment synthesized by petunia flowers, Napoli *et al.* (1990) produced transgenic petunia plants carrying multiple copies of the chalcone synthase (*chs*) gene. This encodes an enzyme that converts coumaroyl-CoA and 3-malonyl-CoA into chalcone, a precursor of anthocyanin pigments. The presence of multiple transgene copies was expected to increase the level of enzyme and result in deeper pigmentation. However, in about 50% of the plants recovered from the experiment, exactly the opposite effect was observed, i.e. the flowers were either pure white or variegated with purple and white sectors. Similar findings were reported by Van der Krol *et al.* (1988) using a transgene encoding another pigment-biosynthesis enzyme, dihydroflavonol-4-reductase. In both cases, it appeared that integration of multiple copies of the transgene led to the suppression of some or all of the transgenes and the cosuppression of homologous endogenous genes.

While troublesome in terms of generating plant lines with high transgene expression levels, cosuppression can also be exploited as a tool for specific gene inactivation. There have been many reports of this nature. For example, transgenic tomatoes have been produced containing a partial copy of the *pg* gene in the sense orientation (Smith *et al.* 1990). As with the antisense *pg* transgenic plants generated previously by the same group (see above), strong inhibition of the endogenous gene was achieved, resulting in fruit with a prolonged shelf-life. Cosuppression has also been demonstrated in animals (Pal-Bhadra *et al.* 1997, Bahramian & Zabl 1999, Dernberg *et al.* 2000) and is probably related to a similar phenomenon called *quelling*, which has been described in fungi (reviewed by Selker 1997, 1999).

The mechanism of cosuppression in plants is complex and can involve silencing at either the transcriptional or post-transcriptional levels (for details, see Box 13.2). One of the most remarkable

Box 13.2 Transgene silencing, cosuppression and RNA interference

Transgene silencing is a complex phenomenon, occurring in all eukaryotes, caused by the introduction of foreign nucleic acid into the cell. Typically, expression of the affected transgene is reduced or abolished, associated with increased methylation at the transgenic locus. An understanding of transgene silencing is important, first, because it is a serious impediment to the use of animal and plant systems for the expression of foreign genes and, second, because if a transgene is homologous to an endogenous gene, the endogenous gene can also be silenced by *cosuppression*. Several different forms of silencing can be distinguished.

spread outwards into flanking host DNA and inactivate linked genes (Jahner & Jaenisch 1985). Many unrelated transgenes in animals and plants have been subject to this type of silencing, suggesting that a specific sequence is not responsible. It is possible that eukaryotic genomes possess mechanisms for scanning and identifying foreign DNA sequences, perhaps based on their unusual sequence context, and then inactivating them by methylation (Kumpatla *et al.* 1998). Prokaryotic DNA may be recognized in this manner, since prokaryotic sequences act as a strong trigger for *de novo* methylation, e.g. in transgenic mice (Clark *et al.* 1997).

Position-dependent silencing and sequence-dependent silencing

These forms of silencing can affect single-copy transgenes and are not, therefore, homology-dependent. Position-dependent silencing occurs where a transgene integrates into a genomic region containing heterochomatin. The repressive chromatin structure and DNA methylation can spread into the transgenic locus from the flanking genomic DNA (Matzke & Matzke 1998); therefore silencing results from a negative position effect (position effects are discussed in more detail in Box 11.1). Single-copy transgenes may also be silenced, even if they integrate into a genomic region that lacks negative position effects. For example, integrated retrovirus vectors often undergo *de novo* silencing associated with increased levels of DNA methylation (Jahner *et al.* 1982) and, indeed, this methylation may

Homology-dependent gene silencing (HDGS)

HDGS is caused by the presence of multiple transgene copies (reviewed by Gallie 1998, Grant 1999, Plasterk & Ketting 2000, Hammond *et al.* 2001). The suppression of transgene expression can occur at the transcriptional or post-transcriptional levels, and homologous endogenous genes are often cosuppressed. Single-copy transgenes homologous to an endogenous gene may, in some cases, also be sufficient to induce cosuppression. In *transcriptional gene silencing* (TGS), no mRNA is produced from the silenced genes and affected loci act as nucleation points for heterochromatin formation and DNA methylation. In *post-transcriptional gene silencing* (PTGS), transcription is actually required for silencing to take place, and it induces the degradation of mRNA, so that very little accumulates in the

continued

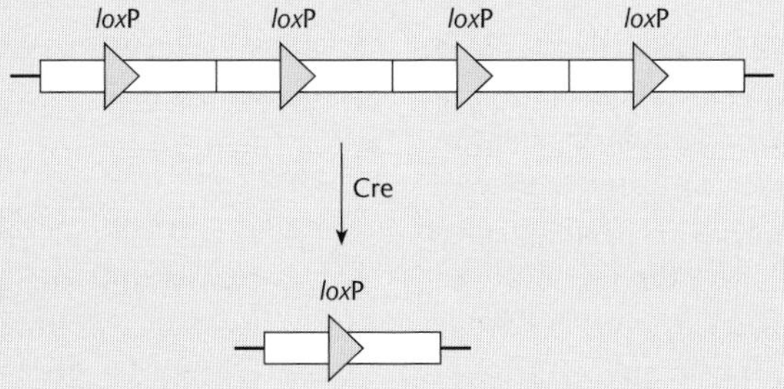

Box 13.2 *continued*

Fig. B13.1 An experiment to demonstrate homology-dependent gene silencing in mammals (also called repeat-induced gene silencing (RIGS)). A transgene construct containing the human β-globin cDNA was modified to contain a single *loxP* site, which is recognized by Cre recombinase. Transgenic mice were generated carrying multiple copies of the transgene, and in these animals the locus was highly methylated and β-globin expression was low. However, when Cre recombinase was expressed, recombination between the *loxP* sites resulted in the excision of all copies of the transgene except one. Reduction in the transgene copy number resulted in increased expression, accompanied by reduced methylation at the transgenic locus. (After Garrick *et al.* 1998.)

cytoplasm. The severity of silencing often correlates with transgene copy number, and this can be demonstrated directly through the use of site-specific recombination, as shown in the figure above.

Different processes may trigger homology-dependent silencing, including the ability of homologous DNA sequences to form secondary structures and the synthesis of aberrant RNA molecules, leading to the production of dsRNA. Both processes are stimulated if the transgenic locus is complex, particularly if transgenes or partial transgenes are arranged as inverted repeats. There may be some cross-talk between the TGS and PTGS mechanisms, but this is currently unclear. In plants, transgenes carried by RNA viruses can induce PTGS, a phenomenon that has been termed virus-induced gene silencing (VIGS). Only replication-competent viruses have this effect, providing further evidence that dsRNA is the trigger for this process. The high-level expression of sense transgenes can also induce PTGS, and it has been suggested that the plant cell may recognize such aberrant transcripts and convert them into dsRNA.

PTGS and RNAi as a common mechanism – the evidence

The involvement of dsRNA in many cases of PTGS in plants suggests that the underlying basis of this phenomenon may be similar to RNA interference (RNAi), which is also a post-transcriptional process. As discussed in the text, both PTGS and RNAi are systemic, i.e. silencing can spread from the site at

which it was induced throughout the entire organism. Recently, small RNA molecules, called *guide RNAs*, have been identified in plants affected by PTGS and animals subjected to RNAi, which can also spread systemically (Hamilton & Baulcombe 1999, Hammond *et al.* 2000, Parrish *et al.* 2000). It is proposed that these RNAs may assemble with certain proteins to generate a catalytic endonuclease complex that cleaves target mRNA molecules efficiently and in a homology-dependent fashion. In cases of PTGS, guide RNAs may form from aberrant transcripts generated by integrated transgenes or directly from dsRNA viral genomes.

What is silencing for?

It is possible that both homology-dependent and sequence-dependent silencing have evolved as a defence against 'invasive' nucleic acids (Yoder *et al.* 1997, Jones *et al.* 1998, Jensen *et al.* 1999, Li *et al.* 1999). This has been supported by the recent isolation of mutants in several organisms that show deficiencies in PTGS or RNAi. Animals impaired for RNAi show increased rates of transposon mobilization, whereas plants impaired for PTGS are more susceptible to viral infection. Interestingly, similar gene products have been identified in diverse organisms, providing further evidence for a link between PTGS and RNAi. A comprehensive discussion of this exciting new area of research is outside the scope of this book, but the interested reader can consult several excellent reviews on the subject (Plasterk & Ketting 2000, Hammond *et al.* 2001).

aspects of post-transcriptional gene silencing (PTGS) is that it is a systemic phenomenon, suggesting that a diffusible signal is involved. This can be demonstrated by grafting a non-silenced transgenic scion on to a silenced transgenic host. The silencing effect is able to spread into the graft, and the systemic effect works even if the two transgenic tissues are separated by up to 30 cm of wild-type stem (Palauqui *et al.* 1997, Voinnet *et al.* 1998).

PTGS in plants can be induced not only by integrated transgenes but also by RNA viruses, as long as there is a region of homology between the virus genome and an integrated gene. For example, the virus may carry a sequence that is homologous to an endogenous gene or to a transgene integrated into the host genome. The effect also works if the plant is transformed with a cDNA construct corresponding to part of the virus genome, as demonstrated by Angell and Baulcombe (1997). The rationale behind this experiment was to transform plants with a cDNA construct corresponding to a chimeric potato virus X (PVX) genome containing the *gusA* reporter gene. Expression of the transgene was expected to generate very high levels of β-glucuronidase (GUS) activity, because, after transcription of the transgene, the resulting viral RNA would be amplified by the virus's own replication system. However, disappointingly, all of the transgenic plants produced extremely low levels of viral RNA and GUS activity. The plants also showed an absence of PVX symptoms and were resistant to PVX infection. The virus-induced silencing effect only worked using replication-competent vectors, suggesting that the double-stranded RNA (dsRNA) intermediate involved in viral replication was the trigger for silencing (see Box 13.2).

Such is the efficiency with which PVX RNA can silence homologous genes in the plant genome that PVX vectors have been used very successfully to generate functional knockouts in plants (reviewed by Baulcombe 1999). For example, Burton *et al.* (2000) described the infection of tobacco plants with PVX vectors containing a cDNA sequence putatively encoding a cellulose synthase. The inoculated plants showed a dwarf phenotype, and levels of cellulose in affected leaves were reduced by 25%. On the basis of this evidence, the investigators concluded that the cDNA did indeed encode such an enzyme and was capable of cosuppressing the endogenous cellulose synthase gene.

RNA interference

RNA interference (RNAi) is a novel phenomenon that has the potential to become an extremely powerful tool for gene silencing in any organism. The process was discovered by Fire *et al.* (1998) while investigating the use of antisense and sense RNA for gene inhibition in the nematode worm *Caenorhabditis elegans*. In one experiment, they introduced both sense and antisense RNA into worms simultaneously and observed a striking and specific inhibitory effect, which was approximately 10-fold more efficient than either single RNA strand alone.

Further investigation showed that only a few molecules of dsRNA were necessary for RNAi in *C. elegans*, indicating that, like the action of ribozymes, the effect was catalytic rather than stoichiometric. Interference could be achieved only if the dsRNA was homologous to the exons of a target gene, indicating that it was a post-transcriptional process. The phenomenon of RNAi appears to be quite general, and this strategy has been used more recently for gene silencing in many other organisms, including *Drosophila* (Kennerdell & Carthew 2000), mice (Wianny & Zernicka-Goetz 2000) and plants (Waterhouse *et al.* 1998, Chuang & Meyerowitz 2000). In *C. elegans*, RNAi is now the standard procedure for gene inactivation, and it is becoming increasingly favoured in *Drosophila* and plants, due to its potency and specificity. Direct injection of the RNA is unnecessary. The use of a construct containing adjacent sense and antisense transgenes producing hairpin RNA (e.g. Chuang & Meyerowitz 2000, Tavernarakis *et al.* 2000) or a single transgene with dual opposing promoters (Wang *et al.* 2000) provides a stable source of dsRNA and hence the potential for permanent gene inactivation. Like PTGS in plants, RNAi is also systemic. RNAi-mediated silencing can be achieved in *C. elegans* by injecting dsRNA into any part of the body (Fire *et al.* 1998), but, more remarkably, simply placing worms in a solution containing dsRNA or feeding them on bacteria that synthesize dsRNA is sufficient to trigger the effect (Tabara *et al.* 1998, Timmons & Fire 1998). The similarities between RNAi and PTGS in plants suggest a common molecular basis, which we discuss briefly in Box 13.2.

Gene inhibition at the protein level

Intracellular antibodies

Antibodies bind with great specificity to particular target antigens and have therefore been exploited in many different ways as selective biochemical agents. Examples discussed in this book include the immunological screening of cDNA expression libraries (p. 109) and the isolation of recombinant proteins by immunoaffinity chromatography (p. 76). The microinjection of antibodies into cells has been widely used to block the activity of specific proteins, but the limitation of this approach is that the inhibitory effect is transient (reviewed by Morgan & Roth 1988). Specific inhibitory effects can also be achieved by microinjecting cells with RNA from hybridoma cell lines (Valle *et al.* 1982, Burke & Warren 1984). Such experiments provided the first evidence that non-lymphoid cells can synthesize and assemble functional antibodies.

To achieve long-term inhibition of specific proteins, cells can be transformed with cDNA constructs that allow the expression of intracellular antibodies (sometimes termed *intrabodies*) (Richardson & Marasco 1995). An important consideration here is that antibodies are large multimeric proteins with, in addition to antigen binding, various effector functions that are non-essential for intracellular protein inhibition. The strategy for expressing intracellular antibodies has been radically simplified using modified antibody forms, such as single-chain Fv (scFv) fragments (Fig. 13.10). These comprise the antigen-binding variable domains of the immunoglobulin heavy and light chains, linked by a flexible peptide arm. Such fragments retain the specificity of the parent monoclonal antibody, but are encoded by a single, relatively small transgene. Further modifications to the expression construct allow the antibody to be targeted to particular intracellular compartments, such as the nucleus, mitochondria or cytosol. It should be noted, however, that antibodies are normally folded and assembled in the endoplasmic reticulum (ER) and Golgi apparatus and are generally less stable in cell compartments outside the secretory pathway.

Due to their long half-life in the ER, intracellular antibodies have been particularly useful for the inhibition of cell-surface receptors, which pass through this compartment *en route* to the plasma membrane. For example, the cell-surface presentation of functional interleukin-2 (IL2) receptors was completely abolished in Jurkat cells stably expressing an anti-IL2Rα scFv fragment in the ER, rendering these cells insensitive to exogenously applied IL2 (Richardson *et al.* 1995). More recently, the same result has been achieved using lentivirus vectors expressing the scFv fragment, demonstrating how intracellular antibodies can be valuable for gene therapy (Richardson *et al.* 1998). Intracellular antibodies have also been used to abolish the activity of oncogenes (Beerli *et al.* 1994, Cochet *et al.* 1998, Caron de Fromentel *et al.* 1999) and to confer virus resistance by inhibiting replication (reviewed by Rondon & Marasco 1997). Functional antibodies, both full-sized immunoglobulins and fragments, can also be expressed in plants. Hiatt *et al.* (1989) were the first to demonstrate the expression of plant recombinant antibodies, dubbed '*plantibodies*', and subsequent experiments have shown that this strategy can be used, as in animal cells, to combat viral diseases by targeting specific viral proteins (Conrad & Fielder 1998). Antibodies expressed in plants have also been used to interfere with physiological processes in the plant, e.g. antibodies against abscisic acid have been used to disrupt signalling by this hormone in tobacco (Artsaenko *et al.* 1995). Note that plants can also be used as bioreactors to produce recombinant antibodies for therapeutic use (Chapter 14).

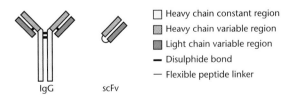

□ Heavy chain constant region
▨ Heavy chain variable region
■ Light chain variable region
— Disulphide bond
— Flexible peptide linker

Fig. 13.10 Comparison of a normal immunoglobulin molecule with a single-chain Fv fragment.

Dominant-negative mutants

In diploid organisms, most loss-of-function mutations generate recessive or semidominant (dosage-related) phenotypes, because the remaining wild-type copy of the gene provides enough gene product for normal or near-normal activity. However, some loss-of-function mutations are fully dominant over the wild-type allele, because the mutant gene product interferes with the activity of the wild-type protein. Such mutants are known as *dominant negatives*, and principally affect proteins that form dimers or larger multimeric complexes.

The deliberate overexpression of dominant-negative transgenes can be used to swamp a cell with mutant forms of a particular protein, causing all functional molecules to be mopped up into inactive complexes. The microinjection of DNA constructs or *in vitro*-synthesized dominant-negative RNA into *Xenopus* embryos has been widely used to examine the functions of cell-surface receptors in development, since many of these are dimeric (e.g. see Amaya *et al.* 1991, Hemmati-Brivanlou & Melton 1992). Dominant-negative proteins stably expressed in mammalian cells have been used predominantly to study the control of cell growth and proliferation. A dominant-negative ethylene receptor from *Arabidopsis* has been shown to confer ethylene insensitivity in transgenic tomato and petunia. The effects of transgene expression included delayed fruit ripening and flower senescence (Wilkinson *et al.* 1997).

Transgenic technology for functional genomics

In the last 5 years, the complete sequences of the genomes of several important model eukaryotic species have been published: the yeast *S. cerevisiae* (Mewes *et al.* 1997), the nematode worm *C. elegans* (*C. elegans* Sequencing Consortium 1998), the fruit fly *Drosophila melanogaster* (Adams *et al.* 2000), the model plant *Arabidopsis thaliana* (*Arabidopsis* Genome Initiative 2000) and, most recently, the human genome itself (International Human Genome Sequencing Consortium 2001, Venter *et al.* 2001). With the wealth of information that has been generated by these genome projects, the next important step is to find out what all the newly discovered genes actually do. This is the burgeoning field of *functional genomics*, which aims to determine the function of all the transcribed sequences in the

genome (the *transcriptome*) and all the proteins that are made (the *proteome*).

In other parts of the book, we have already discussed a number of functional genomics techniques, including microarrays (p. 116) and variants of the yeast two-hybrid system (p. 169). However, perhaps the most important way to establish a gene's function is to see what happens when that gene is either mutated or inappropriately expressed in the context of the whole organism. This chapter has focused on the development of novel transgenic strategies for the analysis of gene function in animals and plants, but such approaches are only applicable to individual genes. For functional genomics, technologies must be available that allow the high-throughput analysis of gene function, which is the only way we can begin to understand what the 10 000–40 000 genes in the genomes of higher eukaryotic cells are for. Such technologies are discussed below.

Insertional mutagenesis

Traditional techniques for generating mutations involve the use of radiation or chemical mutagens. These tend to generate point mutations, and isolating the genes corresponding to a particular mutant phenotype can be a laborious task, particularly in the large genomes of vertebrates and plants.

An alternative is to use DNA as a mutagen. The genomes of most animals and plants contain *transposable elements* – DNA elements that have the ability to jump from site to site in the genome, occasionally interrupting genes and causing mutations. Some of the gene-silencing methods discussed in the previous section appear to be based on defence strategies, many involving DNA methylation, that exist to resist the movement of such elements in the genome. However, if these endogenous transposable elements *can* be mobilized at a sufficient frequency, they can be used to deliberately interrupt functional genes and generate insertional mutants. Importantly, populations of animals or plants carrying transposable elements can be set up for saturation mutagenesis, such that a suitable number of transposition events is induced to theoretically interrupt each gene in the genome at least once somewhere in the population (e.g. see Walbot 2000).

Endogenous transposable elements are not the only sequences that cause insertional mutagenesis. Foreign DNA introduced into a cell may occasionally integrate into an existing gene and disrupt its expression. Insertional mutants have therefore been identified as by-products of other gene-transfer experiments. As an example, we consider the study of Yokoyama *et al.* (1993). They introduced a transgene encoding the enzyme tyrosinase into the male pronuclei of fertilized mouse eggs and recovered a number of founders that went on to produce transgenic lines. In one line, they found a unique and unexpected phenotype – situs inversus (the reversal of left–right asymmetry among the developing organs) and inversion of embryonic turning. This phenotype had nothing to do with tyrosinase activity, since the other lines were not similarly affected, but represented integration of the transgene into a locus required for correct specification of the left–right axis. Generally, aberrant results such as these are not reported unless they warrant investigation in their own right, so insertional mutants are likely to be produced in many unrelated transgenic experiments. A number of serendipitous cases of insertional mutagenesis in transgenic mice, caused by either pronuclear microinjection or retroviral transfer of foreign DNA, have been reviewed (Rijkers *et al.* 1994).

Gene tagging

The use of transposable elements or transgenes for insertional mutagenesis is advantageous over traditional mutagenesis methods, because the interrupted locus becomes 'tagged' with a unique sequence. Cloning genes tagged with endogenous transposons can sometimes be complicated if there are many copies of the element in the genome, but the use of heterologous transposons or transgenes can provide a truly unique sequence tag that identifies the interrupted gene. A genomic library can therefore be generated from the mutant, and the interrupted gene can be identified using hybridization or the polymerase chain reaction (PCR) to identify clones containing the tag. Such clones will also contain flanking DNA from the interrupted genomic locus.

Compared with chromosome walking and similar techniques (p. 107), tagging is a simple procedure that homes in directly on the mutant gene. Several

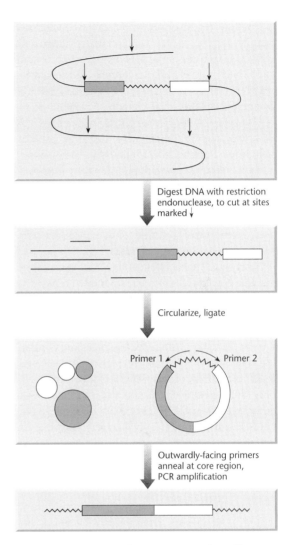

Fig. 13.11 Inverse PCR. The core region is indicated by the wavy line. Restriction sites are marked with arrows, and the left and right regions which flank the core region are represented by closed and open boxes. Primers are designed to hybridize with core sequences and are extended in the directions shown. PCR amplification generates a linear fragment containing left and right flanking sequences.

PCR-based techniques can be used to directly amplify genomic DNA flanking a transposon or transgene tag, thus avoiding the necessity for library construction in phage-λ or cosmid vectors (reviewed by Maes *et al.* 1999). The inverse PCR strategy is shown in Fig. 13.11 (Ochman *et al.* 1988). Genomic DNA from the mutant is digested with a restriction enzyme for which there is no site in the insertional tag. The DNA is thus divided into fragments, one of which will contain the entire insertion sequence flanked by genomic DNA tails. The tails, however, have compatible sticky ends and in a suitably dilute ligation reaction can be self-ligated to generate a circular genomic fragment. PCR primers annealing at the edges of the transposon tag but facing *outwards* can then be used to amplify the flanking genomic sequences. Such PCR strategies have been used to generate fragment libraries of flanking sequences for the rapid identification of genes associated with particular phenotypes (Parinov *et al.* 1999, Tissier *et al.* 1999). If appropriate, these products can be used to screen traditional genomic or cDNA libraries to isolate the uninterrupted clone.

An even greater simplification of this procedure is possible if the origin of replication from a bacterial plasmid and an antibiotic-resistance gene are included in the tag (Perucho *et al.* 1980a). In this case, self-ligation of the construct as above generates an autonomously replicating plasmid containing genomic DNA. If whole preparations of genomic DNA are used to transform competent bacterial cells, only the tag-derived plasmid will replicate, while other genomic circles will be rapidly diluted from the culture. Furthermore, only bacteria carrying the antibiotic-resistance marker will survive selection, allowing pure plasmid DNA corresponding to the flanking regions of the genomic insert to be automatically prepared. This *plasmid-rescue* technique is illustrated in Fig. 13.12.

Both endogenous transposable elements and transgene insertions have therefore been used to isolate genes by tagging in a variety of species. In *D. melanogaster*, P elements have been widely used because of the high transposition frequency and the availability of strains that lack P elements, allowing lines containing individual insertions of interest to be derived by back-crossing (Bingham *et al.* 1981). *Tc*1 elements have been used for transposon tagging in *C. elegans* (e.g. Moerman *et al.* 1986), *Ac–Ds* elements have been widely used in maize (reviewed by Gierl & Saedler 1992) and *Tam* elements have been used in the snapdragon (*Antirrhinum majus*). *Tam* elements are particularly useful because the transposition frequency can be increased more than 800-fold if the temperature at which the plants are raised is reduced to 15°C (Coen *et al.* 1989). Also, it

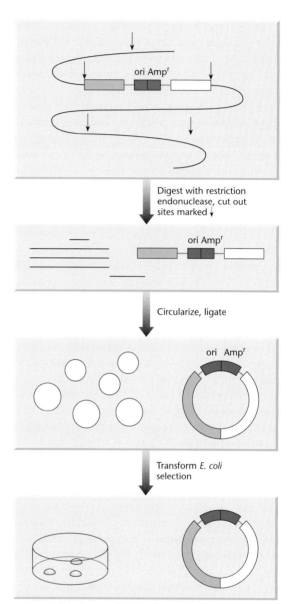

Digest with restriction endonuclease, cut out sites marked ↓

Circularize, ligate

Transform *E. coli* selection

Colonies contain rescued plasmid of interest

Fig. 13.12 The principle of plasmid rescue, a technique for isolating genomic DNA sequences flanking a transposon or transgene integration site.

is notable that approximately 5% of all naturally occurring recessive mutations in the mouse are caused by a particular family of retroviruses. The cloning of several mouse genes has been facilitated by virtue of their linkage to a proviral sequence (e.g. Bowes *et al.* 1993).

The transposition of *Drosophila* P elements appears to be restricted to drosophiloid insects (Handler *et al.* 1993), but other transposable elements are more promiscuous. The *Ac–Ds* elements of maize have been shown to function in a wide range of plants – dicots as well as monocots – and have been extensively used as insertional mutagens and gene tags in heterologous species (reviewed by Sundaresan 1996). T-DNA can also be used as a tag following transfer by *Agrobacterium tumefaciens*. T-DNA mutagenesis and tagging has been used in *Arabidopsis* (Feldmann 1991, Krysan *et al.* 1999) and has recently been extended to other plant systems, such as rice (Jeon *et al.* 2000).

Vectors for insertional mutagenesis and tagging

The development of vectors specifically for insertional mutagenesis and gene tagging arose directly from the use of transposons and the recovery of serendipitous insertional mutants in gene-transfer experiments. The transposable elements of several species have been developed as vectors. Modifications have been carried out in order to control the number of insertion events and to facilitate the cloning and analysis of tagged genes. The use of recombinant P-element derivatives for controlled transposon mutagenesis in *Drosophila* is a good example (reviewed by Cooley *et al.* 1988). Two fly strains are involved. One strain contains the *mutator*, a defective P element carrying useful marker genes, which can be mobilized when provided with a source of transposase. A second strain contains the *jumpstarter*, a *wings-clipped* element, which provides transposase *in trans* but lacks the *cis*-acting elements required for its own mobilization. During a controlled mutagenesis screen, a single jumpstarter element is crossed into a mutator-containing strain, whereupon transposition occurs. In subsequent generations the mutator is stabilized when the chromosome carrying the mutator element segregates from the chromosome bearing the jumpstarter element. The inclusion of a marker gene in the mutator P element allows screening or selection for transformed flies and facilitates subsequent manipulation of the modified locus. The inclusion of a bacterial selectable-marker gene and an *E. coli* origin of replication allows recovery of the tagged gene by plasmid rescue, as discussed above.

Entrapment constructs

Entrapment constructs are genetic tools that combine three important principles of gene-transfer technology: (i) the random integration of transgenes causes insertional mutagenesis and tags the mutated gene; (ii) randomly integrated DNA sequences are subject to variable position effects (Box 11.1); and (iii) reporter genes can be used to assay the activity of regulatory elements to which they become joined (Box 10.1). Entrapment constructs are insertional mutagenesis vectors adapted to provide information about the genomic region into which they integrate. Such vectors contain a reporter gene, whose expression is activated by regulatory elements in the surrounding DNA. When they integrate into the genome, the pattern of reporter-gene expression reveals the activity of nearby genes, allowing investigators to screen for appropriate or interesting insertion events. The corresponding gene can be cloned, because it is tagged by the insertion.

Enhancer traps in Drosophila

The prototype entrapment construct was the *Drosophila* enhancer trap, which originated as an important application of P-element vectors (p. 219). The method employs a *lacZ* reporter construct, in which the reporter gene is transcribed from a minimal promoter, such as a TATA box. Expression from the promoter is weak, because the promoter lacks an enhancer to stimulate its transcriptional activity. P-element-mediated transposition is used to transpose the construct into many different genomic positions in separate fly lines. In some flies, by chance, the construct is transposed to a position where it comes under the influence of an enhancer that activates transcription from the weak promoter (Fig. 13.13). It is often found in practice that, when using a histochemical stain for β-galactosidase activity, the pattern of expression shows cell specificity. Sometimes the pattern of expression is remarkably refined and detailed (O'Kane & Gehring 1987). The pattern of *lacZ* expression is assumed to reflect the cell-type specificity of the enhancer. Presumably, an endogenous gene located within range of the enhancer's effect has the same pattern of expression as the reporter. This assumption is known to be valid in

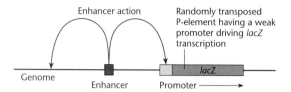

Fig. 13.13 Enhancer trapping. A reporter construct consists of *lacZ* linked to a weak promoter, which requires an enhancer for significant transcriptional activity. P-element-mediated transposition is used to insert the construct at random sites in the fly genome. When the promoter is inserted within the active range of an enhancer, expression of *lacZ* can be detected.

some cases, but the reporter activity does not always exactly match that of an endogenous gene.

Since enhancers are often found a considerable distance from the gene they activate, the enhancer trap cannot be used to directly clone genes by tagging, although it can be used as the basis for a chromosome walk to identify the gene of interest. Enhancer-trap lines have been used to identify and clone novel *Drosophila* genes, but the cell-type-specific expression that is revealed can also be harnessed in other ways. For example, instead of driving expression of a *lacZ* reporter, the principle could be applied to the expression of a toxic gene (such as ricin or

diphtheria toxin), leading to cell death and thus ablation of specific cell lineages in the fly. This has enormous potential, e.g. in studies of the development of the nervous system (O'Kane & Moffat 1992).

In order to facilitate the use of enhancer trapping as a general method for driving cell-specific expression, a modified strategy has been developed. This depends upon the ability of the yeast transcription factor GAL4 to activate transgenes containing its recognition site in the heterologous environment of the fly. The enhancer-trap principle shown in Fig. 13.13 is modified so that the *lacZ* reporter gene is replaced by the coding region for GAL4 (Fig. 13.14). In such flies, GAL4 is now expressed in the pattern dictated by a local enhancer, although this cannot be seen. The pattern of GAL4 expression can be revealed by crossing the enhancer-trap line to flies carrying a reporter transgene in which the *lacZ* gene is coupled to a promoter containing GAL4-binding sites. The beauty of this system is that a bank of fly stocks with different trapped enhancers can be built up, each with a defined pattern of GAL4 expression. Once the patterns of GAL4 expression are known, crosses can be performed to introduce chromosomes containing constructs in which any desired gene is coupled to a GAL4-dependent promoter, giving a particular cell-specific pattern of expression.

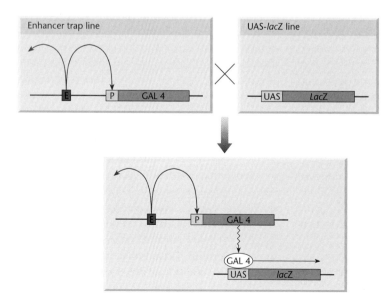

Fig. 13.14 Second-generation enhancer trap, in which the yeast gene for the GAL4 transcription factor is activated by the enhancer. This can be crossed to a responder line in which *lacZ* is driven by a GAL4-dependent promoter.

Enhancer traps in other species

Simple *lacZ* enhancer traps based on the original *Drosophila* model have also been used in mice (Allen *et al.* 1988, Kothary *et al.* 1988). The constructs were introduced into eggs by pronuclear microinjection and the resulting transgenic embryos could be screened for β-galactosidase activity. One disadvantage of this strategy is that, unlike P-element insertion, gene transfer by pronuclear microinjection tends to generate complex, multicopy, transgene arrays, often combined with major structural rearrangements of the genome. These factors complicate the interpretation of reporter expression patterns and hamper efforts to recover the corresponding genes. The use of ES cells and alternative transformation techniques have addressed these problems. ES cells can be transformed by electroporation, resulting in simpler transgenic loci with a lower transgene copy number (Gossler *et al.* 1989), or by retroviral vectors, which generate single-copy insertions (Friedrich & Soriano 1991, Von Melchner *et al.* 1992). Enhancer-trap vectors have also been developed for plants, using T-DNA insertions carrying a *gusA* gene driven by a weak promoter adjacent to the right border repeat (Goldsbrough & Bevan 1991, Topping *et al.* 1991). The activity of endogenous enhancers can be determined by screening the plants for GUS activity. The use of *lacZ* and *gusA* as histological reporters is discussed in Box 13.1.

Gene traps and functional genomics

Although enhancer traps have been widely used in animals and plants, there have been relatively few reports of successful gene-cloning efforts other than in *Drosophila*. Much more use has been made of an alternative type of entrapment construct, the *gene trap*, which is activated only upon direct integration into a functional gene. Depending on the site of integration, the insertion may or may not generate a mutant phenotype, but in either case the gene can be directly cloned by virtue of its unique insertional tag.

The first entrapment vectors in plants were gene traps that employed a promoterless selectable marker providing resistance to the antibiotic neomycin, in an attempt to directly select for T-DNA insertions into functional genes (Andre *et al.* 1986, Teeri *et al.* 1986). A problem with this approach was that selection favoured plants with multiple copies of abnormal T-DNA inserts (Koncz *et al.* 1989). A second selectable marker was therefore incorporated into the vector, allowing transformed plants to be selected for hygromycin resistance, while the trapped neomycin-resistance marker was used as a reporter gene to confirm integration into functional genes. Replacing the neomycin-resistance marker with *gusA* provided the increased versatility of direct visual screening of gene-trap lines, and vectors of this type, based on transposons or T-DNA, have been widely used (Federoff & Smith 1993, Lindsey *et al.* 1993). An alternative type of gene trap incorporates a splice acceptor site, in addition to the promoterless reporter gene, thus making reporter-gene expression dependent on the formation of a transcriptional fusion with the trapped gene (Sundaresan *et al.* 1995). Promoter-trap and splice-trap vectors are compared in Fig. 13.15.

In mice, gene-trap technology has become very sophisticated and can be used, in ES cells, as an excellent high-throughput strategy to identify and characterize new genes. The technique was developed in parallel in a number of laboratories, predominantly based on the splice-trap principle (Gossler *et al.* 1989, Friedrich & Soriano 1991, Skarnes *et al.* 1992, Wurst *et al.* 1995).

Several clever strategies have been employed to address potential pitfalls of the technique and improve the versatility of gene-trap vectors. For example, in order to select ES cells with productive gene-trap insertions, the selectable marker *neo* has been fused to the reporter gene *lacZ* to generate a hybrid marker called β-geo (Gossler *et al.* 1989). A potential disadvantage of the gene-trap strategy is that only in-frame insertions will generate a functional reporter protein. This has been addressed using an internal ribosome entry site (Box 13.3), so that translation of the reporter gene occurs independently of the transcript in which it is embedded (Skarnes *et al.* 1995, Chowdhury *et al.* 1997). The possibility that transiently expressed genes will be overlooked in large-scale screens has been addressed, using a binary system in which Cre recombinase is expressed in the gene-trap construct and activates the *lacZ* reporter

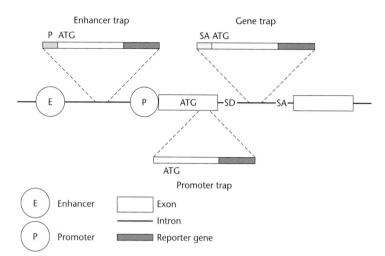

Fig. 13.15 Comparison of the structure and target-integration positions of three different types of entrapment vector – the enhancer trap, a typical gene trap and a promoter-dependent gene trap.

gene, using the RAGE strategy discussed earlier in the chapter (p. 257). Thus, β-galactosidase is expressed constitutively in all cells where Cre was transiently active (Thorey *et al.* 1998). In other vectors, the reporter gene incorporates a nuclear targeting sequence, which carries the enzyme into the nucleus and increases the concentration of the signal (Takeuchi *et al.* 1995). However, relying on endogenous tar-

geting signals provided by the interrupted gene can allow visual screening for proteins that function in specific cell compartments (e.g. Tate *et al.* 1998).

The applicability of gene trapping to high-throughput functional genomics has been demonstrated by Wiles and colleagues, who were able to generate a library of over 12 000 gene-trapped ES cell lines, using retroviral gene-trap vectors (Wiles

---

Box 13.3 Internal ribosome entry sites

The translation of most eukaryotic mRNAs is dependent on the 5′ cap, with which the ribosome associates before scanning for the initiation codon. However, certain RNA viruses and a few endogenous mRNAs contain an *internal ribosome entry site* (IRES), which allows cap-independent translation (reviewed by Mountford & Smith 1995).

An IRES can be very useful in expression vectors because it allows multiple transgenes to be expressed on a single transcript. One way in which this can be exploited is to increase the efficiency of cotransformation. In traditional plasmid-based cotransformation techniques, the transgene and a selectable marker are transcribed independently; therefore many selected cells express the marker but not the transgene. However, if both genes are expressed as a single, dicistronic mRNA, then selection for the marker necessarily identifies cells in which the non-selected transgene is also expressed

(Kaufman *et al.* 1991). Cotransformation with retroviral vectors is also much more efficient if a dicistronic expression system is used in preference to dual-promoter or alternative-splicing vectors, since, in the latter cases, the non-selected gene is often not expressed (e.g. Ghattas *et al.* 1991).

Linking an IRES to a reporter gene also has numerous applications. For example, most transgenes do not confer a convenient phenotype that can be used to track their expression. However, by placing an IRES-controlled reporter gene downstream of the transgene, the expression pattern of the transgene can be established very easily. In this context, IRES elements are particularly useful where transcriptional fusions are generated, because this removes any dependence on in-frame insertions to generate a functional protein. This is important in gene-trap vectors, where the position of integration cannot be controlled (Skarnes *et al.* 1995; see p. 271).

Fig. 13.16 A highly versatile gene-trap vector designed for high-throughput screening. See text for details.

et al. 2000). Such a library of clones can be grown, stored frozen in microtitre plates and systematically analysed to build up a molecular library of flanking sequences. In order to characterize the expression patterns of the trapped genes, it is possible to undertake large-scale screens of mouse embryos or particular tissues of adult mice produced from the ES cells, but prescreening is also possible by exploiting the ability of ES cells to differentiate along different pathways in specific culture media (e.g. Reddy *et al.* 1992, Baker *et al.* 1997, Yang *et al.* 1997). It is also possible to identify ES cells in which particular functional subsets of genes are trapped, e.g. secreted proteins (see below). An elegant gene-trap vector suitable for high-throughput screening has been designed by Zambrowicz *et al.* (1998) (Fig. 13.16). There are two expression cassettes in this vector: the β-geo hybrid marker downstream of a splice acceptor and internal ribosome entry site, and the selectable marker *pur*, conferring resistance to puromycin, under the control of a constitutive promoter but lacking a polyadenylation signal, upstream of a splice donor site. Using this vector, expression of the reporter is dependent on the formation of a transcriptional fusion with the trapped gene, but, since the selectable marker is driven by its own promoter but dependent on the structure of the gene for correct processing, selection for insertions is not biased in favour of genes that are actively expressed in ES cells.

Recently, a *Xenopus* gene trap has also been developed, based on the transgenesis procedure of Kroll and Amaya (1996) (p. 216). In this system, the reporter gene for green fluorescent protein is used (Box 13.1), allowing reporter expression patterns to be analysed in living tadpoles (Bronchain *et al.* 1999). This system could become a powerful tool for functional genomics in *Xenopus* because hundreds or thousands of embryos can be generated in a single day and screened for gene-trap events in real time.

Function-specific trapping

The design of gene-trap constructs can be modified to select for particular classes of genes. For example, Skarnes *et al.* (1995) describe a construct in which the β-geo marker is expressed as a fusion to the transmembrane domain of the CD4 type I protein. If this inserts into a gene encoding a secreted product, the resulting fusion protein contains a signal peptide and is inserted into the membrane of the ER in the correct orientation to maintain β-galactosidase activity. However, if the construct inserts into a different type of gene, the fusion product is inserted into the ER membrane in the opposite orientation and β-galactosidase activity is lost.

CHAPTER 14

Applications of recombinant DNA technology

Introduction

Biotechnology is not new. The making of beer, wine, bread, yoghurt and cheese was practised by ancient civilizations, such as the Babylonians, the Romans and the Chinese. Much, much later came vaccines, the production of basic chemicals (e.g. glycerol, citric acid, lactic acid) and the development of antibiotics. In each of these examples, *existing* properties of microorganisms were exploited. For example, *Penicillium* species naturally make penicillin. What the scientists have done is to increase the yield of penicillin by repeated rounds of mutation and selection, coupled with optimization of the growth medium. Similarly, sexual crosses between related plant species have created high-yielding and disease-resistant varieties of cereals. These improved cereals represent new combinations of genes and alleles already existing in wild strains.

With the development of gene manipulation techniques in the 1970s, there was a major paradigm shift. For the first time microorganisms could be made to synthesize compounds that they had never synthesized before, e.g. insulin production in *E. coli* (Johnson 1983). Soon all sorts of commercially or therapeutically useful proteins were being made in bacteria, principally *E. coli*, and thus the modern biotechnology industry was born. As the techniques developed for manipulating genes in bacteria were extended to plants and animals there was a concomitant expansion of the biotechnology industry to exploit the new opportunities being provided. Today there are many different facets to the commercial exploitation of gene manipulation techniques as shown in Fig. 14.1. Rather than discuss all these topics in detail, for that would take a book in itself, we have chosen to focus on six interdisciplinary themes that reflect both the successes achieved to date and the likely successes in the next decade.

Theme 1: Nucleic acid sequences as diagnostic tools

Introduction to theme 1

Nucleic acid sequences can be used diagnostically in two different ways. The first is to determine whether a particular, relatively long sequence is present in or absent from a test sample. A good example of such an application is the diagnosis of infectious disease. By choosing appropriate probes, one can ascertain in a single step which, if any, microorganisms are present in a sample. Alternatively, a search could be made for the presence of known antibiotic-resistance determinants so that an appropriate therapeutic regime can be instituted. In the second way in which sequences are used diagnostically, the objective is to determine the similarity of sequences from different individuals. Good examples of this approach are prenatal diagnosis of genetic disease and forensic profiling ('DNA fingerprinting').

Detection of sequences at the gross level

Imagine that a seriously ill individual has a disorder of the gastrointestinal tract. A likely cause is a microbial infection and there are a number of candidate organisms (Table 14.1). The question is, which organism is present and to which antibiotics is it susceptible? The sooner one has an answer to these questions, the sooner *effective* therapy can begin. Traditionally, in such a case, a stool specimen would be cultured on a variety of different media and would be examined microscopically and tested with various immunological reagents. A simpler approach is to test the sample with a battery of probes and determine which, if any, hybridize in a simple dot-blot assay. With such a simple format it is possible to vary the stringency of the hybridization reaction to accommodate any sequence differences that might

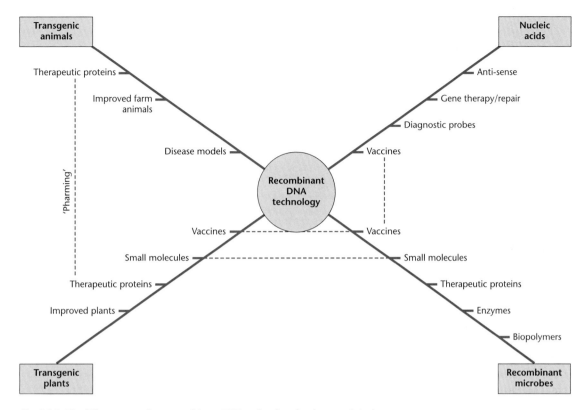

Fig. 14.1 The different ways that recombinant DNA technology has been exploited.

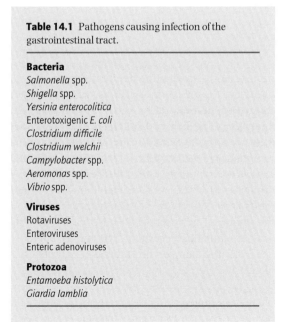

Table 14.1 Pathogens causing infection of the gastrointestinal tract.

Bacteria
Salmonella spp.
Shigella spp.
Yersinia enterocolitica
Enterotoxigenic *E. coli*
Clostridium difficile
Clostridium welchii
Campylobacter spp.
Aeromonas spp.
Vibrio spp.

Viruses
Rotaviruses
Enteroviruses
Enteric adenoviruses

Protozoa
Entamoeba histolytica
Giardia lamblia

exist between the probe and the target. The downside of this approach is that if one wishes to test the sample with 10 different probes, then 10 different dot blots are required; otherwise there is no way of determining which probe has bound to the target. Another disadvantage of this approach is that sufficient target DNA must be present in the sample to enable a signal to be detected on hybridization. Both of these problems can be overcome by using the polymerase chain reaction (PCR).

In a traditional dot-blot assay, the sample DNA is immobilized on a membrane and hybridized with a selection of labelled probes. An alternative is a 'reverse dot blot', where the probes are immobilized on the membrane and hybridized with the sample. In this way, only one hybridization step is required, because each probe occupies a unique position on the membrane (Fig. 14.2). For this approach to work, the sample DNA needs to be labelled and this can be achieved using the PCR. This has the added

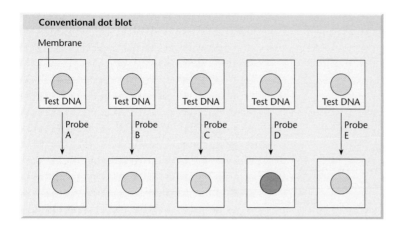

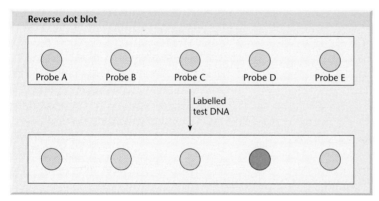

Fig. 14.2 Comparison of conventional dot blot assay with reverse dot blot method. Note that in the latter there is only a single hybridization reaction, regardless of the number of probes used, whereas in the former each probe has to be tested in a separate hybridization step.

benefit of amplifying the sample, thereby minimizing the amount of target DNA required. The downside of this approach is that the amplification step needs to be done with multiple pairs of primers (multiplex PCR), one for each target sequence. Although multiplex PCR is now an established method, considerable optimization is needed for each application (Elnifro *et al.* 2000). The major problems encountered are poor sensitivity or specificity and/or preferential amplification of certain specific targets. The presence of more than one primer pair in the reaction increases the chance of obtaining spurious amplification products because of the formation of primer dimers. Such non-specific products may be amplified more efficiently than the desired target. Clearly, the design of each primer pair is crucial to avoiding this problem. Another important feature is that all primer pairs should enable similar amplification efficiencies for their respective targets.

Despite the difficulties noted above, multiplex PCR has been used successfully in the diagnosis of infectious diseases. For example, Heredia *et al.* (1996) have used the method to simultaneously examine blood for the presence of HIV-1, HIV-2, human T-cell lymphotropic virus (HTLV)-I and HTLV-II. Similarly, Grondahl *et al.* (1999) have used the method to identify which of nine different organisms is responsible for respiratory infections. Once the clinical microbiologist knows the identity of a microorganism in a specimen, he/she can select those antibiotics that might be effective, provided that the organism in question does not carry multiple drug-resistance determinants. The presence of these determinants can also be established using multiplex PCR. Alternatively, if a virus causes the infection, one can monitor the progress of the infection by using quantitative PCR (see p. 23). This approach has been used to monitor cytomegalovirus infections in kidney-transplant patients (Caballero *et al.* 1997). It is worth noting that the development of multiplex PCR technology is being facilitated by the rapid progress in the sequencing of microbial genomes (for an up-to-date list, see http//igweb.integratedgenomics.com/GOLD/), since the data generated enable species-specific genes or sequences to be identified.

Comparative sequence analysis: single-nucleotide polymorphisms (SNPs)

In prenatal diagnosis of genetic disorders, there is a need to determine which alleles of a particular locus are being carried by the fetus, i.e. is the fetus homozygous for the normal or the deleterious allele or is it heterozygous? In forensic DNA profiling the requirement is to match DNA from the perpetrator of a crime with that of a suspect. In each case, a definitive answer could be obtained by sequencing the relevant samples of DNA. While this is possible, it is not practicable for mass screening. An alternative could be to detect hybridization of specific probes. This has been done for the detection of sickle-cell anaemia by Conner *et al.* (1983). They synthesized

two 19-mer oligonucleotides, one of which was complementary to the amino-terminal region of the normal β-globin ($β^A$) gene and the other of which was complementary to the sickle-cell β-globin gene ($β^S$). These oligonucleotides were radiolabelled and used to probe Southern blots. Under appropriate conditions, the probes could distinguish the normal and mutant alleles. The DNA from normal homozygotes only hybridized with the $β^A$ probe and DNA from sickle-cell homozygotes only hybridized with the $β^S$ probe. DNA of heterozygotes hybridized with both probes. These experiments, therefore, showed that oligonucleotide hybridization probes can discriminate between a fully complementary DNA and one containing a single mismatched base. Similar results have been obtained (Fig. 14.3) with a point mutation in the α-antitrypsin gene, which is implicated in pulmonary emphysema (Cox *et al.* 1985).

The single-base changes that occur in the two clinical examples just quoted are examples of single-nucleotide polymorphisms (SNPs, pronounced 'snips'). Many such polymorphisms occur throughout an entire genome and in humans the frequency is about once every 1000 bases. Their distribution is not entirely random. SNPs that alter amino acid

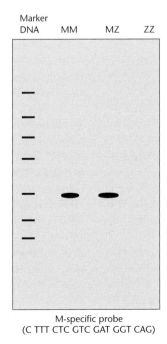
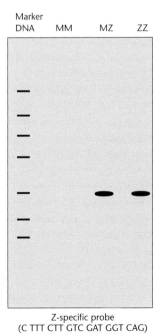

Fig. 14.3 Schematic representation of the use of oligonucleotide probes to detect the normal $α_1$-antitrypsin gene (M) and its Z variant. Human DNA obtained from normal (MM), heterozygous (MZ) and homozygous variant (ZZ) subjects is digested with a restriction endonuclease, electrophoresed and fragments Southern blotted on to a nitrocellulose membrane. The patterns shown were obtained on autoradiography of the filter following hybridization with either the normal (M-specific) or variant (Z-specific) probe.

Marker DNA   MM   MZ   ZZ

M-specific probe
(C TTT CTC GTC GAT GGT CAG)

Marker DNA   MM   MZ   ZZ

Z-specific probe
(C TTT CTT GTC GAT GGT CAG)

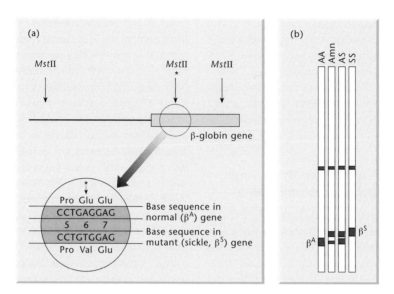

Fig. 14.4 Antenatal detection of sickle-cell genes. Normal individuals are homozygous for the β^A allele, while sufferers from sickle-cell anaemia are homozygous for the β^S allele. Heterozygous individuals have the genotype $\beta^A\beta^S$. In sickle-cell anaemia, the 6th amino acid of β-globin is changed from glutamate to valine. (a) Location of recognition sequences for restriction endonuclease *Mst*II in and around the β-globin gene. The change of A → T in codon 6 of the β-globin gene destroys the recognition site (CCTGAGG) for *Mst*II as indicated by the asterisk. (b) Electrophoretic separation of *Mst*II-generated fragments of human control DNAs (AA, AS, SS) and DNA from amniocytes (Amn). After Southern blotting and probing with a cloned β-globin gene, the normal gene and the sickle gene can be clearly distinguished. Examination of the pattern for the amniocyte DNA indicates that the fetus has the genotype $\beta^A\beta^S$, i.e. it is heterozygous.

sequences occur much less frequently than silent substitutions and SNPs in non-coding regions (Cargill *et al.* 1999). However, they are stably inherited. In some instances, these polymorphisms result in the creation or elimination of a restriction-enzyme site and this can be used diagnostically. A classical example is sickle-cell anaemia (Fig. 14.4), where the mutation from GAG to GTG eliminates restriction sites for the enzymes *Dde*I (CTNAG) and *Mst*II (CCTNAGG) (Chang & Kan 1981, Orkin *et al.* 1982). The mutation has been detected by digesting mutant and normal DNA with the restriction enzyme and performing a Southern-blot hybridization with a cloned β-globin DNA probe.

Many of the SNPs that cause genetic diseases do not lie within a restriction-enzyme site, as is the case in sickle-cell anaemia. However, the restriction fragment length polymorphisms (RFLPs) caused by other SNPs can be used diagnostically, as shown in Fig. 14.5. In this case there is a close linkage between a polymorphic restriction site and the locus of interest and this can be used to trace the inheritance of the gene. When this approach was first developed, a major limitation was the availability of suitable polymorphic markers. Following the sequencing of the entire human genome, an encyclopedia of SNPs is being created (currently 1.4 million) and this will greatly facilitate association studies. Indeed, this approach is now being used to match patients with appropriate drugs (see p. 292). Again, when this approach was developed, RFLPs were detected by genomic Southern blotting. This laborious step can be bypassed by use of the PCR. Enough DNA can be synthesized in the PCR reaction so that after digestion with the restriction enzyme and electrophoresis, the DNA bands are directly visible following staining with ethidium bromide. Another advantage of PCR is that the gel step can be omitted altogether and the SNPs detected directly, using microarrays ('DNA chips') (p. 116).

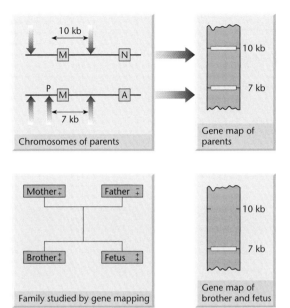

Fig. 14.5 An example of prenatal diagnosis using restriction fragment length polymorphism (RFLP) linkage analysis. The parents are both carriers for a deleterious gene (A): one of their chromosomes carries this determinant, the other its normal allele (N). One of the parental chromosomes carries a polymorphic restriction enzyme site P, which is close enough to A or N for them not to be separated in successive generations. On the chromosome which does not contain this site (–), a particular restriction enzyme cuts out a piece of DNA 10 kb long which contains another locus (M), for which we have a radioactive probe. On the chromosome containing the polymorphism (+), a single base change produces a new site and hence the DNA fragment containing locus M is now only 7 kb. On gene mapping of the parents' DNA using probe M, we see two bands representing either the + or – chromosomes. A previously born child had received the deleterious gene A from both parents and on mapping we find that it has the ++ chromosome arrangement, i.e. only a single 7 kb band. Hence the mutation must be on the + chromosome in both parents. To identify the disease in a fetus in subsequent pregnancies, we shall be looking for an identical pattern, i.e. the 7 kb band only. (Reproduced courtesy of Professor D. Weatherall and Oxford University Press.)

Variable number tardem repeat (VNTR) polymorphisms

In higher eukaryotes, genes and their associated introns occupy only a small proportion of the genome. The intergenic DNA, which makes up the majority of the genome, is composed of a mixture of unique sequences and repetitive sequences. Many of the repetitive DNA sequence elements are arranged in tandem and are known as satellite DNA. Three types of satellite DNA can be distinguished on the basis of the level of repetition and the repeat-unit length (Table 14.2). Not only is satellite DNA dispersed throughout the genome, it is highly variable and provides a valuable tool for genetic individualization. An example of this is shown schematically in Fig. 14.6. In this case the RFLPs detected are due to variations in the number of repeat units (VNTR polymorphisms) between restriction sites, rather than changes in the location of the restriction sites, as discussed in the previous section.

VNTR polymorphisms are of importance to clinical geneticists because a number of important hereditary diseases are associated with alterations in the degree of repetition of microsatellites (for reviews see Bowater & Wells 2000, Gutekunst *et al.* 2000, Usdin & Grabczyck 2000). Probably the best example of such a disease is Huntington's chorea, which is caused by an expansion of a CAG trinucleotide repeat in exon 1 of the gene coding for a protein of unknown function, which has been named huntingtin. Expansion beyond 40 repeat units correlates with the onset and progression of the disease (for review see Reddy *et al.* 1999).

Forensic applications of VNTRs

The existence of VNTR polymorphisms is of great utility in paternity testing and criminal investigations, since they allow ready comparison of DNA samples in the absence of detailed genetic information by the generation of a DNA profile or fingerprint. In principle, a multilocus DNA fingerprint can be generated either by the simultaneous application of several probes, each one specific for a particular locus, or by applying a single DNA probe that simultaneously detects several loci. When DNA profiling was first developed (Jeffreys *et al.* 1985a), multilocus probes were used and these were derived from a tandemly repeated sequence within an intron of the myoglobin gene (Fig. 14.7). These probes can hybridize to other autosomal loci – hence their utility. The first criminal court case to use DNA fingerprinting was in Bristol, UK, in 1987, when a link was shown between a burglary and a rape. In the following

Table 14.2 Classification of satellite DNA.

Type of repeat	Degree of repetition per locus	Number of loci	Repeat-unit length (bp)
Satellite	10^3–10^7	One to two per chromosome	1000–3000
Minisatellite	10–10^3	Thousands per genome	9–100
Microsatellite	10–10^2	Up to 10^5 per genome	1–6

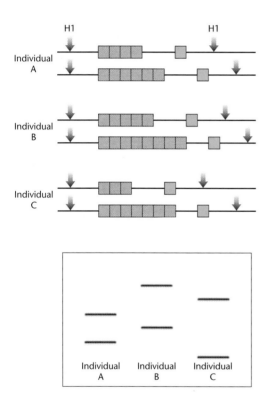

Fig. 14.6 Restriction fragment length polymorphisms caused by a variable number of tandem repeats between the two *Hinf*I restriction sites. The upper part of the diagram shows the DNA structure for three different individuals. The lower part of the diagram shows the pattern obtained on electrophoresis of *Hinf*I cut DNA from the three individuals after hybridization with a probe complementary to the sequence shown in pink.

		A
Core sequence	G G A G G T G G G C A G G A G G	
Probe 33.6	$[(A G G G C T G G A G G)_3]_{18}$	
Probe 33.15	$(A G A G G T G G G C A G G T G G)_{29}$	
	C	
Probe 33.5	$(G G G A G T G G G C A G G A G G)_{14}$	

Fig. 14.7 Probes used for DNA fingerprinting.

Multilocus probes can also be used to prove or disprove paternity and a unique example, which was part of an immigration test case, is shown in Box 14.1. The technique also has application in many other areas, such as pedigree analysis in cats and dogs (Jeffreys & Morton 1987), confirming cell-line authenticity in animal cell cultures (Devor *et al.* 1988, Stacey *et al.* 1992) and monitoring the behaviour and breeding success of bird populations (Burke & Bruford 1987).

In criminal cases, a major disadvantage of multi-locus probes is the complexity of the DNA fingerprint provided. Showing innocence is easy, but proving identity is fraught with problems. The issue boils down to calculations of the probability that two profiles match by chance as opposed to having come from the same person (Lewontin & Hartl 1991). For this reason forensic scientists have moved to the use of single-locus probes and an example is shown in Fig. 14.8. The latest variation of the technique, introduced in the UK in 1999, targets 10 distinct loci, and the likelihood of two people sharing the same profile is less than one in a billion (thousand million). Chance matches are even less likely in the USA, where the FBI routinely examines 13 VNTR loci. Another advantage of single-locus probes is

year, DNA-fingerprinting evidence was used in the USA. It is worth noting that DNA evidence has been used to prove innocence as well as guilt (Gill & Werrett 1987).

Box 14.1 Use of DNA fingerprinting in an immigration test case

In 1984, a Ghanaian boy was refused entry into Britain because the immigration authorities were not satisfied that the woman claiming him as her son was in fact his mother. Analysis of serum proteins and erythrocyte antigens and enzymes showed that the alleged mother and son were related but could not determine whether the woman was the boy's mother or aunt. To complicate matters, the father was not available for analysis nor was the mother certain of the boy's paternity. DNA fingerprints from blood samples taken from the mother and three children who were undisputedly hers, as well as the alleged son, were prepared by Southern blot hybridization to two of the mini-satellite probes shown in Fig. B14.1. Although the father was absent, most of his DNA fingerprint could be reconstructed from paternal-specific DNA fragments present in at least one of the three undisputed siblings but absent from the mother. The DNA fingerprint of the alleged son contained 61 scorable fragments, all of which

were present in the mother and/or at least one of the siblings. Analysis of the data showed the following.
• The probability that either the mother or the father by chance possess all 61 of the alleged son's bands is 7×10^{-22}. Clearly the alleged son is part of the family.
• There were 25 maternal-specific fragments in the 61 identified in the alleged son and the chance probability of this is 2×10^{-5}. Thus the mother and alleged son are related.
• If the alleged mother of the boy in question is in fact a maternal aunt, the chance of her sharing the 25 maternal-specific fragments with her sister is 6×10^{-6}.

When presented with the above data (Jeffreys *et al.* 1985b), as well as results from conventional marker analysis, the immigration authorities allowed the boy residence in Britain. In a similar kind of investigation, a man originally charged with murder was shown to be innocent (Gill & Werrett 1987).

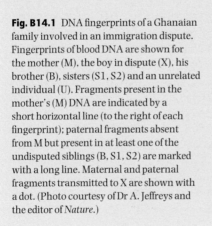

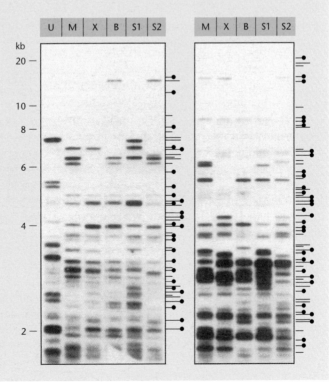

Fig. B14.1 DNA fingerprints of a Ghanaian family involved in an immigration dispute. Fingerprints of blood DNA are shown for the mother (M), the boy in dispute (X), his brother (B), sisters (S1, S2) and an unrelated individual (U). Fragments present in the mother's (M) DNA are indicated by a short horizontal line (to the right of each fingerprint); paternal fragments absent from M but present in at least one of the undisputed siblings (B, S1, S2) are marked with a long line. Maternal and paternal fragments transmitted to X are shown with a dot. (Photo courtesy of Dr A. Jeffreys and the editor of *Nature*.)

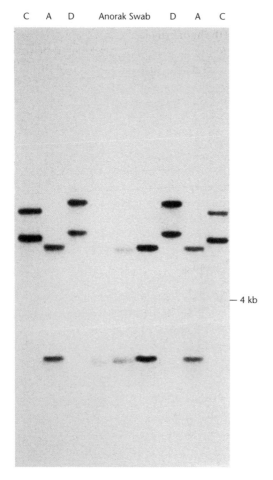

C A D Anorak Swab D A C

— 4 kb

Fig. 14.8 Use of a single locus probe to determine the identity of a rapist. Semen was extracted from an anorak and a vaginal swab. The victim's profile is in track D and that of two suspects in tracks A and C. The profile matches individual A. (Photo courtesy of Dr. P. Gill.)

that it is possible to convert the DNA profile into a numerical format. This enables a database to be established and all new profiles can be matched to that database.

Detection of VNTR polymorphisms requires that an adequate amount of DNA be present in the test sample. This is not a problem in paternity disputes but can be an issue in forensic testing. With the advent of single-locus probes, the amount of DNA required is much less of an issue, since the test loci in the sample can be amplified by PCR. As a result, it now is possible to type DNA from a face-mask

worn by a bank robber, a cigarette-butt discarded at the scene of a crime or the back of a stamp on an envelope used to send a 'poison-pen' or blackmail letter.

Historical genetics

Just as multilocus probes have been used for many applications other than in crime testing, so too have single-locus probes. A good example is the determination of the parentage of grapevines used for wine making. Grapevines are propagated vegetatively, so that individual vines of a cultivar are genetically identical to each other and to the single original seedling from which the cultivar originated. Most of the cultivars in existence in north-eastern France are centuries old and their provenance was not known. However, using 17 microsatellite loci, Bowers *et al.* (1999) were able to show that 16 of the common cultivars have genotypes consistent with their being progeny of a single pair of grape cultivars that were widespread in the region in the Middle Ages.

VNTRs can be found in mitochondrial DNA, as well as nuclear DNA, and these have particular applications. The reasons for this are threefold. First, mitochondrial sequences are passed from mother to child in the egg. Thus, brothers and sisters have identical mitochondrial DNA. Secondly, the small size of mitochondrial DNA (16–20 kb) means that there is less scope for variability, but this is more than compensated for by the copy number (~10 000 copies per cell). That is, mitochondrial DNA is naturally amplified. Thirdly, in very old or degraded specimens, the nuclear DNA may be totally decomposed, but mitochondrial DNA can still be recovered. For example, mitochondrial-DNA analysis was used to confirm that skeletons found in Ekaterinburg, Russia, were the remains of the last tsar and his family (Gill *et al.* 1994). A similar analysis showed that an individual living in Cheddar Gorge in the UK was related to a Stone Age individual whose skull was found nearby. Since bones are more likely than soft tissue to survive in the event of major accidents that involve fire, mitochondrial DNA analysis will play an increasingly important role in identifying victims. Indeed, such an analysis was done in the UK following the 1999

Paddington train crash, in which one carriage was completely incinerated.

Just as the mitochondrion is transmitted maternally, the Y chromosome is transmitted only through male descendants. Because there is only a single copy of the Y chromosome in normal diploid cells, recombination between different Y chromosomes does not occur. Any changes that do occur in the Y chromosome from generation to generation must arise from DNA rearrangements or by accumulation of random mutations. That is, the Y chromosome should be highly conserved. Sykes and Irven (2000) obtained proof of this. They probed a randomly ascertained sample of males with the surname 'Sykes' with four Y-chromosome microsatellites and found that half of them had the same Y haplotype. This suggests that all those with the same haplotype have a common ancestor, even though conventional genealogical analysis suggests otherwise.

Theme 2: New drugs and new therapies for genetic diseases

Introduction to theme 2: proteins as drugs

One of the earliest commercial applications of gene-manipulation techniques was the production in bacteria of human proteins with therapeutic applications. Not surprisingly, the first such products were recombinant versions of proteins already used as therapeutics: human growth hormone and insulin. Prior to the advent of genetic engineering, human growth hormone was produced from pituitary glands removed from cadavers. Not only did this limit the supply of the hormone but, in some cases, it resulted in recipients contracting Creutzfeld–Jakob syndrome. The recombinant approach resulted in unlimited supplies of safe material. This safety aspect has been extended to various clotting factors that were originally isolated from blood but now carry the risk of HIV infection. As the methods for cloning genes became more and more sophisticated, an increasing number of lymphokines and cytokines were identified and significant amounts of them produced for the first time. A number of these were shown to have therapeutic potential and found their way into clinical practice (Table 14.3).

The first generation of protein drugs were exact copies of the human molecules but protein engineering is now being used to develop second-generation molecules with improved properties (see theme 4, p. 299). More recently, macromodifications have been made to proteins, as exemplified by the recently approved drug (Table 14.3) for rheumatoid arthritis, which consists of the tumour necrosis factor receptor fused to the Fc portion of human IgG1.

Transgenic animals and plants as bioreactors: 'pharming'

'Pharming' is the play on words that refers to the use of transgenic animals and plants to produce recombinant therapeutic proteins. As discussed in earlier chapters, recombinant-protein synthesis in animal cells has a number of advantages over microbial expression systems, the most important of which is the authentic post-translational modifications that are performed in animal cells. However, large-scale culture of animal cells is expensive, due to the amount of medium and serum required and the necessity for precise and constant growth conditions. The production of growth hormone in the serum of transgenic mice (Palmiter *et al.* 1982a) (see p. 209) provided the first evidence that recombinant proteins could be produced, continuously, in the body fluids of animals. Five years later, several groups reported the secretion of recombinant proteins in mouse milk. In each case, this was achieved by joining the transgene to a mammary-specific promoter, such as that from the casein gene. The first proteins produced in this way were sheep β-lactoglobulin (Simons *et al.* 1987) and human tissue-plasminogen activator (tPA) (Gordon *et al.* 1987, Pittius *et al.* 1988). There have been over 100 such reports since these early experiments, and a selection is listed in Table 14.4.

Although proteins can be produced at high concentrations in mouse milk (e.g. 50 ng/ml for tPA), the system is not ideal, due to the small volume of milk produced. Therefore, other animals, such as sheep and goats, have been investigated as possible bioreactors. Such animals not only produce large volumes of milk, but the regulatory practices regarding the use of their milk are more acceptable. An early success was Tracy, a transgenic ewe

Table 14.3 Some recombinant proteins that are used therapeutically.

Year	Product	Clinical indication
1982	Human insulin	Diabetes
1985	Human growth hormone	Pituitary dwarfism
1986	Hepatitis B vaccine	Prevention of hepatitis B infection
	Interferon-α_{2a}	Hairy-cell leukaemia
1987	Tissue plasminogen activator	Acute myocardial infarction
1989	Erythropoietin	Anaemia associated with chronic renal failure
1990	Interferon-γ_{1b}	Chronic granulomatous disease
1991	Granulocyte–macrophage-colony-stimulating factor	Bone-marrow transplant
	Granulocyte-colony-stimulating factor	Chemotherapy-induced neutropenia
1992	Human interleukin-2	Renal-cell carcinoma
	Factor VIII	Haemophilia A
1993	Human DNase	Cystic fibrosis
1994	Glucocerebrosidase	Gaucher's disease
1996	Interferon-β_{1a}	Multiple sclerosis
1997	Factor IX	Haemophilia B
	Consensus interferon	Chronic HCV infection
	Platelet growth factor	Chemotherapy-induced thrombocytopenia
	Platelet-derived growth factor β	Lower-extremity diabetic ulcers
1998	Tumour necrosis factor receptor linked to Fc portion of human IgG1	Rheumatoid arthritis
	Glucagon	Hypoglycaemia
1999	Factor VIIa	Haemophilia

Table 14.4 Some recombinant proteins produced in the secretions of animal bioreactors.

System	Species	Product	Reference
Milk	Mouse	Sheep β-lactoglobulin	Simons *et al.* 1987
		Human tissue-plasminogen activator	Gordon *et al.* 1987
		Human urokinase	Meade *et al.* 1990
		Human growth hormone	Devinoy *et al.* 1994
		Human fibrinogen	Prunkard *et al.* 1996
		Human nerve growth factor	Coulibaly *et al.* 1999
		Spider silk	Karatzas *et al.* 1999
	Rabbit	Human erythropoietin	Massoud *et al.* 1996
	Sheep	Human α_1-antitrypsin	Wright *et al.* 1991
	Goat	Human tissue-plasminogen activator	Ebert *et al.* 1991
Blood serum	Rabbit	Human α_1-antitrypsin	Massoud *et al.* 1991
	Pig	Recombinant antibodies	Lo *et al.* 1991, Weidle *et al.* 1991
Urine	Mouse	Human growth hormone	Kerr *et al.* 1998
Semen	Mouse	Human growth hormone	Dyck *et al.* 1999

producing extremely high levels (30 g/l) of human α_1-antitrypsin (AAT) in her milk (Wright *et al.* 1991). Artificially inseminated eggs were microinjected with a DNA construct containing an *AAT* gene fused to a β-lactoglobin promoter. These eggs were implanted into surrogate mothers, of which 112 gave birth. Four females, including Tracy, and one male were found to have incorporated intact copies of the gene and all five developed normally. Over the lactation period, sheep can produce 250–800 l of milk, so the production potential is significant.

Using similar protocols, Ebert *et al.* (1991) have demonstrated the production of a variant of human tPA in goat milk. Of 29 offspring, one male and one female contained the transgene. The transgenic female underwent two pregnancies and one out of five offspring was transgenic. Milk collected over her first lactation contained only a few milligrams of tPA per litre, but improved expression constructs have since resulted in an animal generating several grams per litre of the protein. Recombinant human antithrombin III, which is used to prevent blood clots forming in patients that have undergone heart-bypass operations, was the first protein expressed in transgenic animal milk to reach commercial production, and is currently marketed by Genzyme Transgenics Corporation.

The production of foreign proteins in secreted body fluids has the obvious advantage that transgenic animals can be used as a renewable source of the desirable molecule. In addition to milk, other production systems have been investigated, including serum (Massoud *et al.* 1991), semen (Dyck *et al.* 1999) and urine (Kerr *et al.* 1998). In each case, an important consideration is whether the protein is stable and whether it folds and assembles correctly. The assembly of complex proteins comprising up to three separate polypeptides has been demonstrated in milk, e.g. fibrinogen (Prunkard *et al.* 1996), collagen (John *et al.* 1999) and various immunoglobulins (e.g. Castilla *et al.* 1998). Other abundantly secreted fluids that are likely to be exploited for recombinant-protein expression in the future include the albumen of hens' eggs and silkworm cocoons. There has already been some success with the latter, using both microinjection (e.g. Nagaraju *et al.* 1996) and infection of silkworm larvae with baculovirus vectors (Tamura *et al.* 1999, Yamao *et al.* 1999) (see

Chapter 10). The use of animals as bioreactors has been extensively reviewed (Clark 1998, Rudolph 1999, Wall 1999, Houdebine 2000).

Plants as bioreactors

Plants are a useful alternative to animals for recombinant-protein production because they are inexpensive to grow and scale-up from laboratory testing to commercial production is easy. Therefore, there is much interest in using plants as production systems for the synthesis of recombinant proteins and other speciality chemicals. There is some concern that therapeutic molecules produced in animal expression systems could be contaminated with small quantities of endogenous viruses or prions, a risk factor that is absent from plants. Furthermore, plants carry out very similar post-translational modification reactions to animal cells, with only minor differences in glycosylation patterns (Cabanes-Macheteau *et al.* 1999). Thus plants are quite suitable for the production of recombinant human proteins for therapeutic use.

A selection of therapeutic proteins that have been expressed in plants is listed in Table 14.5. The first such report was the expression of human growth hormone, as a fusion with the *Agrobacterium* nopaline synthase enzyme, in transgenic tobacco and sunflower (Barta *et al.* 1986). Tobacco has been the most frequently used host for recombinant-protein expression although edible crops, such as rice, are now becoming popular, since recombinant proteins produced in such crops could in principle be administered orally without purification. The expression of human antibodies in plants has particular relevance in this context, because the consumption of plant material containing recombinant antibodies could provide passive immunity (i.e. immunity brought about without stimulating the host immune system). Antibody production in plants was first demonstrated by Hiatt *et al.* (1989) and During *et al.* (1990), who expressed full-size immunoglobulins in tobacco leaves. Since then, many different types of antibody have been expressed in plants, predominantly tobacco, including full-size immunoglobulins, Fab fragments and single-chain Fv fragments (scFvs). For example, a fully humanized antibody against herpes simplex virus-2 (HSV-2) has been expressed

Table 14.5 A selection of pharmaceutical recombinant human proteins expressed in plant systems.

Species	Recombinant human product	Reference
Tobacco, sunflower (plants)	Growth hormone	Barta *et al.* 1986
Tobacco, potato (plants)	Serum albumin	Sijmons *et al.* 1990
Tobacco (plants)	Epidermal growth factor	Higo *et al.* 1993
Rice (plants)	α-Interferon	Zhu *et al.* 1994
Tobacco (cell culture)	Erythropoietin	Matsumoto *et al.* 1995
Tobacco (plants)	Haemoglobin	Dieryck *et al.* 1997
Tobacco (cell culture)	Interleukins-2 and 4	Magnuson *et al.* 1998
Tobacco (root culture)	Placental alkaline phosphatase	Borisjuk *et al.* 1999
Rice (cell culture)	α_1-Antitrypsin	Terashima *et al.* 1999
Tobacco (seeds)	Growth hormone	Leite *et al.* 2000
Tobacco (chloroplasts)	Growth hormone	Staub *et al.* 2000

in soybean (Zeitlin *et al.* 1998). Even secretory IgA (sIgA) antibodies, which have four separate polypeptide components, have been successfully expressed in plants. This experiment involved the generation of four separate transgenic tobacco lines, each expressing a single component, and the sequential crossing of these lines to generate plants in which all four transgenes were stacked (Ma *et al.* 1995). Plants producing recombinant sIgA against the oral pathogen *Streptococcus mutans* have been generated (Ma *et al.* 1998), and these plant-derived antibodies ('plantibodies') have recently been commercially produced as the drug CaroRx™, marketed by Planet Biotechnology Inc. A number of other biotechnology companies are bringing antibody-expressing transgenic plants into commercial production (see Fischer & Emans 2000).

The impact of genomics

Many of the drugs currently on the market treat the *symptoms* of the disease rather than the *cause* of the disease. This is analogous to reversing a mutant phenotype by selecting a mutation at a second site. Not surprisingly, many of these drugs have side-effects quite separate from those (e.g. toxicity) caused by their metabolism. Now that the first drafts of the human genome sequence are available, it will be possible to convert disease phenotypes into nucleotide changes in specific genes (Bailey *et al.*

2001). These genes will then become targets for new small-molecule drugs. Already drugs have been developed based on such genotype–phenotype correlations. For example, Wettereau *et al.* (1998) identified a molecule that normalizes atherogenic lipoprotein levels caused by a genetic deletion of a microsomal triglyceride transfer protein.

The identification of genetic changes associated with particular disease phenotypes offers a number of novel approaches to the development of therapies. As well as using such changes as novel targets for small-molecule drug design, there is an opportunity to use the techniques described in Chapter 11 to generate animals with the exact same genetic defect and which can be used as models to test new drug candidates (*disease modelling*). Furthermore, where drugs cannot be developed to treat a particular disorder, there might be an opportunity to correct the disease by further modification to the genome (*gene therapy*). Finally, it is likely that, in the near future, transgenic animals could be used to provide healthy organs for humans requiring transplants (*xenotransplantation*) (Box 14.2). These topics are discussed in more detail below.

Transgenic animals as models of human disease

Mammals have been used as models for human disease for many years, since they can be exploited to

Box 14.2 Xenotransplantation

Transplantation is widely used to treat organ failure but there is a shortage of organ donors resulting in long waiting times and many unnecessary deaths. In the future, transgenic animals could be used to supply functional organs to replace failing human ones. This process is termed *xenotransplantation*. There is vigorous debate concerning the ethics of xenotransplantation but, ethics aside, the technique remains limited by the phenomenon of hyperacute rejection, which is caused by the host immune system. Hyperacute rejection is dependent both on antibodies raised against the foreign organ and the activation of the host complement system. In both cases, the major trigger for rejection appears to be a disaccharide group (Gal-α(1,3)-Gal) which is present in pigs but not in primates (Cooper *et al.* 1994, Sandrin *et al.* 1994).

Transgenic strategies have been investigated to avoid hyperacute rejection, including the expression of complement-inactivating protein on the cell surface (Cozzi & White 1995, reviewed by Pearse *et al.* 1997), the expression of antibodies against the disaccharide group (Vanhove *et al.* 1998) and attempts to inhibit the expression of α(1,3) galactosyltransferase, the enzyme that forms this particular carbohydrate linkage, an enzyme that is present in pigs but not in primates. In the latter case,

the simplest strategy would be to knock out the gene by homologous recombination. Gene targeting has not yet been achieved in pigs, although the success of gene targeting/nuclear transfer in sheep (Chapter 11) suggests that knockout pigs could be produced in the next few years. Alternative procedures include introducing genes encoding other carbohydrate-metabolizing enzymes, so that the Gal-α(1,3)-Gal groups are modified into some other less immunogenic moiety (e.g. Sandrin *et al.* 1995, Cohney *et al.* 1997, Osman *et al.* 1997). Once hyperacute rejection has been overcome, there may be further problems, including delayed rejection, involving natural killer cells and macrophages (Bach *et al.* 1996), and the requirement for T-cell tolerance (Bracy *et al.* 1998, Kozlowski *et al.* 1998, Yang *et al.* 1998). There are also concerns that endogenous pig retroviruses could be activated following transplantation, perhaps even recombining with human retroviruses to produce potent new hybrids with unknown properties. Detailed studies have so far shown no evidence of such a phenomenon (e.g. Heneine *et al.* 1998, Patience *et al.* 1998). For recent reviews on the prospects of xenotransplantation (particularly porcine-to-human), the reader should consult Lambrigts *et al.* (1998), Sandrin and McKenzie (1999) and Logan (2000).

carry out detailed analyses of the molecular basis of disease and to test newly developed therapeutics prior to clinical trials in humans. Before the advent of transgenic animal technology, however, models of inherited diseases (i.e. diseases with a genetic basis) were difficult to come by. They could be obtained as spontaneously occurring mutants, suitable mutant animals identified in mutagenesis screens and susceptible animal strains obtained by selective breeding. Gene manipulation now offers a range of alternative strategies to create *specific* disease models (see reviews by Smithies 1993, Bedell *et al.* 1997, Petters & Sommer 2000).

Some of the earliest transgenic disease models were mice predisposed to particular forms of cancer, because the germ line contained exogenously derived

oncogenes (e.g. Sinn *et al.* 1987). This exemplifies so-called gain-of-function diseases, which are caused by a dominantly acting allele and can be modelled simply by adding that allele to the normal genome, e.g. by microinjection into eggs. Other gain of function diseases that have been modelled in this way include the Gerstmann–Straussler–Scheinker (GSS) syndrome, a neurodegenerative disease caused by a dominantly acting mutated prion protein gene. In one patient suffering from this disease, a mutation was identified in codon 102 of the prion protein gene. Transgenic mice were created carrying this mutant form of the gene in addition to the wild-type locus and were shown to develop a similar neurodegenerative pathology to their human counterparts (Hsiao *et al.* 1990). Other examples of gain-of-function disease models include Alzheimer's disease, which was modelled by overexpression of the amyloid precursor protein (Quon *et al.* 1991), and the triplet-repeat disorder spinocerebellar ataxia type 1 (Burright *et al.* 1995). Simple transgene addition can also be used to model diseases caused by dominant negative alleles, as recently shown for the premature ageing disease, Werner's syndrome (Wang *et al.* 2000).

Recessively inherited diseases are generally caused by loss of function, and these can be modelled by gene knockout. The earliest report of this strategy was a mouse model for hypoxanthine-guanine phosphoribosyltransferase (HPRT) deficiency, generated by disrupting the gene for HRPT (Kuehn *et al.* 1987). A large number of genes have been modelled in this way, including those for cystic fibrosis (Dorin *et al.* 1992, Snouwaert *et al.* 1992), fragile-X syndrome (Dutch–Belgian Fragile X Consortium, 1994), β-thalassaemia (Skow *et al.* 1983, Ciavattia *et al.* 1995) and mitochondrial cardiomyopathy (Li *et al.* 2000). Gene targeting has been widely used to model human cancers caused by the inactivation of tumour suppressor genes, such as *TP53* and *RB1* (reviewed by Ghebranious & Donehower 1998, Macleod & Jacks 1999).

While the studies above provide models of single-gene defects in humans, attention is now shifting towards the modelling of more complex diseases, which involve multiple genes. This is a challenging area of research but there have been some encouraging early successes. In many cases, the crossing of different modified mouse lines has led to interesting discoveries. For example, *undulated* mutant mice lack the gene encoding the transcription factor Pax-1, and *Patch* mutant mice are heterozygous for a null allele of the platelet-derived growth-factor gene. Hybrid offspring from a mating between these two strains were shown to model the human birth defect spina bifida occulta (Helwig *et al.* 1995). In other cases, such crosses have pointed the way to possible novel therapies. For example, transgenic mice overexpressing human α-globin and a mutant form of the human β-globin gene that promotes polymerization provide good models of sickle-cell anaemia (Trudel *et al.* 1991). However, when these mice are crossed to those ectopically expressing human fetal haemoglobin in adulthood, the resulting transgenic hybrids show a remarkable reduction in disease symptoms (Blouin *et al.* 2000). Similarly, crossing transgenic mice overexpressing the anti-apoptotic protein Bcl-2 to rds mutants, which show inherited slow retinal degeneration, resulted in hybrid offspring in which retinal degeneration was strikingly reduced. This indicates that Bcl-2 could possibly be used in gene therapy to treat the equivalent human retinal-degeneration syndrome (Nir *et al.* 2000).

The most complex diseases involve many genes, and transgenic models would be difficult to create. However, it is often the case that such diseases can be reduced to a small number of 'major genes' with severe effects and a larger number of minor genes. Thus, it has been possible to create mouse models of Down's syndrome, which in humans is generally caused by the presence of three copies of chromosome 21. Trisomy for the equivalent mouse chromosome 16 is a poor model because the two chromosomes do not contain all the same genes. However, a critical region for Down's syndrome has been identified by studying Down's patients with partial deletions of chromosome 21. The generation of yeast artificial chromosome (YAC) transgenic mice carrying this essential region provides a useful model of the disorder (Smith *et al.* 1997) and has identified increased dosage of the *Dyrk1a* (*minibrain*) gene as an important component of the learning defects accompanying the disease. Animal models of Down's syndrome have been reviewed (Kola & Hertzog 1998, Reeves *et al.* 2001).

Gene transfer to humans – gene therapy

The scope of gene therapy

Gene therapy is any procedure used to treat disease by modifying the genetic information in the cells of the patient. In essence, gene therapy is the antithesis of the disease modelling discussed above. Whereas disease modelling takes a healthy animal and uses gene-manipulation techniques to induce a specific disease, gene therapy takes a diseased animal (or human) and uses gene-manipulation techniques in an attempt to correct the disorder and return the individual to good health. Gene transfer can be carried out in cultured cells, which are then reintroduced into the patient, or DNA can be transferred to the patient *in vivo*, directly or using viral vectors.

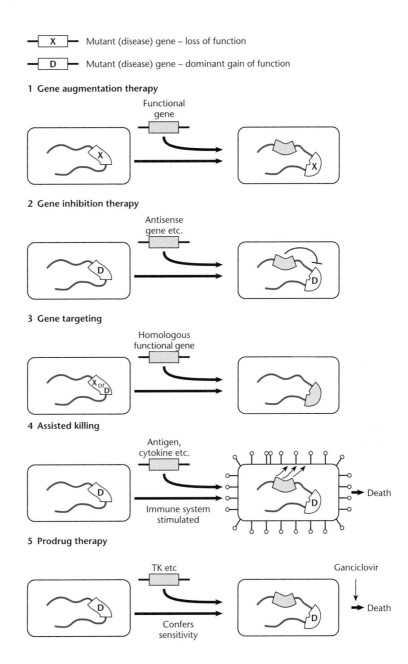

Fig. 14.9 Overview of gene-therapy strategies.

The *ex vivo* approach can be applied only to certain tissues, such as bone marrow, in which the cells are amenable to culture. Gene therapy can be used to treat diseases caused by mutations in the patient's own DNA (inherited disorders, cancers), as well as infectious diseases, and is particularly valuable in cases where no conventional treatment exists or where that treatment is inherently risky. Strategies include the following (Fig. 14.9):

- Gene-augmentation therapy (GAT), where DNA is added to the genome with the aim of replacing a missing gene product.
- Gene targeting to correct mutant alleles.
- Gene-inhibition therapy, using techniques such as antisense RNA expression or the expression of intracellular antibodies to treat dominantly acting diseases.
- The targeted ablation of specific cells.

Therapeutic gene transfer effectively generates transgenic human cell clones and, for this reason, only somatic cells can be used as targets. The prospect of germ-line transgenesis in humans raises serious ethical concerns and, with the rapid advances in technology allowing germ-line transformation and nuclear transfer in numerous mammals, these concerns will need to be addressed in the very near future (Johnson 1998). As an alternative to permanent gene transfer, transient gene therapy can be achieved using oligonucleotides, which can disrupt gene expression at many levels but do not permanently change the genetic material of the cell (Pollock & Gaken 1995).

The tools and techniques for gene therapy are essentially similar to those used for gene transfer to any animal cells. Transfection, direct delivery or transduction (see Chapter 11) can be used to introduce DNA into cells. Viral vectors are most popular because of their efficiency of gene transfer *in vivo*. However, extreme precautions need to be taken to ensure the safety of such vectors, avoiding potential problems, such as the production of infectious viruses by recombination and the pathological effects of viral replication. A number of viral vectors have been developed for gene therapy, including those based on oncoretroviruses, lentiviruses, adenovirus, adeno-associated virus, herpes virus and a number of hybrid vectors combining advantageous elements of different parental viruses (Robbins *et al.* 1998, Reynolds *et al.* 1999). The risks associated with viral vectors have promoted research into other delivery methods, the most popular of which include direct injection of DNA into tissues (e.g. muscle), the injection of liposome–DNA complexes into the blood and direct transfer by particle bombardment. Although inherently much safer than viruses, such procedures show a generally low efficiency (Scheule & Cheng 1996, Tseng & Huang 1998).

Gene-augmentation therapy for recessive diseases

The first human genetic-engineering experiment was one of *gene marking*, rather than gene therapy, and was designed to demonstrate that an exogenous gene could be safely transferred into a patient and that this gene could subsequently be detected in cells removed from the patient. Both objectives were

met. Tumour-infiltrating lymphocytes (cells that naturally seek out cancer cells and then kill them by secreting proteins such as tumour necrosis factor (TNF)) were isolated from patients with advanced cancer. The cells were then genetically marked with a neomycin-resistance gene and injected back into the same patient (Rosenberg *et al.* 1990).

The first clinical trial using a therapeutic gene-transfer procedure involved a 4-year-old female patient, Ashanthi DeSilva, suffering from severe combined immune deficiency, resulting from the absence of the enzyme adenosine deaminase (ADA). This disease fitted many of the ideal criteria for gene-therapy experimentation. The disease was life-threatening (therefore making the possibility of unknown treatment-related side-effects ethically acceptable), but the corresponding gene had been cloned and the biochemical basis of the disease was understood. Importantly, since ADA functions in the salvage pathway of nucleotide biosynthesis (p. 177), cells in which the genetic lesion had been corrected had a selective growth advantage over mutant cells, allowing them to be identified and isolated *in vitro*. Conventional treatment for ADA deficiency involves bone-marrow transplantation from a matching donor. Essentially the same established procedure could be used for gene therapy, but the bone-marrow cells would be derived from the patient herself and would be genetically modified *ex vivo* (Fig. 14.10). Cells from the patient were subjected to leucophaeresis and mononuclear cells were isolated. These were grown in culture under conditions that stimulated T-lymphocyte activation and growth and then transduced with a retroviral vector carrying a normal *ADA* gene as well as the neomycin-resistance gene. Following infusion of these modified cells, both this patient and a second, who began treatment in early 1991, showed an improvement in their clinical condition as well as in a battery of *in vitro* and *in vivo* immune-function studies (Anderson 1992). However, the production of recombinant ADA in these patients is transient, so each must undergo regular infusions of recombinant T lymphocytes. Research is ongoing into procedures for the transformation of bone-marrow stem cells, which would provide a permanent supply of corrected cells.

Gene-augmentation therapies for a small number of recessive single-gene diseases are now undergoing

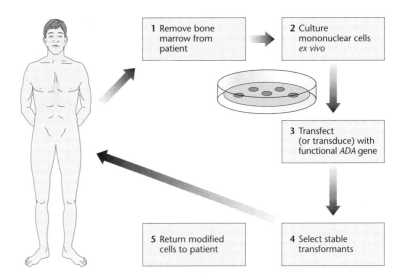

Fig. 14.10 Procedure for *ex vivo* gene therapy, based on the treatment for ADA deficiency.

clinical trials. We consider cystic fibrosis (CF) as an example. CF is a disorder that predominantly affects the lungs, liver and pancreas. The disease is caused by the loss of a cAMP-regulated membrane-spanning chloride channel. This results in an electrolyte imbalance and the accumulation of mucus, often leading to respiratory failure. CF is a recessive disorder suggesting that the loss of function could be corrected by introducing a functional copy of the gene. Indeed, epithelial cells isolated from CF patients can be restored to normal by transfecting them with the cloned cystic-fibrosis transmembrane regulator (*CFTR*) cDNA. Unlike ADA deficiency, the cells principally affected by CF cannot be cultured and returned to the patient, so *in vivo* delivery strategies must be applied. Targeted delivery of the *CFTR* cDNA to affected cells has been achieved using adenoviral vectors, which have a natural tropism for the epithelial lining of the respiratory system. Recombinant viruses carrying the *CFTR* cDNA have been introduced into patients using an inhaler (Zabner *et al.* 1993, Hay *et al.* 1995, Knowles *et al.* 1995). The *CFTR* cDNA has also been introduced using liposomes (e.g. Caplen *et al.* 1995). While such treatments have resulted in *CFTR* transgene expression in the nasal epithelium, there were neither consistent changes in chloride transport nor reduction in the severity of CF symptoms, i.e. they have been largely ineffective.

Gene-therapy strategies for cancer

Cancer gene therapy was initially an extension of the early gene-marking experiments. The tumour-infiltrating leucocytes were transformed with a gene for TNF in addition to the neomycin-resistance gene, with the aim of improving the efficiency with which these cells kill tumours by increasing the amount of TNF they secrete. Although TNF is highly toxic to humans at levels as low as 10 µg/kg body weight, there have been no side-effects from the gene therapy and no apparent organ toxicity from secreted TNF (Hwu *et al.* 1993). One alternative strategy is to transform the tumour cells themselves, making them more susceptible to the immune system through the expression of cytokines or a foreign antigen. Another is to transform fibroblasts, which are easier to grow in culture, and then co-inject these together with tumour cells to provoke an immune response against the tumour. A number of such 'assisted killing' strategies have been approved for clinical trials (see review by Ockert *et al.* 1999).

Direct intervention to correct cancer-causing genes is also possible. Dominantly acting genes (oncogenes) have been targeted using antisense technology, either with antisense transgenes, oligonucleotides (see Carter & Lemoine 1993, Nellen & Lichtenstein 1993) or ribozymes (Welch *et al.* 1998, Muotri *et al.* 1999). An early report of cancer gene therapy with antisense oligonucleotides was that of Szczylik et al. (1991) for the treatment of chronic myeloid leukaemia. They used two 18-mers specific for the BCR–ABL gene junction generated by the chromosomal translocation that causes this particular cancer, and showed that colony formation was suppressed in cells removed from cancer patients. Cancers caused by loss of tumour-suppressor gene function have been addressed by replacement strategies, in which a functional copy of the appropriate gene is delivered to affected cells (e.g. see Cai et al. 1993, Harper et al. 1993, Smith et al. 1993, Hahn et al. 1996). A further strategy, known as prodrug activation therapy, involves the activation of a particular enzyme specifically in cancer cells, which converts a non-toxic 'prodrug' into a toxic product, so killing the cancer cells. This can be achieved by driving the expression of a so-called 'suicide gene' selectively in cancer cells. An example is the HSV thymidine kinase gene, in combination with the prodrug ganciclovir. Thymidine kinase converts ganciclovir into a nucleoside analogue, which is incorporated into DNA and blocks replication by inhibiting the DNA polymerase. Activation of the enzyme specifically in cancer cells can be achieved by preferential delivery to dividing cells through the use of onco-retroviruses (e.g. Moolten 1986, Culver et al. 1992, Klatzmann et al. 1996). Another way is to use transcriptional regulatory elements that are active only in cancer cells (e.g. Harris et al. 1994, Su et al. 1996).

The importance of SNPs

As noted earlier (p. 277), SNPs are single-base variations that occur about once every 1000 bases along the human genome. With the availability of the human genome sequence, a high-resolution SNP map is being developed and this will facilitate disease therapy in a number of ways (Rothberg 2001). Two examples will be cited here: understanding polygenic disorders and pharmacogenomics.

The disease studies cited above have all focused on single-gene disorders. However, many of the common diseases of humans are polygenic in origin and attempts to map genes for these complex conditions have generally failed. The availability of SNP maps is providing a new tool for use in genetic association studies to identify genes for polygenic disorders, and success has already been achieved with hypertension (Geller et al. 2000), non-insulin-dependent diabetes (Deeb et al. 2000) and cardiovascular disease (Mason et al. 1999).

Pharmacogenomics is the term used to describe the identification and elucidation of genetic variations that will have an impact on the efficacy of drugs. SNPs are essential to pharmacogenomics (McCarthy & Hilfiker 2000), for they provide an easy means of determining the genotype of the patient. For example, the 2-adrenergic receptor agonists are the most widely used agents in the treatment of asthma and several polymorphisms have been described within the target genes. Several studies have shown associations between SNPs in these genes and response to therapy. One study (Buscher et al. 1999) found that homozygotes for one allele were up to 5.3 times more likely to respond to albuterol than homozygotes for the other allele. Another striking example is the response of Alzheimer's patients to the drug tacrine. Approximately 80% of patients not carrying the ApoE4 allele improved after tacrine treatment, whereas 60% of patients with the allele deteriorated after treatment (Poirier et al. 1995). One scenario for the future is that, before prescribing a drug, patients will be genotyped by SNP analysis to determine the most effective therapy.

Another use for pharmacogenomics is the prevention of adverse drug reactions resulting from drug metabolism. For example, between 3 and 10% of the Caucasian population fail to metabolize the adrenergic-blocking drug debrisoquine and treatment results in severe hypotension. In Afro-Americans the frequency of this 'poor metabolizer' condition is 5% and in Asians it is only 1%. Affected individuals are homozygous for a mutant cytochrome P450 gene (CYP2D6) and they also fail to metabolize over 20% of all commonly prescribed drugs, including codeine (Gonzalez et al. 1988). The same gene also has alleles that cause an elevated-metabolizer phenotype, which has been correlated with increased susceptibility to cancer. Clearly, before selecting patients for clinical trials of new drugs, it would make sense to screen candidates for their drug-metabolizing phenotype and most large pharmaceutical companies now do this.

Theme 3: Combating infectious disease

Introduction to theme 3

The usual way of treating bacterial and infectious diseases is with antibiotics. As is well known, certain microbes quickly develop resistance to the antibiotics in current use and this means that new antibiotics are required. The traditional way of obtaining new antibiotics is the screening of new microbial isolates from nature. An alternative way will be described in theme 5: combinatorial biosynthesis (p. 306). Another way is to identify new cellular targets and screen chemical libraries for inhibitory activities. A number of methods for identifying key genes involved in pathogenesis have been developed and these are described later in this section. In contrast to bacteria and fungi, viruses are not sensitive to antibiotics and few therapies have been available. However, this could change with the development of antisense drugs, as described in the previous section (p. 291). Where suitable therapies do exist, it can be advantageous to know the identity of the pathogen as soon as possible. Conventional laboratory procedures take several days, but PCR methodology offers a more rapid identification, as was described on p. 274. For many pathogens, prevention is much better than cure, and hence vaccines are of great value. Gene-manipulation techniques have greatly facilitated the development of new vaccines, as described below.

Novel routes to vaccines

An effective vaccine generates humoral and/or cell-mediated immunity, which prevents the development of disease upon exposure to the corresponding pathogen. This is accomplished by presenting pertinent antigenic determinants to the immune system in a fashion which mimics that in natural infections. Conventional viral vaccines consist of inactivated, virulent strains or live, attenuated strains, but they are not without their problems. For example, many viruses have not been adapted to grow to high titre in tissue culture, e.g. hepatitis B virus. There is a danger of vaccine-related disease when using inactivated virus, since replication-competent virus may remain in the inoculum. Outbreaks of foot-and-mouth disease in Europe have been attributed to this

cause. Finally, attenuated virus strains have the potential to revert to a virulent phenotype upon replication in the vaccinee. This occurs about once or twice in every million people who receive live polio vaccine. Recombinant DNA technology offers some interesting solutions.

Given the ease with which heterologous genes can be expressed in various prokaryotic and eukaryotic systems, it is not difficult to produce large quantities of purified immunogenic material for use as a subunit vaccine. A whole series of immunologically pertinent genes have been cloned and expressed but, in general, the results have been disappointing. For example, of all the polypeptides of foot-and-mouth disease virus, only VP1 has been shown to have immunizing activity. However, polypeptide VP1 produced by recombinant means was an extremely poor immunogen (Kleid *et al.* 1981). Perhaps it is not too surprising that subunit vaccines produced in this way do not generate the desired immune response, for they lack authenticity. The hepatitis B vaccine, which is commercially available (Valenzuela *et al.* 1982), differs in this respect, for expression of the surface antigen in yeast results in the formation of virus-like particles. A similar phenomenon is seen with a yeast Ty vector carrying a gene for HIV coat protein (Adams *et al.* 1987). These subunit vaccines also have another disadvantage. Being inert, they do not multiply in the vaccinee and so they do not generate the effective cellular immune response essential for the recovery from infectious disease.

Recombinant bacterial vaccines

An alternative approach to the development of live vaccines is to start with the food-poisoning organism *Salmonella typhimurium*. This organism can be attenuated by the introduction of lesions in the *aro* genes, which encode enzymes involved in the biosynthesis of aromatic amino acids, *p*-aminobenzoic acid and enterochelins. Whereas doses of 10^4 wild-type *S. typhimurium* reproducibly kill mice, *aro* mutants do not kill mice when fed orally, even when doses as high as 10^{10} organisms are used. However, the mutant strains can establish self-limiting infections in the mice and can be detected in low numbers in organs such as the liver and spleen. Such attenuated strains of *S. typhimurium* are particularly attractive as carriers of heterologous antigens because they can be delivered orally and because they can stimulate humoral, secretory and cellular immune responses in the host (Charles & Dougan 1990). Already a wide range of heterologous antigens have been expressed in such vaccine strains (Hackett 1993).

The use of the BCG vaccine strain as an alternative vector has many advantages, including its known safety, low cost, widespread use as a childhood vaccine and ability of a single dose to induce long-lasting protection. Several recombinant BCG strains have been constructed that stably express foreign genes (Stover et al. 1991) and preliminary results from animal studies are very encouraging. An alternative vector is the human oral commensal *Streptococcus gordonii* (Fischetti et al. 1993).

A different procedure for attenuating a bacterial pathogen to be used as a vaccine has been proposed by Kaper et al. (1984). They attenuated a pathogenic strain of *Vibrio cholera* by deletion of DNA sequences encoding the A_1 subunit of the cholera enterotoxin. A restriction-endonuclease fragment encoding the A_1, but not the A_2 or B, sequence was deleted *in vitro* from cloned cholera-toxin genes. The mutation was then recombined into the chromosome of a pathogenic strain. The resulting strain produces the immunogenic but non-toxic B subunit of cholera toxin but is incapable of producing the A subunit. This strain has been found to be safe and immunogenic in carefully controlled clinical trials in a number of countries (Cryz 1992).

Yet another approach to vaccine preparation has been developed by Pizza et al. (2000), based on whole-genome sequencing. They wanted to develop a vaccine against group B meningococci but were hampered by the fact that there was considerable sequence variation in surface-exposed proteins. Starting with the entire genome sequence of a group B strain of *Neisseria meningitidis*, they identified 350 proteins as potential protective antigens. All 350 candidate antigens were expressed in *E. coli*, purified and used to immunize mice. The sera from the mice allowed the identification of proteins that are surface-exposed in meningococci, that are conserved across a range of strains and that induce a bactericidal antibody response, a property known to correlate with vaccine efficacy in humans.

Recombinant viruses as vaccines

Recombinant viruses can be used as vectors to express heterologous antigens, and thus function as live vaccines. The first animal virus to be exploited in this way was vaccinia, which had been used previously as a non-recombinant vaccine providing cross-protection against variola virus, the causative agent of smallpox. Contributing to its success as a live vaccine were its stability as a freeze-dried preparation, its low production cost and the ability to administer the vaccine by simple dermal abrasion. Vaccinia remains the most widely explored recombinant viral vaccine, and many antigens have been expressed using this vector, essentially using the methods described on p. 197 (Ulaeto & Hruby 1994, Moss 1996). The most successful recombinant-virus vaccination campaign to date involved the use of recombinant vaccinia virus expressing rabies-virus glycoprotein. This was administered to the wild population of foxes in central Europe by providing a bait consisting of chicken heads spiked with the virus. The epidemiological effects of vaccination were most evident in eastern Switzerland, where two decades of rabies came to a sudden end after only three vaccination campaigns (Brochier et al. 1991, Flamand et al. 1992), and in Belgium, where the disease was all but eliminated (Brochier et al. 1995).

One disadvantage of this approach in humans is the unacceptably high risk of adverse reactions to the vaccine. For use in humans, vaccinia must be further attenuated to make it replication-deficient and to minimize the likelihood of replication-competent viruses arising by recombination (e.g. Cooney et al. 1991, Sutter & Moss 1992, Tartaglia et al. 1992). Highly attenuated vaccinia derivatives (Table 14.6) have been used to express a range of viral, bacterial and parasite antigens, with some reaching clinical trials (reviewed by Moss 1996, Paoletti 1996). Other poxviruses, such as canarypox, have also been developed as potential vaccines (e.g. see Perkus et al. 1995, Fries et al. 1996, Myagkikh et al. 1996).

As well as vaccinia, adenovirus and alphavirus vectors have also been investigated as potential recombinant vaccines. Adenovirus has been used to express various antigens, including HSV and rabies virus glycoproteins (Gallichan et al. 1993, Xiang et al. 1996). Among the alphaviruses, Semliki Forest

Table 14.6 Immune response to heterologous antigens expressed by vaccinia virus recombinants.

Antigen	Neutralizing antibodies	Cellular immunity	Animal protection
Rabies virus glycoprotein	+	+	+
Vesicular stomatitis glycoprotein	+	+	+
Herpes simplex virus glycoprotein D	+	+	+
Hepatitis B surface antigen	+	+	+
Influenza virus haemagglutinin	+	+	+
Human immunodeficiency virus envelope	+	+	not determined

Reproduced with permission from Tartaglia & Paoletti (1988).

Table 14.7 A selection of recombinant vaccines against animal viruses produced in plants.

Antigen	Host-plant system	Reference
Herpes virus B surface antigen	Tobacco	Mason *et al.* 1992
Rabies glycoprotein	Tomato	McGarvey *et al.* 1995
Norwark virus coat protein	Tobacco, potato	Mason *et al.* 1996
Foot-and-mouth virus VP1	*Arabidopsis*	Carrillo *et al.* 1998
Cholera toxin B subunit	Potato	Arakawa *et al.* 1998
Human cytomegalovirus glycoprotein B	Tobacco	Tackaberry *et al.* 1999

virus has been used to express the gp160 protein of HIV (Berglund *et al.* 1997) and Sindbis virus has been used to express antigens from Japanese encephalitis virus (Pugachev *et al.* 1995).

Recombinant proteins expressed by viruses have been shown to be immunogenetic. However, stronger stimulation of the immune response can often be achieved by presenting the antigen on the surface of the virus attached to a host virus-derived carrier protein. The advantage of this strategy is that the recombinant antigen is presented as multiple copies. This strategy is borrowed from the surface display of foreign antigens on bacteria that are used as live vaccines (see above) but is much safer. Indeed, while a number of conventional viral vaccines, such as vaccinia, have been developed as surface-display systems (e.g. see Katz & Moss 1997, Katz *et al.* 1997), even plant viruses can be used for this purpose, as discussed below (Porta & Lomonossoff 1996, Johnson *et al.* 1997).

Plants as edible vaccines

Plants have been explored as a cheap, safe and efficient production system for subunit vaccines (Table 14.7), with the added advantage that orally administered vaccines can be ingested by eating the plant, therely eliminating the need for processing and purification (reviewed by Mason & Arntzen 1995, Walmsey & Arntzen 2000). The earliest demonstration was the expression of a surface antigen from the bacterium *S. mutans* in tobacco. This bacterium is the causative agent of dental caries, and it was envisaged that stimulation of a mucosal immune response would prevent the bacteria colonizing the teeth and therefore protect against tooth decay (Curtis & Cardineau, 1990).

A number of edible transgenic plants have been generated expressing antigens derived from animal viruses. For example, rabies glycoprotein has been expressed in tomato (McGarvey *et al.* 1995), hepatitis

B virus antigen in lettuce (Ehsani *et al.* 1997) and cholera antigen in potato (Arakawa *et al.* 1997). As well as animal virus antigens, autoantigens associated with diabetes have also been produced (Ma *et al.* 1997, Porceddu *et al.* 1999). Plants have also been infected with recombinant viruses expressing various antigen epitopes on their surfaces. Cowpea mosaic virus (CMV) has been extensively developed as a heterologous antigen-presenting system (Porta *et al.* 1994, Lomonossoff & Hamilton 1999). There have been some recent successes in vaccination trials using recombinant CMV vectors expressing epitopes of HIV gp41 (McLain *et al.* 1995, 1996a,b) and canine parvovirus (Dalsgaard *et al.* 1997). The first clinical trials using a plant-derived vaccine were conducted in 1997 and involved the ingestion of transgenic potatoes expressing the B subunit of the *E. coli* heat-labile toxin, which causes diarrhoea. This resulted in a successful elicitation of mucosal immunity in test subjects (Tacket *et al.* 1994).

DNA vaccines

The immune system generates antibodies in response to the recognition of proteins and other large molecules carried by pathogens. In each of the examples above, the functional component of the vaccine introduced into the host is the protein responsible for the elicitation of the immune response. The introduction of DNA into animals does not generate an immune response *against the DNA molecule*, but, if that DNA is expressed to yield a protein, that protein can stimulate the immune system. This is the basis of DNA vaccination, as first demonstrated by Ulmer *et al.* (1993). DNA vaccines generally comprise a bacterial plasmid carrying a gene encoding the appropriate antigen under the control of a strong promoter that is recognized by the host cell. The advantages of this method include its simplicity, its wide applicability and the ease with which large quantities of the vaccine can be produced. The DNA may be administered by injection, using liposomes or by particle bombardment. In the original demonstration, Ulmer and colleagues introduced DNA corresponding to the influenzavirus nucleoprotein and achieved protection against influenza infection. Since then, many DNA vaccines have been used to target viruses (e.g. measles (Cardoso *et al.* 1996);

HIV (Wang *et al.* 1993, Fuller *et al.* 1997, Hinkula *et al.* 1997); Ebola virus (Xu *et al.* 1998)), other pathogens (e.g. tuberculosis (Huygen *et al.* 1996)) and even the human cellular prion protein in mice (Krasemann *et al.* 1996).

The DNA-vaccination approach has several additional advantages. These include the following:
• Certain bacterial DNA sequences have the innate ability to stimulate the immune system (see Klinman *et al.* 1997, Roman *et al.* 1997).
• Other genes encoding proteins influencing the function of the immune response can be co-introduced along with the vaccine (e.g. Kim *et al.* 1997).
• DNA vaccination can be used to treat diseases that are already established as a chronic infection (e.g. Mancini *et al.* 1996).

In principle, DNA vaccination has much in common with gene therapy (discussed above), since both processes involve DNA transfer to humans, using a similar selection of methods. However, while the aim of gene therapy is to alleviate disease, by either replacing a lost gene or blocking the expression of a dominantly acting gene, the aim of DNA vaccination is to prevent disease, by causing the expression of an antigen that stimulates the immune system.

Selecting targets for new antimicrobial agents

In attempting to develop new antimicrobial agents, including ones that are active against intractable pathogens such as the malarial parasite, it would be useful to know which genes are both essential for virulence and unique to the pathogen. Once these genes have been identified, chemical libraries can be screened for molecules that are active against the gene product. Two features of this approach deserve further comment. First, inhibition of the gene product could attenuate the organism's virulence but would not result in death of the organism *in vitro*; that is, no effect would be seen in whole-organism inhibition assays and so molecules only active *in vivo* would be missed. Secondly, target genes can be selected on the basis that there are no human counterparts. Thus, active molecules will be less likely to be toxic to humans. A number of different approaches have been developed for identifying virulence determinants, and details are presented below for three of these: *in vivo* expression technology (IVET), differential fluorescence induction and signature-tagged mutagenesis.

In vivo expression technology (IVET)

IVET was developed to positively select those genes that are induced specifically in a microorganism when it infects an animal or plant host (Mahan *et al.* 1993). The basis of the system is a plasmid carrying a promoterless operon fusion of the *lacYZ* genes fused to the *purA* or *thyA* genes downstream of a unique *BglII* cloning site (Fig. 14.11). This operon fusion was constructed in a suicide-delivery plasmid. Cloning of pathogen DNA into the *BglII* site results in the construction of a pool of transcriptional fusions driven by promoters present in the

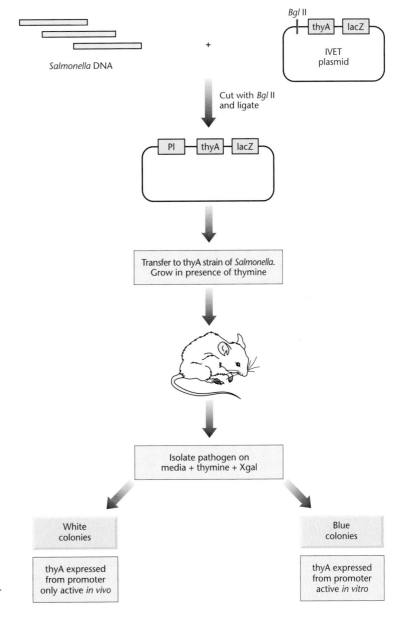

Fig. 14.11 The basic principle of IVET. See text for details.

cloned DNA. The pool of fusions is transferred to a strain of the pathogen carrying a *purA* or *thyA* deletion and selection made for integration into the chromosome. The recombinant pathogen is then used to infect a test animal. Fusions that contain a promoter that is active in the animal allow transcription of the *purA* or *thyA* gene and hence bacterial survival. When the surviving pathogens are reisolated from the test animal, they are tested *in vitro* for their levels of β-galactosidase. Clones that contain fusions to genes that are *specifically* induced in the test animal show little *lacZ* expression on laboratory media and DNA sequencing can be used to identify the genes to which these promoters belong.

The IVET system was initially developed for use in a murine typhoid model. Since it was first described, IVET has been used with a wide variety of Gram-positive and Gram-negative organisms, including 'difficult' organisms, such as *Mycobacterium*. One disadvantage of the original IVET system is that it inherently selects for promoters that are constitutively and highly expressed *in vivo*. An alternative IVET system uses *tnpR* operon fusions in place of *purA* or *thyA*. The *tnpR* locus encodes a site-specific resolvase and is used as the selection gene. Expression of this gene results in resolvase synthesis and the deletion of a resolvase-specific reporter. The advantage of this system is that it can be used to identify those genes that are expressed only transiently during animal infection (Camilli & Mekalanos 1995, Camilli 1996). Handfield and Levesque (1999) have described a number of other modifications to the IVET technology.

Differential fluorescence induction

This technique represents a different way of identifying environmentally controlled promoters and was originally developed to facilitate identification of *Salmonella* genes that are differentially expressed with macrophages (Valdivia & Falkow 1997). Fragments of DNA from *Salmonella* were fused to a promotorless gene for green fluorescent protein (GFP) and returned to the *Salmonella*, which was then used to infect macrophages. Cells that became fluorescent were recovered using a fluorescence-activated cell sorter (FACS) and grown on media in the absence of macrophages. Bacteria that were non-fluorescent in the extracellular environment were sorted and used for a second round of macrophage infection. Bacteria still capable of generating fluorescent macrophages were found to contain GFP fusions that were up-regulated by the macrophage's intracellular environment.

Signature-tagged mutagenesis

This technique is a variation on the use of transposon mutagenesis that has been applied to the identification of bacterial virulence genes. The basic principle is to create a large number of different transposon-generated mutants of a pathogenic organism and to identify those mutants that can survive *in vitro* but not *in vivo* – that is, it is a whole-genome scan for habitat-specific genes. Although one could test each mutant individually, this would be very laborious and would use large numbers of animals. By using signature-tagged mutagenesis, one can test large numbers of different mutants *simultaneously in the same animal*. This is achieved by tagging each transposon mutant with a different DNA sequence or tag.

The use of sequence-tagged mutagenesis was first described for the identification of virulence genes from *S. typhimurium* in a mouse model of typhoid fever (Hensel *et al.* 1995), as shown in Fig. 14.12. The tags comprise different sequences of 40 bp $[NK]_{20}$, where N = A, C, G or T and K = G or T. The arms were designed so that the amplification of the tags by PCR with specific primers would produce probes with 10 times more label in the central region than in each arm. The double-stranded tags were ligated into a Tn5 transposon and transferred from *E. coli* to *S. typhimurium* by conjugation. A library of over 1500 exconjugants resulting from transposition events was stored in microtitre dishes. Of these exconjugants, 1152 were selected and prepared as 12 pools of 96 mutants. Each pool was injected into the peritoneum of a different mouse and infection allowed to proceed. Bacteria were then recovered from each mouse by plating spleen homogenates on culture media. DNA was extracted from the recovered bacteria and the tags in this DNA were amplified by PCR. Those tags present in the initial pool of bacteria but missing from the recovered media represent mutations in genes essential for

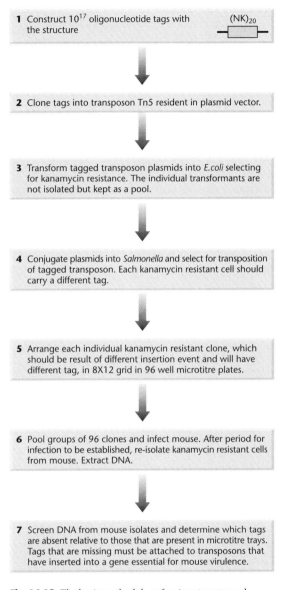

Fig. 14.12 The basic methodology for signature tagged mutagenesis.

The flow chart steps read:

1 Construct 10^{17} oligonucleotide tags with the structure $(NK)_{20}$

2 Clone tags into transposon Tn5 resident in plasmid vector.

3 Transform tagged transposon plasmids into *E.coli* selecting for kanamycin resistance. The individual transformants are not isolated but kept as a pool.

4 Conjugate plasmids into *Salmonella* and select for transposition of tagged transposon. Each kanamycin resistant cell should carry a different tag.

5 Arrange each individual kanamycin resistant clone, which should be result of different insertion event and will have different tag, in 8X12 grid in 96 well microtitre plates.

6 Pool groups of 96 clones and infect mouse. After period for infection to be established, re-isolate kanamycin resistant cells from mouse. Extract DNA.

7 Screen DNA from mouse isolates and determine which tags are absent relative to those that are present in microtitre trays. Tags that are missing must be attached to transposons that have inserted into a gene essential for mouse virulence.

virulence. In this way 28 different mutants with attenuated virulence were identified and some of these mutants were in genes not previously identified.

The principle of signature-tagged mutagenesis has been extended to the analysis of pathogenicity determinants in a wide range of bacteria (for a review, see Handfield & Levesque 1999) and to fungi (Brookman & Denning 2000).

Environomics

The techniques described above to identify bacterial virulence genes can be applied to other microbe–host interactions, e.g. protozoan infections of humans, bacterial or fungal pathogenesis of plants, or even beneficial rhizosphere–microbe interactions (Rainey 1999). Essentially, these techniques are methods for scanning the entire microbial genome for genes that are expressed under particular environmental conditions. This approach has been termed 'environomics'.

Theme 4: Protein engineering

Introduction to theme 4

One of the most exciting aspects of recombinant DNA technology is that it permits the design, development and isolation of proteins with improved operating characteristics and even completely novel proteins. The simplest example of protein engineering involves site-directed mutagenesis to alter key residues, as originally shown by Winter and colleagues (Winter *et al.* 1982, Wilkinson *et al.* 1984). From a detailed knowledge of the enzyme tyrosyl-tRNA synthetase from *Bacillus stearothermophilus*, including its crystal structure, they were able to predict point mutations in the gene that should increase the enzyme's affinity for the substrate ATP. These changes were introduced and, in one case, a single amino acid change improved the affinity for ATP by a factor of 100. Using a similar approach, the stability of an enzyme can be increased. Thus Perry and Wetzel (1984) were able to increase the thermostability of T4 lysozyme by the introduction of a disulphide bond. However, although new cysteine residues can be introduced at will, they will not necessarily lead to increased thermal stability (Wetzel *et al.* 1988).

Improving therapeutic proteins with single amino acid changes

As noted earlier (p. 283), many recombinant proteins are now being used therapeutically. With some of them, protein engineering has been used to generate second-generation variants with improved pharmacokinetics, structure, stability and bioavailability (Bristow 1993). For example, in the neutral solutions used for therapy, insulin is mostly assembled as zinc-containing hexamers. This self-association may limit absorption. By making single amino acid substitutions, Brange *et al.* (1988) were able to generate insulins that are essentially monomeric at pharmaceutical concentrations. Not only have these insulins preserved their biological activity, but they are also absorbed two to three times faster. Similarly, replacing an asparagine residue with glutamine altered the glycosylation pattern of TPA. This in turn significantly increased the circulatory half-life, which in the native enzyme is only 5 min (Lau *et al.* 1987). Proteins can also be engineered to be resistant to oxidative stress, as has been shown with AAT (see Box 14.3).

Improving enzymes: subtilisin as a paradigm for protein engineering

Proof of the power of gene manipulation coupled with the techniques of *in vitro* (random and site-directed) mutagenesis as a means of generating improved enzymes is provided by the work done on subtilisin over the past 15 years (for review, see Bryan 2000). Every property of this serine protease has been altered, including its rate of catalysis, substrate specificity, pH-rate profile and stability to oxidative, thermal and alkaline inactivation. In the process, well over 50% of the 275 amino acids of subtilisin have been changed. At some positions in the molecule, the effects of replacing the usual amino acid with all the other 19 natural amino acids have been evaluated.

Box 14.3 Oxidation-resistant variants of α_1-antitrypsin (AAT)

Cumulative damage to lung tissue is thought to be responsible for the development of emphysema, an irreversible disease characterized by loss of lung elasticity. The primary defence against elastase damage is AAT, a glycosylated serum protein of 394 amino acids. The function of AAT is known because its genetic deficiency leads to a premature breakdown of connective tissue. In healthy individuals there is an association between AAT and neutrophil elastase followed by cleavage of AAT between methionine residue 358 and serine residue 359 (see Fig. B14.2).

After cleavage, there is negligible dissociation of the complex. Smokers are more prone to emphysema, because smoking results in an increased concentration of leucocytes in the lung and consequently increased exposure to neutrophil elastase. In addition, leucocytes liberate oxygen free radicals and these can oxidize methionine-358 to methionine sulphoxide. Since methionine sulphoxide is much bulkier than methionine, it does not fit into the active site of elastase. Hence oxidized AAT is a poor inhibitor. By means of site-directed mutagenesis, an oxidation-resistant mutant of AAT has been constructed by replacing methionine-358 with valine (Courtney *et al.* 1985). In a laboratory model of inflammation, the modified AAT was an effective inhibitor of elastase and was not inactivated by oxidation. Clinically, this could be important, since intravenous replacement therapy with plasma concentrates of AAT is already being tested on patients with a genetic deficiency in AAT production.

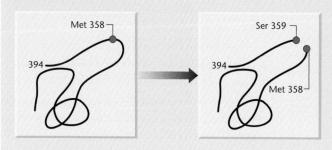

Fig. B14.2 The cleavage of α_1-antitrypsin on binding to neutrophil elastase.

Many of the changes described above were made to improve the ability of subtilisin to hydrolyse protein when incorporated into detergents. However, serine proteases can be used to synthesize peptides and this approach has a number of advantages over conventional methods (Abrahmsen *et al.* 1991). A problem with the use of subtilisin for peptide synthesis is that hydrolysis is strongly favoured over aminolysis, unless the reaction is undertaken in organic solvents. Solvents, in turn, reduce the half-life of subtilisin. Using site-directed mutagenesis, a number of variants of subtilisin have been isolated with greatly enhanced solvent stability (Wong *et al.* 1990, Zhong *et al.* 1991). Changes introduced included the minimization of surface changes to reduce solvation energy, the enhancement of internal polar and hydrophobic interactions and the introduction of conformational restrictions to reduce the tendency of the protein to denature. Designing these changes requires an extensive knowledge of the enzyme's structure and function. Chen and Arnold (1991, 1993) have provided an alternative solution. They utilized random mutagenesis combined with screening for enhanced proteolysis in the presence of solvent (dimethyl formamide) and substrate (casein).

The engineering of subtilisin has now gone one step further, in that it has been modified such that aminolysis (synthesis) is favoured over hydrolysis, even in aqueous solvents. This was achieved by changing a serine residue in the active site to cysteine (Abrahmsen *et al.* 1991). The reasons for this enhancement derive mainly from the increased affinity and reactivity of the acyl intermediate for the amino nucleophile (Fig. 14.13). These engineered 'peptide ligases' are in turn being used to synthesize novel glycopeptides. A glycosyl amino acid is used in peptide synthesis to form a glycosyl peptide ester,

which will react with another *C*-protected peptide in the presence of the peptide ligase to form a larger glycosyl peptide.

Methods for engineering proteins: the rational approach

Given the range of modifications made to subtilisin, the question is not 'what modifications are possible?' (for review, see Arnold 2001) but 'how do I achieve the modifications that I wish to make?' There are two general techniques: the rational approach and 'directed evolution'. There are two variations on the rational approach. The first of these makes use of comparisons between related proteins. For example, barnase and binase are ribonucleases that have 85% sequence identity but differ in their thermal stability. Using site-directed mutagenesis, Serrano *et al.* (1993) tested the impact on thermostability of barnase of all 17 amino acid differences between barnase and binase. They observed effects between +1.1 and −1.1 kcal/mol and a multiple mutant combining six of the substitutions displayed a stability increase of 3.3 kcal/mol over the wild-type enzyme.

The second variation on the rational approach requires that the three-dimensional structure of the protein be known. Based on an analysis of this structure, particular residues are selected for mutagenesis and the biological effects then evaluated. That this approach works was clearly shown by the early work on lysozyme and subtilisin, but the chances of success are not predictable. One disadvantage of this approach is illustrated by the results of Spiller *et al.* (1999), who were investigating the thermostability of an esterase using directed evolution. When they evaluated the effect of certain mutations, they concluded that no amount of rational analysis of the crystal structure would have led them to predict the results obtained.

Protein engineering through directed evolution

Directed evolution involves repeated rounds of random mutagenesis, followed by selection for the improved property of interest. Any number of methods can be used to introduce the mutations, including mutagenic base analogues, chemical

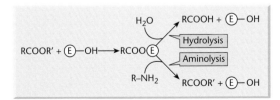

Fig. 14.13 The aminolysis (synthetic) and hydrolysis reactions mediated by an acylated protease.

mutagenesis, error-prone PCR and spiked synthetic oligonucleotides. The key element in the process is the ability to screen large numbers of mutants. An example is the isolation of a more thermostable subtilisin. Up to 1000 mutant clones are gridded out on replica plates and, once they are grown, one plate is incubated at an elevated temperature long enough to inactivate the wild-type enzyme. When an assay for hydrolytic activity is subsequently performed, only mutants with stability greater than that of the wild type will display measurable activity. Once stable mutants have been identified, the replicate colony can be grown to identify the mutation. Once stabilizing single amino acid changes have been identified, building a highly stable subtilisin can be accomplished by combining individual amino acids into the same molecule (Pantoliano *et al.* 1988, 1989, Zhao & Arnold 1999).

One of the disadvantages of the screening method described above is that it is labour-intensive and the maximum feasible number of mutants that can be examined in a single screen is 10^4–10^5. By combin-ing mutagenesis with phage display, up to 10^9 different mutants can *theoretically* be screened in a single experiment. The constraint with this approach is that phage display is really best suited to selecting proteins with altered binding characteristics, rather than the other properties one might wish to engineer. Despite this, a number of groups (Atwell & Wells 1999, Demartis *et al.* 1999, Olsen *et al.* 2000) were able to select variant proteases with novel substrate specificities.

Gene families as aids to protein engineering

An alternative approach to directed evolution is 'DNA shuffling', which is also known as 'molecular breeding' (Minshull & Stemmer 1999, Ness *et al.* 2000). This method can only be adopted if the target protein belongs to a known protein family. If it does, the genes for the different family members are isolated and artificial hybrids created (Fig. 14.14). As an example of this approach, Ness *et al.* (1999)

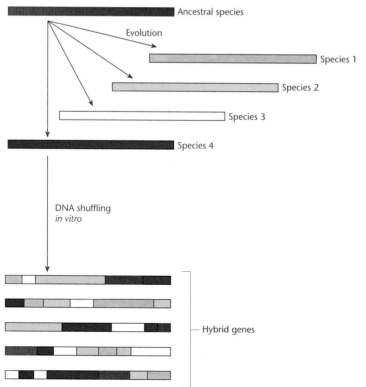

Fig. 14.14 Schematic representation of gene shuffling.

started with the genes for 26 members of the sub-tilisin family and created a library of chimeric proteases. When this library was screened for four distinct enzyme properties, variants were found that were significantly improved over any of the parental enzymes for each individual property. In a similar way, Powell *et al.* (2000) were able to select retroviruses with increased stability for use as gene-therapy vectors.

As a final twist on the methods for engineering proteins, Lehmann *et al.* (2000) have combined the use of protein families with a rational approach. They started with a family of mesophilic phytases whose amino acid sequences had been determined. Using these data they constructed a 'consensus' phytase sequence and found that enzyme with this sequence was much more thermostable than any of the parent enzymes. Refinement of the approach combined with site-directed mutagenesis resulted in the isolation of an enzyme whose unfolding temperature was 27–34°C higher than that of the parent enzymes. Whereas all the methods described earlier tend to yield mutants with single amino acid changes, this method generated variants with multiple amino acid changes spread throughout the molecule.

Theme 5: Metabolic engineering

Introduction to theme 5

When the large-scale production of penicillin was begun in the 1940s, the yields of penicillin were measured in micrograms per litre of culture. Demand for the antibiotic was outstripping supply and higher-yielding strains were badly needed. Since nothing was known about the biosynthetic pathway, a pro-gramme of strain improvement was set in place that involved random mutation and screening. The best strain from each cycle of improvement then became the starting-point for the next round of selection. In this way the yield of penicillin was steadily increased until it reached the tens of grams per litre that can be achieved today. As each new antibiotic was discov-ered, the same process of strain improvement was applied. In every case, the biochemical and genetic basis of the beneficial mutations was not known. Only when the details of gene regulation and metabolic-pathway regulation (allosteric control)

had been elucidated could we even begin to under-stand how antibiotic yields might have been improved.

Based on our current knowledge of metabolic regulation we can predict that the changes in the improved strains described above will involve all of the following:

- Removal of rate-limiting transcriptional and allo-steric controls.
- Kinetic enhancement of rate-limiting enzymes.
- Genetic blockage of competing pathways.
- Enhanced carbon commitment to the primary metabolic pathway from central metabolism.
- Modification of secondary metabolic pathways to enhance energy metabolism and availability of enzymatic cofactors.
- Enhanced transport of the compound out of the cell.

Today, if one starts with a wild-type strain and wants to turn it into an overproducer, then the approach would be to use recombinant DNA tech-nology to make these desired changes in a rational way, as exemplified below by phenylalanine. There are many other examples of rational 'metabolic engineering' and these have been reviewed by Chotani *et al.* (2000).

Designed overproduction of phenylalanine

Phenylalanine is a key raw material for the synthesis of the artificial sweetener aspartame. Phenylalanine can be synthesized chemically but is too expensive if made this way. In the 1980s bacterial strains that overproduced phenylalanine were developed, using the traditional mutation and selection method. At the same time, a programme of rational strain devel-opment was instituted at G.D. Searle, the company who owned the patent for aspartame. The starting-point for this programme was an analysis of the biosynthetic pathway (Fig. 14.15). Removing feed-back inhibition of key steps is an essential first step. In the case of a phenylalanine producer, it is essential to knock out any feedback inhibition of the pathway from chorismate to phenylalanine and this was achieved by selecting strains resistant to pheny-lalanine analogues. The conversion of erythrose-4-phosphate (E4P) and phosphoenol pyruvate (PEP) to DAHP is also subject to feedback inhibition, but, since there are three different enzymes here, each

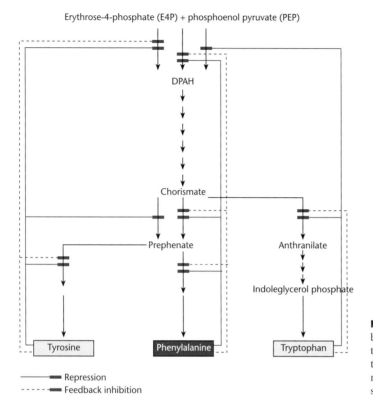

Erythrose-4-phosphate (E4P) + phosphoenol pyruvate (PEP)

DPAH

Chorismate

Prephenate

Anthranilate

Indoleglycerol phosphate

Tyrosine

Phenylalanine

Tryptophan

━━━■ Repression
- - - - ■ Feedback inhibition

Fig. 14.15 The regulation of the biosynthesis of aromatic amino acids. Note that indoleglycerol phosphate is converted to tryptophan via indole but that the indole normally is not released from the tryptophan synthase complex.

inhibited by a different aromatic end-product, all that is necessary for a phenylalanine overproducer is to clone the tryptophan-sensitive enzyme and have it overexpressed. To overcome repression of enzyme synthesis, the existing promoters were removed and replaced with one that could be controlled more easily in industrial-scale fermentations.

The above changes removed the natural control circuits. The next step was to remove competing pathways, i.e. the synthesis of tyrosine and tryptophan. This was easily achieved by making a tyrosine and tryptophan double auxotroph. Note that stable (non-reverting) auxotrophs can best be made by deleting part or all of the relevant genes. This is a task that is easy using recombinant DNA technology. Once all the control circuits and competing pathways had been removed, attempts were made to increase the carbon flux through the biosynthetic pathway. Surprisingly, overexpressing all the genes in the pathway did not enhance the yield of phenylalanine. One explanation was that the supply of precursors (E4P and/or PEP) was rate-limiting. This

was confirmed when cloning transketolase (to enhance E4P levels) and eliminating pyruvate kinase (to enhance PEP levels) enhanced yields.

New routes to small molecules

Recombinant DNA technology can be used to develop novel routes to small molecules. Good examples are the microbial synthesis of the blue dye indigo (Ensley *et al.* 1983) and the black pigment melanin (Della-Cioppa *et al.* 1990). Neither compound is produced in bacteria. The cloning of a single gene from *Pseudomonas putida* – that encoding naphthalene dioxygenase – resulted in the generation of an *E. coli* strain able to synthesize indigo in a medium containing tryptophan (Fig. 14.16). Similarly, cloning a tyrosinase gene in *E. coli* led to conversion of tyrosine to dopaquinone, which spontaneously converts to melanin in the presence of air. To overproduce these compounds, one generates a strain of *E. coli* that overproduces either tryptophan or tyrosine, rather than phenylalanine, as described

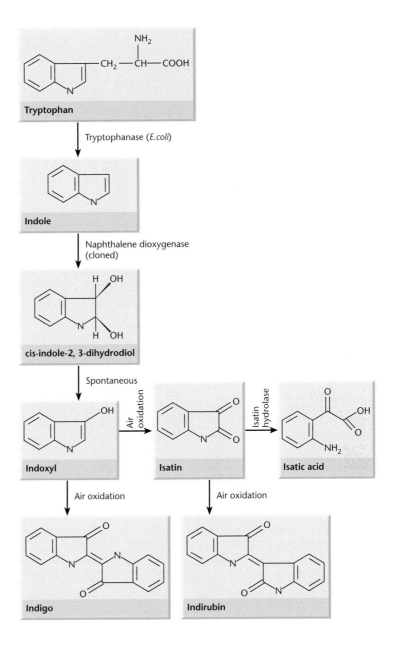

Fig. 14.16 The biosynthesis of indigo in *E. coli* and the formation of alternative end products.

above. With both indigo and melanin, yields are improved by increasing the levels of cofactors. Also, in the case of indigo biosynthesis, it is necessary to engineer the tryptophan synthase gene. The reason for this is that indole is an intermediate in the biosynthesis of tryptophan (Fig. 14.15). However, normally it is not free in the cytoplasm but remains trapped within the tryptophan synthase complex. By modifying the *trpB* gene, encoding the subunit of

tryptophan synthase, it was possible for the indole to be released for conversion by the dioxygenase (Murdock *et al.* 1993).

One disadvantage of the new route to indigo is that one of the intermediates in its synthesis, indoxyl, can undergo an alternative spontaneous oxidation to isatin and indirubin. The latter compound is an isomer of indigo with similar dyeing properties, but instead of being blue it is a deep

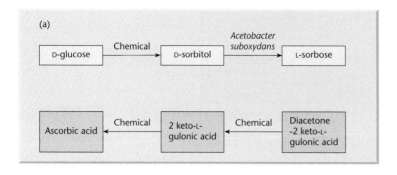

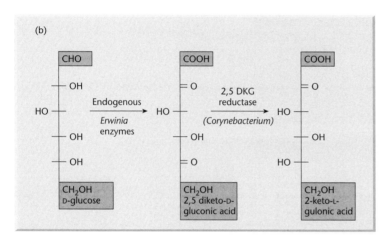

Fig. 14.17 Simplified route to vitamin C (ascorbic acid) developed by cloning in *Erwinia* the *Corynebacterium* gene for 2,5-diketogluconic acid reductase. (a) Classical route to vitamin C. (b) The simplified route to 2-ketogulonic acid, the immediate precursor of vitamin C.

burgundy colour. To make textile-quality indigo, there must be no indirubin present. Screening soil microorganisms with the capacity to degrade indole resulted in the identification of an enzyme, isatin hydrolase, that can degrade isatin to isatic acid. After cloning the gene for isatin hydrolase in the indigo overproducing strains, the indigo product obtained performed as well as chemically produced material.

A slightly different approach to that above has yielded a new route to vitamin C. The conventional process starts with glucose and comprises one mirobiological and four chemical steps (Fig. 14.17). By cloning in *Erwinia* a single gene – that from *Corynebacterium* encoding 2,5-diketogluconic acid reductase – the process can be simplified to a single microbiological and a single chemical step (Anderson *et al.* 1985). After observations of unexpectedly low yields of 2-ketogulonic acid in the recombinant

strain, it was found that 2-ketogulonic acid was converted to L-idonic acid by an endogenous 2-ketoaldonate reductase. Cloning, deletion mutagenesis and homologous recombination of the mutated reductase gene into the chromosome were some of the several steps taken to develop an organism capable of accumulating large amounts (120 g/l) of 2-ketogulonic acid (Lazarus *et al.* 1990). So far, attempts to manufacture vitamin C directly from glucose have been unsuccessful. However, enzymes that can convert 2-ketogulonic acid to ascorbic acid have been identified and the objective now is to clone these activities into *Erwinia* (Chotani *et al.* 2000).

Combinatorial biosynthesis

A number of widely used antibiotics and immuno-suppressants belong to a class of molecules known

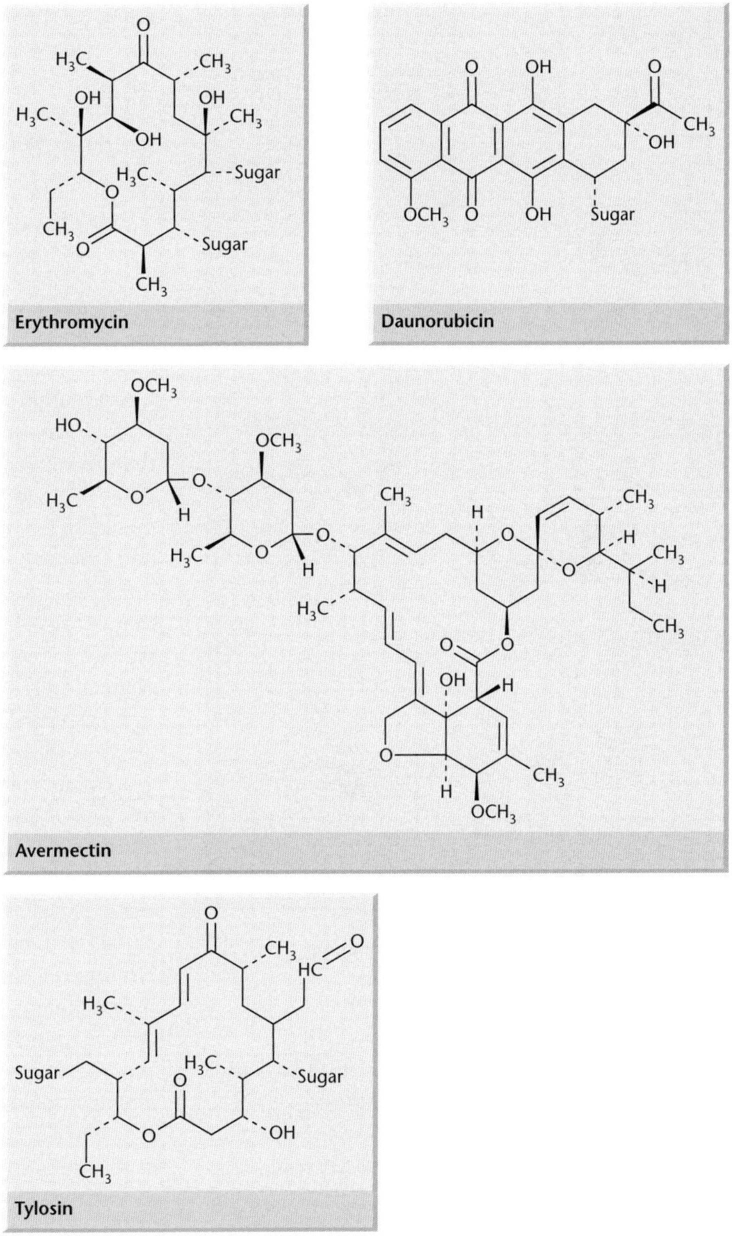

Fig. 14.18 Some examples of polyketides.

as polyketides. These molecules, which are synthesized by actinomycetes, have a fairly complex structure (Fig. 14.18). The genes involved in the biosynthesis of polyketides are clustered, thereby facilitating the cloning of all of the genes controlling the synthetic pathway. The first cluster (the *act* genes) to be cloned was that for actinorhodin. When parts of the *act* gene cluster were introduced into streptomycetes making related polyketides, completely new antibiotics were produced (Hopwood *et al.* 1985). For example, introducing the *actVA* gene from *Streptomyces coelicolor* into a strain that makes medermycin leads to the synthesis of mederrhodinA (Fig. 14.19). This approach has been repeated many times since

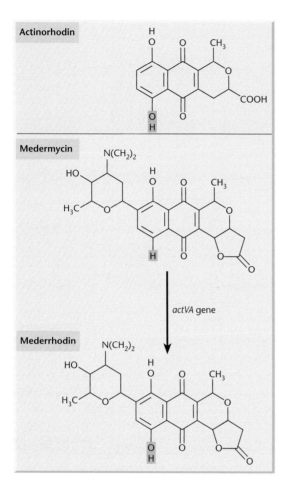

Fig. 14.19 The formation of the new antibiotic mederrhodin from medermycin by the actVA gene product.

with other polyketides (for review, see Baltz 1998) and is known as *combinatorial biosynthesis*.

Once a number of polyketide biosynthetic gene clusters had been cloned and sequenced, new insights were gained on the mechanism of synthesis. In particular, two enzymic modes of synthesis were discovered. In particular, polyketide synthesis takes place on an enzyme complex in a manner analogous to fatty acid synthesis. Furthermore, there are two types of complex. In type II complexes, the different enzymic activities are encoded by separate subunits. In contrast, in type I synthesis all the different enzyme activities are encoded by a single, very large gene. Clearly, the polyketide synthases are prime candidates for DNA shuffling, as described in the

previous theme (p. 302), and this approach has been widely adopted (Baltz 1998). However, novel polyketides can also be generated by simply changing the *order* of the different activities in type I synthases (McDaniel *et al.* 1999).

Engineering metabolic control over recombinant pathways

When a recombinant cell overproduces a protein or the components of a biosynthetic pathway, there is a marked reduction in metabolic activity, coupled with a retardation of growth (Kurland & Dong 1996). These phenotypic characteristics result from the fact that the constant demands of production placed upon the cell interfere with its changing requirements for growth. To alleviate this problem, Farmer and Liao (2000) designed a dynamic control circuit that is able to sense the metabolic state of the cell and thereby regulate the expression of a recombinant pathway. This approach is termed 'metabolic control engineering'.

An essential component of a dynamic controller is a signal that will reflect the metabolic state of the cell. Acetyl phosphate was selected as the signal, since it is known to be a regulator of various operons influenced by nutrient availability. A component of the Ntr regulon in *E. coli*, NRI, is capable of sensing the acetyl phosphate level in the cell. When phosphorylated by acetyl phosphate, it is capable of binding to the *glnAp2* promoter and activating transcription (Fig. 14.20). To reconstruct this control module, the NRI-binding site and the *glnAp2* promoter were inserted into a plasmid vector upstream of a cloning site.

When the *lacZ* gene was placed under the control of the *glnAp2* promoter, there was no significant β-galactosidase synthesis until late in the exponential phase of growth, just as expected. A similar result was obtained when the *lacZ* gene was replaced with a construct encoding two different metabolic enzymes. Finally, as a real test of the system, an engineered construct encoding a pathway for the synthesis of the carotenoid lycopene was placed under the control of the *glnAp2* promoter. This was a particularly interesting test because one of the precursors of lycopene, pyruvate, is also an immediate precursor of acetyl phosphate. Once again, product

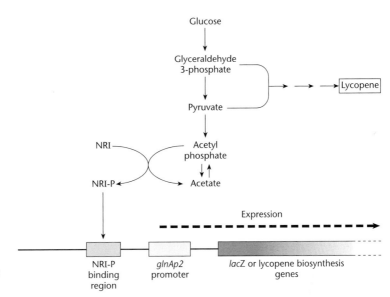

Fig. 14.20 Construction of a metabolic control circuit. See text for details.

(lycopene) formation was controlled by the metabolic state of the cell. Although metabolic control engineering is still at an early stage, it represents a significant degree of sophistication of control compared with the use of simple controllable promoters.

Metabolic engineering in plant cells

Plants synthesize an incredibly diverse array of useful chemicals. Most are products of *secondary metabolism* – that is, biochemical pathways that are not involved in the synthesis of essential cellular components but which synthesize more complex molecules that provide additional functions. Examples of these functions are attraction of pollinators and resistance to pests and pathogens. In many cases, these secondary metabolites have specific and potent pharmacological properties in humans: well-known examples include caffeine, nicotine, morphine and cocaine.

Plants have long been exploited as a source of pharmaceutical compounds, and a number of species are cultivated specifically for the purpose of extracting drugs and other valuable molecules. We discussed above how gene transfer to bacteria and yeast can be used to produce novel chemicals, so in theory it would be possible to transfer the necessary components from these useful plants into microbes for large-scale production. However, the secondary metabolic pathways of plants are so extensive and complex that, in most cases, such a strategy would prove impossible. Fortunately, advances in plant transformation have made it possible to carry out metabolic engineering in plants themselves, and large-scale plant cell cultures can be used in the same manner as microbial cultures for the production of important phytochemicals (reviewed by Verpoorte 1998, Verpoorte *et al.* 2000).

The secondary metabolic pathways of most plants produce the same basic molecular skeletons, but these are 'decorated' with functional groups in a highly specific way, so that particular compounds may be found in only one or a few plant species. Furthermore, such molecules are often produced in extremely low amounts, so extraction and purification can be expensive. For example, the Madagascar periwinkle *Catharanthus roseus* is the source of two potent anti-cancer drugs called vinblastine and vincristine. These terpene indole alkaloids are too complex to synthesize in the laboratory and there are no alternative natural sources. In *C. roseus*, these molecules are produced in such low amounts that over 1 ha of plants must be harvested to produce a single gram of each drug, with a commercial value of over $1 million.

It would be much more convenient to produce such drugs in fermenters containing cultured plant cells, and this has been achieved for a number of compounds, two of which (paclitaxel and shikonin) have reached commercial production (see Verpoorte *et al.* 2000). However, cell-suspension cultures often do not produce the downstream products made by the parent plant. This applies to vinblastine and vincristine from *C. roseus*, and also to other important drugs, such as morphine, codeine and hyoscyamine. Part of the reason for this is the complexity of secondary metabolism. The entire pathway is not completed in a single cell, but is often segregated into different cell types, with consequent shuttling of intermediates between cells. Within the cell, different stages of the pathway are also compartmentalized, so that intermediates must be transported between organelles. As in bacteria, knowledge of the target biosynthetic pathway is therefore essential for metabolic engineering, but in plants only a few secondary pathways are understood in sufficient

detail. Examples include the phenylpropanoid and flavonoid pathways, which yield anthocyanins (plant pigments) and phytoalexins (antimicrobial compounds), and the terpene indole alkaloid biosynthetic pathway, which generates important alkaloids, such as vinblastine and vincristine. As more plant genomes are sequenced, we are likely to learn much more about such pathways.

All terpene indole alkaloids derive from a single universal precursor called strictosidine. This is formed by the convergence of two pathways, the iridoid pathway (culminating in secologanin) and the terpenoid pathway (culminating in tryptamine). Strictosidine is formed by the condensation of secologanin and tryptamine, catalysed by the enzyme strictosidine synthase, and is then further modified in later steps to produce the valuable downstream alkaloids (Fig. 14.21). The conversion of tryptophan

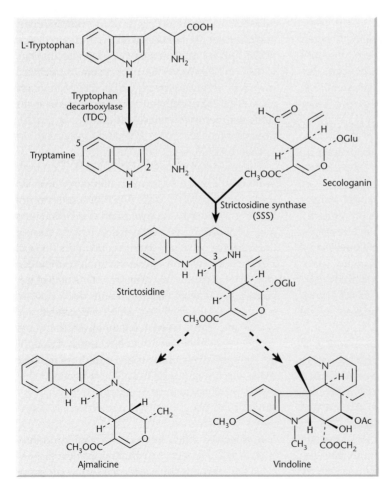

Fig. 14.21 Abbreviated pathway of terpene indole alkaloid biosynthesis, showing the conversion of tryptophan into tryptamine by the enzyme tryptophan decarboxylase, the condensation of tryptamine and secologanin into strictosidine, and the later diversification of strictosidine into valuable akaloids, such as ajmalicine and vindoline (a precursor of both vinblastine and vincristine).

to tryptamine is a rate-limiting step in the terpenoid pathway, and this has been addressed by overexpressing the enzyme tryptophan decarboxylase in *C. roseus* cell-suspension cultures. However, while transformed cultures produced much higher levels of tryptamine, no downstream alkaloids were synthesized (Goddijn *et al.* 1995, Canel *et al.* 1998). The simultaneous overexpression of the next enzyme in the pathway, strictosidine synthase, did increase the levels of the alkaloid ajmalicine, a useful sedative, in some cultures, but did not result in the synthesis of vinblastine or vincristine (Canel *et al.* 1998).

Thus it seems that single-step engineering may remove known bottlenecks only to reveal the position of the next. The limited success of single-gene approaches has resulted in the development of alternative strategies for the coordinated regulation of entire pathways using transcription factors. Using the yeast one-hybrid system (Vidal *et al.* 1996a), a transcription factor called ORCA2 has been identified that binds to response elements in the genes for tryptophan decarboxylase, strictosidine synthase and several other genes encoding enzymes in the same pathway. A related protein, ORCA3, has been identified using insertional vectors that activate genes adjacent to their integration site. By bringing the expression of such transcription factors under the control of the experimenter, entire metabolic pathways could be controlled externally (see review by Memelink *et al.* 2000).

Apart from the modification of endogenous metabolic pathways to produce more (or less) of a specific endogenous compound, plants can also be engineered to produce heterologous or entirely novel molecules. An example is the production of the alkaloid scopolamine, an anticholinergic drug, in *Atropa belladonna* (Hashimoto *et al.* 1993, Hashimoto & Yamada 1994). Scopolamine is produced in *Hyoscyamus niger* but not in *A. belladonna*, which accumulates the immediate precursor hyoscyamine. *H. niger* converts hyoscyamine into scopolamine using the enzyme hyoscyamine-6-hydroxylase (H6H), which is absent in *A. belladonna*. Hashimoto and colleagues isolated a complementary DNA (cDNA) encoding H6H from *H. niger* and expressed it in *A. belladonna*. The transgenic *A. belladonna* plants produced scopolamine because they were able to extend the metabolic pathway beyond its endogenous end-point. Another example of the production of novel chemicals in plants is the diversion of carbon backbones from fatty acid synthesis to the formation of polyhydroxyalkanoates, which form biodegradable thermoplastics (Steinbuchel & Fuchtenbusch 1998). In this case, the foreign genes derive not from other plants, but from bacteria.

Theme 6: Plant breeding in the twenty-first century

Introduction to theme 6

We have discussed several uses for transgenic plants earlier in this chapter, i.e. as bioreactors producing recombinant proteins (p. 285) and novel metabolites (see previous section). Transgenic plants can potentially express any foreign gene, whether that gene is derived from bacteria, yeast, other plants or even animals. The scope for exploitation and improvement is virtually limitless, and gene-manipulation techniques have therefore given the biotechnology industry a new lease of life. In the following sections, we consider the development of plant biotechnology in two key areas: improvement of agronomic traits and modification of production traits.

Improving agronomic traits

The initial focus of plant biotechnology was on improving agronomic traits, i.e. the protection of crops against pests, pathogens and weeds, and thus increasing yields. Major crop losses are caused every year by these so-called 'biotic' constraints, as well as physical (or 'abiotic') factors, such as flooding, drought, soil quality, etc. The aims of the biotechnology industry went hand in hand with those of conventional breeders, but offered the possibility of importing useful genes from distant species that could not be used for breeding. It has been found that, in many cases, single genes transferred from another organism can provide high levels of protection.

Herbicide resistance

Herbicides generally affect processes that are unique to plants, e.g. photosynthesis or amino acid biosynthesis (see Table 14.8). Both crops and weeds share

Table 14.8 Mode of action of herbicides and method of engineering herbicide-resistant plants.

Herbicide	Pathway inhibited	Target enzyme	Basis of engineered resistance to herbicide
Glyphosate	Aromatic amino acid biosynthesis	5-Enol-pyruvyl shikimate-3-phosphate (EPSP) synthase	Overexpression of plant EPSP gene or introduction of bacterial glyphosate-resistant *aroA* gene
Sulphonylurea	Branched-chain amino acid biosynthesis	Acetolactate synthase (ALS)	Introduction of resistant *ALS* gene
Imidazolinones	Branched-chain amino acid biosynthesis	ALS	Introduction of mutant *ALS* gene
Phosphinothricin	Glutamine biosynthesis	Glutamine synthetase	Overexpression of glutamine synthetase or introduction of the *bar* gene, which detoxifies the herbicide
Atrazine	Photosystem II	Q_B	Introduction of mutant gene for Q_B protein or introduction of gene for glutathione-*S*-transferase, which can detoxify atrazines
Bromoxynil	Photosynthesis		Introduction of nitrilase gene, which detoxifies bromoxynil

these processes, and developing herbicides that are selective for weeds is very difficult. An alternative approach is to modify crop plants so that they become resistant to broad-spectrum herbicides, i.e. incorporating selectivity into the plant itself rather than relying on the selectivity of the chemical. Two approaches to engineering herbicide resistance have been adopted. In the first, the target molecule in the cell either is rendered insensitive or is overproduced. In the second, a pathway that degrades or detoxifies the herbicide is introduced into the plant. An example of each strategy is considered below.

Glyphosate is a non-selective herbicide that inhibits 5-enol-pyruvylshikimate-3-phosphate (EPSP) synthase, a key enzyme in the biosynthesis of aromatic amino acids in plants and bacteria. A glyphosate-tolerant *Petunia hybrida* cell line obtained after selection for glyphosate resistance was found to overproduce the EPSP synthase as a result of gene amplification. A gene encoding the enzyme was subsequently isolated and introduced into petunia plants under the control of a cauliflower mosaic virus (CaMV) 35S promoter. Transgenic plants expressed increased levels of EPSP synthase in their chloroplasts and were significantly more tolerant to glyphosate (Shah *et al.* 1986). An alternative approach

to glyphosate resistance has been to introduce a gene encoding a mutant EPSP synthase. This mutant enzyme retains its specific activity but has decreased affinity for the herbicide. Transgenic tomato plants expressing this gene under the control of an opine promoter were also glyphosate-tolerant (Comai *et al.* 1985). Following on from this early research, several companies have introduced glyphosate tolerance into a range of crop species, with soybean and cotton the first to reach commercialization (Nida *et al.* 1996, Padgette *et al.* 1996). Currently, nearly three-quarters of all transgenic plants in the world are resistant to glyphosate (James 2000).

Phosphinothricin (PPT) is an irreversible inhibitor of glutamine synthetase in plants and bacteria. Bialaphos, produced by *Streptomyces hygroscopicus*, consists of PPT and two alanine residues. When these residues are removed by peptidases the herbicidal component PPT is released. To prevent self-inhibition of growth, bialaphos-producing strains of *S. hygroscopicus* also produce the enzyme phosphinothricin acetyltransferase (PAT), which inactivates PPT by acetylation. The *bar* gene that encodes the acetylase has been introduced into potato, tobacco and tomato cells using *Agrobacterium*-mediated transformation. The resultant plants were

(a)   (b)

Fig. 14.22 Evaluation of phosphinothricin in transgenic tobacco plants under field conditions. (a) Untransformed control plants. (b) Transgenic plants. (Photographs courtesy of Dr J. Botterman and the editor of *Biotechnology*.)

resistant to commercial formulations of PPT and bialaphos in the laboratory (De Block *et al.* 1987) and in the field (De Greef *et al.* 1989) (Fig. 14.22). More recently, it has been shown that bialaphos-resistant transgenic rice plants that were inoculated with the fungi causing sheath blight disease and subsequently treated with the herbicide were completely protected from infection (Uchimiya *et al.* 1993). This agronomically important result depends on the observation that bialaphos is toxic to fungi as well as being a herbicide. PPT resistance is widely used in plants as a selectable marker (see p. 231); however, it has also been introduced into a number of different crops for weed control, including sugar cane and rice (Gallo-Meagher & Irvine 1996, Oard *et al.* 1996). Thus far, PPT-resistant transgenic plants have not been commercially released.

The benefit of herbicide-resistant transgenic crops is the increased yield and seed quality as competing weed species are eliminated. However, there was initial concern that this would come with an associated penalty of increased herbicide use, which could have a serious impact on the environment. Contrary to these predictions, the introduction of herbicide-resistant plants has actually reduced chemical use by up to 80% in many areas, as farmers adopt better weed-control policies and switch to herbicides with low use rates. A further risk is that transgenes for herbicide tolerance could spread to weed species, resulting in a new breed of 'superweeds' (Kling 1996). It is too early to say whether this will be a problem. Although a range of herbicide-resistant

transgenic crops are being tested, only crops resistant to glyphosate or bromoxynil are currently grown on a commercial scale. The benefits and risks of herbicide-resistant crops and strategies to combat gene transfer to weeds are discussed in recent reviews (Gressel 1999, 2000).

Virus resistance

Major crop losses occur every year as a result of viral infections, e.g. tobacco mosaic virus (TMV) causes losses of over $50 million per annum in the tomato industry. There is a useful phenomenon known as cross-protection, in which infection of a plant with one strain of virus protects against superinfection with a second, related strain. The mechanism of cross-protection is not fully understood, but it is believed that the viral coat protein is important. Powell-Abel *et al.* (1986) developed transgenic plants that express the TMV coat protein and which had greatly reduced disease symptoms following virus infection. Since that observation, the principle of heterologous coat-protein expression has been extended to many different plants and viruses (reviewed by Beachy *et al.* 1990). In the case of resistance to TMV, the coat protein must be expressed in the epidermis and in the vascular tissue, through which the virus spreads systemically (Clark *et al.* 1990). Transgenic squash containing multiple viral coat-protein genes and demonstrating resistance to cucumber mosaic virus, water-melon mosaic virus 2 and zucchini yellow mosaic virus was the first virus-resistant transgenic crop to reach commercial production (Tricoli *et al.* 1995).

While the heterologous coat-protein approach can be successful, it has been demonstrated that, in many cases, the effect of transgene expression is mediated at the RNA rather than the protein level. This can be proved by generating transgenic plants carrying coat-protein genes that cannot be translated to yield functional protein, as first shown by Lindbo and Dougherty (1992) using the tobacco etch virus coat-protein gene. The transgene RNA apparently interferes with viral replication (a phenomenon called RNA-mediated viral resistance (RMVR)). This requires homology between the transgene and the target virus, and involves high-level transgene transcription but low-level accumulation of the transcript. This has much in common with post-transcriptional gene silencing (discussed in more detail in Chapter 13), which can lead to transgene silencing and the cosuppression of homologous endogenous genes, as well as viral resistance. The interested reader can consult several excellent reviews covering this and related phenomena (Waterhouse *et al.* 1998, Grant 1999, Plasterk & Ketting 2000, Hammond *et al.* 2001).

A different method of minimizing the effects of plant virus infection was developed by Gehrlach *et al.* (1987). They generated plants that expressed the satellite RNA of tobacco ringspot virus and such plants were resistant to infection with tobacco ringspot virus itself. Another potential method of inducing resistance to viruses is the production of antiviral proteins in transgenic plants. American pokeweed produces an antiviral protein called dianthrin that functions as a ribosome-inactivating protein. The cDNA for this protein has been cloned (Lin *et al.* 1991) and expressed in *Nicotiana benthamiana* (a relative of tobacco), providing resistance against African cassava mosaic virus (ACMV) (Hong *et al.* 1996). Interestingly in this experiment, the dianthrin gene was expressed under the control of an ACMV promoter, such that the antiviral protein was expressed only upon viral infection. In this manner, the toxic effects of constitutive transgene expression were avoided.

Antibodies specific for virion proteins have also been used to protect plants from viruses. In the first demonstration of this approach, Tavladoraki *et al.* (1993) expressed a single-chain Fv fragment (scFv) specific for ACMV in transgenic *N. benthamiana*, and demonstrated resistance to viral infection. Other groups have generated transgenic tobacco plants expressing antibodies specific for TMV, resulting in reduced infectivity. Voss *et al.* (1995) expressed full-size IgGs, while Zimmermann *et al.* (1998) expressed scFv fragments. Targeting scFv fragments to the plasma membrane also provides protection against virus infection (Schillberg *et al.* 2001).

Resistance to microbial pathogens

Progress has also been made in developing resistance to plant-pathogenic fungi which are traditionally controlled by appropriate farming practices (e.g. crop rotation) and the application of expensive and environmentally harmful fungicides. A straightforward approach is to engineer plants with antifungal proteins from heterologous species. This was first demonstrated by Broglie *et al.* (1991) who showed that expression of bean chitinase can protect tobacco and oil-seed rape from post-emergent damping off caused by *Rhizoctonia solani*. Plants synthesize a wide range of so-called 'pathogenesis-related proteins' (PR proteins), such as chitinase, which are induced by microbial infection. They also synthesize antifungal peptides called defensins and other antifungal proteins. As the genes for more of these proteins have been cloned and characterized, the number of transgenic plants constitutively expressing such proteins continues to rise. For example, tobacco osmotin has been expressed in transgenic potato, providing resistance to *Phytophthora infestans* (Liu *et al.* 1994), and in transgenic rice, providing resistance to *R. solani* (Lin *et al.* 1995). Instead of using a protein to provide direct protection, a metabolic-engineering strategy can be utilized. Phytoalexins are alkaloids with antifungal activity, and transforming plants with genes encoding the appropriate biosynthetic enzymes can increase their synthesis. Hain *et al.* (1993) generated tomato plants expressing the grapevine gene for stilbene synthase, and these plants demonstrated increased resistance to infection by *Botrytis cinerea*. Similarly, Anzai *et al.* (1989) have used a bacterial gene facilitating tabtoxin detoxification to protect tomato plants against *Pseudomonas syringae* infection.

An alternative to the use of antifungal proteins or metabolites is to manipulate the hypersensitive response, which is a physiological defence mechanism used by plants to repel attacking pathogens. Resistance occurs only in plants carrying a resistance gene (*R*) that corresponds to an avirulence (*avr*) gene in the pathogen. Elicitors (signalling molecules) released by pathogens are detected by the plant and activate a range of defence responses, including cell death, PR-gene expression, phytoalexin synthesis and the deposition of cellulose at the site of invasion, forming a physical barrier. Importantly, the hypersensitive response is systemic, so that neighbouring cells can pre-empt pathogen invasion. A recently developed strategy is to transfer avirulence genes from the pathogen to the plant, under the control of a pathogen-inducible promoter. This has been demonstrated in tomato plants transformed with the *avr9* gene from *Chladosporium fulvum*, resulting in resistance to a range of fungal, bacterial and viral diseases (Keller *et al.* 1999, reviewed by Melchers & Stuiver 2000).

Resistance to insects and other pests

Insect pests represent one of the most serious biotic constraints to crop production. For example, more than a quarter of all the rice grown in the world is lost to insect pests, at an estimated cost of nearly US$50 billion. This is despite an annual expenditure of approximately US$1.5 billion on insecticides for this crop alone. Insect-resistant plants are therefore desirable not only because of the potential increased yields, but also because the need for insecticides is eliminated and, following on from this, the undesirable accumulation of such chemicals in the environment is avoided. Typical insecticides are non-selective, so they kill harmless and beneficial insects as well as pests. For these reasons, transgenic plants have been generated expressing toxins that are selective for particular insect species.

Research is being carried out on a wide range of insecticidal proteins from diverse sources. However, all commercially produced insect-resistant transgenic crops express toxin proteins from the Gram-positive bacterium *Bacillus thuringiensis* (Bt). Unlike other *Bacillus* species, *Bt* produces crystals during sporulation, comprising one or a small number of ~130 kDa protoxins called crystal proteins. These proteins are potent and highly specific insecticides. The specificity reflects interactions between the crystal proteins and receptors in the insect midgut. In susceptible species, ingested crystals dissolve in the alkaline conditions of the gut and the protoxins are activated by gut proteases. The active toxins bind to receptors on midgut epithelial cells, become inserted into the plasma membrane and form pores that lead to cell death (and eventual insect death) through osmotic lysis. Approximately 150 distinct Bt toxins have been identified and each shows a unique spectrum of activity.

Bt toxins have been used as topical insecticides since the 1930s, but never gained widespread use, because they are rapidly broken down on exposure to daylight and thus have to be applied several times during a growing season. Additionally, only insects infesting the exposed surface of sprayed plants are killed. These problems have been addressed by the expression of crystal proteins in transgenic plants. Bt genes were initially introduced into tomato (Fischhoff *et al.* 1987) and tobacco (Barton *et al.* 1987, Vaeck *et al.* 1987) and later into cotton (Perlak *et al.* 1990), resulting in the production of insecticidal proteins that protected the plants from insect infestation. However, field tests of these plants revealed that higher levels of the toxin in the plant tissue would be required to obtain commercially useful plants (Delannay *et al.* 1989). Attempts to increase the expression of the toxin gene in plants by use of different promoters, fusion proteins and leader sequences were not successful. However, examination of the bacterial *cry1Ab* and *cry1Ac* genes indicated that they differed significantly from plant genes in a number of ways (Perlak *et al.* 1991). For example, localized AT-rich regions resembling plant introns, potential plant polyadenylation signal sequences, ATTTA sequences that can destabilize mRNA and rare plant codons were all found. The elimination of undesirable sequences and modifications to bring codon usage into line with the host species resulted in greatly enhanced expression of the insecticidal toxin and strong insect resistance of the transgenic plants in field tests (Koziel *et al.* 1993). By carrying out such enhancements, Perlak and colleagues expressed a modified *cry3A* gene in potato to provide resistance against Colorado beetle

(Perlak *et al.* 1993). In 1995, this crop became the first transgenic insect-resistant crop to reach commercial production, as NewLeaf™ potato marketed by Monsanto. The same company also released the first commercial transgenic, insect-resistant varieties of cotton (Bollgard™, expressing *cry1Ac* and protected against tobacco bollworm) and maize (YieldGard™, expressing *cry1Ab* and resistant to the European corn-borer). Many other biotechnology companies have now produced Bt-transgenic crop plants resistant to a range of insects.

Although Bt-transgenic plants currently dominate the market, there are many alternative insecticidal proteins under investigation. Two types of protein are being studied in particular: proteins that inhibit digestive enzymes in the insect gut (proteinase and amylase inhibitors) and lectins (carbohydrate-binding proteins). Research into these alternatives is driven in part by the fact that some insects are not affected by any of the known Bt crystal proteins. Homopteran insects, mostly sap-sucking pests such as planthoppers, fall into this category, but have been shown to be susceptible to lectins such as *Galanthus nivalis* agglutinin (GNA). This lectin has been expressed in many crops, including potato (Shi *et al.* 1994, Gatehouse *et al.* 1996), rice (Bano-Maqbool & Christou 1999), tomato and tobacco (reviewed by Schuler *et al.* 1998).

Modification of production traits

Production traits can include many characteristics, such as the colour and smell of flowers and the taste, consistency and nutritional composition of food. While modifying flower colour would seem a harmless pursuit (experiments in this area are discussed on p. 261), the genetic modification of food for human consumption is the subject of intensive current debate. Opinions as to the benefits and risks of the latter differ widely according to region, ethical standpoint and financial/food security of the nation. There is no doubt, however, that gene manipulation has the potential to greatly increase the quality of food. Examples include modifying the starch or oil content, facilitating the accumulation of vitamins and minerals in foods that are normally poor in such nutrients, and altering the properties of fruits and vegetables to delay ripening, increase the levels of

soluble solids and facilitate processing. Several examples are discussed below and Box 14.4 describes the recently developed terminator technology, which provides a mechanism by which producers of transgenic crops could dictate where and when their modified plants are grown.

Delayed ripening

The earliest example of quality improvement by gene transfer was the introduction of antisense constructs targeting the polygalacturonase (*pg*) gene in tomato to delay ripening (Sheehy *et al.* 1988, Smith *et al.* 1988), as discussed in Chapter 13 (p. 260). Other strategies to delay ripening include: cosuppression of the *pg* gene using sense constructs (Smith *et al.* 1990), the use of antisense RNA to suppress the expression of two key enzymes required for ethylene synthesis (Hamilton *et al.* 1990, Oeller *et al.* 1991) and the degradation of a key intermediate in the ethylene biosynthetic pathway (Klee *et al.* 1991; see review by Fray & Grierson 1993).

Starch and oil modification

Much research has been carried out into strategies for the modification of the starch and oil content of seeds. In each case, it is essential to have a thorough understanding of the normal biosynthetic pathways. Starch is the major storage carbohydrate in higher plants. It consists of two components: amylose, which is a linear polymer, and amylopectin, which is a branched polymer. The physicochemical, nutritional and textural properties of food are significantly influenced by the nature of the starch. For example, cooked amylose produces fibre-like structures that are resistant to digestion, serve as dietary fibre and require less insulin for metabolism. Amylopectin, in contrast, is waxy and viscous. The food and other industries use a wide range of different starches. These are obtained by sourcing the starch from different plant varieties coupled with chemical and enzymatic modification. The genetic modification of plants offers a new approach to creating novel starches with new functional properties.

The enzyme ADP-glucose pyrophosphorylase (GP) catalyses the first step in starch biosynthesis in plants. Stark *et al.* (1992) cloned the gene for an *E.*

Box 14.4 Protecting your investment: terminator technology

The production of any marketable commodity requires a great deal of investment in terms of research and development. Most commodities are not self-renewing, so manufacturers can recover their investments through long-term sales. Transgenic plants, however, are capable of reproduction. Farmers have to make an initial outlay for seed purchase, but, by saving a proportion of seeds from each harvest, they need never return to the manufacturer again. For producers who have invested heavily in development, this is unacceptable. So how can the producers protect their investment?

Until recently, this was achieved through the use of contracts that obliged farmers not to save any seeds. However, such contracts are difficult to enforce, particularly in developing countries. A novel strategy is to modify the plants so that the seeds are sterile, thereby forcing the farmer to return to the manufacturer year after year for fresh supplies of seeds. This strategy, which has been termed *technology protection* by the biotechnology companies and *terminator technology* by disgruntled consumers and opponents, is perhaps the most controversial potential use of genetically modified plants. In essence, the technology works rather like the anti-copying devices now incorporated into video recorders – the industry is protecting its investment by preventing unauthorized duplication of its products. The principle of terminator technology is the expression of a toxic transgene, the terminator gene, at a critical stage of embryonic development, thus killing the embryo and rendering the seeds incapable of germination. Several variants on this technology were described in a patent application, granted in March 1998, to Pine Lands Corporation, which has since been bought by Monsanto. One of the variations is discussed below.

In this example, the terminator gene encodes a ribosomal-inactivating protein and is expressed under the control of a promoter that is active in late embryogenesis, such as the promoter of the *lea* (late embryogenesis abundant) gene. Seeds produced from such plants are sterile, but, since seed storage products (e.g. starch and oil) accumulate in early and mid-embryonic development, the seeds are nutritionally unimpaired. In order to maintain a stock of fertile plants, the terminator gene is controlled by inducible site-specific recombination, as described in detail in Chapter 13. The gene is rendered inactive by inserting a blocking element between the gene and its promoter. This blocking element is flanked by *loxP* sites. The transgenic plant also contains a *cre* recombinase gene, which is controlled by a tetracycline-inducible promoter, and a third transgene encoding the tetracycline repressor protein, which is constitutively expressed. In the absence of tetracycline, the repressor prevents *cre* expression, the terminator gene is not activated and the plants can be grown as normal, allowing the producer to grow plants and produce seeds in unlimited quantity. Before distribution, however, the seeds are soaked in tetracycline, which causes the repressor to release the *cre* promoter, thus inducing the *cre* gene, leading to excision of the blocking fragment and activation of the terminator gene. Since the terminator gene is controlled by an embryonic promoter, it is not switched on until the following generation of developing seeds.

The advent of terminator technology was met by a public outcry, particularly from farmers and from environmentalists concerned that terminator genes could spread from the transgenic crops into wild plants, with unknown consequences. Perhaps in response to this, Monsanto has since pledged not to implement the technology now at its disposal.

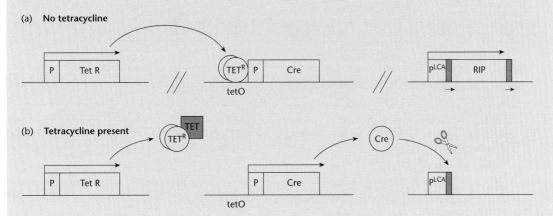

Oliver *et al.* (Pine Land Corporation) United States Patent 5 723 765, 'Control of Plant Gene Expression', 3 March 1998.

coli mutant GP which was deficient in allosteric regulation. When this gene was inserted into potato plants, the tubers accumulated higher levels of starch but had normal starch composition and granule size. Since transgenic plants expressing the wild-type *E. coli* GP had normal levels of starch, the allosteric regulation of GP, and not its absolute level, must influence starch levels. A different method of modulating GP activity has been used by Muller-Rober *et al.* (1992). They used an antisense construct to reduce the level of plant GP to 2–5% of wild type. These plants had very low levels of starch (also 2–5% of wild type) but had a very much higher number of tubers. These tubers had six- to eightfold more glucose and sucrose and much lower levels of patatin and other storage proteins. Visser *et al.* (1991) have modified the starch composition of potato plants. They demonstrated that potato plants expressing an antisense gene to the granule-bound starch synthase have little or no amylose compared with wild-type potato, where it can be as much as 20%.

Whereas some plants accumulate starch as a carbon reserve, others accumulate high levels of triacylglycerols in seeds or mesocarp tissues. Higher plants produce over 200 kinds of fatty acids, some of which are of food value. However, many are likely to have industrial (non-food) uses of higher value than edible fatty acids (Kishore & Somerville 1993, Murphy 1999). Thus there is considerable interest in using gene-manipulation techniques to modify the fatty acid content of plants. As a first step towards this goal, a number of genes encoding desaturases have been cloned, e.g. the MS desaturase, which converts linoleic acid to α-linolenic acid (Arondel *et al.* 1992).

In most plants the Δ9 stearoyl acyl carrier protein (ACP) desaturase catalyses the first desaturation step in seed-oil biosynthesis, converting stearoyl-ACP to oleoyl-ACP. When antisense constructs were used to reduce the levels of the enzyme in developing rape-seed embryos, the seeds had greatly increased stearate levels (Knutzon *et al.* 1992). The significance of this result is that high stearate content is of value in margarine and confectionery fats.

Most oil-seed crops accumulate triacylglycerols with C_{16} to C_{22} acyl chains. However, plants that accumulate medium-chain triacylglycerols (MCT), with C_8 to C_{12} acyl chains, would be very useful, for these lipids are used in detergent synthesis. Some plants, such as the California bay, accumulate MCT because of the presence of a medium-chain specific thioesterase. When the gene for this thioesterase was expressed in rape-seed and *Arabidopsis*, the transgenic plants accumulated high levels of MCT (Voelker *et al.* 1992).

Improving vitamin and mineral content

The vitamin and mineral content of plants varies from species to species, and from tissue to tissue in a particular plant. Milled cereal grains – basically the endosperm component of the seed – are particularly deficient in essential vitamins and minerals, and yet they represent the staple food for much of the world's population. In developed countries, where diet is varied, this is seldom a problem. However, in poorer nations, where cereal grains are often the only food available, this can cause major health problems. Iron deficiency is the most widespread nutrient deficiency in the world, reflecting the combination of low iron in cereal grains and the high level of phytic acid, which reduces the efficiency of iron absorption from the gut. Many cereal grains also lack β-carotene, which is required in the diet for vitamin A biosynthesis. Vitamin A deficiency in early childhood leads to blindness, an avoidable consequence of poor diet in many parts of Asia and Africa.

A number of attempts to increase the nutritional value of cereals have been reported, and we discuss two examples of genetically engineered rice below, since this cereal is the staple diet of at least two-thirds of the world's population. Lucca *et al.* (2001) described three different routes to improving the iron content of rice grain, which could be used singly or in combination. The first route was to transform rice with a gene encoding ferritin, an iron-storage protein. The overexpression of bean ferritin in rice endosperm led to the accumulation of iron in a bioavailable form, resulting in transgenic rice grains with twice the normal levels of iron. The second route was to transform rice with a gene from the fungus *Aspergillus fumigatus* encoding the enzyme phytase, which metabolizes phytic acid. Since the transgenic grain has lower levels of phytic acid, iron absorption from the gut should be more efficient. The third approach was to transform rice with a construct causing the overexpression of an endogenous cysteine-rich metallothionein-like protein, since cysteine peptides have been shown to have the opposite effect to phytic acid, i.e. they enhance iron absorption. The resulting rice grains had over five times the normal level of cysteine residues and a 130-fold increase in phytase activity, theoretically sufficient to eliminate all phytic acid from the grains.

Burkhardt *et al.* (1997) described rice plants transformed with the phytoene synthase gene from the daffodil (*Narcissus pseudonarcissus*). The enzyme encoded by this gene represents the first of four steps in the β-carotene biosynthesis pathway, and the plants were shown to accumulate the immediate product of the recombinant enzyme, phytoene. Further work by the same group (Ye *et al.* 2000) resulted in rice plants expressing several daffodil enzymes simultaneously in the endosperm, and recapitulated the entire heterologous pathway. Hence, the transgenic rice grains were highly enriched for β-carotene (provitamin A) and were consequently golden in colour.

Epilogue: from genes to genomes

Principles of Gene Manipulation tells the story of the development of genetic-engineering technology, a story that began in the mid-1970s. When the first edition of this book was published in 1980, basic techniques, such as cloning, sequencing and expression analysis, were all in their infancy and many of today's most widely used procedures, including the PCR, had not been invented. In those days, the cloning and characterization of a single gene was an elaborate task that required a great deal of inventiveness and hard work. The sequencing and analysis of entire cellular genomes was an impossible goal, as far-fetched as space travel was to people 100 years ago.

Over the last 25 years, however, the technology of gene manipulation has rapidly increased in sophistication. Through the sharing of information between laboratories, standard molecular-biology procedures have become more robust and reliable, new techniques have evolved, and many of the more laborious procedures that scientists had to carry out in the early years have been streamlined or automated. Biotechnology companies now take on most of the burden of developing and refining the tools and protocols, producing high-quality enzymes, specialist molecular-biology kits and premade custom genomic or cDNA libraries. Most importantly, there has been a drive to develop high-throughput versions of many of the techniques discussed in this book, allowing more and more samples to be processed in parallel, producing more data in less time than ever before. Sequencing has been at the forefront of this paradigm shift, with the result that the sequencing of entire genomes is now commonplace. As this sixth edition of *Principles of Gene Manipulation* goes to press, over 40 bacterial genomes plus those of four model eukaryotes have been completely sequenced and the first draft of the entire human genome sequence has been published (International Human Genome Sequencing Consortium 2001, Venter *et al.* 2001).

These extraordinary achievements have ushered in a new era in molecular biology, which is analysis at the whole-genome level. More data are emerging from genome-sequencing projects and the random cloning of partial cDNAs (known as expressed sequence tags (ESTs)) than can be dealt with using traditional single-gene analysis methods. A new breed of high-throughput functional-analysis techniques has been developed, some of which are discussed briefly in Chapters 6, 9 and 13 of this book. Simple northern blots and reverse transcription (RT)-PCR experiments are being replaced by the use of DNA chips containing hundreds or thousands of individual sequences that can be hybridized with a suitable probe simultaneously. Instead of analysing single proteins by western blots, two-dimensional gel electrophoresis is being used to simultaneously study every single protein expressed under various conditions. The creation and functional analysis of individual mutants is being superseded by systematic approaches to mutating and characterizing every single gene in the genome. Techniques are also being developed for the analysis of all possible protein–protein interactions in the cell.

While the impact of the new genomics is discussed in this book, as least as far as it affects trends in gene-manipulation technology, a full account of its development, application and potential requires a book of its own. The story continues in our sister text, *Principles of Genome Analysis and Genomics* (third edition).

References

Aaij C. & Borst P. (1972) The gel electrophoresis of DNA. *Biochim. Biophys. Acta* **269**, 192–200.

Abrahmsen L., Tom J., Burnier J. *et al.* (1991) Engineering subtilisin and its substrates for efficient ligation of peptide bonds in aqueous solution. *Biochemistry* **30**, 4151–9.

Adams M.D., Kelley J.M., Gocayne J.D. *et al.* (1991) Complementary DNA sequencing: expressed sequence tags and human genome project. *Science* **252**, 1651–6.

Adams M.D., Dubnick M., Kerlavage A.R. *et al.* (1992) Sequence identification of 2,375 human brain genes. *Nature* **355**, 632–4.

Adams M.D., Celniker S.E., Holt R.A. *et al.* (2000) The genome sequence of *Drosophila melanogaster*. *Science* **287**, 2185–95.

Adams S.E., Dawson K.M., Gull K., Kingsman S.M. & Kingsman A.J. (1987) The expression of hybrid HIV: Ty virus-like particles in yeast. *Nature* **329**, 68–70.

Adati N., Ito T., Koga C. *et al.* (1995) Differential display analysis of gene expression in developing embryos of *Xenopus laevis*. *Biochim. Biophys. Acta* **1262**, 43–51.

Agapov E.V., Frolov I., Lindenbach B.D. *et al.* (1998) Noncytopathic Sindbis virus RNA vectors for heterologous gene expression. *Proc. Nat. Acad. Sci. USA* **95**, 12989–94.

Ahlquist P. & Janda M. (1984) cDNA cloning and *in vitro* transcription of the complete brome mosaic virus genome. *Mol. Cell. Biol.* **4**, 2876–82.

Ahlquist P., French R., Janda M. & Loesch-Fries S. (1984) Multicomponent RNA plant virus infection derived from cloned viral cDNA. *Proc. Nat. Acad. Sci. USA* **81**, 7066–70.

Ahlquist P., French R. & Bujarski J.J. (1987) Molecular studies of brome mosaic virus using infectious transcripts from cloned cDNA, *Adv. Virus Res.* **32**, 215–42.

Albert H., Dale E.C., Lee E. & Ow D.W. (1995) Site specific integration of DNA into wild type and mutant *lox* sites placed in the plant genome. *Plant J.* **7**, 649–59.

Allen N.D., Cran D.G., Barton S.C. *et al.* (1988) Transgenes as probes for active chromosomal domains in mouse development. *Nature* **333**, 852–5.

Al-Shawi R., Kinnaird J., Burke J. & Bishop J.O. (1990) Expression of a foreign gene in a line of transgenic mice is modulated by a chromosomal position effect. *Mol. Cell. Biol.* **10**, 1192–8.

Altschul S.F., Boguski M.S., Gish W. & Woolton J.C. (1994) Issues in searching molecular databases. *Nature Genet.* **6**, 119–29.

Alwine J.C., Kemp D.J., Parker B.A. *et al.* (1979) Detection of specific RNAs or specific fragments of DNA by fractionation in gels and transfer to diazobenzyloxymenthyl paper. *Methods Enzymol.* **68**, 220–42.

Amaya E., Musci T.J. & Kirschner M.W. (1991) Expression of a dominant negative mutant of the FGF receptor disrupts mesoderm formation in *Xenopus* embryos. *Cell* **66**, 257–70.

An G.H., Costa M.A. & Ha S.B. (1990) Nopaline synthase promoter is wound-inducible and auxin-inducible. *Plant Cell* **2**, 225–33.

Anderson S., Marks C.B., Lazarus R. *et al.* (1985) Production of 2-keto-L-gulonate, an intermediate in L-ascorbate synthesis by a genetically modified *Erwinia herbicola*. *Science* **230**, 144–9.

Andre D., Colau D., Schell J., van Montagu M. & Hernalsteens, J.-P. (1986) Gene tagging in planta by a T-DNA insertion that generates APH (3′) II plant gene fusions. *Mol. Gen. Genet.* **204**, 512–18.

Andrulis I.L. & Siminovitch L. (1981) DNA-mediated gene transfer of beta-aspartylhydroxamate resistance into Chinese hamster ovary cells. *Proc. Nat. Acad. Sci. USA* **78**, 5724–8.

Angell S.M. & Baulcombe D.C. (1997) Consistent gene silencing in transgenic plants expressing a replicating potato virus X RNA. *EMBO J.* **16**, 3675–84.

Anzai H., Yoneyama K. & Yamaguchi I. (1989) Transgenic tobacco resistant to a bacterial disease by the detoxification of a pathogenic toxin. *Mol. Gen. Genet.* **219**, 492–4.

Aoyama T. & Chua N.-H. (1997) A glucocorticoid-mediated transcriptional induction system in transgenic plants. *Plant J.* **11**, 605–12.

Arabidopsis Genome Initiative (2000) Analysis of the genome sequence of the flowering plant *Arabidopsis thaliana*. *Nature* **408**, 796–815.

Arakawa T., Chong D.K.X., Merritt J.L. & Langridge W.H.R. (1997) Expression of cholera toxin B subunit oligomers in transgenic potato plants. *Transgenic Res.* **6**, 403–13.

Arakawa T., Chong D.K. & Langridge W.H. (1998) Efficacy of a food plant-based oral cholera toxin B subunit vaccine. *Nature Biotechnol.* **16**, 292–7.

Arber W. (1965) Host specificity of DNA produced by *Escherichia coli*. V. The role of methionine in the production of host specificity. *J. Mol. Biol.* **11**, 247–56.

Arber W. & Dussoix D. (1962) Host specificity of DNA produced by *Escherichia coli*. I. Host controlled modification of bacteriophage λ. *J. Mol. Biol.* **5**, 18–36.

Arencibia A.D., Carmona E.R., Tellez P. *et al.* (1998) An efficient protocol for sugarcane (*Saccharum* spp L) transformation mediated by *Agrobacterium tumefaciens*. *Transgenic Res.* **7**, 213–22.

Arnold D., Feng L., Kim J. & Heintz N. (1994) A strategy for the analysis of gene expression during neural development. *Proc. Nat. Acad. Sci. USA* **91**, 9970–4.

Arnold F.H. (2001) Combinatorial and computational challenges for biocatalyst design. *Nature* **409**, 253–7.

Arondel V., Lemieux B., Hwang I. *et al.* (1992) Map-based cloning of a gene controlling omega-3 fatty acid desaturation in *Arabidopsis*. *Science* **258**, 1353–5.

Artsaenko O., Peisker M., zur Nieden U. *et al.* (1995) Expression of a single chain Fv antibody against abscisic acid creates a wilty phenotype in transgenic tobacco. *Plant J.* **8**, 745–50.

Ashburner M. (1989) Drosophila: *A Laboratory Handbook*. Cold Spring Harbor Laboratory Press, Cold Spring Harbor, New York.

Atchison R.W., Casto B.C. & Hammond W.M. (1965) Adenovirus-associated defective viral particles. *Science* **149**, 754–6.

Atwell S. & Wells J.A. (1999) Selection for improved subtiligases by phage display. *Proc. Nat. Acad. Sci. USA* **96**, 9497–502.

Austin S. & Nordstrom K. (1990) Partition-mediated incompatibility of bacterial plasmids. *Cell* **60**, 351–4.

Axel R., Fiegelson P. & Schutz G. (1976) Analysis of the complexity and diversity of mRNA from chicken oviduct and liver. *Cell* **11**, 247–54.

Bach F.H., Winkler H., Ferran C., Hancock W.W. & Robson S.C. (1996) Delayed xenograft rejection. *Immunol. Today* **17**, 379–84.

Backman K. & Boyer H.W. (1983) Tetracycline resistance determined by pBR322 is mediated by one polypeptide. *Gene* **26**, 197–203.

Bagdasarian M., Bagdasarian M.M., Coleman S. & Timmis K.N. (1979) New vector plasmids for gene cloning in *Pseudomonas*. In *Plasmids of Medical, Environmental and Commercial Importance*, eds Timmis K.N. & Pühler A., pp. 411–22. Elsevier/North-Holland Biomedical Press, Amsterdam.

Bagdasarian M., Lurz R., Rückert B. *et al.* (1981) Specific-purpose plasmid cloning vectors. II. Broad host range, high copy number, RSF1010-derived vectors, and a host–vector system for gene cloning in *Pseudomonas*. *Gene* **16**, 237–47.

Bagdasarian M.M., Amann E., Lurz R., Ruckert B. & Bagdasarian M. (1983) Activity of the hybrid *trp–lac* (*tac*) promoter of *Escherichia coli* in *Pseudomonas putida*: construction of broad-host-range, controlled-expression vectors. *Gene* **26**, 273–82.

Bagga J.S. & Wilusz J. (1999) Northwestern screening of expression libraries. *Methods Mol. Biol.* **118**, 245–56.

Baguisi A., Behboodi E., Melican D.T. *et al.* (1999) Production of goats by somatic cell nuclear transfer. *Nature Biotechnol.* **17**, 456–61.

Bahramian M.B. & Zabl H. (1999) Transcriptional and posttranscriptional silencing of rodent alpha1 (I) collagen by a homologous transcriptionally self-silenced transgene. *Mol. Cell. Biol.* **19**, 274–83.

Bailey D., Zanders E. & Dean P. (2001) The end of the beginning for genomic medicine. *Nature Biotechnol.* **19**, 207–9.

Baim S.B., Labow M.A., Levine A.J. & Shenk T. (1991) A chimeric mammalian transactivator based on the lac repressor that is regulated by temperature and isopropyl beta-D-thiogalactopyranoside. *Proc. Nat. Acad. Sci. USA* **88**, 5072–6.

Baker R.K., Haendel M.A., Swanson B.J. *et al.* (1997) *In vitro* pre-selection of gene-trapped embryonic stem cell clones for characterizing novel developmentally regulated genes in the mouse. *Devel. Biol.* **185**, 201–14.

Balbás P., Soberon X., Merino E. *et al.* (1986) Plasmid vector pBR322 and its special-purpose derivatives – a review. *Gene* **50**, 3–40.

Ballivet M., Nef P., Coutourier S. *et al.* (1988) Electrophysiology of a chick neuronal nicotinic acetylcholine receptor expressed in *Xenopus* oocytes after cDNA injection. *Neuron* **1**, 847–52.

Baltz R.H. (1998) Genetic manipulation of antibiotic-producing *Streptomyces*. *Trends Microbiol.* **6**, 76–83.

Baneyx F. (1999) Recombinant protein expression in *Escherichia coli*. *Curr. Opin. Biotechnol.* **10**, 411–21.

Banfi S., Borsani G., Rossi E. *et al.* (1996) Identification and mapping of human cDNAs homologous to *Drosophila* mutant genes through EST database searching. *Nature Genet.* **13**, 167–74.

Bano-Maqbool S. & Christou P. (1999) Multiple traits of agronomic importance in transgenic indica rice plants: analysis of transgene integration patterns, expression levels and stability. *Mol. Breeding* **5**, 471–80.

Barnard E.A., Houghton M., Miledi R., Richards B.M. & Sumikawa K. (1982) Molecular genetics of the acetylcholine receptor and its insertion and organization in the membrane. *Biol. Cell* **45**, 383.

Barnes W.M. (1980) DNA cloning with single-stranded phage vectors. In *Genetic Engineering*, eds Setlow J.K. & Hollaender A., Vol. 2, pp. 185–200. Plenum Press, New York.

Barnes W.M. (1994) PCR amplification of up to 35 kbp DNA with high fidelity and high yield from lambda bacteriophage templates. *Proc. Nat. Acad. Sci. USA* **91**, 2216–20.

Barta A., Sommergruber K., Thompson D. *et al.* (1986) The expression of a nopaline synthase-human growth hormone chimaeric gene in transformed tobacco and sunflower callus tissue. *Plant Mol. Biol.* **6**, 347–57.

Barth S., Huhn M., Matthey B. *et al.* (2000) Compatible-solute supported periplasmic expression of functional recombinant proteins under stress conditions. *Appl. Environ. Microbiol.* **66**, 1572–9.

Barton K.A., Whitely H.R. & Yang N.S. (1987) *Bacillus thuringiensis* delta endotoxin expressed in transgenic

Nicotiana tabacum provides resistance to lepidopteran insects. *Plant Physiol.* **85**, 1103–9.

Bass S., Greener R. & Wells J.A. (1990) Hormone phage: an enrichment method for variant proteins with altered binding properties. *Proteins* **8**, 309–14.

Bates P.F. & Swift R.A. (1983) Double *cos* site vectors: simplified cosmid cloning. *Gene* **26**, 137–46.

Bates S., Cashmore A.M. & Wilkins B.M. (1998) IncP plasmids are unusually effective in mediating conjugation of *Escherichia coli* and *Saccharomyces cerevisiae*: involvement of the Tra2 mating system. *J. Bacteriol.* **180**, 6538–43.

Baubonis W. & Saur B. (1993) Genomic targeting with purified Cre recombinase. *Nucl. Acids Res.* **21**, 2025–9.

Baulcombe D.C. (1999) Fast forward genetics based on virus-induced gene silencing. *Curr. Opin. Plant Biol.* **2**, 109–13.

Baulcombe D.C., Chapman S. & SantaCruz S.S. (1995) Jellyfish green fluorescent protein as a reporter for virus-infections. *Plant J.* **7**, 1045–53.

Beach L.R. & Palmiter R.D. (1981) Amplification of the metallothionein-I gene in cadmium-resistant mouse cells. *Proc. Nat. Acad. Sci. USA* **78**, 2110–14.

Beachy R., Loesch-Fries S. & Tumer N. (1990) Coat-protein mediated resistance against virus infection. *Ann. Rev. Phytopathol.* **28**, 451–74.

Bechtold N., Ellis J. & Pelletier G. (1993) In planta *Agrobacterium*-mediated gene transfer by infiltration of adult *Arabidopsis thaliana* plants. *C. R. Acad. Sci. Paris Life Sci.* **316**, 1194–9.

Beck E. & Bremer E. (1980) Nucleotide sequence of the gene *ompA* encoding the outer membrane protein II of *Escherichia coli* K-12. *Nucl. Acids Res.* **8**, 3011–24.

Becker D.M., Fikes J.D. & Guarente L. (1991) A cDNA encoding a human CCAAT-binding protein cloned by functional complementation in yeast. *Proc. Nat. Acad. Sci. USA* **88**, 1968–72.

Bedell M.A., Jenkins N.A. & Copeland N.G. (1997) Mouse models of human disease. Part II. Recent progress and future directions. *Genes Devel.* **11**, 11–43.

Beerli R.R., Wels W. & Hynes N.E. (1994) Intracellular expression of single chain antibodies reverts ErbB-2 transformation. *J. Biol. Chem.* **269**, 23931–6.

Beggs J.D. (1978) Transformation of yeast by a replicating hybrid plasmid. *Nature* **275**, 104–9.

Belfort M. & Roberts R.J. (1997) Homing endonucleases: keeping the house in order. *Nucl. Acids Res.* **25**, 3379–88.

Bell G.I., Merryweather J.P., Sanchez-Pescador R. *et al.* (1984) Sequence of a cDNA clone encoding human preproinsulin-like growth factor II. *Nature* **310**, 775–7.

Belshaw P.J., Ho S.N., Crabtree G.R. & Schreiber S.L. (1996) Controlling protein association and subcellular localization with a synthetic ligand that induces hetero-dimerization of proteins. *Proc. Nat. Acad. Sci. USA* **93**, 4604–7.

Bender M.A., Palmer T.D., Gelinas R.E. & Miller A.D. (1987) Evidence that the packaging signal of Moloney murine leukemia virus extends into the gag region. *J. Virology* **61**, 1639–46.

Bender W., Spierer P. & Hogness D.S. (1983) Chromosomal walking and jumping to isolate DNA from the *Ace* and *rosy* loci and the bithorax complex in *Drosophila melanogaster. J. Mol. Biol.* **168**, 17–33.

Bendig M.M. & Williams J.G. (1983) Replication and expression of *Xenopus laevis* globin genes injected into fertilized *Xenopus* eggs. *Proc. Nat. Acad. Sci. USA* **80**, 6197–201.

Benihoud K., Yeh P. & Perricaudet M. (1999) Adenovirus vectors for gene delivery. *Curr. Opin. Biotechnol.* **10**, 440–7.

Benson D.A., Boguski M., Lipman D.J. & Ostell J. (1996) GenBank. *Nucl. Acids Res.* **24**, 1–5.

Benson D.A., Boguski M., Lipman D.J. & Ostell J. (1997) GenBank. *Nucl. Acids Res.* **25**, 1–6.

Bentley D.R., Todd C., Collins C. *et al.* (1992) The development and application of automated gridding for the efficient screening of yeast and bacterial ordered libraries. *Genomics* **12**, 534–41.

Benton W.D. & Davis R.W. (1977) Screening λgt recombinant clones by hybridization to single plaques *in situ*. *Science* **196**, 180–2.

Berg P. (1981) Dissections and reconstructions of genes and chromosomes. *Science* **213**, 296–303.

Berger J., Hauber J., Hauber R., Geiger R. & Cullen B.R. (1988) Secreted placental alkaline phosphatase: a powerful new quantitative indicator of gene expression in eukaryotic cells. *Gene* **66**, 1–10.

Berges H., Joseph-Liauzun E. & Fayet O. (1996) Combined effects of the signal sequence and the major chaperone proteins on the export of human cytokines in *Escherichia coli. Appl. Environ. Microbiol.* **62**, 55–60.

Berglund P., Quesada-Rolander M., Putkonen P. *et al.* (1997) Outcome of immunization of cynomolgus monkeys with recombinant Semliki Forest virus encoding human immunodeficiency virus type 1 envelope protein and challenge with a high dose of SHIV-4 virus. *AIDS Res. Hum. Retroviruses* **13**, 1487–95.

Berglund P., Tubulekas I. & Liljestrom O. (1996) Alphaviruses as vectors for gene delivery. *Trends Biotechnol.* **14**, 130–4.

Berkner K.L. (1992) Expression of heterologous sequences in adenoviral vectors. *Curr. Top. Microbiol. Immunol.* **158**, 39–67.

Berns K.I., Pinkerton T.C., Thomas G.F. & Hoggan M.D. (1975) Detection of adeno-associated virus (AAV)-specific nucleotide sequences in DNA isolatd from latently infected Detroit 6 cells. *Virology* **68**, 556–60.

Bertholet C., Drillien R. & Wittek R. (1985) One hundred base pairs of 5′ flanking sequence of vaccinia virus late gene are sufficient to temporally regulate transcription. *Proc. Nat. Acad. Sci. USA* **82**, 2096–100.

Betancourt O.H., Attal J., Theron M.C., Puissant C. & Houdebine L.M. (1993) Efficiency of introns from various origins in fish cells. *Mol. Marine Biol. Biotechnol.* **2**, 181–8.

Bett A.J., Prevec L. & Graham F.L. (1993) Packaging capacity and stability of human adenovirus type 5 vectors. *J. Virol.* **67**, 5911–21.

Bevan M. (1984) Binary *Agrobacterium* vectors for plant transformation. *Nucl. Acids Res.* **12**, 8711–21.

Bevan M., Barnes W. & Chilton M.D. (1983a) Structure and transcription on the nopaline synthase gene region of T-DNA. *Nucl. Acids Res.* **11**, 369–85.

Bevan M.W., Flavell R.B. & Chilton M.D. (1983b) A chimaeric antibiotic resistance gene as a selectable market for plant cell transformation. *Nature* **304**, 184–7.

Bibb M.J., Ward J.M. & Hopwood D.A. (1978) Transformation of plasmid DNA into *Streptomyces* at high frequency. *Nature* **274**, 398–400.

Bibb M.J., Schottel J.L. & Cohen S.N. (1980) A DNA cloning system for interspecies gene transfer in antibiotic-producing *Streptomyces*. *Nature* **284**, 526–31.

Bickley J. & Hopkins D. (1999) Inhibitors and enhancers of PCR. In *Analytical Molecular Biology: Quality and Validation*, eds Saunders G.C. & Parkes H.C., pp. 81–102. Royal Society of Chemistry, London.

Bilan R., Futterer J. & Sautter C. (1999) Transformation of cereals. *Genet. Eng.* **21**, 113–57.

Bills M.M., Medd J.M., Chappel R.J. & Adler B. (1993) Construction of a shuttle vector for use between *Pasteurella multocida* and *Escherichia coli*. *Plasmid* **30**, 268–73.

Bingham P.M., Levis R. & Rubin G.M. (1981) Cloning of DNA sequences from the white locus of *D. melanogaster* by a novel and general method. *Cell* **25**, 693–704.

Bingham P.M., Kidwell M.G. & Rubin G.M. (1982) The molecular basis of P–M hybrid dysgenesis: the role of the P element, a P-strain-specific transposon family. *Cell* **29**, 995–1004.

Bingle L.E.H. & Thomas C.M. (2001) Regulatory circuits for plasmid survival. *Curr. Opin. Microbiol.* **4**, 194–200.

Birch D.E. (1996) Simplified hot start PCR. *Nature* **381**, 445–6.

Birch R.G. & Bower R. (1994) Principles of gene transfer using particle bombardment. In *Particle Bombardment Technology for Gene Transfer*, eds Yang N.-Y. & Christou P., pp. 3–37. Oxford University Press, Oxford.

Birnboim H.C. & Doly J. (1979) A rapid alkaline extraction procedure for screening recombinant plasmid DNA. *Nucl. Acids Res.* **7**, 1513–23.

Birren B. & Lai E. (1994) Rapid pulsed field separation of DNA molecules up to 250 kb. *Nucl. Acids Res.* **22**, 5366–70.

Birren B.W., Lai E., Clark S.M., Hood L. & Simon M.I. (1988) Optimized conditions for pulsed field gel electrophoretic separations of DNA. *Nucl. Acids Res.* **16**, 7563–82.

Bishop J.O. & Smith P. (1989) Mechanism of chromosomal integration of microinjected DNA. *Mol. Biol. Med.* **6**, 283–98.

Blatny J.M., Brautaset T., Winther-Larsen H.C., Haughan K. & Valla S. (1997) Construction and use of a versatile set of broad-host-range cloning and expression vectors based on the RK2 replicon. *Appl. Environ. Microbiol.* **63**, 370–9.

Blattner F.R., Williams B.G., Blechl A.E. *et al.* (1977) Charon phages: safer derivatives of bacteriophage lambda for DNA cloning. *Science* **196**, 161–9.

Blobel G. & Dobberstein B. (1975) Transfer of proteins across membranes. I. Presence of proteolytically processed and unprocessed nascent immunoglobulin light chains on membrane-bound ribosomes of murine myeloma. *J. Cell Biol.* **67**, 835–51.

Blochlinger K. & Diggelmann H. (1984) Hygromycin B phosphotransferase as a selectable marker for DNA transfer experiments with higher eucaryotic cells. *Mol. Cell. Biol.* **4**, 2929–31.

Bloom K.S. & Carbon J. (1982) Yeast centromere DNA is a unique and highly ordered structure in chromosomes and small circular minichromosomes. *Cell* **29**, 305–17.

Blouin M.J., Beauchemin H., Wright A. *et al.* (2000) Genetic correction of sickle cell disease: insights using transgenic mouse models. *Nature Med.* **6**, 177–82.

Blum P., Velligan M., Lin N. & Matin A. (1992) DnaK-mediated alterations in human growth hormone protein inclusion bodies. *Biotechnology* **10**, 301–4.

Bochmann H., Gehrisch S. & Jaross W. (1999) The gene structure of the human growth factor bound protein GRB2. *Genomics* **56**, 203–7.

Boder E.T. & Wittrup K.D. (1997) Yeast surface display for screening combinatorial polypeptide libraries. *Nature Biotechnol.* **15**, 553–8.

Boeke J.D., Vovis G.F. & Zinder N.D. (1979) Insertion mutant of bacteriophage fl sensitive to *Eco*RI. *Proc. Nat. Acad. Sci. USA* **76**, 2699–702.

Boeke J.D., Lacroute F. & Fink G.R. (1984) A positive selection for mutants lacking orotidine-5'-phosphate decarboxylase activity in yeast: 5-fluoro-orotic acid resistance. *Mol. Gen. Genet.* **197**, 345–6.

Boeke J.D., Xu H. & Fink G.R. (1988) A general method for the chromosomal amplification of genes in yeast. *Science* **239**, 280–2.

Boguski M.S. (1995) The turning point in genome research. *Trends Biochem. Sci.* **20**, 295–332.

Boguski M.S., Lowe T.M.J. & Toltoshev C.M. (1993) dbEST – database for 'expressed sequence tags'. *Nature Genet.* **4**, 332–3.

Bohl D., Naffakh N. & Heard J.M. (1997) Long term control of erythropoietin secretion by doxycycline in mice transplanted with engineered primary myoblasts. *Nature Med.* **3**, 299–305.

Bolivar F., Rodriguez R.L., Betlach M.C. & Boyer H.W. (1977a) Construction and characterization of new cloning vehicles. I. Ampicillin-resistant derivatives of the plasmid pMB9. *Gene* **2**, 75–93.

Bolivar F., Rodriguez R.L., Greene P.J. *et al.* (1977b) Construction and characterization of new cloning vehicles. II. A multipurpose cloning system. *Gene* **2**, 95–113.

Bolker M., Bohnert H.U., Braun K.H., Gorl J. & Kahmann R. (1995) Tagging pathogenicity genes in *Ustilago maydis* by restriction enzyme-mediated integration (REMI). *Mol. Gen. Genet.* **248**, 547–52.

Bomhoff G.H., Klapwijk F.M., Kester M.C.M. *et al.* (1976) Octopine and nopaline synthesis and breakdown genetically controlled by a plasmid of *Agrobacterium tumefaciens*. *Mol. Gen. Genet.* **145**, 177–81.

Bonifer C. (1999) Long distance chromatin mechanisms controlling tissue-specific gene locus activation. *Gene* **238**, 277–89.

Bonifer C. (2000) Developmental regulation of eukaryotic gene loci: which *cis*-regulatory information is required? *Trends Genet.* **16**, 310–15.

Bonifer C., Vidal M., Grosveld F. & Sippel A.E. (1990) Tissue-specific and position independent expression of complete gene domain for chicken lysozyme in transgenic mice. *EMBO J.* **9**, 2843–8.

Borisjuk N.V., Borisjuk L.G., Logendra S. *et al.* (1999) Production of recombinant proteins in plant root exudates. *Nature Biotechnol.* **17**, 466–9.

Bork P. (1996) Go hunting in sequence databases but watch out for the traps. *Trends Genet.* **12**, 425–7.

Boshart M., Weber F., Jahn G. *et al.* (1985) A very strong enhancer is located upstream of an immediate early gene of human cytomegalovirus. *Cell* **41**, 521–30.

Bosselman R.A., Hsu R.Y., Boggs T. *et al.* (1989) Germline transmission of exogenous genes in the chicken. *Science* **243**, 533–5.

Bostian K.A., Ellio O., Bussey H. *et al.* (1984) Sequence of the prepro-toxin ds RNA gene of Type 1 killer yeast: multiple processing events produce a two-component toxin. *Cell* **36**, 741–51.

Botstein D. & Davis R.W. (1982) Principles and practice of recombinant DNA research with yeast. In *The Molecular Biology of the Yeast Saccharomyces*, eds Strathern J.N., Jones E.W. & Broach J.R. Cold Spring Harbor Press, Cold Spring Harbor, New York.

Botstein D. & Fink G.R. (1988) Yeast: an experimental organism for modern biology. *Science* **240**, 1439–43.

Boviatsis E.J., Park J.S., Sena-Esteves M. *et al.* (1994) Long-term survival of rats harboring brain neoplasms treated with ganciclovir and a herpes simplex virus vector that retains an intact thymidine kinase gene. *Cancer Res.* **54**, 5745–51.

Bowater R.P. & Wells R.D. (2000) The intrinsically unstable life of DNA triplet repeats associated with human hereditary disorders. *Prog. Nucl. Acid Res. Mol. Biol.* **66**, 159–202.

Bower R. & Birch R.G. (1992) Transgenic sugarcane plants via microprojectile bombardment. *Plant J.* **2**, 409–416.

Bowers J., Boursiquot J.-M., This P. *et al.* (1999) Historical genetics. The parentage of chardonnay, gamay, and other wine grapes of northeastern France. *Science* **285**, 1562–5.

Bowes C., Li T., Frankel W.N. *et al.* (1993) Localization of a retroviral element within the *rd* gene coding for the β-subunit of cGMP phosphodiesterase. *Proc. Nat. Acad. Sci. USA* **90**, 2955–9.

Boyce F.M. & Bucher N.L.R. (1996) Baculovirus-mediated gene transfer into mammalian cells. *Proc. Nat. Acad. Sci. USA* **93**, 2348–52.

Boyko W.L. & Ganschow R.E. (1982) Rapid identification of *Escherichia coli* transformed by pBR322 carrying inserts at the *Pst*I site. *Anal. Biochem.* **122**, 85–8.

Boynton J.E., Gillham N.W., Harris E.H. *et al.* (1988) Chloroplast transformation in *Chlamydomonas* with high-velocity microprojectiles. *Science* **240**, 1534–8.

Bracy J.L., Sachs D.H. & Iacomini J. (1998) Inhibition of xenoreactive natural antibody production by retroviral gene therapy. *Science* **281**, 1845–7.

Bradley A., Evans M., Kaufman M.H. & Robertson E. (1984) Formation of germ line chimeras from embryo-derived teratocarcinoma cell lines. *Nature* **309**, 255–6.

Brady G., Funk A., Mattern J., Schutz G. & Brown R. (1985) Use of gene transfer and a novel cosmid rescue strategy to isolate transforming sequences. *EMBO J.* **4**, 2583–8.

Brandes C., Plautz J.D., Stanewsky R. *et al.* (1996) Novel features of *Drosophila* period transcription revealed by real-time luciferase reporting. *Neuron* **16**, 687–92.

Brandon E.P., Idzerda R.L. & McKnight G.S. (1995a) Targeting the mouse genome: a compendium of knock-outs. Part I. *Curr. Biol.* **5**, 625–34.

Brandon E.P., Idzerda R.L. & McKnight G.S. (1995b) Targeting the mouse genome: a compendium of knock-outs. Part II. *Curr. Biol.* **5**, 758–65.

Brandon E.P., Idzerda R.L. & McKnight G.S. (1995c) Targeting the mouse genome: a compendium of knock-outs. Part III. *Curr. Biol.* **5**, 873–81.

Brange J., Ribel U., Hansen J.F. *et al.* (1988) Monomeric insulins obtained by protein engineering and their medical implications. *Nature* **333**, 679–82.

Brenner S.E. (1995) Network sequence retrieval. *Trends Genet.* **11**, 247–8.

Breter H.J., Knoop M.-T. & Kirchen H. (1987) The mapping of chromosomes in *Saccharomyces cerevisiae*. I. A cosmid vector designed to establish, by cloning *cdc* mutants, numerous start loci for chromosome walking in the yeast genome. *Gene* **53**, 181–90.

Briddon R.W., Pinner M.S., Stanley J. & Markham P.G. (1990) Geminivirus coat protein gene replacement alters insect specificity. *Virology* **177**, 85–94.

Briggs R. & King T.J. (1952) Transplantation of living nuclei from blastula cells into enucleated frog's eggs. *Proc. Nat. Acad. Sci. USA* **38**, 455–63.

Brinster R.L., Chen H.Y., Trumbauer M. *et al.* (1981) Somatic expression of Herpes thymidine kinase in mice following injection of a fusion gene into eggs. *Cell* **27**, 223–31.

Brinster R.L., Chen H.Y., Warren R., Sarthy A. & Palmiter R.D. (1982) Regulation of metallothionein-thymidine kinase fusion plasmids injected into mouse eggs. *Nature* **296**, 39–42.

Brisson N., Paszkowski J., Penswick J.R. *et al.* (1984) Expression of a bacterial gene in plants by using a viral vector. *Nature* **310**, 511–14.

Bristow A.F. (1993) Recombinant-DNA-derived insulin analogues as potentially useful therapeutic agents. *Trends Biotechnol.* **11**, 301–5.

Brocard J., Warot X., Wendling O. *et al.* (1997) Spatio-temporally controlled site-specific somatic mutagenesis in the mouse. *Proc. Nat. Acad. Sci. USA* **94**, 14559–63.

Brochier B., Kieny M.P., Costy F. *et al.* (1991) Large-scale eradication of rabies using recombinant vaccinia-rabies vaccine. *Nature* **354**, 520–2.

Brochier B., Costy F. & Pastoret P.P. (1995) Elimination of fox rabies from Belgium using a recombinant vaccinia–rabies vaccine: an update. *Vet. Microbiol.* **46**, 269–79.

Broglie K., Chet I., Holliday M. *et al.* (1991) Transgenic plants with enhanced resistance of the fungal pathogen *Rhizoctonia solanii. Science* **254**, 1194–7.

Bron S. & Luxen E. (1985) Segregational instability of pUB110-derived recombinant plasmids in *Bacillus subtilis. Plasmid* **14**, 235–44.

Bron S., Bosma P., Van Belkum M. & Luxen E. (1988) Stability function in the *Bacillus subtilis* plasmid pTA1060. *Plasmid* **18**, 8–16.

Bron S., Peijnenburg A., Peeters B., Haima P. & Venema G. (1989) Cloning and plasmid (in)stability in *Bacillus subtilis.* In *Genetic Transformation and Expression*, eds Butler O.O., Harwood C.R. & Moseley B.E.B., pp. 205–19. Intercept, Andover.

Bronchain O.J., Hartley K.O. & Amaya E. (1999) A gene trap approach in *Xenopus. Curr. Biol.* **9**, 1195–8.

Brookman J.L. & Denning D.W. (2000) Molecular genetics in *Aspergillus fumigatus. Curr. Opin. Microbiol.* **3**, 468–74.

Broome S. & Gilbert W. (1978) Immunological screening method to detect specific translation products. *Proc. Natl. Acad. Sci. USA* **75**, 2746–9.

Brosius J. (1984) Toxicity of an overproduced foreign gene product in *Escherichia coli* and its use in plasmid vectors for the selection of transcription terminators. *Gene* **27**, 161–72.

Brough D.E., Lizonova A., Hsu C., Kulesa V.A. & Kovesdi I. (1996) A gene transfer vector–cell line system for complete functional complementation of adenovirus early regions E1 and E4. *J. Virol.* **70**, 6497–501.

Bryan P.N. (2000) Protein engineering of subtilisin. *Biochim. Biophys. Acta* **1543**, 203–22.

Buchanan-Wollaston V., Passiatore J.E. & Channon F. (1987) The *mob* and *ori* T functions of a bacterial plasmid promote its transfer to plants. *Nature* **328**, 172–5.

Buchholz F., Ringrose L., Angrand P.O., Rossi F. & Stewart A.F. (1996) Different thermostabilities of FLP and Cre recombinases: implications for applied site-specific recombination. *Nucl. Acids Res.* **24**, 4256–62.

Buchholz F., Angrand P.-O. & Stewart A.F. (1998) Improved properties of FLP recombinase evolved by cycling mutagenesis. *Nature Biotechnol.* **16**, 657–62.

Buchschacher G.L.J. & Panganiban A.T. (1992) Human immunodeficiency virus vectors for inducible expression of foreign genes. *J. Virol.* **66**, 2731–9.

Buller R.M., Janik J.E., Sebring E.D. & Rose J.A. (1981) Herpes simplex virus types 1 and 2 completely help adenovirus associated virus replication. *J. Virol.* **40**, 241–7.

Bundock P. & Hooykaas P.J.J. (1996) Integration of *Agrobacterium tumefaciens* T-DNA in the *Saccharomyces cerevisiae* genome by illegitimate recombination. *Proc. Nat. Acad. Sci. USA* **93**, 15272–5.

Bundock P., Den Dulk-Ras A., Beijersbergen A. & Hooykaas P.J.J. (1995) Trans-kingdom T-DNA transfer from *Agrobacterium tumefaciens* to *Saccharomyces cerevisiae. EMBO J.* **14**, 3206–14.

Burke B. & Warren G. (1984) Microinjection of messenger RNA for an anti-Golgi antibody inhibits intracellular transport of a viral membrane protein. *Cell* **36**, 847–56.

Burke D.T., Carle G.F. & Olson M.V. (1987) Cloning of large segments of exogenous DNA into yeast by means of artificial chromosome vectors. *Science* **236**, 806–13.

Burke T. & Bruford M.W. (1987) DNA fingerprinting in birds. *Nature* **327**, 149–52.

Burkhardt P.K., Beyer P., Wunn J. *et al.* (1997) Transgenic rice (*Oryza sativa*) endosperm expressing daffodil (*Narcissus pseudonarcissus*) phytoene synthase accumulates phytoene, a key intermediate of provitamin A biosynthesis. *Plant J.* **11**, 1071–8.

Burnette W.N. (1981) Western blotting: electrophoretic transfer of proteins from sodium dodecyl sulphate–polyacrylamide gels to unmodified nitrocellulose and radiographic detection with antibody and radioiodinated protein A. *Anal. Biochem.* **112**, 195–203.

Burright E.N., Clark H.B., Servadio A. *et al.* (1995) SCA1 transgenic mice – a model for neurodegeneration caused by CAG trinucleotide expansion. *Cell* **82**, 937–48.

Burton R.A., Gibeaut D.M., Bacic A. *et al.* (2000) Virus-induced silencing of a plant cellulose synthase gene. *Plant Cell* **12**, 691–705.

Buscher R., Herrmann V. & Insel P.A. (1999) Human adrenoreceptor polymorphisms. Evolving recognition of clinical importance. *Trends Pharmacol. Sci.* **20**, 94–9.

Caballero O.L. & 13 others (1997) Highly-sensitive single-step PCR protocol for diagnosis and monitoring of human cytomegalovirus infection in renal transplant recipients. *J. Clin. Microbiol.* **35**, 283–9.

Cabanes-Macheteau M., Fitchette-Laine A.C., Loutelier-Bourhis C. *et al.* (1999) N-glycosylation of a mouse IgG expressed in transgenic tobacco plants. *Glycobiology* **9**, 365–72.

Cai D.W., Mukhopadhyay T., Lui T., Fujiwara T. & Roth J.A. (1993) Stable expression of the wild type p53 gene in human lung cancer cells after retrovirus mediated gene transfer. *Hum. Gene Ther.* **4**, 617–24.

Camilli A. (1996) Noninvasive techniques for studying pathogenic bacteria in the whole animal. *Trends Microbiol.* **4**, 295–6.

Camilli A. & Mekalanos J.J. (1995) Use of recombinase gene fusions to identify *Vibrio cholerae* genes induced during infection. *Mol. Microbiol.* **18**, 671–683.

Caminci P., Kvam C., Kiamura A. *et al.* (1996) High efficiency full length cDNA cloning by biotinylated CAP trapper. *Genomics* **37**, 327–36.

Campbell K.H.S., McWhir J., Richie W.A. & Wilmut I. (1996) Sheep cloned by nuclear transfer from a cultured cell line. *Nature* **380**, 64–7.

Canel C., Lopez-Cardoso M.I., Whitmer S. *et al.* (1998) Effects of over expression of strictosidine synthase and tryptophan decarboxylase on alkaloid production by cell cultures of *Catharanthus roseus*. *Planta* **205**, 414–19.

Canfield V., Emanuael J.R., Spickofsky N., Levenson R. & Margolskee R.F. (1990) Ouabain resistant mutants of the rat Na, K-ATPase α2 isoform identified using an episomal expression vector. *Mol. Cell. Biol.* **10**, 1367–72.

Cangelosi G.A., Best E.A., Martinetti G. & Nester E.W. (1991) Genetic analysis of *Agrobacterium*. *Methods Enzymol.* **204**, 384–97.

Canosi U., Morelli G. & Trautner T.A. (1978) The relationship between molecular structure and transformation efficiency of some *S. aureus* plasmids isolated from *B. subtilis*. *Mol. Gen. Genet.* **166**, 259–67.

Canosi U., Iglesias A. & Trautner T.A. (1981) Plasmid transformation in *Bacillus subtilis*: effects of insertion of *Bacillus subtilis* DNA into plasmid pC194. *Mol. Gen. Genet.* **181**, 434–40.

Capecchi M.R. (1980) High efficiency transformation by direct microinjection of DNA into cultured mammalian cells. *Cell* **22**, 479–88.

Caplen N.J., Alton E.W.F.W., Middleton P.G. *et al.* (1995) Liposome-mediated CFTR gene transfer to the nasal epithelium of patients with cystic fibrosis. *Nature Med.* **1**, 39–46.

Cardoso A.I., Blixenkrone-Moller M., Fayolle J. *et al.* (1996) Immunization with plasmid DNA encoding for the measles virus hemagglutinin and nucleoprotein leads to humoral and cell-mediated immunity. *Virology* **225**, 293–9.

Cargill M. & many others (1999) Characterization of single-nucleotide polymorphisms in coding regions of human genes. *Nature Genet.* **22**, 231–8.

Cariello N., Swenberg J. & Skopek T. (1991) Fidelity of *Thermococcus litoralis* DNA polymerase (Vent™) in PCR determined by denaturing gradient gel electrophoresis. *Nucl. Acids Res.* **19**, 4193–8.

Caron de Fromentel C., Gruek N., Venot C. *et al.* (1999) Restoration of transcriptional activity of p53 mutants in human tumour cells by intracellular expression of anti-p53 single chain Fv fragments. *Oncogene* **18**, 551–7.

Carrillo C., Wigdorovitz A., Oliveros J.C. *et al.* (1998) Protective immune response to foot and mouth disease virus with VP1 expressed in transgenic plants. *J. Virol.* **72**, 1688–90.

Carroll M.W. & Moss B. (1995) *E. coli* β-glucuronidase (GUS) as a marker for recombinant vaccinia viruses. *Biotechniques* **19**, 352–5.

Carter G. & Lemoine N.R. (1993) Antisense technology for cancer therapy: does it make sense? *Br. J. Cancer* **67**, 869–76.

Carter P., Bedouelle H. & Winter G. (1985) Improved oligonucleotide site-directed mutagenesis using M13 vectors. *Nucl. Acids Res.* **13**, 4431–43.

Cartier M., Chang M. & Stanners C. (1987) Use of the *Escherichia coli* gene for asparagine synthetase as a selective marker in a shuttle vector capable of dominant transfection and amplification in animal cells. *Mol. Cell. Biol.* **7**, 1623–8.

Castanie M.P., Berges H., Oreglia J., Prere M.F. & Fayet O. (1997) A set of pBR322-compatible plasmids allowing the testing of chaperone-assisted folding of proteins over-expressed in *Escherichia coli*. *Anal. Biochem.* **254**, 150–2.

Castilla J., Pintado B., Sola I., Sanchez-Morgado J.M. & Enjuanes L. (1998) Engineering passive immunity in transgenic mice secreting virus-neutralizing antibodies in milk. *Nature Biotechnol.* **16**, 34954.

C. elegans Sequencing Consortium (1998) Genome sequence of the nematode *C. elegans*: a platform for investigating biology. *Science* **282**, 2012–18.

Cepko C.L., Roberts B.E. & Mulligan R.C. (1984) Construction and applications of a highly transmissible murine retrovirus shuttle vector. *Cell* **37**, 1053–62.

Cereghino J.L. & Cregg J.M. (1999) Applications of yeast in biotechnology: protein production and genetic analysis. *Curr. Opin. Biotechnol.* **10**, 422–7.

Cereghino J.L. & Cregg J.M. (2000) Heterologous protein expression in the methylotrophic yeast *Pichia pastoris*. *FEMS Microbiol. Rev.* **24**, 45–66.

Cereghino J.L., Helinski D.R. & Toukdarian A.E. (1994) Isolation and characterisation of DNA-binding mutants of a plasmid replication initiation protein utilizing an *in vivo* binding assay. *Plasmid* **31**, 89–99.

Cesarini G., Muesing M.A. & Polisky B. (1982) Control of Col E1 DNA replication: the *rop* gene product negatively affects transcription from the replication primer promoter. *Proc. Nat. Acad. Sci. USA* **79**, 6313–17.

Cesarini G., Helmer-Citterich M. & Castagnoli L. (1991) Control of Col E1 plasmid replication by antisense RNA. *Trends Genet.* **7**, 230–5.

Chada K., Magram J. & Constantini F. (1986) An embryonic pattern of expression of a human fetal globin gene in transgenic mice. *Nature* **319**, 685–9.

Chakrabarti S., Breaching K. & Moss B. (1985) Vaccinia virus expression vector: coexpression of β-galactosidase provides visual screening of recombinant plaques. *Mol. Cell. Biol.* **5**, 3403–9.

Chakrabarti S., Robert-Guroff M., Wong-Staal F., Gallo R.C. & Moss B. (1986) Expression of the HTLV-III envelope gene by a recombinant vaccinia virus. *Nature* **320**, 535–7.

Chalfie M., Tu Y., Euskirchen G., Ward W.W. & Prasher D.C. (1994) Green fluorescent protein as a marker for gene expression. *Science* **263**, 802–5.

Champoux J.J. (1995) Roles of ribonuclease H in reverse transcription. In *Reverse Transcriptase*, eds Skalka, A.M. & Goff, S.P., pp. 103–17. Cold Spring Harbor Press, Cold Spring Harbor, New York.

Chan A.W.S., Homan E.J., Ballou L.U., Burns J.C. & Brennel R.D. (1998) Transgenic cattle produced by reverse transcribed gene transfer in oocytes. *Proc. Nat. Acad. Sci. USA* **95**, 14028–33.

Chan A.W.S., Luetjen C.M., Dominko T. *et al.* (2000) Foreign DNA transmission by ICSI: injection of spermatozoa bound with exogenous DNA results in embryonic GFP expression and live Rhesus monkey births. *Mol. Hum. Reprod.* **6**, 26–33.

Chan A.W.S., Chong K.Y., Martinovich C., Simerly C. & Shatten G. (2001) Transgenic monkeys produced by retroviral gene transfer into mature oocytes. *Science* **291**, 309–12.

Chan M.-T., Lee T.-M. & Chang H.-H. (1992) Transformation of indica rice (*Oryza sativa* L.) mediated by *Agrobacterium tumefaciens*. *Plant Cell Physiol.* **33**, 577–83.

Chan M.-T., Chang H.-H., Ho S.-L., Tong W.-F. & Yu S.-M. (1993) *Agrobacterium*-mediated production of transgenic rice plants expressing a chimeric α-amylase promoter/β-glucuronidase gene. *Plant Mol. Biol.* **22**, 491–506.

Chaney W.G., Howard D.R., Pollard J.W., Sullustio, S. & Stanley P. (1986) High frequency transfection of CHO cells using polybrene. *Somat. Cell Mol. Genet.* **12**, 237–44.

Chang A.C.Y., Nunberg J.H., Kaufman R.K. *et al.* (1978) Phenotypic expression in *E. coli* of a DNA sequence coding for mouse dihydrofolate reductase. *Nature* **275**, 617–24.

Chang J.C. & Kan Y.W. (1981) Antenatal diagnosis of sickle-cell anaemia by direct analysis of the sickle mutation. *Lancet* **ii**, 1127–9.

Chang L.M.S. & Bollum F.J. (1971) Enzymatic synthesis of oligodeoxynucleotides. *Biochemistry* **10**, 536–42.

Chang S. & Cohen S.N. (1979) High-frequency transformation of *Bacillus subtilis* protoplasts by plasmid DNA. *Mol. Gen. Genet.* **168**, 111–15.

Chang S.S., Park S.K., Kim B.C. *et al.* (1994) Stable genetic transformation of *Arabidopsis thaliana* by *Agrobacterium* inoculation in planta. *Plant J.* **5**, 551–8.

Chapman A.B., Costello M.A., Lee R. & Ringold G.M. (1983) Amplification and hormone-regulated expression of a mouse mammary tumor virus-Ecogpt fusion plasmid in mouse 3T6 cells. *Mol. Cell. Biol.* **3**, 1421–9.

Chapman S., Kavanagh T. & Baulcombe D. (1992) Potato virus-X as a vector for gene-expression in plants. *Plant J.* **2**, 549–57.

Charles I. & Dougan G. (1990) Gene expression and the development of live enteric vaccines. *Trends Biotechnol.* **8**, 117–21.

Charpentier E., Gerbaud G. & Courvalin P. (1999) Conjugative mobilization of the rolling-circle plasmid pIP823 from *Listeria monocytogenes* BM4293 among Gram-positive and Gram-negative bacteria. *J. Bacteriol.* **181**, 3368–74.

Chater K.F. & Hopwood D.A. (1983) Streptomyces genetics. In *Biology of the Actinomycetes*, eds Goodfellow M. Mordarski M. & Williams S.T., pp. 229–85. Academic Press, London.

Chatoo B.B., Sherman F., Azubalis D.A. *et al.* (1979) Selection of *lys2* mutants of the yeast *Saccharomyces cerevisiae* by the utilisation of α-aminoadipate. *Genetics* **93**, 51–65.

Chatterjee P.K. & Coren J.S. (1997) Isolating large nested deletions in bacterial and P1 artificial chromosomes by *in vivo* P1 packaging of products of Cre-catalysed recombination between the endogenous and a transported *loxP* site. *Nucl. Acids Res.* **25**, 2205–12.

Chee M., Yang R., Hubbell E. *et al.* (1996) Accessing genetic information with high-density DNA arrays. *Science* **274**, 610–14.

Chen C. & Okayama H. (1987) High efficiency transformation of mammalian cells by plasmid DNA. *Mol. Cell. Biol.* **7**, 2745–51.

Chen C. & Okayama H. (1988) Calcium phosphate-mediated gene transfer: a highly efficient transfection system for stably transforming cells with plasmid DNA. *Biotechniques* **6**, 632.

Chen C.Y., Oppermann H. & Hitzeman R.A. (1984) Homologous versus heterologous gene expression in the yeast *Saccharomyces cerevisiae*. *Nucl. Acids Res.* **12**, 8951–70.

Chen K. & Arnold F.H. (1991) Enzyme engineering for non-aqueous solvents: random mutagenesis to enhance activity of subtilisin E in polar organic media. *Biotechnology* **9**, 1073–7.

Chen K. & Arnold F.H. (1993) Tuning the activity of an enzyme for unusual environments: sequential random mutagenesis of subtilisin E for catalysis in dimethylformamide. *Proc. Nat. Acad. Sci. USA* **90**, 5618–22.

Chen Z. (1996) Simple modifications to increase specificity of the 5′-RACE procedure. *Trends Genet.* **12**, 87–8.

Cheng M., Fry J.E., Pang S.Z. *et al.* (1997) Genetic transformation of wheat mediated by *Agrobacterium tumefaciens*. *Plant Physiol.* **115**, 971–80.

Cheng S., Fockler C., Barmes W.M. & Higuchi R. (1994a) Effective amplification of long targets from cloned inserts and human genome DNA. *Proc. Nat. Acad. Sci. USA* **91**, 5695.

Cheng S., Chang S.Y., Gravitt P. & Respess R. (1994b) Long PCR. *Nature* **369**, 684–5.

Cheung V.G. & Nelson S.F. (1996) Whole genome amplification using a degenerate oligonucleotide primer allows hundrends of genotypes to be performed on less than one nanogram of genomic DNA. *Proc. Nat. Acad. Sci. USA* **93**, 14676–9.

Chiang T.-R. & McConlogue L. (1988) Amplification of heterologous ornithine decarboxylase in Chinese hamster ovary cells. *Mol. Cell. Biol.* **8**, 764–9.

Chien C.-T., Bartel P.L., Sternglanz R. & Fields S. (1991) The two-hybrid system: a method to identify and clone genes for proteins that interact with a protein of interest. *Proc. Nat. Acad. Sci. USA* **88**, 9578–82.

Chinery S.A. & Hinchliffe E. (1989) A novel class of vector for yeast transformation. *Curr. Genet.* **16**, 21–5.

Chiu M.I., Katz H. & Berlin V. (1994) RAPT1, a mammalian homolog of yeast Tor, interacts with the FKBP12/rapamycin complex. *Proc. Nat. Acad. Sci. USA* **91**, 12574–8.

Chiu W., Niwa Y., Zeng W. *et al.* (1996) Engineered GFP as a vital reporter in plants. *Curr. Biol.* **3**, 325–30.

Cho R.J., Campbell M.J., Winzeler E.A. *et al.* (1998) A genome-wide transcriptional analysis of the mitotic cell cycle. *Mol. Cell* **2**, 65–73.

Chong S. & 12 others (1997) Single-column purification of free recombinant proteins using a self-cleavable affinity tag derived from a protein splicing element. *Gene* **192**, 271–81.

Chong S., Montello G.E., Zhang A. *et al.* (1998) Utilizing the C-terminal cleavage activity of a protein splicing element to purify recombinant protein in a single chromatographic step. *Nucl. Acids Res.* **26**, 5109–15.

Chotani G., Dodge T., Hsu A. *et al.* (2000) The commercial production of chemicals using pathway engineering. *Biochim. Biophys. Acta* **1543**, 434–55.

Chowdhury K., Bonaldo P., Torres M., Stoykova A. & Gruss P. (1997) Evidence for the stochastic integration of gene trap vectors into the mouse germline. *Nucl. Acids Res.* **25**, 1531–6.

Christensen A.H. & Quail P.H. (1996) Ubiquitin promoter-based vectors for high-level expression of selectable and/or screenable marker genes in monocotyledonous plants. *Transgenic Res.* **5**, 213–18.

Christou P. & Swain W.F. (1990) Cotransformation frequencies of foreign genes in soybean cell cultures. *Theor. Appl. Genet.* **90**, 97–104.

Christou P., McCabe D.E. & Swain W.F. (1988) Stable transformation of soybean callus by DNA-coated gold particles. *Plant Physiol.* **87**, 671–4.

Christou P., Ford T.L. & Kofron M. (1991) Production of transgenic rice (*Oryza sativa* L.) plants from agronomically important indica and japonica varieties via electric discharge particle acceleration of exogenous DNA into immature zygotic embryos. *Biotechnology* **9**, 957–62.

Chu G. & Sharp P.A. (1981) SV40 DNA transfection of cells in suspension: analysis of the efficiency of transcription and translation of T-antigen. *Gene* **13**, 197–202.

Chu G., Vollrath D. & Davis R. (1986) Separation of large DNA molecules by contour clamped homogenous electric fields. *Science* **234**, 1582–5.

Chu S., DeRisi J., Eisen M. *et al.* (1998) The transcriptional program of sporulation in budding yeast. *Science* **282**, 699–705.

Chuang C.F. & Meyerowitz E.M. (2000) Specific and heritable genetic interference by double-stranded RNA in *Arabidopsis thaliana*. *Proc. Nat. Acad. Sci. USA* **97**, 4985–90.

Ciavattia D.J., Ryan T.M., Farmer S.C. & Townes T.M. (1995) Mouse model of human beta-zero thalassemia: targeted deletion of the mouse beta major- and beta minor-globin genes in embryonic stem cells. *Proc. Nat. Acad. Sci. USA* **92**, 9259–63.

Cibelli J.B., Stice S.L., Golueke P.J. *et al.* (1998) Cloned transgenic claves produced from nonquiescent fetal fibroblasts. *Science* **280**, 1256–8.

Clark A.J. (1998) The mammary gland as a bioreactor: expression processing and production of recombinant proteins. *J. Mammary Gland Biol. Neop.* **3**, 337–49.

Clark A.J., Archibald A.L., McClenaghan M. *et al.* (1993) Enhancing the efficiency of transgene expression. *Phil. Trans. R. Soc. London B* **339**, 225–32.

Clark A.J., Cowper A., Wallace R., Wright G. & Simmons J.P. (1992) Rescuing transgene expression by co-integration. *Biotechnology* **10**, 1450–4.

Clark A.J., Harold G. & Yull F.E. (1997) Mammalian cDNA and prokaryotic reporter sequences silence adjacent transgenes in transgenic mice. *Nucl. Acids Res.* **25**, 1009–14.

Clark W., Register J., Eichholtz D. *et al.* (1990) Tissue-specific expression of the TMV coat protein in transgenic tobacco plants affects the level of coat protein-mediated virus protection. *Virology* **179**, 640–7.

Clarke D.L., Johansson C.B., Wilbertz J. *et al.* (2000) Generalized potential of adult neural stem cells. *Science* **288**, 1660–3.

Clarke L. & Carbon J. (1976) A colony bank containing synthetic Col E1 hybrid plasmids representative of the entire *E. coli* genome. *Cell* **9**, 91–9.

Clarke L. & Carbon J. (1980) Isolation of a yeast centromere and construction of functional small circular chromosomes. *Nature* **287**, 504–9.

Clough S.J. & Bent A. (1998) Floral dip: a simplified method for *Agrobacterium*-mediated transformation of *Arabidopsis thaliana*. *Plant J.* **16**, 735–43.

Cochet O., Kenigsberg M., Delumeau I. *et al.* (1998) Intracellular expression of an antibody fragment neutralizing p21 ras promotes tumor regression. *Cancer Res.* **58**, 1170–6.

Cochran M.A., Puckett C. & Moss B. (1985) *In vitro* mutagenesis of the promoter region for a vaccinia virus gene: evidence for tandem early and late regulatory signals. *J. Virol.* **54**, 30–7.

Cockett M.I., Bebbington C.R. & Yarranton G.T. (1990) High level expression of tissue inhibitor of metalloproteinases in Chinese hamster ovary cells using glutamine synthetase gene amplification. *Biotechnology* **8**, 662–7.

Coen E.S., Robbins T.P., Almeida J., Hudson A. & Carpenter R. (1989) Consequences and mechanisms of transposition in *Antirrhinum majus*. In *Mobile DNA* eds Berg D.E. & Howe M.M., pp. 41–55. American Society for Microbiology, Washington DC.

Cohen S.N., Chang A.C.Y. & Hsu L. (1972) Nonchromosomal antibiotic resistance in bacteria: genetic transformation of *Escherichia coli* by R-factor DNA. *Proc. Nat. Acad. Sci. USA* **69**, 2110–14.

Cohney S., McKenzie I.F.C., Patton K. *et al.* (1997) Down-regulation of Galα (1,3) Gal expression by α1,2–fucosyltransferase: further characterization of α1,2–fucosyltransferase transgenic mice. *Transplantation* **64**, 495–500.

Colas P. & Brent R. (1998) The impact of the two-hybrid and related methods on biotechnology. *Trends Biotechnol.* **16**, 355–63.

Colas P., Cohen B., Jessen T. *et al.* (1996) Genetic selection of peptide aptamers that recognize and inhibit cyclin-dependent kinase 2. *Nature* **380**, 548–50.

Colbère-Garapin F., Horodniceanu F., Kourilsky P. & Garapin A.C. (1981) A new dominant hybrid selective marker for higher eukaryotic cells. *J. Mol. Biol.* **150**, 1–14.

Coleclough C. & Erlitz E.L. (1985) Use of primer restriction end adapters in a novel cDNA cloning strategy. *Gene* **34**, 305–14.

Collart F.R. & Huberman E. (1987) Amplification of the IMP dehydrogenase gene in Chinese hamster cells resistant to mycophenolic acid. *Mol. Cell. Biol.* **7**, 3328–31.

Collas P. & Robl J.M. (1990) Relationship between nuclear remodeling and development in nuclear transplant rabbit embryos. *Biol. Reprod.* **45**, 455–65.

Collins F.S. & Weissman S.M. (1984) Directional cloning of DNA fragments at a large distance from an initial probe: a circularization method. *Proc. Nat. Acad. Sci. USA* **81**, 6812–16.

Collins F.S., Drumm M.L., Cole J.L. *et al.* (1987) Construction of a general human chromosome jumping library, with application to cystic fibrosis. *Science* **235**, 1046–9.

Collins J. & Brüning H.J. (1978) Plasmids usable as gene-cloning vectors in an *in vitro* packaging by coliphage λ: 'cosmids'. *Gene* **4**, 85–107.

Collins J. & Hohn B. (1979) Cosmids: a type of plasmid gene-cloning vector that is packageable *in vitro* in bacteriophage λ heads. *Proc. Nat. Acad. Sci. USA* **75**, 4242–6.

Colman A. (1984) Translation of eukaryotic messenger RNA in *Xenopus* oocytes. In: *Transcription and Translation – A Practical Approach*, eds Hames BD & Higgens SJ, pp. 271–302. IRL Press, Oxford.

Colman A., Lane C., Craig R. *et al.* (1981) The influence of topology and glycosylation on the fate of heterologous secretory proteins made in *Xenopus* oocytes. *Eur. J. Biochem.* **113**, 339–48.

Comai L., Facciotti D., Hiatt W.R. *et al.* (1985) Expression in plants of a mutant *aroA* gene from *Salmonella typhimurium* confers tolerance to glyphosate. *Nature* **317**, 741–4.

Condreay J.P., Witherspoon S.M., Clay W.C. & Kost T.A. (1999) Transient and stable gene expression in mammalian cells transduced with a recombinant baculovirus vector. *Proc. Nat. Acad. Sci. USA* **96**, 127–32.

Cone R.D. & Mulligan R.C. (1984) High-efficiency gene transfer into mammalian cells: generation of helper-free recombinant retrovirus with broad mammalian host range. *Proc. Nat. Acad. Sci. USA* **81**, 6349–53.

Conner B.J., Reyes, A.A., Morin C. *et al.* (1983) Detection of sickle cell βˢ-globin allele by hybridization with synthetic oligonucleotides. *Proc. Nat. Acad. Sci. USA* **80**, 278–82.

Conrad B., Savchenko R.S., Breves R. & Hofeweister J. (1996) A T7 promoter-specific, inducible protein expression system for *Bacillus subtilis*. *Mol. Gen. Genet.* **250**, 230–6.

Conrad M. & Topal M.D. (1989) DNA and spermidine provide a switch mechanism to regulate the activity of restriction enzyme Nae I. *Proc. Nat. Acad. Sci. USA* **86**, 9707–11.

Conrad U. & Fielder U. (1998) Compartment-specific accumulation of recombinant immunoglobulins in plant cells: an essential tool for antibody production and immunomodulation of physiological functions and pathogen activity. *Plant Mol. Biol.* **38**, 101–9.

Contente S. & Dubnau D. (1979) Characterization of plasmid transformation in *Bacillus subtilis*: kinetic properties and the effect of DNA conformation. *Mol. Gen. Genet.* **167**, 251–8.

Cooke N.E., Coit D., Weiner R.I., Baxter J.D. & Martial J.A. (1980) Structure of cloned DNA complementary to rat prolactin messenger RNA. *J. Biol. Chem.* **255**, 6502–10.

Cooley L., Berg C. & Spradling A. (1988) Controlling P element insertional mutagenesis. *Trends Genet.* **4**, 254–8.

Cooney E.L., Collier A.C., Greenberg P.D. *et al.* (1991) Safety and immunological response to a recombinant vaccinia virus vaccine expressing HIV envelope glycoprotein. *Lancet* **337**, 567–72.

Cooper D.K., Koren E. & Oriol R. (1994) Oligosaccharides and discordant xenotransplantation. *Immunol. Rev.* **141**, 31–58.

Cormack B.P., Bertram G., Egerton M. *et al.* (1997) Yeast-enhanced green fluorescent protein (yEGFP): a reporter of gene expression in *Candida albicans*. *Microbiology* **143**, 303–11.

Cornelis P. (2000) Expressing genes in different *Escherichia coli* compartments. *Curr. Opin. Biotechnol.* **11**, 450–4.

Cosloy S.D. & Oishi M. (1973) Genetic transformation in *Escherichia coli* K12. *Proc. Nat. Acad. Sci. USA* **70**, 84–7.

Coulibaly S., Besenfelder U., Fleischmann M. *et al.* (1999) Human nerve growth factor beta (hNGF-β): mammary gland specific expression and production in transgenic rabbits. *FEBS Lett.* **444**, 111–16.

Coulson A. (1994) High-performance searching of bio-sequence databases. *Trends Biotechnol.* **12**, 76–80.

Courtney M., Jallat S., Terrier L.-H. *et al.* (1985) Synthesis in *E. coli* of alpha₁-antitrypsin variants of therapeutic potential for emphysema and thrombosis. *Nature* **313**, 149–51.

Cox D.W., Woo S.L.C. & Mansfield T. (1985) DNA restriction fragments associated with alpha-antitrypsin indicate a single origin for deficiency allele PIZ. *Nature* **316**, 79–81.

Cozzi E. & White D.J.G. (1995) The generation of transgenic pigs as potential organ donors for humans. *Nature Med.* **1**, 964–6.

Craig N.L. (1988) The mechanism of conservative site-specific recombination. *Ann. Rev. Genet.* **22**, 77–105.

Cryz S.J. (1992) Live attenuated vaccines for human use. *Curr. Opin. Biotechnol.* **3**, 298–302.

Cullen B.R. & Malim M.H. (1992) Secreted placental alkaline phosphatase as a eukaryotic reporter gene. *Methods Enzymol.* **216**, 362–8.

Culver K.W., Ram Z., Wallbridge S. *et al.* (1992) *In vivo* gene transfer with retroviral vector-producer cells for treatment of experimental brain tumors. *Science* **256**, 1150–2.

Cunningham T.P., Montelaro R.C. & Rushlow K.E. (1993) Lentivirus envelope sequences and proviral genomes are stabilized in *Escherichia coli* when cloned in low-copy-number plasmid vectors. *Gene* **124**, 93–8.

Curtis R.I. & Cardineau C.A. (1990) Oral immunisation by transgenic plants. World Patent Application, WO 90/02484.

Cwirla S.E., Peters E.A., Barrett R.W. & Dower W.J. (1990) Peptides on phage: a vast library of peptides for identifying ligands. *Proc. Nat. Acad. Sci. USA* **87**, 309–14.

Dahme M., Bartsch U., Martini R. *et al.* (1997) Disruption of the mouse L1 gene leads to malformations of the nervous system. *Nature Genet.* **17**, 346–9.

Dalbadie-McFarland G., Cohen L.W., Riggs A.D. *et al.* (1982) Oligonucleotide-directed mutagenesis as a general and powerful method for studies of protein functions. *Proc. Nat. Acad. Sci. USA* **79**, 6409–13.

Dale E.C. & Ow D.W. (1991) Gene transfer with subsequent removal of the selection gene from the host genome. *Proc. Nat. Acad. Sci. USA* **23**, 10558–62.

Dalmay T., Hamilton A., Mueller E. & Baulcombe D.C. (2000) Potato virus X amplicons in *Arabidopsis* mediate genetic and epigenetic gene silencing. *Plant Cell* **12**, 369–79.

Dalsgaard K., Uttenthal A., Jones T.D. *et al.* (1997) Plant-derived vaccine protects target animals against a virus disease. *Nature Biotechnol.* **15**, 248–52.

Dang C. & Jayasena S.D. (1996) Oligonucleotide inhibitors of Taq DNA polymerase facilitate detection of low copy number targets by PCR. *J. Mol. Biol.* **264**, 268–78.

Danos O. & Mulligan R.C. (1988) Safe and efficient generation of recombinant retroviruses with amphotropic and ecotropic host ranges. *Proc. Nat. Acad. Sci. USA* **85**, 6460–4.

Datta S.K., Peterhans A., Datta K. & Potrykus I. (1990) Genetically-engineered fertile Indica rice recovered from protoplasts. *Biotechnolcgy* **8**, 571–4.

Datta K., Vasquez A., Tu J. *et al.* (1998) Constitutive and tissue-specific differential expression of the *cryIA* (b) gene in transgenic rice plants conferring resistance to rice insect pest. *Theor. Appl. Genet.* **97**, 20–30.

Daubert S., Shepherd R.J. & Gardner R.C. (1983) Insertional mutagenesis of the cauliflower mosaic-virus genome. *Gene* **25**, 201–8.

Davidson B.L., Allen E.D., Kozarsky K.F., Wilson J.M. & Roessler B.J. (1993) A model system for *in vivo* gene transfer into the central nervous system using an adenoviral vector. *Nature Genet.* **3**, 219–23.

Davidson B.L., Stein C.S., Heth J.A. *et al.* (2000) Recombinant adeno-associated virus type 2, 4, 5 vectors: transduction of variant cell types and regions in the mammalian central nervous system. *Proc. Nat. Acad. Sci. USA* **97**, 3428–32.

Davies K., Young B.D., Elles R.G., Hill M.E. & Williamson R. (1981) Cloning of a representative genomic library of the human X chromosome after sorting by flow cytometry. *Nature* **293**, 374–76.

Davis B. & MacDonald R.J. (1988) Limited transcription of rat elastase I transgene repeats in transgenic mice. *Genes Devel.* **2**, 13–22.

Davis G.D., Elisee C., Newham D.M. & Harrison R.G. (1999) New fusion protein systems designed to give soluble expression in *Escherichia coli. Biotechnol. Bioeng.* **65**, 382–8.

Davison J., Chevalier N. & Brunel F. (1989) Bacteriophage T7 RNA polymerase-controlled specific gene expression in *Pseudomonas. Gene* **83**, 371–5.

Dawson W.O., Lewandowski D.J., Hilf M.E. *et al.* (1989) A tobacco mosaic virus-hybrid expresses and loses an added gene. *Virology* **172**, 285–292.

De Benedetti A. & Rhoads R.E. (1991) A novel BK virus-based episomal vector for expression of foreign genes in mammalian cells. *Nucl. Acids Res.* **19**, 1925–31.

De Block M. & Debrouwer D. (1991) Two T-DNAs cotransformed into *Brassica napus* by a double *Agrobacterium tumefaciens* infection are mainly integrated at the same locus. *Theor. Appl. Genet.* **82**, 257–63.

De Block M., Herrera-Estrella L., Van Montagu M., Schell J. & Zambryski P. (1984) Expression of foreign genes in regenerated plants and their progeny. *EMBO J.* **3**, 1681–9.

De Block M., Schell J. & Van Montagu M. (1985) Chloroplast transformation by *Agrobacterium tumefaciens. EMBO J.* **4**, 1367–72.

De Block M., Botterman J., Vandewiele M. *et al.* (1987) Engineering herbicide resistance in plants by expression of a detoxifying enzyme. *EMBO J.* **6**, 2513–18.

De Boer H.A. & Hui A.S. (1990) Sequences within ribosome binding site affecting messenger RNA translatability and method to direct ribosomes to single messenger RNA species. *Methods Enzymol.* **185**, 103–14.

De Boer H.A., Hui A., Comstock L.J., Wong E. & Vasser M. (1983b) Portable Shine–Dalgarno regions: a system for a systematic study of defined alterations of nucleotide sequences within *E. coli* ribosome binding sites. *DNA* **2**, 231–41.

Dedieu J.F., Vigne E., Torrent C. *et al.* (1997) Long term gene delivery into the liver of immunocompetent mice with E1/E4 defective adenoviruses. *J. Virol.* **71**, 4626–37.

Deeb S.S. & 8 others (2000) A pro 12A1a substitution in PPARgamma2 associated with decreased receptor activity, lower body mass index and improved insulin sensitivity. *Nature Genet.* **20**, 284–7.

De Grado M., Castan P. & Berenguer J. (1999) A high-transformation-efficiency cloning vector for *Thermus thermophilus*. *Plasmid* **42**, 241–5.

De Greef W., Delon R., De Block M., Leemans J. & Botterman J. (1989) Evaluation of herbicide resistance in transgenic crops under field conditions. *Biotechnology* **7**, 61–4.

De Greve H., Leemans J., Hernalsteens J.P. *et al.* (1982a) Regeneration of normal fertile plants that express octopine synthase from tobacco crown galls after deletion of tumor-controlling functions. *Nature* **300**, 752–5.

De Greve H., Phaese P., Seurwick J. *et al.* (1982b) Nucleotide sequence and transcript map of the *Agrobacterium tumefaciens* Ti plasmid-encoded octopine synthase gene. *J. Mol. Appl. Genet.* **1**, 499–511.

De Groot M.J.A., Bundock P., Hooykaas P.J.J. & Beijersbergen A.G.M. (1998) *Agrobacterium tumefaciens*-mediated transformation of filamentous fungi. *Nature Biotechnol.* **16**, 839–42.

Delannay X., La Valle B.J., Proksch R.K. *et al.* (1989) Field performance of transgenic tomato plants expressing the *Bacillus thuringiensis* var. *kurstaki* insect control protein. *Biotechnology* **7**, 1265–9.

De La Pena A., Lorz H. & Schell J. (1987) Transgenic plants obtained by injecting DNA into young floral tillers. *Nature* **325**, 274–6.

Della-Cioppa G., Garger S.J., Sverlow G.G., Turpen T.J. & Grill L.K. (1990) Melanin production in *Escherichia coli* from a cloned tyrosinase gene. *Biotechnology* **8**, 634–8.

Delneri D., Brancia F.L. & Oliver S.G. (2001) Towards a truly integrative biology through the functional genomics of yeast. *Curr. Opin. Biotechnol.* **12**, 87–91.

Del Solar G., Giraldo R., Ruiz-Echevarria M.J., Espinosa M. & Diaz-Orejas R. (1998) Replication and control of circular plasmids. *Microbiol. Mol. Biol. Rev.* **62**, 434–64.

Demartis S., Huber A., Viti F. *et al.* (1999) A strategy for the isolation of catalytic activities from repertoires of enzymes displayed on phage. *J. Mol. Biol.* **286**, 617–33.

Deng C. & Capecchi M.R. (1992) Re-examination of gene targeting frequency as a function of the extent of homology between the targeting vector and the target locus. *Mol. Cell. Biol.* **12**, 3365–71.

Deng G. & Wu R. (1981) An improved procedure for utilizing terminal transferase to add homopolymers to the 3′ termini of DNA. *Nucl. Acids Res.* **9**, 4173–88.

Denhardt D.T. (1966) A membrane-filter technique for the detection of complementary DNA. *Biochem. Biophys. Res. Commun.* **23**, 641–6.

Depicker A., Stachel S., Dhaese P., Zambryski P. & Goodman H.M. (1982) Nopaline synthase: transcript mapping and DNA sequence. *J. Mol. Appl. Genet.* **1**, 561–74.

Depicker A., Herman L., Jacobs A., Schell J. & Van Montagu M. (1985) Frequencies of simultaneous transformation with different T-DNAs and their relevance to the *Agrobacterium*/plant cell interaction. *Mol. Gen. Genet.* **201**, 477–84.

Deretic V., Chandrasekharappa S., Gill J.F., Chaterjee D.K. & Chakrabarty A.M. (1987) A set of cassettes and improved vectors for genetic and biochemical characterization of *Pseudomonas* genes. *Gene* **57**, 61–72.

De Risi J., Vishwanath R.L. & Brown P.O. (1997) Exploring the metabolic and genetic control of gene expression on a genomic scale. *Science* **278**, 690–6.

Dernberg A.F., Zalevsky J., Colaiacovo M.P. & Villeneuve A.M. (2000) Transgene-mediated cosuppression in the *C. elegans* germ line. *Genes Devel.* **14**, 1578–83.

De Ruyter P.G.G.A., Kuipers O.P. & De Vos W.M. (1996) Controlled gene expression systems for *Lactococcus lactis* with the food-grade inducer nisin. *Appl. Environ. Microbiol.* **62**, 3662–7.

De Saint Vincent B.R., Delbruck S., Eckhart W. *et al.* (1981) The cloning and reintroduction into animal cells of a functional CAD gene, a dominant amplifiable genetic marker. *Cell* **27**, 267–77.

Deshayes A., Herrera-Estrella L. & Caboche M. (1985) Liposome-mediated transformation of tobacco mesophyll protoplasts by an *Escherichia coli* plasmid. *EMBO J.* **4**, 2731–7.

Dessaux Y. & Petit A. (1994) Opines as screenable markers for plant transformation. In *Plant Molecular Biology Manual*, 2nd edn, eds Gelvin S.B. & Schilperoort R.A., Section 3, pp. 1–12. Kluwer Academic Publishers.

Devinoy E., Thepot D., Stinnakre M.G. *et al.* (1994) High level production of human growth hormone in the milk of transgenic mice: the upstream region of the rabbit whey acidic protein (WAP) gene targets transgene expression to the mammary gland. *Transgenic Res.* **3**, 79–89.

Devlin J.J., Panganiban L.C. & Devlin P.E. (1990) Random peptide libraries: a source of specific protein binding molecules. *Science* **249**, 404–6.

Devlin P.E., Drummond R.J., Toy P. *et al.* (1988) Alteration of amino-terminal codons of human granulocyte-colony-stimulating factor increases expression levels and allows efficient processing by methionine aminopeptidase in *Escherichia coli*. *Gene* **65**, 13–22.

Devor E.J., Ivanovich A.K., Hickok J.M. & Todd R.D. (1988) A rapid method for confirming cell line identity: DNA fingerprinting with a minisatellite probe from M13 bacteriophage. *Biotechniques* **6**, 200–1.

De Vos W.M., Venema G., Canosi U. & Trautner T.A. (1981) Plasmid transformation in *Bacillus subtilis*: fate of plasmid DNA. *Mol. Gen. Genet.* **181**, 424–33.

De Vos W.M., Kleerebezem M. & Kuipers O.P. (1997) Expression systems for industrial Gram-positive bacteria with low guanine and cytosine content. *Curr. Opin. Biotechnol.* **8**, 547–53.

De Wet J.R., Fukushima H., Dewji N.N. *et al.* (1984) Chromogenic immunodetection of human serum albumin and a-L-glucosidase clones in a human hepatoma cDNA expression library. *DNA* **3**, 437–47.

De Wet J.R., Wood K.V., De Luca M., Helinski D.R. & Subramani S. (1987) Firefly luciferase gene: structure and expression in mammalian cells. *Mol. Cell. Biol.* **7**, 725–37.

De Zoeten G.A., Penswick J.R., Horisberger M.A. *et al.* (1989) The expression, localization, and effect of a human interferon in plants. *Virology* **172**, 213–22.

D'Halluin K., Bossut M., Bonne E. *et al.* (1992) Transformation of sugar beet (*Beta vulgaris* L.) and evaluation of herbicide resistance in transgenic plants. *Biotechnology* **10**, 309–14.

Dieffenbach C.W. & Dvesler G.S. (eds) (1995) *PCR Primer: A Laboratory Manual*. Cold Spring Harbor Laboratory Press, Cold Spring Harbor, New York.

Dieryck W., Pagnier J., Poyart C. *et al.* (1997) Human haemoglobin from transgenic tobacco. *Nature* **386**, 29–30.

Di Maio D., Triesman R. & Maniatis T. (1982) A bovine papillomavirus vector which propagates as an episome in both mouse and bacterial cells. *Proc. Nat. Acad. Sci. USA* **79**, 4030–4.

DiMaio J.J. & Shillito R.D. (1989) Cryopreservation technology for plant cell cultures. *J. Tissue Cult. Methods* **12**, 163–9.

Dirks W., Wirth M. & Hauser H. (1993) Dicistronic transcription units for gene expression in mammalian cells. *Gene* **129**, 247–9.

Dittmar M.T., Simmons G., Donaldson Y. *et al.* (1997) Biological characterization of human immunodeficiency virus type 1 clones derived from different organs of an AIDS patient by long-range PCR. *J. Virol.* **71**, 5140–7.

Dobson M.J., Tuite M.F., Roberts N.A. *et al.* (1982) Conservation of high efficiency promoter sequences in *Saccharomyces cerevisiae*. *Nucl. Acids Res.* **10**, 2625–37.

Doetschman T., Gregg R.G., Maeda N. *et al.* (1987) Targeted correction of a mutant *HPRT* gene in mouse embryonic stem cells. *Nature* **330**, 57657–8.

Dong J., Teng W., Buchholz W.G. & Hall T.C. (1996) *Agrobacterium*-mediated transformation of Javanica rice. *Planta* **199**, 612–17.

Donson J., Kearney C.M., Hilf M.E. & Dawson W.O. (1991) Systemic expression of a bacterial gene by a tobacco mosaic virus-based vector. *Proc. Nat. Acad. Sci. USA* **88**, 7204–8.

Dorin J.R., Dickinson P., Alton E.W. *et al.* (1992) Cystic-fibrosis in the mouse by targeted insertional mutagenesis. *Nature* **359**, 211–15.

Dorin J.R., Stevenson B.J., Fleming S. *et al.* (1994) Long term survival of the exon 10 insertional cystic fibrosis mutant mouse is a consequience of low-level residual wild type *Cftr* gene expression. *Mammalian Genome* **5**, 465–72.

Dower W.J., Miller J.F. & Ragsdale C.W. (1988) High efficiency transformation of *E. coli* by high voltage electroporation. *Nucl. Acids Res.* **16**, 6127–45.

Drummond M.H. & Chilton M.-D. (1978) Tumor-inducing (Ti) plasmids of *Agrobacterium* share extensive regions of DNA homology. *J. Bacteriol.* **136**, 1178–83.

Dubendorff J.W. & Studier F.W. (1991) Controlling basal expression in an inducible T7 expression system by blocking the target T7 promoter with *lac* repressor. *J. Mol. Biol.* **219**, 45–59.

Dubnau D. (1999) DNA uptake in bacteria. *Ann. Rev. Microbiol.* **53**, 217–44.

DuBridge R.B. & Calos M.P. (1988) Recombinant shuttle vectors for the study of mutation in mammalian cells. *Mutagenesis* **3**, 1–9.

Dueschle U., Kammerer W., Genz R. & Bujard H. (1986) Promoters of *Escherichia coli*: a hierarchy of *in vivo* strength indicates alternate structure. *EMBO J.* **5** 2987–94.

Dugaiczyk A., Boyer H.W. & Goodman H.M. (1975) Ligation of *Eco*RI endonuclease-generated DNA fragments into linear and circular structures. *J. Mol. Biol.* **96**, 171–84.

Duguid J.R., Rohwer R.G. & Seed B. (1988) Isolation of cDNAs of scrapie-modulated RNAs by subtractive hybridization of a cDNA library. *Proc. Nat. Acad. Sci. USA* **85**, 5738–42.

Dujon B. (1996) The yeast genome project: what did we learn? *Trends Genet.* **12**, 263–70.

Dujon B. & 107 others (1994) Complete DNA sequence of yeast chromosome XI. *Nature* **369**, 371–8.

Dunham R.A. (1999) Utilization of transgenic fish in developing countries: potential benefits and risks. *J. World Aquacult. Soc.* **30**, 1–11.

Dunn J.J. & Studier F.W. (1983) Complete nucleotide sequence of bacteriophage T7 DNA and the locations of T7 genetic elements. *J. Mol. Biol.* **166**, 477–535.

During K., Hippe J., Kreuzaler F. & Schell J. (1990) Synthesis and self-assembly of a functional monoclonal antibody in transgenic *Nicotiana tabacum*. *Plant Mol. Biol.* **15**, 281–93.

Dussoix D. & Arber W. (1962) Host specificity of DNA produced by *Escherichia coli*. II. Control over acceptance of DNA from infecting phage λ. *J. Mol. Biol.* **5**, 37–49.

Dutch–Belgian Fragile X Consortium (1994) *Fmr1* knock-out mice: a model to study fragile X mental retardation. *Cell* **78**, 23–33.

Duttweiler H.M. & Gross D.S. (1998) Bacterial growth medium that significantly increases the yield of recombinant plasmid. *Biotechniques* **24**, 438–44.

Dworkin M.B. & Dawid I.B. (1980) Use of a cloned library for the study of abundant poly(A)$^+$ RNA during *Xenopus laevis* development. *Dev. Biol.* **76**, 449–64.

Dyck M.K., Gagne D., Ouellet M. *et al.* (1999) Seminal vesicle production and secretion of growth hormone into seminal fluid. *Nature Biotechnol.* **17**, 1087–90.

Ebert K.M., Selgrath J.P., Ditullio P. *et al.* (1991) Transgenic production of a variant of human tissue-type plasminogen activator in goat milk: generation of transgenic goats and analysis of expression. *Biotechnology* **9**, 835–8.

Eckert K.A. & Kunkel T.A. (1990) High fidelity DNA synthesis by the *Thermus aquaticus* DNA polymerase. *Nucl. Acids Res.* **18**, 3739–44.

Eckhardt T. (1978) A rapid method for the identification of plasmid deoxyribonucleic acid in bacteria. *Plasmid* **1**, 584–8.

Edery I., Chu L.L., Soneberg N. & Pelletier J. (1995) An efficient strategy to isolate full length cDNAs based on an mRNA cap retention procedure (CAPture). *Mol. Cell Biol.* **15**, 3363–71.

Efrat S., Lieser M., Wu Y. *et al.* (1994) Ribozyme-mediated attenuation of pancreatic β-cell glucokinase expression in transgenic mice results in impaired glucose-induced insulin secretion. *Proc. Nat. Acad. Sci. USA* **91**, 2051–5.

Efstratiadis A., Kafatos F.C., Maxam A.M. & Maniatis T. (1976) Enzymatic *in vitro* synthesis of globin genes. *Cell* **7**, 279–88.

Eguchi Y., Itoh T. & Tomizawa J. (1991) Antisense RNA. *Ann. Rev. Biochem.* **60**, 631–652.

Ehrlich S.D. (1977) Replication and expression of plasmids from *Staphylococcus aureus* in *Bacillus subtilis*. *Proc. Nat. Acad. Sci. USA* **74**, 1680–2.

Ehrlich S.D. (1978) DNA cloning in *Bacillus subtilis*. *Proc. Nat. Acad. Sci. USA* **75**, 1433–6.

Ehrlich S.D., Noirot P., Petit M.A. *et al.* (1986) Structural instability of *Bacillus subtilis* plasmids. In *Genetic Engineering*, eds Setlow J.K. & Hollaender A., Vol. 8, pp. 71–83. Plenum, New York.

Ehsani P., Khabiri A. & Domansky N.N. (1997) Polypeptides of hepatitis B surface antigen in transgenic plants. *Proc. Nat. Acad. Sci. USA* **190**, 107–11.

Eichholtz D.A., Rogers S.G., Horsch R.B. *et al.* (1987) Expression of mouse dihydrofolate reductase gene confers methotrexate resistance in transgenic petunia plants. *Somat. Cell Mol. Genet.* **13**, 67–76.

Ellenberg J., Lippincott-Schwartz J. & Presley J.F. (1999) Dual colour imaging with GFP variants. *Trends Cell Biol.* **9**, 52–6.

Elnifro E.M., Ashshi A.M., Cooper R.J. & Klapper P.E. (2000) Multiplex PCR: optimization and application in diagnostic virology. *Clin. Microbiol. Rev.* **13**, 559–570.

Emmerman M. & Temin H.M. (1984) Genes with promoters in retrovirus vectors can be independently suppressed by an epigenetic mechanism. *Cell* **39**, 449–67.

Endean D. & Smithies O. (1989) Replication of plasmid DNA in fertilized *Xenopus* eggs is sensitive to both the topology and size of the injected template. *Chromosoma* **97**, 307–14.

Engler G., Depicker A., Maenhaut R. *et al.* (1981) Physical mapping of DNA base sequence homologies between an octopine and a nopaline Ti plasmid of *Agrobacterium tumefaciens*. *J. Mol. Biol.* **152**, 183–208.

English J.J. & Baulcombe D.C. (1997) The influence of small changes in transgene transcription on homology-dependent virus resistance and gene silencing. *Plant J.* **12**, 1311–18.

English J.J., Mueller E. & Baulcombe D.C. (1996) Suppression of virus accumulation in transgenic plants exhibiting silencing of nuclear genes. *Plant Cell* **8**, 179–88.

Ensley B.D., Ratzkin B.J., Osslund T.D. *et al.* (1983) Expression of naphthalene oxidation genes in *Escherichia coli* results in biosynthesis of indigo. *Science* **222**, 167–9.

Erickson R.P., Lai L.W. & Grimes J. (1993) Creating a conditional mutation of *Wnt-1* by antisense transgenesis provides evidence that *Wnt-1* is not essential for spermatogenesis. *Devel. Biol.* **14**, 274–81.

Erlich H.A. (ed.) (1989) *PCR Technology: Principles and Applications for DNA Amplification*. Stockton Press.

Erlich H.A., Gelfand D.H. & Sakai R.K. (1988) Specific DNA amplification. *Nature* **331**, 461–2.

Ernst J.F. (1988) Codon usage and gene expression. *Trends Biotechnol.* **6**, 196–9.

Eskin B. & Linn S. (1972) The deoxyribonucleic modification and restriction enzymes of *Escherichia coli* B. II. Purification, subunit structure, and catalytic properties of the restriction endonuclease. *J. Biol. Chem.* **247**, 6183–91.

Estrem S.T., Gaal T., Ross W. & Gourse R.L. (1998) Identification of an UP element consensus sequence for bacterial promoters. *Proc. Nat. Acad. Sci. USA* **95**, 9761–6.

Estruch J.J., Chriqui D., Grossmann K., Schell J. & Spena A. (1991a) The plant oncogene rolC is responsible for the release of cytokinins from glucoside conjugates. *EMBO J.* **10**, 2889–95.

Estruch J.J., Schell J. & Spena A. (1991b) The protein encoded by the rolB plant oncogene hydrolyzes indole glucosides. *EMBO J.* **10**, 3125–8.

Etchegaray J.-P. & Inouye M. (1999) Translational enhancement by an element downstream of the initiation codon in *Escherichia coli*. *J. Biol. Chem.* **274**, 10079–85.

Etkin L.D., Pearman B. & Ansah-Yiadom R. (1987) Replication of injected DNA templates in *Xenopus* embryos. *Exp. Cell Res.* **169**, 468–77.

Evans G.A., Lewis K. & Rothenberg B.E. (1989) High efficiency vectors for cosmid microcloning and genomic analysis. *Gene* **79**, 9–20.

Evans M.J. & Kaufman M.H. (1981) Establishment in culture of pluripotential cells from mouse embryos. *Nature* **292**, 154–6.

Evans P.D., Cook S.N., Riggs P.D. & Noren C.J. (1995) LITMUS: multipurpose cloning vectors with a novel system for bidirectional *in vitro* transcription. *Biotechniques* **19**, 130–5.

Fabret C., Quentin Y., Guiseppi A. *et al.* (1995) Analysis of errors in finished DNA sequences: the surfactin operon of *Bacillus subtilis* as an example. *Microbiology* **141**, 345–50.

Farmer W.R. & Liao J.C. (2000) Improving lycopene production in *Escherichia coli* by engineering metabolic control. *Nature Biotechnol.* **18**, 533–7.

Federoff N.V. & Smith D.L. (1993) A versatile system for detecting transposition in *Arabidopsis*. *Plant J.* **3**, 273–89.

Feldmann K.A. (1991) T-DNA insertion mutagenesis in *Arabidopsis*: mutational spectrum. *Plant J.* **1**, 71–82.

Feldmann K.A. & Marks M.D. (1987) *Agrobacterium*-mediated transformation of germinating seeds of *Arabidopsis thaliana* – a non-tissue culture approach. *Mol. Gen. Genet.* **208**, 1–9.

Felgner J.H., Kumar R., Sridhar C.N. *et al.* (1994) Enhanced gene delivery and mechanism studies with a novel series of cationic lipid formulations. *J. Biol. Chem.* **269**, 2550–61.

Felgner P.L., Gadek T.R., Holm M. *et al.* (1987) Lipofection: a highly efficient lipid-mediated DNA-transfection procedure. *Proc. Nat. Acad. Sci. USA* **84**, 7413–17.

Feng Y.Q., Seibler J., Alami R. *et al.* (1999) Site-specific chromosomal integration in mammalian cells: highly efficient CRE recombinase-mediated cassette exchange. *J. Mol. Biol.* **292**, 779–85.

Ferretti L. & Sgaramella V. (1981) Temperature dependence of the joining by T4 DNA ligase of termini produced by type II restriction endonucleases. *Nucl. Acids Res* **9**, 85–93.

Festenstein R., Tolaini M., Carbella P. *et al.* (1996) Locus control region function and heterochromatin-induced position effect variegation. *Science* **271**, 1123–5.

Fickett J.W. (1996) Finding genes by computer: the state of the art. *Trends Genet.* **12**, 316–20.

Fiel R., Brocard J., Mascrez B. *et al.* (1996) Ligand-activated site-specific recombination in mice. *Proc. Nat. Acad. Sci. USA* **93**, 10887–90.

Fiering S., Epner E., Robinson K. *et al.* (1995) Targeted deletion of 5′HS2 of the murine beta-globin LCR reveals that it is not essential for proper regulation of the beta-globin locus. *Genes Devel.* **9**, 2203–13.

Figge J., Wright C., Collins C.J., Roberts T.M. & Livingston D.M. (1988) Stringent regulation of stably integrated chloramphenicol acetyl transferase genes by *E. coli* lac repressor in monkey cells. *Cell* **52**, 713–22.

Fincham J.R.S. (1989) Transformation in fungi. *Microbiol. Rev.* **53**, 148–70.

Finer J.J. & McMullen M.D. (1990) Transformation of cotton (*Gossypium hirsutum* L.) via particle bombardment. *Plant Cell Rep.* **8**, 586–9.

Finer J.J., Vain P., Jones M.W. & McMullen M.D. (1992) Development of the particle inflow gun for DNA delivery to plant cells. *Plant Cell Rep.* **11**, 323–8.

Fire A., Xu S., Montgomery M.K. *et al.* (1998) Potent and specific genetic interference by double stranded RNA in *Caenorhabditis elegans*. *Nature* **391**, 806–11.

Firsheim W. & Kim P. (1997) Plasmid partitioning and replication in *Escherichia coli*: is the membrane the key? *Mol. Microbiol.* **23**, 1–10.

Fischer R. & Emans N. (2000) Molecular farming for pharmaceutical proteins. *Transgenic Res.* **9**, 279–99.

Fischetti V.A., Medaglini D., Oggioni M. & Pozzi G. (1993) Expression of foreign proteins on Gram-positive commensal bacteria for mucosal vaccine delivery. *Curr. Opin. Biotechnol.* **4**, 603–10.

Fischhoff D.A., Bowdish K.S., Perlak F.J. *et al.* (1987) Insect tolerant transgenic tomato plants. *Biotechnology* **5**, 807–13.

Fitch M.M.M., Manshardt R.M., Gonsalves D., Slightom J.L. & Sandford J.C. (1990) Stable transformation of papaya via microprojectile bombardment. *PCR* **9**, 189–94.

Flamand A., Coulon P., Lafay F. *et al.* (1992) Eradication of rabies in Europe. *Nature* **360**, 115–16.

Fleischmann R.D., Adams M.D., White O. *et al.* (1995) Whole-genome random sequencing and assembly of *Haemophilus influenzae* Rd. *Science* **269**, 496–512.

Foord O.S. & Rose E.A. (1994) Long-distance PCR. *PCR Methods Appl.* **3**, S149–S161.

Forman M.D., Stack R.F., Masters P.S., Hauer C.R. & Baxter S.M. (1998) High level, context dependent misincorporation of lysine for arginine in *Saccharomyces cerevisiae* a1 homeodomain expressed in *Escherichia coli*. *Protein Sci.* **7**, 500–3.

Forrester W.C., Takegawa S., Papayannopoulou T., Atamatoyannopoulos G. & Groudine M. (1987) Evidence for a locus activation region: the formation of developmentally stable hypersensitive sites in globin expressing hybrids. *Nucl. Acids Res.* **83**, 10159–77.

Fournier R.E. & Ruddle F.H. (1977) Microcell-mediated transfer of murine chromosomes into mouse, Chinese hamster, and human somatic cells. *Proc. Nat. Acad. Sci. USA* **74**, 319–23.

Franconi A., Roggero P., Pirazzi P. *et al.* (1999) Functional expression in bacteria and plants of an scFv antibody fragment against tospoviruses. *Immunotechnology* **4**, 189–201.

Franke C.A., Rice C.M., Strauss J.H. & Hruby D.E. (1985) Neomycin resistance as a dominant selectable marker for selection and isolation of vaccinia virus recombinants. *Mol. Cell. Biol.* **5**, 1918–24.

Franklin F.C.H., Bagdasarian M., Bagdasarian M.M. & Timmis K.N. (1981) Molecular and functional analysis of the TOL plasmid pWWO from *Pseudomonas putida* and cloning of genes for the entire regulated aromatic ring *meta* cleavage pathway. *Proc. Nat. Acad. Sci. USA* **78**, 7458–62.

Fraser M.J. (1992) The baculovirus infected insect cell as a eukaryotic gene expression system. *Curr. Top. Mol. Biol.* 131–72.

Fray R.G. & Grierson D. (1993) Molecular genetics of tomato fruit ripening. *Trends Genet.* **9**, 438–43.

French R., Janda M. & Ahlquist P. (1986) Bacterial gene inserted in an engineered RNA virus: efficient expression in monocotyledonous plant cells. *Science* **231**, 1294–7.

Frengen E., Weichenhan D., Zhao B. *et al.* (1999) A modular, positive selection bacterial artificial chromosome vector with multiple cloning sites. *Genomics* **58**, 250–3.

Fridell Y.W.C. & Searles L.L. (1991) Vermilion as a small selectable marker gene for *Drosophila* transformation. *Nucl. Acids Res.* **19**, 5082.

Friedrich G. & Soriano P. (1991) Promoter traps in embryonic stem cells: a genetic screen to identify and mutate developmental genes in mice. *Genes Devel.* **5**, 1513–23.

Fries L.F., Tartaglia J., Taylor J. *et al.* (1996) Human safety and immunogenicity of a canarypox rabies glycoprotein recombinant vaccine: an alternative poxvirus vector system. *Vaccine* **14**, 428–34.

Frischauf A.-M., Lehrach H., Poustka A. & Murray N. (1983) Lambda replacement vectors carrying polylinker sequences. *J. Mol. Biol.* **170**, 827–42.

Frijters A.C.J., Zhang Z. Van Damme M. *et al.* (1997) Construction of a bacterial artificial chromosome library containing large *Eco*RI and *Hind* III genomic fragments of lettuce. *Theor. Appl. Genet.* **94**, 390–9.

Frohman M.A. & Martin G.R. (1989) Rapid amplification of cDNA ends using nested primers. *Technique* **1**, 165–70.

Frohman M.A., Dush M.K. & Martin G. (1988) Rapid production of full-length cDNAs from rare transcripts: amplification using a single gene-specific oligonucleotide primer. *Proc. Nat. Acad. Sci. USA* **85**, 8998–9002.

Frolov I. & Schlesinger S. (1994) Translation of Sindbis virus mRNA: effects of sequences downstream of the initiating codon. *J. Virol.* **68**, 8111–17.

Frolov I., Agapov E., Hoffman T.A. Jr *et al.* (1999) Selection of RNA replicons capable of persistent non-cytopathic replication in mammalian cells. *J. Virol.* **73**, 3854–65.

Fromm M.E., Morrish F., Armstrong C. *et al.* (1990) Inheritance and expression of chimeric genes in the progeny of transgenic maize plants. *Biotechnology* **8**, 833–44.

Fuerst T.R., Niles E.G., Studier W.F. & Moss B. (1986) Eukaryotic transient-expression system based on recombinant vaccinia virus that synthesizes bacteriophage T7 RNA polymerase. *Proc. Nat. Acad. Sci. USA* **83**, 8122–6.

Fuerst T.R., Earl P.L. & Moss B. (1987) Use of hybrid vaccinia virus T and RNA polymerase system for expression of target genes. *Mol. Cell. Biol.* **7**, 2538–44.

Fuh G. & Sidhu S.S. (2000) Efficient phage display of polypeptides fused to the carboxy-terminus of the M13 gene-3 minor coat protein. *FEBS Lett.* **480**, 231–4.

Fuh G., Pisabarro M.T., Li Y. *et al.* (2000) Analysis of PDZ domain ligand interactions using carboxy-terminal phage display. *J. Biol. Chem.* **275**, 21486–91.

Fukunaga R., Sokawa Y. & Nagata S. (1984) Constitutive production of human interferons by mouse cells with bovine papillomavirus as vector. *Proc. Nat. Acad. Sci. USA* **81**, 5086–90.

Fukushige S. & Sauer B. (1992) Targeted genomic integration with a positive selection *lox* recombination vector allows highly reproducible gene expression in mammalian cells. *Proc. Nat. Acad. Sci. USA* **89**, 7905–9.

Fuller D.J., Corb M.M., Barnett S., Steimer K. & Haynes J.R. (1997) Enhancement of immunodeficiency virus-specific immune responses in DNA-immunized rhesus macaques. *Vaccine* **15**, 924–6.

Fulton A.M., Adams S.E., Mellor J., Kingsman S.M. & Kingsman A.J. (1987) The organisation and expression of the yeast transposon, Ty. *Microbiol. Sci.* **4**, 180–5.

Futterer J. & Hohn T. (1996) Translation in plants – rules and exceptions. *Plant Mol. Biol.* **32**, 159–89.

Gabizon R. & Taraboulos A. (1997) Of mice and (mad) cows – transgenic mice help to understand prions. *Trends Genet.* **13**, 264–9.

Gallichan W.S., Johnson D.C., Graham F.L. & Rosenthal K.L. (1993) Mucosal immunity and protection after intranasal immunization with recombinant adenovirus expressing herpes simplex virus glycoprotein B. *J. Infect. Dis.* **168**, 622–9.

Gallie D.R. (1996) Translational control of cellular and viral mRNAs. *Plant Mol. Biol.* **32**, 145–58.

Gallie D.R. (1998) Controlling gene expression in transgenics. *Curr. Opin. Plant Biol.* **1**, 166–72.

Gallie D.R., Gay P. & Kado C.I. (1988) Specialized vectors for members of Rhizobiaceae and other Gram-negative bacteria. In *Vectors. A Survey of Molecular Cloning Vectors and Their Uses*, eds Rodriguez R.L. & Denhardt D.T., pp. 333–42. Butterworths, London.

Gallo-Meagher M. & Irvine J.E. (1996) Herbicide resistant transgenic sugarcane plants containing the *bar* gene. *Crop Sci.* **36**, 1367–74.

Gallwitz D. & Sures I. (1980) Structure of a split yeast gene: complete nucleotide sequence of the actin gene in *Saccharomyces cerevisiae*. *Proc. Nat. Acad. Sci. USA* **77**, 2546–50.

Gao G.P., Yang Y.P. & Wilson J.M. (1996) Biology of adenovirus vectors with E1 and E4 deletions for liver-directed gene therapy. *J. Virol.* **70**, 8934–43.

Garcia T., Benhamou B., Gofflo D. *et al.* (1992) Switching agonistic, antagonistic and mixed transcriptional responses to 11b-substituted progestins by mutation of the progesterone receptor. *Mol. Endocrinol.* **6**, 2071–8.

Garfinkel D.J., Boeke J.D. & Fink G.R. (1985) Ty element transposition: reverse transcription and virus-like particles. *Cell* **42**, 507–17.

Garrick D., Fiering S., Martin D.I. & Whitelaw E. (1998) Repeat-induced gene silencing in mammals. *Nature Genet.* **18**, 56–9.

Gatehouse A.M.R., Down R.E., Powell K.S. *et al.* (1996) Transgenic potato plants with enhanced resistance to the peach-potato aphid *Myzus persicae*. *Entomol. Exp. Appl.* **79**, 295–307.

Gatz C. & Quail P.H. (1988) Tn-10-encoded Tet repressor can regulate an operator-containing plant promoter. *Proc. Nat. Acad. Sci. USA* **85**, 1394–7.

Gatz C., Kaiser A. & Wendenburg R. (1991) Regulation of a modified CaMV 35S promoter by the Tn-10-encoded Tet repressor in transgenic tobacco. *Mol. Gen. Genet.* **227**, 229–37.

Gatz C., Frohberg C. & Wendenburg R. (1992) Stringent repression and homogeneous de-repression by tetracycline of a modified CaMV 35S promoter in intact transgenic tobacco plants. *Plant J.* **2**, 397–404.

Gehrlach W., Llewellyn D. & Haseloff J. (1987) Construction of a plant disease resistance gene from the satellite RNA of tobacco ringspot virus. *Nature* **328**, 802–5.

Geiger H., Sick S., Bonifer C. & Muller A.M. (1998) Globin gene expression is reprogrammed in chimeras generated by injecting adult haemopoietic stem cells into mouse blastocysts. *Cell* **93**, 1055–65.

Geller D.S. & 9 others (2000) Activating mineralcorticoid receptor mutation in hypertension exacerbated by pregnancy. *Science* **289**, 119–23.

Genilloud O., Garrido M.C. & Moreno F. (1984) The transposon Tn5 carries a bleomycin resistance determinant. *Gene* **32**, 225–33.

Gerard G.F. & D'Allesio J.M. (1993) Reverse transcriptase (EC2.7.7.49): the use of cloned Moloney murine leukemia virus reverse transcriptase to synthesize DNA from RNA. *Methods Mol. Biol.* **16**, 73–94.

Gerard R.D. & Gluzman Y. (1985) A new host cell system for regulated simian virus 40 DNA replication. *Mol. Cell. Biol.* **5**, 3231–40.

Gerhold D. & Caskey C.T. (1996) It's the genes! EST access to the human genome content. *Bioessays* **18**, 973–81.

Gershoni J.M. & Palade G.E. (1982) Electrophoretic transfer of proteins from sodium dodecyl sulfate-polyacrylamide gels to a positively charged membrane filter. *Anal. Biochem.* **124**, 396–405.

Gething M.-J. & Sambrook J. (1981) Cell-surface expression of influenza haemogglutinin from a cloned DNA copy of the RNA gene. *Nature* **293**, 620–5.

Geyer C.R., Colman Lerner A. & Brent R. (1999) Mutagenesis by peptide aptamers identifies genetic network members and pathway connections. *Proc. Nat. Acad. Sci. USA* **96**, 8567–72.

Ghattas I.R., Sanes J.R. & Majors J.E. (1991) The encephalomyocarditis virus internal ribosome entry site allows efficient coexpression of two genes from a recombinant provirus in cultured cells and in embryos. *Mol. Cell Biol.* **11**, 5848–59.

Ghebranious N. & Donehower L.A. (1998) Mouse models in tumor suppression. *Oncogene* **17**, 3385–400.

Gierl A. & Saedler H. (1992) Plant transposable elements and gene tagging. *Plant Mol. Biol.* **19**, 39–49.

Gilbert D.M. & Cohen S.N. (1987) Bovine papillomavirus plasmids replicate randomly in mouse fibroblasts throughout S phase of the cell cycle. *Cell* **50**, 59–68.

Gill P. & Werrett D.J. (1987) Exclusion of a man charged with murder by DNA fingerprinting. *Forensic Sci. Int.* **35**, 145–8.

Gill P., Ivanov P.L., Kimpton C. *et al.* (1994) Identification of the remains of the Romanov family by DNA analysis. *Nature Genet.* **6**, 130–5.

Gillam S., Astell C.R. & Smith M. (1980) Site-specific mutagenesis using oligodeoxyribonucleotides: isolation of a phenotypically silent φX174 mutant, with a specific nucleotide deletion, at very high efficiency. *Gene* **12**, 129–37.

Girgis S.I., Alevizaki M., Denny P., Ferrier G.J.M. & Legon S. (1988) Generation of DNA probes for peptides with highly degenerate codons using mixed primer PCR. *Nucl. Acids Res.* **16**, 10371.

Giri A. & Narassu M.L. (2000) Transgenic hairy roots: recent trends and applications. *Biotechnol. Adv.* **18**, 1–22.

Glazer A.N. & Mathies R.A. (1997) Energy-transfer fluorescent reagents for DNA analyses. *Curr. Opin. Biotechnol.* **8**, 94–102.

Gluzman Y. (1981) SV40-transformed simian cells support the replication of early SV40 mutants. *Cell* **23**, 175–82.

Goddijn O.J.M., Pennings E.J.M., Van der Helm P., Verpoorte R. & Hoge J.H.C. (1995) Overexpression of a tryptophan decarboxylase cDNA in *Catharanthus roseus* crown gall calli results in increased tryptamine levels but not in increased terpenoid indole alkaloid production. *Transgenic Res.* **4**, 315–23.

Goff S.P. & Berg P. (1976) Construction of hybrid viruses containing SV40 and λ phage DNA segments and their propagation in cultured monkey cells. *Cell* **9**, 695–705.

Goldberg D.A., Posakony J.W. & Maniatis T. (1983) Correct developmental expression of a cloned alcohol dehydrogenase gene transduced into the *Drosophila* germ line. *Cell* **34**, 59–73.

Goldin A.L. (1991) Expression of ion channels in oocytes. In *Methods in Cell Biology*, Vol. 36. *Xenopus laevis: Practical Uses in Cell and Molecular Biology*, eds Kay KB & Peng HB, pp. 487–509. Academic Press, New York.

Golds T.J., Maliga P. & Koop H.U. (1993) Stable plastid transformation in PEG-treated protoplasts of *Nicotiana tabacum*. *Bio/Technology* **11**, 95–7.

Goldsbrough A. & Bevan M. (1991) New patterns of gene activity in plants detected using an *Agrobacterium* vector. *Plant Mol. Biol.* **16**, 263–9.

Goldsbrough A.P., Lastrella C.N. & Yoder J.I. (1993) Transposition mediated repositioning and subsequent elimination of marker genes from transgenic tomato. *Biotechnology* **11**, 1286–92.

Golic K. (1991) Site-specific recombination between homologous chromosomes in *Drosophila*. *Science* **252**, 958–61.

Gonzalez F.J. & 8 others (1988) Characterization of the common genetic defect in humans deficient in debrisoquine metabolism. *Nature* **331**, 442–6.

Gordon J.W. & Ruddle F.H. (1981) Integration and stable germ line transmission of genes injected into mouse pronuclei. *Science* **214**, 1244–6.

Gordon K., Lee E., Vitale J.A. *et al.* (1987) Production of human tissue plasminogen activator in transgenic mouse milk. *Biotechnology* **5**, 1183–7.

Gordon M.P., Farrand S.K., Sciaky D. *et al.* (1979) In *Molecular Biology of Plants, Symposium, University of Minnesota*, ed. Rubenstein I. Academic Press, London.

Gordon-Kamm W.J., Spencer T.M., Maugano M.L. *et al.* (1990) Transformation of maize cells and regeneration of fertile transgenic plants. *Plant Cell* **2**, 603–18.

Gorlach J., Volrath S., Knauf-Beiter G. *et al.* (1996) Benzothiadiazole, a novel class of inducers of systemic acquired resistance, activates gene expression and disease resistance in wheat. *Plant Cell* **8**, 629–43.

Gorman C.M., Merlino G.T., Willingham M.C., Pastan I. & Howard B. (1982a) The Rous sarcoma virus long terminal repeat is a strong promoter when introduced into a variety of eukaryotic cells by DNA-mediated transfection. *Proc. Nat. Acad. Sci. USA* **79**, 6777–81.

Gorman C.M., Moffat L.F. & Howard B.H. (1982b) Recombinant genome which expresses chloramphenicol acetyl transferase in mammalian cells. *Mol. Cell. Biol.* **2**, 1044–51.

Gorman C.M., Gies D., McCray G. & Huang M. (1989) The human cytomegalovirus major immediate early promoter can be *trans*-activated by adenovirus early proteins. *Virology* **171**, 377–85.

Gorman C.M., Gies D.R. & McCray G. (1990) Transient production of proteins using an adenovirus transformed cell line. *DNA Protein Eng. Technol.* **2**, 3–10.

Gormley E.P. & Davies J. (1991) Transfer of plasmid RSF1010 by conjugation from *Escherichia coli* to *Streptomyces lividans* and *Mycobacterium smegmatis*. *J. Bacteriol.* **173**, 6705–8.

Gorziglia M.I., Kadan M.J., Yei S. *et al.* (1996) Elimination of both E1 and E2a from adenovirus vectors further improves prospects for *in vivo* human gene therapy. *J. Virol.* **70**, 4173–8.

Gossen M. & Bujard H. (1992) Tight control of gene expression in mammalian cells by tetracycline-responsive promoters. *Proc. Nat. Acad. Sci. USA* **89**, 5547–51.

Gossen M., Freundlieb S., Bender G. *et al.* (1995) Transcriptional activation by tetracyclines in mammalian cells. *Science* **268**, 1766–9.

Gossler A., Joyner A.L., Rossant J. & Skarnes W.C. (1989) Mouse embryonic stem cells and reporter constructs to detect developmentally regulated genes. *Science* **244**, 463–5.

Gottesmann M.E. & Yarmolinsky M.D. (1968) The integration and excision of the bacteriophage lambda genome. *Cold Spring Harbor Symp. Quant. Biol.* **33**, 735–47.

Gourse R.L., Gaal T., Aiyar S.E. *et al.* (1998) Strength and regulation without transcription factors: lessons from bacterial rRNA promoters. *Cold Spring Harbor Symp. Quant. Biol.* **63**, 131–9.

Gowland P.C. & Hardmann D.J. (1986) Methods for isolating large bacterial plasmids. *Microbiol. Sci.* **3**, 252–4.

Graham F.L., Smiley J., Russell W.C. & Nairn R. (1977) Characteristics of a human cell line transformed by DNA from human adenovirus type 5. *J. Gen. Virol.* **36**, 59–74.

Graham F.L. & Van der Erb A.J. (1973) A new technique for the assay of infectivity of human adenovirus 5 DNA. *Virology* **52**, 546–50.

Graham M.W., Craig S. & Waterhouse P.M. (1997) Expression patterns of vascular-specific promoters RolC and Sh in transgenic potatoes and their use in engineering PLRV-resistant plants. *Plant Mol. Biol.* **33**, 729–35.

Grant S.R. (1999) Dissecting the mechanisms of post-transcriptional gene silencing: divide and conquer. *Cell* **96**, 303–6.

Green P., Pines O. & Inouye M. (1986) The role of antisense RNA in gene regulation. *Ann. Rev. Biochem.* **55**, 569–97.

Gregory R.J., Cheng S.H., Rich D.P. *et al.* (1990) Expression and characterization of the cystic fibrosis transmembrane conductance regulator. *Nature* **347**, 382–6.

Gressel J. (1999) Tandem constructs: preventing the rise of superweeds. *Trends Biotechnol.* **17**, 361–6.

Gressel J. (2000) Molecular biology of weed control. *Transgenic Res.* **9**, 355–82.

Gribskov M. & Burgess R.R. (1983) Overexpression and purification of the sigma subunit of *Escherichia coli* RNA polymerase. *Gene* **26**, 109–18.

Grimsley N., Hohn T., Davies J.W. & Hohn B. (1987) *Agrobacterium*-mediated delivery of infectious maize streak virus into maize plants. *Nature* **325**, 177–9.

Grondahl B., Puppe W., Hoppe A. *et al.* (1999) Rapid identification of nine microorganisms causing acute respiratory tract infection by single-tube multiplex reverse transcription-PCR: feasibility study. *J. Clin. Microbiol.* **37**, 1–7.

Gronenborn B. & Messing J. (1978) Methylation of single-stranded DNA *in vitro* introduces new restriction endonuclease cleavage sites. *Nature* **272**, 375–7.

Gronenborn B., Gardner R.C., Schaefer S. & Shepherd R.J. (1981) Propagation of foreign DNA in plants using cauliflower mosaic-virus as vector. *Nature* **294**, 773–776.

Grosveld F. (1999) Activation by locus control regions? *Curr. Opin. Genet. Devel.* **9**, 152–7.

Grosveld F., Van Assendelft G.B., Greaves D.R. & Kollias G. (1987) Position independent, high level expression of the human β-globin gene in transgenic mice. *Cell* **51**, 975–85.

Grunstein M. & Hogness D.S. (1975) Colony hybridization: a method for the isolation of cloned DNAs that contain a specific gene. *Proc. Nat. Acad. Sci. USA* **72**, 3961–5.

Gruss A. & Ehrlich S.D. (1989) The family of highly inter-related single-related deoxyribonucleic acid plasmids. *Microbiol. Rev.* **53**, 231–41.

Gryczan T.J., Contente S. & Dubnau D. (1980) Molecular cloning of heterologous chromosomal DNA by recombination between a plasmid vector and a homologous resident plasmid in *Bacillus subtilis*. *Mol. Gen. Genet.* **177**, 459–67.

Gu H., Marth J.D., Orban P.C., Mossmann H. & Rajewsky K. (1994) Deletion of a DNA polymerase beta gene segment in T-cells using cell type-specific gene targeting. *Science* **265**, 103–6.

Gu S.Y., Huang T.M., Ruan L. *et al.* (1996) First EBV vaccine trial in humans using recombinant vaccinia virus expressing the major membrane antigen. *Devel. Biol. Stand.* **84**, 171–7.

Guarente L. (1987) Regulatory proteins in yeast. *Ann. Rev. Genet.* **21**, 425–52.

Gubler U. & Hoffman B.J. (1983) A simple and very efficient method for generating cDNA libraries. *Gene* **26**, 263–9.

Guerineau F. & Mullineaux P. (1989) Nucleotide sequence of the sulfonamide resistance plasmid R46. *Nucl. Acids Res.* **14**, 4370.

Guhathakurta A., Viney I. & Summers D. (1996) Accessory proteins impose site selectivity during ColE1 dimer resolution. *Mol. Microbiol.* **20**, 613–20.

Gumport R.I. & Lehman I.R. (1971) Structure of the DNA ligase adenylate intermediate: lysine (ε-amino) linked AMP. *Proc. Nat. Acad. Sci. USA* **68**, 2559–63.

Gurdon J.B. (1986) Nuclear transplantation in eggs and oocytes. *J. Cell Sci. Suppl.* **4**, 287–318.

Gurdon J.B. (1991) Nuclear transplantation in *Xenopus*. *Methods Cell Biol.* **36**, 299–309.

Gurdon J.B., Lane C.D., Woodland H.R. & Marbaix G. (1971) Use of frog eggs and oocytes for the study of messenger RNA and its translation in living cells. *Nature* **233**, 177–82.

Gutekunst C.A., Norflus F. & Hersch S.M. (2000) Recent advances in Huntington's disease. *Curr. Opin. Neurol.* **13**, 445–50.

Guzman L.M., Belin D., Carson M.J. & Beckwith J. (1995) Tight regulation, modulation, and high-level expression by vectors containing the arabinose PBAD promoter. *J. Bacteriol.* **177**, 4121–30.

Hacia J.G. (1999) Resequencing and mutational analysis using oligonucleotide microarrays. *Nature Genet. Suppl.* **21**, 42–7.

Hackett J. (1993) Use of *Salmonella* for heterologous gene expression and vaccine delivery systems. *Curr. Opin. Biotechnol.* **4**, 611–15.

Haenlin M., Steller H., Pirotta V. & Momer E. (1985) A 43kb cosmid P transposon rescues the fs(1)K10 morphogenetic locus and three adjacent *Drosophila* developmental mutants. *Cell* **40**, 827–37.

Hager L.J. & Palmiter R.D. (1981) Transcriptional regulation of mouse liver metallothionein I gene by glucocorticoids. *Nature* **291**, 340–2.

Hagio T., Hirabayashi T., Machii H. & Tomotsune H. (1995) Production of fertile transgenic barley (*Hordeum vulgare* L.) plants using the hygromycin resistance marker. *Plant Cell Rep.* **14**, 329–34.

Hahn S., Hoar E.T. & Guarente L. (1985) Each of three 'TATA elements' specifies a subset of the transcription initiation sites at the CYC-1 promoter of *Saccharomyces cerevisiae*. *Proc. Nat. Acad. Sci. USA* **82**(24), 8562–6.

Hahn S.A., Schutte M., Hoque A.T.M.S. *et al.* (1996) *DPC4*, a candidate tumor suppressor gene at human chromosome 18q21.1. *Science* **271**, 350–3.

Haima P., Bron S. & Venema G. (1987) The effect of restriction on shotgun cloning and plasmid stability in *Bacillus subtilis* Marburg. *Mol. Gen Genet.* **209**, 335–42.

Haima P., Bron S. & Venema G. (1988) A quantitative analysis of shotgun cloning in *Bacillus subtilis* protoplasts. *Mol. Gen. Genet.* **213**, 364–9.

Haima P., Bron S. & Venema G. (1990) Novel plasmid marker rescue transformation system for molecular cloning in *Bacillus subtilis* enabling direct selection of recombinants. *Mol. Gen. Genet.* **223**, 185–91.

Hain R., Reif H.J., Krause E. *et al.* (1993) Disease resistance results from foreign phytoalexin expression in a novel plant. *Nature* **361**, 153–6.

Hall C.V., Jacob P.E., Ringold G.M. & Lee F. (1983) Expression and regulation of *Escherichia coli lacZ* gene fusions in mammalian cells. *J. Mol. Appl. Genet.* **2**, 101–9.

Hall R.D., RiksenBruinsma T., Weyens G.J. *et al.* (1996) A high efficiency technique for the generation of transgenic sugar beets from stomatal guard cells. *Nature Biotechnol.* **14**, 1133–8.

Hamil J.D., Parr A.J., Rhodes M.J.C., Robins R.J. & Walton N.J. (1987) New routes to plant secondary products. *Biotechnology* **5**, 800–4.

Hamilton A.J. & Baulcombe D.C. (1999) A species of small antisense RNA in posttranscriptional gene silencing in plants. *Science* **286**, 950–2.

Hamilton A.J., Lycett G.W. & Grierson D. (1990) Antisense gene that inhibits synthesis of the hormone ethylene in transgenic plants. *Nature* **346**, 284–7.

Hamilton C.M. (1997) A binary-BAC system for plant transformation with high molecular weight DNA. *Gene* **200**, 107–16.

Hamilton C.M., Frary A., Lewis C. & Tanksley S.D. (1996) Stable transfer of high molecular weight DNA into plant chromosomes. *Proc. Nat. Acad. Sci. USA* **93**, 9975–9.

Hamilton C.M., Frary A., Xu Y.M., Tanksley S.D. & Zhang H.B. (1999) Construction of tomato genomic DNA libraries in a binary-BAC (BIBAC) vector. *Plant J.* **18**, 223–9.

Hammer R.E., Krumlauf R., Camper S.A., Brinster R.L. & Tilghman S.M. (1987) Diversity of alpha-fetoprotein gene expression in mice is generated by a combination of separate enhancer elements. *Science* **235**, 53–8.

Hammerschmidt W. & Sugden B. (1988) Identification and characterization of ori$_{lyt}$, a lytic origin of DNA replication of Epstein–Barr virus. *Cell* **55**, 427–33.

Hammond S.M., Bernstein E., Beach D. & Hannon G.J. (2000) An RNA-directed nuclease mediates posttranscriptional gene silencing in *Drosophila* cells. *Nature* **404**, 293–6.

Hammond S.M., Caudy A.A. & Hannon G.J. (2001) Post-transcriptional gene silencing by double-stranded RNA. *Nature Rev. Genet.* **2**, 110–19.

Hammond-Kosack K.E., Staskowicz B.J., Jones J.D.G. & Baulcombe D.C. (1995) Functional expression of a fungal avirulence gene from a modified potato virus X genome. *Mol. Plant–Microbe Interactions* **8**, 181–5.

Hanahan D. (1983) Studies on transformation of *Escherichia coli* with plasmids. *J. Mol. Biol.* **166**, 557–80.

Hanahan D. & Meselson M. (1980) Plasmid screening at high colony density. *Gene* **10**, 63–7.

Hanahan D., Jessee J. & Bloom F.R. (1991) Plasmid transformation of *Escherichia coli* and other bacteria. *Methods Enzymol.* **204**, 63–113.

Handfield M. & Levesque R.C. (1999) Strategies for isolation of *in vivo* expressed genes from bacteria. *FEMS Microbiol. Rev.* **23**, 69–91.

Handler A.M., Gomez S.P. & Obrochta D.A. (1993) A functional analysis of the P element gene transfer vector in insects. *Arch. Insect Biochem. Physiol.* **22**, 373–84.

Hanks M., Wurst W., Anson-Cartwright L., Auerbach A.B. & Joyner A.L. (1995) Rescue of the en-1 mutant phenotype by replacement of en-1 with en-2. *Science* **269**, 679–82.

Harbers K., Jahner D. & Jaenisch R. (1981) Microinjection of cloned retroviral genomes into mouse zygotes: integration and expression in the animal. *Nature* **293**, 540–3.

Harper J.W., Adami G.R., Wei N., Keyomarsi K. & Ellege S.J. (1993) The p21 Cdk-interacting protein Cip1 is a potent inhibitor of G1 cyclin-dependent kinases. *Cell* **75**, 805–16.

Hartl D.L. (1996) EST! EST!! EST!!! *Bioessays* **18**, 1021–3.

Hartl D.L., Nurminsky D.I., Jones R.W. & Lozovskaya E.R. (1994) Genome structure and evolution in *Drosophila*– applications of the framework P1 map. *Proc. Nat. Acad. Sci. USA* **91**, 6824–9.

Hartman S.C. & Mulligan R.C. (1988) Two dominant-acting selectable markers for gene transfer studies in mammalian cells. *Proc. Nat. Acad. Sci. USA* **85**, 8047–51.

Harris J.D., Gutierrez A.A., Hurst H.C., Sikora K. & Lemoine, N.R. (1994) Gene therapy for cancer using tumour-specific prodrug activation. *Gene Ther.* **1**, 170–5.

Haseloff J. & Amos B. (1995) GFP in plants. *Trends Genet.* **11**, 328–9.

Haseloff J., Siemering K.R., Prasher D.C. & Hodge S. (1997) Removal of a cryptic intron and subcellular localization of green fluorescent protein are required to mark transgenic *Arabidopsis* plants brightly. *Proc. Nat. Acad. Sci. USA* **94**, 2122–7.

Haseloff J., Dormand E.L. & Brand A.H. (1999) Live imaging with green fluorescent protein. *Methods Mol. Biol.* **122**, 241–59.

Hashemzadeh-Bonehi L., Mehraein-Ghomi F., Mitsopoulos C., Hennessey E.S. & Broome-Smith J.K. (1998) Importance of using lac rather than ara promoter vectors for modulating the levels of toxic gene products in *Escherichia coli*. *Mol. Microbiol.* **30**, 676–8.

Hashimoto T. & Yamada Y. (1994) Alkaloid biogenesis: molecular aspects. *Ann. Rev. Plant Physiol. Plant Mol. Biol.* **45**, 257–85.

Hashimoto T., Yun D.-J. & Yamada Y. (1993) Production of tropane alkaloids in genetically engineered root cultures. *Phytochemistry* **32**, 713–18.

Hasty P., Ramirez-Solis R., Krumlauf R. & Bradley A. (1991a) Introduction of subtle mutation into the *Hox-2.6* locus in embryonic stem cells. *Nature* **350**, 243–6.

Hasty P., Rivera-Perez J. & Bradley A. (1991b) The length of homology required for gene targeting in embryonic stem cells. *Mol. Cell. Biol.* **11**, 5586–91.

Haughn G.W., Smith J., Mazur B. & Somerville C. (1988) Transformation with mutant *Arabidopsis* acetolactate synthase gene renders tobacco resistant to sulfonylurea herbicides. *Mol. Gen. Genet.* **211**, 266–71.

Hawley D.K. & McClure W.R. (1983) Compilation and analysis of *Escherichia coli* promoter DNA sequences. *Nucl. Acids Res.* **11**, 2237–55.

Hay J.G., McElvaney N.G., Herena J. & Crystal R.G. (1995) Modification of nasal epithelial potential differences of individuals with cystic fibrosis consequent to local administration of a normal CFTR cDNA adenovirus gene transfer vector. *Hum. Gene Ther.* **6**, 1487–96.

Hayes R.J., Petty I.T.D., Coutts R.H.A. & Buck K.W. (1988) Gene amplification and expression in plants by a replicating geminivirus vector. *Nature* **334**, 179–82.

Hayes R.J., Coutts R.H.A. & Buck K.W. (1989) Stability and expression of bacterial genes in replicating gemini-virus vectors in plants. *Nucl. Acids Res.* **17**, 2391–403.

Hayes W. (1968) *The Genetics of Bacteria and their Viruses*, 2nd edn. Blackwell Scientific Publications, Oxford.

Hayford M.B., Medford J.I., Hoffmann N.L., Rogers S.G. & Klee H.J. (1988) Development of a plant transformation selection system based on expression of genes encoding gentamicin acetyltransferases. *Plant Physiol.* **86**, 1216–22.

Hayman G.T. & Bolen P.L. (1993) Movement of shuttle plasmids from *Escherichia coli* into yeasts other than *Saccharomyces cerevisiae* using trans-kingdom conjugation. *Plasmid* **30**, 251–7.

Haynes J.R., McCabe D.E., Swain W.F., Widera G. & Fuller J.T. (1996) Particle-mediated nucleic acid immunization. *J. Biotechnol.* **44**, 37–42.

Hedgepeth J., Goodman H.M. & Boyer H.W. (1972) DNA nucleotide sequence restricted by the RI endonuclease. *Proc. Nat. Acad. Sci. USA* **69**, 3448–52.

Heim R. & Tsein R. (1996) Engineering green fluorescent protein for improved brightness, longer wavelengths and fluorescence energy resonance transfer. *Curr. Biol.* **6**, 178–82.

Heinemann J.A. & Sprague G.F. (1989) Bacterial conjugative plasmids mobilize DNA transfer between bacteria and yeast. *Nature* **340**, 205–9.

Heinkoff S. (1990) Position effect variegation after 60 years. *Trends Genet.* **6**, 422–6.

Heinrich J.C., Tabler M. & Louis C. (1983) Attenuation of white gene expression in transgenic *Drosophila melanogaster*: possible role of catalytic antisense RNA. *Devel. Genet.* **14**, 258–65.

Helfman D.M. & Hughes S.H. (1987) Use of antibodies to screen cDNA expression libraries prepared in plasmid vectors. *Methods Enzymol.* **152**, 451–7.

Helfman D.M., Feramisco J.R., Fiddes J.C., Thomas G.P. & Hughes S.H. (1983) Identification of clones that encode chicken tropomyosin by direct immunological screening of a cDNA expression library. *Proc. Nat. Acad. Sci. USA* **80**, 31–5.

Hellens R., Mullineaux P. & Klee H. (2000a) A guide to *Agrobacterium* binary Ti vectors. *Trends Plant Sci.* **5**, 446–51.

Hellens R.P., Edwards E.A., Leyland N.R., Bean S. & Mullineaux P.M. (2000b) pGreen: a versatile and flexible binary Ti vector for *Agrobacterium*-mediated plant transformation. *Plant Mol. Biol.* **42**, 819–32.

Heller R., Song K., Villaret D. *et al.* (1990) Amplified expression of tumor necrosis factor receptor in cells transfected with Epstein–Barr virus shuttle vector cDNA libraries. *J. Biol. Chem.* **264**, 5708–17.

Heller R.A., Schena M., Chai A. *et al.* (1997) Discovery and analysis of inflamatory disease-related genes using cDNA microarray. *Proc. Nat. Acad. Sci. USA* **94**, 2150–5.

Helmann J.D. (1995) Compilation and analysis of *Bacillus subtilis* sigma A-dependent promoter sequences: evidence for extended contact between RNA polymerase and upstream promoter DNA. *Nucl. Acids Res.* **23**, 2351–60.

Helmer G., Casadaban M., Bevan M., Kayes L. & Chilton M.-D. (1984) A new chimeric gene as a marker for plant transformation: the expression of *Escherichia coli* β-galactosidase in sunflower and tobacco cells. *Biotechnology* **2**, 520–7.

Helwig U., Imai K., Schmahl W. *et al.* (1995) Interaction between undulated and patch leads to an extreme form of spina-bifida in double-mutant mice. *Nature Genet.* **11**, 60–3.

Hemmati-Brivanlou A. & Melton D.A. (1992) A truncated activin receptor inhibits mesoderm induction and formation of axial structures in *Xenopus* embryos. *Nature* **359**, 609–14.

Hendy S., Chen Z.C., Barker H. *et al.* (1999) Rapid production of single-chain Fv fragments in plants using a potato virus X episomal vector. *J. Immunol. Method.* **231**, 137–46.

Heneine W., Tibell A. & Switzer W.M. (1998) No evidence of infection with porcine endogenous retrovirus in recipients of porcine islet-cell xenografts. *Lancet* **352**, 695–9.

Hensel M., Shea J.E., Gleeson C. *et al.* (1995) Simultaneous identification of bacterial virulence genes by negative selection. *Science* **269**, 400–3.

Heredia A., Soriano V., Weiss S.H. *et al.* (1996) Development of a multiplex PCR assay for the simultaneous detection and discrimination of HIV-1, HIV-2, HTLV-I and HTLV-II. *Clin. Diagn. Virol.* **7**, 85–92.

Hermonat O.L. & Muzyczka N. (1984) Use of adeno-associated virus as a mammalian DNA cloning vector: transduction of neomycin resistance into mammalian tissue culture cells. *Proc. Nat. Acad. Sci. USA* **81**, 6466–70.

Hernalsteens J.P., Thiatoong L., Schell J. & Vanmontagu M. (1984) An agrobacterium-transformed cell-culture from the monocot *Asparagus officinalis*. *EMBO J.* **3**, 3039–41.

Herrera-Estrella L., Depicker A., Van Montagu M. & Schell J. (1983a) Expression of chimaeric genes transferred into plant cells using a Ti-plasmid-derived vector. *Nature* **303**, 209–13.

Herrera-Estrella L., DeBlock M., Messens E. *et al.* (1983b) Chimeric genes as dominant selectable markers in plant cells. *EMBO J.* **2**, 987–95.

Herskowitz I. (1974) Control of gene expression in bacteriophage lambda. *Ann. Rev. Genet.* **7**, 289–324.

Herweijer H., Latendresse J.S., Williams P. *et al.* (1995) A plasmid-based self-amplifying Sindbis virus vector. *Hum. Gene Ther.* **6**, 1161–7.

Heyman J.A. & 17 others (1999) Genome-scale cloning and expression of individual open reading frames using topoisomerase I-mediated ligation. *Genome Res.* **9**, 383–92.

Hiatt A.H., Cafferkey R. & Bowdish K. (1989) Production of antibodies in transgenic plants. *Nature* **342**, 76–8.

Hicks J.B., Strathern J.N., Klar A.J.S. & Dellaporta S.L. (1982) Cloning by complementation in yeast: the mating type genes. In *Genetic Engineering*, eds Setlow J.K. & Hollaender A., pp. 219–48. Plenum Press, New York.

Hiei Y., Ohta S., Komari T. & Kumashiro T. (1994) Efficient transformation of rice (*Oryza sativa* L) mediated by *Agrobacterium* and sequence analysis of the boundaries of the T-DNA. *Plant J.* **6**, 241–82.

Higo K., Saito Y. & Higo H. (1993) Expression of a chemically synthesized gene for human epidermal growth factor under the control of cauliflower mosaic virus 35S promoter in transgenic tobacco. *Biosci. Biotechnol. Biochem.* **57**, 1477–81.

Higo K.E., Otaka E. & Osawa S. (1982) Purification and characterization of 30S ribosomal proteins from *Bacillus subtilis*: correlation to *Escherichia coli* 30S proteins. *Mol. Gen. Genet.* **185**, 239–44.

Higuchi R., Krummel B. & Saiki R.K. (1988) A general method of *in vitro* preparation and specific mutagenesis of DNA fragments: study of protein and DNA interactions. *Nucl. Acids Res.* **16**, 7351–67.

Higuchi R., Dollinger G., Walsh P.S. & Griffith R. (1992) Simultaneous amplification and detection of specific DNA sequences. *Biotechnology* **10**, 413–17.

Higuchi R., Fockler C., Dollinger G. & Watson R. (1993) Kinetic PCR analysis: real-time monitoring of DNA amplification reactions. *Biotechnology* **11**, 1026–30.

Hille J., Verhweggen F., Roelvink P. *et al.* (1986) Bleomycin resistance: a new dominant selectable marker for plant cell transformation. *Plant Mol. Biol.* **7**, 171–6.

Hinkula J., Lundholm P. & Wahren B. (1997) Nucleic acid vaccination with HIV regulatory genes: a combination of HIV-1 genes in separate plasmids induces strong immune responses. *Vaccine* **15**, 874–8.

Hinnebusch J. & Tilly K. (1993) Linear plasmids and chromosomes in bacteria. *Mol. Microbiol.* **10**, 917–22.

Hinnen A., Hicks J.B. & Fink G.R. (1978) Transformation of yeast. *Proc. Nat. Acad. Sci. USA* **75**, 1929–33.

Hirayama T., Ishida C., Kuromori T. *et al.* (1997) Functional cloning of a cDNA encoding Mei2-like protein from *Arabidopsis thaliana* using a fission yeast pheromone receptor deficient mutant. *FEBS Lett.* **413**, 16–20.

Hirsch V.M., Fuerst T.R., Sutter G. *et al.* (1996) Patterns of viral replication correlate with outcome in simian immunodeficiency virus (SIV)-infected macaques: effect of prior immunization with a trivalent SIV vaccine in modified vaccinia virus Ankara. *J. Virol.* **70**, 3741–52.

Hoekma A., Hirsch P.R., Hooykass P.J.J. & Schiperoort R.A. (1983) A binary plant vector strategy based on separation of *vir*- and T-regions of the *Agrobacterium tumefaciens* Ti-plasmid. *Nature* **303**, 179–83.

Hoekstra W.P.M., Bergmans J.E.N. & Zuidweg E.M. (1980) Role of *rec*BC nuclease in *Escherichia coli* transformation. *J. Bacteriol.* **143**, 1031–2.

Hoffman C., Sandig V., Jennigs G. *et al.* (1995) Efficient gene transfer into human hepatocytes by baculovirus vectors. *Proc. Nat. Acad. Sci. USA* **92**, 10099–103.

Hoffman F. (1996) Laser microbeams for the manipulation of plant cells and subcellular structures. *Plant Sci.* **113**, 1–11.

Hofmann A., Nolan G.P. & Blau H.M. (1996) Rapid retroviral delivery of tetracycline-inducible genes in a single autoregulatory cassette. *Proc. Nat. Acad. Sci. USA* **93**, 5185–90.

Hogan B. & Lyons K. (1988) Gene targeting: getting nearer the mark. *Nature* **336**, 304–5.

Hoheisel J.D., Maier E., Mott E. *et al.* (1993) High resolution cosmid and P1 maps spanning the 14 Mb genome of the fission yeast *S. pombe*. *Cell* **73**, 109–20.

Hohn B. (1975) DNA as substrate for packaging into bacteriophage lambda, *in vitro*. *J. Mol. Biol.* **98**, 93–106.

Hohn B. & Murray K. (1977) Packaging recombinant DNA molecules into bacteriophage particles *in vitro*. *Proc. Nat. Acad. Sci. USA* **74**, 3259–63.

Hollister J.R., Shaper J.H. & Jarvis D.L. (1998) Stable expression of mammalian β1,4-galactosyltransferase extends the N-glycosylation pathway in insect cells. *Glycobiology* **8**, 473–80.

Holm P.B., Olsen O., Scnorf M., Brinch-Pedersen H. & Knudsen S. (2000) Transformation of barley by microinjection into isolated zygote protoplasts. *Transgenic Res.* **9**, 21–32.

Holmes D.L. & Stellwagen N.C. (1990) The electric field dependence of DNA mobilities in agarose gels: a reinvestigation. *Electrophoresis* **11**, 5–15.

Hols P., Baulard, A., Garmyn D. *et al.* (1992) Isolation and characterization of genetic expression and secretion signals from *Enterococcus faecalis* through the use of broad-host-range alpha-amylase probe vectors. *Gene* **118**, 21–30.

Hong Y.G., Saunders K., Hartley M.R. & Stanley J. (1996) Resistance to geminivirus infection by virus-induced expression of dianthin in transgenic plants. *Virology* **220**, 119–27.

Hooykaas P.J.J. & Beijersbergen A.G.M. (1994) The virulence system of *Agrobacterium tumefaciens*. *Ann. Rev. Phytopathol.* **32**, 157–79.

Hooykaas P.J.J. & Schilperoort R.A. (1992) *Agrobacterium* and plant genetic engineering. *Plant Mol. Biol.* **19**, 15–38.

Hopwood D.A. (1999) Forty years of genetics with *Streptomyces*: from *in vivo* through *in vitro* to *in silico*. *Microbiology* **145**, 2183–202.

Hopwood D.A., Malpartida F., Kieser H.M. *et al.* (1985) Production of 'hybrid' antibiotics by genetic engineering. *Nature* **314**, 642–4.

Horland P., Flick J., Johnston M. & Sclafani R.A. (1989) Galactose as a gratuitous inducer of *GAL* gene expression in yeasts growing on glucose. *Gene* **83**, 57–64.

Horsch R.B., Fraley R.T., Rogers S.G. *et al.* (1984) Inheritance of functional genes in plants. *Science* **223**, 496–8.

Horsch R.B., Fry J.E., Hoffmann N.L. *et al.* (1985) A simple and general method for transferring genes into plants. *Science* **227**, 1229–31.

Horvath H., Huang J.T., Wong O. *et al.* (2000) The production of recombinant proteins in transgenic barley grains. *Proc. Nat. Acad. Sci. USA* **97**, 1914–19.

Houdebine L.M. (1997) *Transgenic Animals: Generation and Use.* Harwood Academic Publishers, Switzerland.

Houdebine L.M. (2000) Transgenic animal bioreactors. *Transgenic Res.* **9**, 305–20.

Hsiao C.L. & Carbon J. (1981) Characterization of a yeast replication origin (ars2) and construction of stable minichromosomes containing cloned yeast centromere DNA (CEN 3). *Gene* **15**, 157–66.

Hsiao K.K., Scott M., Foster D. *et al.* (1990) Spontaneous neurodegeneration in transgenic mice with mutant prion protein. *Science* **250**, 1587–90.

Hu M.C.-T. & Davidson N. (1987) The inducible lac operator–repressor system is functional in mammalian cells. *Cell* **48**, 555–6.

Hu S.-L., Kosowski S.P. & Dalrymple J.M. (1986) Expression of AIDS virus envelope gene by a recombinant vaccinia virus. *Nature* **320**, 537–40.

Hu W. & Chen C. (1995) Expression of *Aequorea* green fluorescent protein in plant cells. *FEBS Lett.* **369**, 331–4.

Huang H.C. & Brown D.D. (2000) Overexpression of *Xenopus laevis* growth hormone stimulates growth of tadpoles and frogs. *Proc. Nat. Acad. Sci. USA* **97**, 190–4.

Huang J. & Schreiber S.L. (1997) A yeast genetic system for selecting small molecule inhibitors of protein–protein interactions in nanodroplets. *Proc. Nat. Acad. Sci. USA* **94**, 13396–401.

Huang R. & Reusch R.N. (1995) Genetic competence in *Escherichia coli* requires poly-beta-hydroxybutyrate/calcium polyphosphate membrane complexes and certain divalent cations. *J. Bacteriol.* **177**, 486–90.

Huang R.Y. & Kowalski D. (1993) A DNA unwinding element and an ARS consensus comprise a replication origin within a yeast chromosome. *EMBO J.* **12**, 4521–31.

Hubank M. & Schatz D.G. (1994) Identifying differences in mRNA expression by representational difference analysis of cDNA. *Nucl. Acids Res.* **22**, 5640–8.

Huber M.C., Kruger G. & Bonifer C. (1996) Genomic position effects lead to an inefficient reorganization of nucleosomes in the 5′-regulatory region of the chicken lysozyme locus in transgenic mice. *Nucl. Acids Res.* **24**, 1443–53.

Hui A., Hayflick J., Dinkelspiel K. & De Boer H.A. (1984) Mutagenesis of the three bases preceding the start codon of the β-galactosidase mRNA and its effect on translation in *Escherichia coli*. *EMBO J.* **3**, 623–9.

Humphries P., Old R., Coggins L.W. *et al.* (1978) Recombinant plasmids containing *Xenopus laevis* structural genes derived from complementary DNA. *Nucl. Acids Res.* **5**, 905–24.

Huygen K., Content J., Montgomery D.L. *et al.* (1996) Immunogenicity and protective efficacy of a tuberculosis DNA vaccine. *Nature Med.* **8**, 893–8.

Hwu P., Yannelli J., Kriegler M. *et al.* (1993) Functional and molecular characterization of tumor-infiltrating lymphocytes transduced with tumor necrosis factor-alpha cDNA for the gene therapy of cancer in humans. *J. Immunol.* **150**, 4101–15.

Ihssen P.E., McKay L.R., McMillan I. & Phillips R.B. (1990) Ploidy manipulation and gynogenesis in fishes: cytogenetic and fisheries applications. *Trans. Am. Fish Soc.* **119**, 698–717.

Iida A., Morikawa H. & Yamada Y. (1990) Stable transformation of cultured tobacco cells by DNA-coated gold particles accelerated by gas pressure driven particle gun. *Appl. Microbiol. Biotechnol.* **33**, 560–3.

Ikawa M., Yamada S., Nakanishi T. & Okabe M. (1999) Green fluorescent protein as a vital marker in mammals. *Curr. Top. Devel. Biol.* **44**, 1–20.

Ikemura T. (1981a) Correlation between the abundance of *Escherichia coli* transfer RNAs and the occurrence of the respective codons in its protein genes. *J. Mol. Biol.* **146**, 1–21.

Ikemura T. (1981b) Correlation between the abundance of *Escherichia coli* transfer RNAs and the occurrence of the respective codons in its protein genes: a proposal for a synonymous codon choice that is optimal for the *E. coli* translational system. *J. Mol. Biol.* **151**, 389–409.

Imler J.L. *et al.* (1996) Novel complementation cell lines derived from human lung carcinoma A549 cells support the growth of E1-deleted adenovirus vectors. *Gene Ther.* **3**, 75–84.

Ingmer H. & Cohen S.N. (1993) Excess intracellular concentration of pSC101 RepA protein interferes with both plasmid DNA replication and partitioning. *J. Bacteriol.* **175**, 7834–41.

Innis M.A., Gelfand D.H., Sninsky J.J. & White T.J. (eds) (1990) *PCR Protocols: A Guide to Methods and Applications.* Academic Press, New York.

Inoue H., Nojima H. & Okayama H. (1990) High efficiency transformation of *Escherichia coli* with plasmids. *Gene* **96**, 23–8.

International Human Genome Sequencing Consortium (2001) Initial sequencing and analysis of the human genome. *Nature* **409**, 860–921.

Ioannou P.A., Amemiya C.T., Garnes J. *et al.* (1994) A new bacteriophage P1-derived vector for the propagation of large human DNA fragments. *Nature Genet.* **6**, 84–9.

Ish-Horowicz D. & Burke J.F. (1981) Rapid and efficient cosmid cloning. *Nucl. Acids Res.* **9**, 2989–98.

Ishida Y., Saito H., Ohta S. *et al.* (1996) High efficiency transformation of maize (*Zea mays* L) mediated by *Agrobacterium tumefaciens*. *Nature Biotechnol.* **14**, 745–50.

Iyengar A., Muller F. & Maclean N. (1996) Regulation and expression of transgenes in fish – a review. *Transgenic Res.* **5**, 147–66.

Izumi M., Miyazawa H., Kamakura T. *et al.* (1991) Blasticidin S-resistance gene (*bsr*): a novel selectable marker for mammalian cells. *Exp. Cell Res.* **197**, 229–33.

Jackson D.A., Symons R.H. & Berg P. (1972) Biochemical method for inserting new genetic information into DNA of Simian virus 40: circular SV40 DNA molecules containing lambda phage genes and the galactose operon of *Escherichia coli*. *Proc. Nat. Acad. Sci. USA* **69**, 2904–9.

Jacobs E., Dewerchin M. & Boeke J.D. (1988) Retrovirus-like vectors for *Saccharomyces cerevisiae*: integration of foreign genes controlled by efficient promoters into yeast chromosomal DNA. *Gene* **67**, 259–69.

Jaenisch R. (1988) Transgenic animals. *Science* **240**, 1468–74.

Jaenisch R. & Mintz B. (1974) Simian virus 40 DNA sequences in DNA of healthy adult mice derived from preimplantation blastocysts injected with viral DNA. *Proc. Nat. Acad. Sci. USA* **71**, 1250–4.

Jahner D. & Jaenisch R. (1985) Retrovirus-induced *de novo* methylation of flanking host sequences correlates with gene inactivity. *Nature* **315**, 594–7.

Jahner D., Stuhlmann H., Stewart C.L. *et al.* (1982) *De novo* methylation and expression of retroviral genomes during mouse embryogenesis. *Nature* **298**, 623–8.

Jakobovits A., Moore A.L., Green L.L. *et al.* (1993) Germ line transmission and expression of a human-derived yeast artificial chromosome. *Nature* **362**, 255–8.

James C. (2000) *Global Status of Commercial Transgenic Crops: 2000*. ISAAA Briefs No. 21, ISAAA, Ithaca, New York.

James H.A. & Gibson I. (1998) The therapeutic potential of ribozymes. *Blood* **91**, 371–82.

James P., Halladay J. & Craig E.A. (1996) Genomic libraries and a host strain designed for highly efficient two-hybrid selection in yeast. *Genetics* **144**, 1425–36.

Jannière L., Bruand C. & Ehrlich S.D. (1990) Structurally stable *Bacillus subtilis* cloning vectors. *Gene* **81**, 53–61.

Jefferson R.A., Burgess S.M. & Hirsh D. (1986) β-Glucuronidase from *Escherichia coli* as a gene-fusion marker. *Proc. Nat. Acad. Sci. USA* **83**, 8447–51.

Jefferson R.A., Kavanagh T.A. & Bevan M.W. (1987a) GUS fusions: β-glucuronidase as a sensitive and versatile gene fusion marker in higher plants. *EMBO J.* **6**, 3901–7.

Jefferson R.A., Klass M., Wolf N. & Hirsh D. (1987b) Expression of chimeric genes in *Caenorhabditis elegans*. *J. Mol. Biol.* **193**, 41–6.

Jeffreys A.J. & Morton D.B. (1987) DNA fingerprints of cats and dogs. *Animal Genet.* **18**, 1–15.

Jeffreys A.J., Wilson V. & Thein S.L. (1985a) Individual-specific 'fingerprints' of human DNA. *Nature* **316**, 76–9.

Jeffreys A.J., Brookfield J.F.Y. & Semenoff R. (1985b) Positive identification of an immigration test case using human DNA fingerprints. *Nature* **317**, 577–9.

Jelinsky S. & Samson L. (1999) Global response of *Saccharomyces cerevisiae* to an alkylating agent. *Proc. Nat. Acad. Sci. USA* **96**, 1486–91.

Jen G.C. & Chilton M.D. (1986) The right border region of pTiT37 T-DNA is intrinsically more active than the left border region in promoting T-DNA transformation. *Proc. Nat. Acad. Sci. USA* **83**, 3895–9.

Jensen J.S., Marcker K.A., Otten L. & Schell J. (1986) Nodule-specific expression of a chimaeric soybean leghaemoglobin gene in transgenic *Lotus corniculatus*. *Nature* **321**, 669–74.

Jensen S., Gassama M.P. & Heidmann T. (1999) Taming of transposable elements by homology-dependent gene silencing. *Nature Genet.* **21**, 209–12.

Jeon J.-S., Lee S., Jung K.-H. *et al.* (2000) T-DNA insertional mutagenesis for functional genomics in rice. *Plant J.* **22**, 561–70.

Jespers L.S., Messens J.H., De Keyser A. *et al.* (1995) Surface expression and ligand-based selection of cDNAs

fused to filamentous phage gene VI. *Bio/Technology* **13**, 378–82.

Johanning F.W., Conry R.M., LoBuglio A.F. *et al.* (1995) A Sindbis virus mRNA polynucleotide vector achieves prolonged and high level heterologous gene expression *in vivo*. *Nucl. Acids Res.* **23**, 1495–501.

John D.C.A., Watson R., Kind A.J. *et al.* (1999) Expression of an engineered form of recombinant procollagen in mouse milk. *Nature Biotechnol.* **17**, 385–9.

Johnson I.S. (1983) Human insulin from recombinant DNA technology. *Science* **219**, 632–7.

Johnson J., Lin T. & Lomonossoff G. (1997) Presentation of heterologous peptides on plant viruses: genetics, structure and function. *Ann. Rev. Phytopathol.* **35**, 67–86.

Johnson M. (1998) Cloning humans? *Bioessays* **19**, 737–9.

Johnston M. (1987) A model fungal gene regulatory mechanism: the GAL genes of *Saccharomyces cerevisiae*. *Microbiol. Rev.* **51**, 458–76.

Jones A.L., Thomas C.L. & Maule A.J. (1998) *De novo* methylation and cosuppression induced by a cytoplasmically-replicating plant RNA virus. *EMBO J.* **17**, 6385–93.

Jones I.M., Primrose S.B., Robinson A. & Ellwood D.C. (1980) Maintenance of some Col E1-type plasmids in chemostat culture. *Mol. Gen. Genet.* **180**, 579–84.

Jones J.D.G., Gilbert D.E., Grady K.L. & Jorgensen R.A. (1987) T-DNA structure and gene expression in petunia plants transformed by *Agrobacterium tumefaciens* C58 derivatives. *Mol. Gen. Genet.* **207**, 478–85.

Jones K. & Murray K. (1975) A procedure for detection of heterologous DNA sequences in lambdoid phage by *in situ* hybridization. *J. Mol. Biol.* **51**, 393–409.

Jorgensen R.A., Snyder C. & Jones J.D.G. (1987) T-DNA is organized predominantly in inverted repeat structures in plants transformed with *Agrobacterium tumefaciens* C58 derivatives. *Mol. Gen. Genet.* **207**, 471–7.

Joyner A.L. (ed.) (1998) *Gene Targeting: A Practical Approach*, 2nd edn. Oxford University Press, Oxford.

Julius D., Blair L.C., Brake A.J., Sprague G.F. & Thorner J. (1983) Yeast alpha-factor is processed from a larger precursor polypeptide: the essential role of a membrane-bound dipeptidyl amino-peptidase. *Cell* **32**, 839–52.

Julius D., Scheckman R. & Thorner J. (1984) Glycosylation and processing of prepro-alpha-factor through the yeast secretory pathway. *Cell* **36**, 309–18.

Kado C.I. (1998) *Agrobacterium*-mediated horizontal gene transfer. *Genet. Eng.* **20**, 1–24.

Kahl G. & Schell J.S. (eds) (1982) *Molecular Biology of Plant Tumours*, pp. 211–67. Academic Press, New York.

Kanalas J.J. & Suttle D.P. (1984) Amplification of the UMP synthase gene and enzyme overproduction in pyrazofurin-resistant rat hepatoma cells: molecular cloning of a cDNA for UMP synthase. *J. Biol. Chem.* **259**, 1848–53.

Kane S.E., Troen B.R., Gal S. *et al.* (1988) Use of a cloned multidrug resistance gene for amplification and over-production of major excreted protein, a transformation regulated secreted acid protease. *Mol. Cell. Biol.* **8**, 3316–21.

Kanevski I.F., Thakur S., Cosowsky L. *et al.* (1992) Tobacco lines with high copy number of replicating recombinant geminivirus vectors after biolistic DNA delivery. *Plant J.* **2**, 457–63.

Kao C.M., Katz L. & Khosia C. (1994) Engineered biosynthesis of a complete macrolone in a heterologous host. *Science* **265**, 509–12.

Kaper J.B., Lockman H., Baldini M.M. & Levine M.M. (1984) A recombinant live oral cholera vaccine. *Biotechnology* **1**, 345–9.

Kapust R.B. & Waugh, D.S. (1999) *Escherichia coli* maltose-binding protein is uncommonly effective at promoting the solubility of polypeptides to which it is fused. *Protein Sci.* **8**, 1668–74.

Karatzas C., Zhou J.F., Huang Y. *et al.* (1999) Production of recombinant spider silk (biosteel™) in the milk of transgenic animals. In *Transgenic Animal Research Conference Tahoe City USA*, p. 34.

Karess R.E. & Rubin G.M. (1984) Analysis of P transposable element functions in *Drosophila*. *Cell* **38**, 135–46.

Karn J., Brenner S., Barnett L. & Cesareni G. (1980) Novel bacteriophage λ cloning vector. *Proc. Nat. Acad. Sci. USA* **77**, 5172–6.

Karn J., Matthews H.W., Gait M.J. & Brenner S. (1984) A new selective cloning vector, λ2001, with sites for *Xba*I, *Bam*HI, *Hind*III, *Eco*RI, *Sst*I and *Xho*I. *Gene* **32**, 217–24.

Katagiri F., Lam E. & Chua N.H. (1989) Two tobacco DNA-binding proteins with homology to the nuclear factor CREB. *Nature* **340**, 727–30.

Katakura Y., Ametani A., Totsuka M., Nagafuchi S. & Kaminogawa S. (1999) Accelerated secretion of mutant beta-lactoglobulin in *Saccharomyces cerevisiae* resulting from a single amino acid substitution. *Biochim. Biophys. Acta* **1432**, 302–12.

Katsuki M., Sato M., Kimura M. *et al.* (1988) Conversion of normal behaviour to shiverer by myelin basic protein antisense cDNA in transgenic mice. *Science* **241**, 593–5.

Katz E. & Moss B. (1997) Immunogenicity of recombinant vaccinia viruses that display the HIV type 1 envelope glycoprotein on the surface of infectious virions. *AIDS Res. Hum. Retroviruses* **13**, 1497–500.

Katz E., Wolffe,E.J. & Moss B. (1997) The cytoplasmic and transmembrane domains of the vaccinia virus B5R protein target a chimeric human immunodeficiency virus type 1 glycoprotein to the outer envelope of nascent vaccinia virions. *J. Virol.* **71**, 3178–87.

Kaufman P.D. & Rio D.C. (1991) Germline transformation of *Drosophila melanogaster* by purified P element transposase. *Nucl. Acids Res.* **19**, 6336.

Kaufman R.J. (1990a) Strategies for obtaining high level expression in mammalian cells. *Technique* **2**, 221–36.

Kaufman R.J. (1990b) Vectors used for expression in mammalian cells. *Methods Enzymol.* **185**, 487–511.

Kaufman R.J., Wasley L.C., Spiliotes A.T. *et al.* (1985) Coamplification and coexpression of human tissue-type plasminogen activator and murine dihydrofolate reductase sequences in Chinese hamster ovary cells. *Mol. Cell. Biol.* **5**, 1730–59.

Kaufman R.J., Murtha P., Ingolia D.E., Yeung C.-Y. & Kellems R.E. (1986) Selection and amplification of heterologous genes encoding adenosine deaminase in mammalian cells. *Proc. Nat. Acad. Sci. USA* **83**, 3136–40.

Kaufman R.J., Wasley L.C. & Dorner A.J. (1988) Synthesis, processing, and secretion of recombinant human factor VIII expressed in mammalian cells. *J. Biol. Chem.* **263**, 6352–62.

Kaufman R.J., Davies M.V., Wasley L.C. & Michnick D. (1991) Improved vectors for stable expression of foreign genes in mammalian cells by use of the untranslated leader sequence from EMC virus. *Nucl. Acids Res.* **19**, 4485–90.

Kawasaki E.S. (1990) Amplification of RNA. In *PCR Protocols: A Guide to Methods and Applications*, eds Innis M.A., Gelfand D.H., Sninsky J.J. & White T.J., pp. 21–7. Academic Press, New York.

Keggins K.M., Lovett P.S. & Duvall E.J. (1978) Molecular cloning of genetically active fragments of *Bacillus* DNA in *Bacillus subtilis* and properties of the vector plasmid pUB110. *Proc Nat. Acad. Sci. USA* **75**, 1423–7.

Keilty S. & Rosenberg M. (1987) Constitutive function of a positively regulated promotor reveals new sequences essential for activity. *J. Biol. Chem.* **262**, 6389–95.

Keller H., Pamboukdjian N., Ponchet M. *et al.* (1999) Pathogen-induced elicitin production in transgenic tobacco generates a hypersensitive response and non-specific disease resistance. *Plant Cell* **11**, 223–35.

Kelly T.J. & Smith H.O. (1970) A restriction enzyme from *Hemophilus influenzae*. II. Base sequence of the recognition site. *J. Mol. Biol.* **51**, 393–409.

Kennerdell J.R. & Carthew R.W. (2000) Heritable gene silencing in *Drosophila* using double stranded RNA. *Nature Biotechnol.* **17**, 896–8.

Kerr D.E., Liang F., Bondioli K.R. *et al.* (1998) The bladder as a bioreactor: urothelium production and secretion of growth hormone into urine. *Nature Biotechnol.* **16**, 75–9.

Khan M.S. & Maliga P. (1999) Fluorescent antibiotic resistance marker for tracking plastid transformation in higher plants. *Nature Biotechnol.* **17**, 910–15.

Kieser T. & Hopwood D.A. (1991) Genetic manipulation of *Streptomyces*: integrating vectors and gene replacement. *Methods Enzymol.* **204**, 430–58.

Kieser T., Bibb M.J., Buttner M.J. *et al.* (eds) (2000) *Practical Streptomyces Genetics*. John Innes Foundation. Norwich, UK.

Kilby N.J., Smith M.R. & Murray J.A. (1993) Site specific recombinases: tools for genome engineering. *Trends Genet.* **9**, 413–21.

Kim J.J., Bagarazzi M.L., Trivedi N. *et al.* (1997) Engineering of *in vivo* immune responses to DNA immunization via codelivery of costimulatory molecule genes. *Nature Biotechnol.* **15**, 641–6.

Kim L., Mogk A. & Schumann W. (1996) A xylose-inducible *Bacillus subtilis* integration vector and its application. *Gene* **181**, 71–6.

Kim S.K. & Wold B.J. (1985) Stable reduction of thymidine kinase activity in cells expressing high levels of antisense RNA. *Cell* **42**, 129–38.

Kim U.-J., Birren B.W., Slepak T. *et al.* (1996) Construction and characterization of a human bacterial artificial chromosome library. *Genomics* **34**, 213–18.

Kimelman D. & Kirchner M. (1989) An antisense mRNA directs the covalent modification of the transcripts encoding fibroblast growth factor in *Xenopus* oocytes. *Cell* **59**, 687–96.

King L.A. & Possee R.D. (1992) *The Baculovirus Expression System: A Laboratory Guide.* Chapman & Hall, London.

King T.J. & Briggs R. (1956) Serial transplantation of embryonic nuclei. *Cold Spring Harbor Symp. Quant. Biol.* **21**, 271–90.

Kishore G.M. & Somerville C.R. (1993) Genetic engineering of commercially useful biosynthetic pathways in transgenic plants. *Curr. Opin. Biotechnol.* **4**, 152–8.

Kitts P.A. & Possee R.D. (1993) A method for producing recombinant baculovirus expression vectors at high frequency. *Biotechniques* **14**, 810–17.

Klatzmann D., Herson S., Cherin P. *et al.* (1996) Gene therapy for metastatic malignant melanoma: evaluation of tolerance to intratumoral injection of cells producing recombinant retroviruses carrying the herpes simplex virus type 1 thymidine kinase gene, to be followed by ganciclovir administration. *Hum. Gene Ther.* **7**, 155–67.

Klee H.J., Hayford M.B., Kretzmer K.A., Barry G.F. & Kishore G.M. (1991) Control of ethylene synthesis by expression of a bacterial enzyme in transgenic tomato plants. *Plant Cell* **3**, 1187–93.

Kleid D.G., Yansura D., Small B. *et al.* (1981) Cloned viral protein vaccine for foot-and-mouth disease: responses in cattle and swine. *Science* **214**, 1125–9.

Klein T.M. & Fitzpatrick-McElligott (1993) Particle bombardment: a universal approach for gene transfer to cells and tissues. *Curr. Opin. Biotechnol.* **4**, 583–90.

Klein T.M., Wolf E.D., Wu R. & Sanford J.C. (1987) High-velocity micro-projectiles for delivering nucleic acids into living cells. *Nature* **327**, 70–3.

Klein T.M., Fromm M.E., Weissinger A. *et al.* (1988a) Transfer of foreign genes into intact maize cells with high velocity microprojectiles. *Proc. Nat. Acad. Sci. USA* **85**, 4305–9.

Klein T.M., Harper E.C., Svab Z. *et al.* (1988b) Stable genetic transformation of intact *Nicotiana* cells by the particle bombardment process. *Proc. Nat. Acad. Sci. USA* **85**, 8502–5.

Kling J. (1996) Could transgenic supercrops one day breed superweeds? *Science* **274**, 180–1.

Klinman D.M., Yamshchikov G. & Ishigatsubo Y. (1997) Contribution of CpG motifs to the immunogenicity of DNA vaccines. *J. Immunol.* **158**, 3635–9.

Kloti A., Iglesias V.A., Wunn J. *et al.* (1993) Gene transfer by electroporation into intact scutellum cells of wheat embryos. *Plant Cell Rep.* **12**, 671–5.

Knowles M.R., Hohneker K.W., Zhou Z. *et al.* (1995) A controlled study of adenoviral-vector-mediated gene transfer in the nasal epithelium of patients with cystic fibrosis. *N. Engl. J. Med.* **333**, 823–31.

Knutzon D.S., Thompson G.A., Radke S.E. *et al.* (1992) Modification of *Brassica* seed oil by antisense expression of a stearoyl-acyl carrier protein desaturase gene. *Proc. Nat. Acad. Sci. USA* **89**, 2624–8.

Kohler R.H., Cao J., Zipfel W.R., Webb W.W. & Hanson M. (1997) Exchange of protein molecules through connections between higher plant plastids. *Science* **276**, 2039–42.

Kok M., Rekik M., Witholt B. & Harayama S. (1994) Conversion of pBR322-based plasmids into broad-host-range vectors by using the Tn3 transposition mechanism. *J. Bacteriol.* **176**, 6566–71.

Kola I. & Hertzog P.J. (1998) Down syndrome and mouse models. *Curr. Opin. Genet. Devel.* **8**, 316–21.

Kollias G., Wrighton N., Hurst J. & Grosveld F. (1986) Regulated expression of human $^A\gamma$-, β-, and hybrid γβ-globin genes in transgenic mice: manipulation of the developmental expression patterns. *Cell* **46**, 89–94.

Komari T., Hiei Y., Saito Y., Murai N. & Kumashiro T. (1996) Vectors carrying two separate T-DNAs for cotransformation of higher plants mediated by *Agrobacterium tumefaciens* and segregation of transformants free from selection markers. *Plant J.* **10**, 165–74.

Koncz C., Olsson O., Langridge W.H.R., Schell J. & Szalay A.A. (1987) Expression and assembly of functional bacterial luciferase in plants. *Proc. Nat. Acad. Sci. USA* **84**, 131–5.

Koncz C., Martini N., Mayerhofer R. *et al.* (1989) High frequency T-DNA mediated gene tagging in plants. *Proc. Nat. Acad. Sci. USA* **86**, 8467–71.

Koonin E.V., Bork P. & Sander C. (1994) Yeast chromosome III: new gene functions. *EMBO J.* **13**, 493–503.

Koonin E.V., Mushegian A.R. & Rudd K.E. (1996) Sequencing and analysis of bacterial genomes. *Curr. Biol.* **6**, 404–16.

Koop H.U., Steinmuller K., Wagner H. *et al.* (1996) Integration of foreign sequences into the tobacco plastome via polyethylene glycol-mediated protoplast transformation. *Planta* **199**, 193–201.

Kost T.A. & Condreay J.P. (1999) Recombinant baculoviruses as expression vectors for insect and mammalian cells. *Curr. Opin. Biotechnol.* **10**, 428–33.

Kotewicz M.L., Sampson C.M., D'Alessio J.M. & Gerard G.F. (1988) Isolation of cloned Moloney murine leukemia virus reverse transcriptase lacking ribonuclease H activity. *Nucl. Acids Res.* **16**, 265–77.

Kothary R., Clapoff S., Brown A. *et al.* (1988) A transgene containing *lacZ* inserted into *dystonia* locus is expressed in neural tube. *Nature* **335**, 435–7.

Kotin R.M., Siniscalco M., Samulski R.J. *et al.* (1990) Site specific integration by adeno-associated virus. *Proc. Nat. Acad. Sci. USA* **87**, 2211–15.

Kovach M.E., Elzer P.H., Hill D.S. *et al.* (1995) Four new derivatives of the broad-host-range cloning vector pBBR1MCS, carrying different antibiotic resistance cassettes. *Gene* **166**, 175–6.

Kozak M. (1986) Point mutations define a sequence flanking the AUG initiator codon that modulates translation by eukaryotic ribosomes. *Cell* **44**, 283–92.

Kozak M. (1999) Initiation of translation in prokaryotes and eukaryotes. *Gene* **234**, 187–208.

Koziel M.G., Beland G.L., Bowman C. *et al.* (1993) Field performance of elite transgenic maize plants expressing an insecticidal protein derived from *Bacillus thuringiensis*. *Biotechnol.* **11**, 194–200.

Koziel M.G., Carozzi N.B. & Desai N. (1996) Optimizing expression of transgenes with an emphasis on post-transcriptional events. *Plant Mol. Biol.* **32**, 393–405.

Kozlowski T., Monroy R., Xu Y. *et al.* (1998) Anti-Gal (alpha) 1–3Gal antibody response to porcine bone marrow in unmodified baboons and baboons conditioned for tolerance induction. *Transplantation* **66**, 176–82.

Kramer B., Kramer W. & Fritz H.-J. (1984a) Different base/base mismatches are corrected with different efficiencies by the methyl-directed DNA mismatch-repair system of *E. coli*. *Cell* **38**, 879–88.

Kramer R.A., Cameron J.R. & Davis R.W. (1976) Isolation of bacteriophage λ containing yeast ribosomal RNA genes: screening by *in situ* RNA hybridization to plaques. *Cell* **8**, 227–32.

Kramer W., Drutsa V., Jansen H.-W. *et al.* (1984b) The gapped duplex DNA approach to oligonucleotide-directed mutation construction. *Nucl. Acids Res.* **12**, 9441–56.

Krasemann S., Groschup M., Hunsmann G. & Bodemer W. (1996) Induction of antibodies against human prion proteins (PrP) by DNA-mediated immunization of PrP[0/0] mice. *J. Immunol. Methods* **199**, 109–18.

Kroll K.L. & Amaya E. (1996) Transgenic *Xenopus* embryos from sperm nuclear transplantations reveal FGF signaling requirements during gastrulation. *Development* **122**, 3173–83.

Kruger S. (1988) EcoRII can be activated to cleave refractory DNA recognition sites. *Nucl. Acids Res.* **16**, 3997–4008.

Krysan P.J., Young J.C. & Sussman M.R. (1999) T-DNA as an insertional mutagen in *Arabidopsis*. *Plant Cell* **11**, 2283–90.

Kubo M. & Kakimoto T. (2001) The *CYTOKININ HYPERSENSITIVE* genes of *Arabidopsis* negatively regulate the cytokinin signaling pathway for cell division and chloroplast development. *Plant J.* **23**, 385–94.

Kudla B. & Nicolas A. (1992) A multisite integrative cassette for the yeast *Saccharomyces cerevisiae*. *Gene* **119**, 49–56.

Kuehn M.R., Bradley A., Robertson E.J. & Evans M.J. (1987) A potential model for Lesch–Nyhan syndrome through introduction of HPRT mutations in mice. *Nature* **326**, 295–8.

Kuhn R., Schwenk F., Auget M. & Rajewsky K. (1995) Inducible gene targeting in mice. *Science* **269**, 1427–9.

Kuipers O.P., Beerthuyzen M.M., De Ruyter P.G.G.A., Luesink E.J. & De Vos W.M. (1995) Autoregulation of nisin biosynthesis in *Lactococcus lactis* by signal transduction. *J. Biol. Chem.* **270**, 27299–304.

Kumagai M.H., Turpen T.H., Weinzettl N. *et al.* (1993) Rapid, high-level expression of biologically-active alpha-trichosanthin in transfected plants by an RNA viral vector. *Proc. Nat. Acad. Sci. USA* **90**, 427–30.

Kumar A. *et al.* (1995) Potato plants expressing antisense and sense *S*-adenosylmethionine decarboxylase (SAMDC) transgenes show altered levels of polyamines and ethylene: antisense plants display abnormal phenotypes. *Plant J.* **9**, 147–58.

Kumpatla S.P., Chandrasekharah M.B., Iyer L.M., Li G. & Hall T.C. (1998) Genome intruder scanning and modulation systems and transgene silencing. *Trends Plant Sci.* **3**, 97–104.

Kunik T., Tzfira T., Kapulnik Y. *et al.* (2001) Genetic transformation of HeLa cells by *Agrobacterium*. *Proc. Nat. Acad. Sci. USA* **98**, 1871–6.

Kunkel L.M. (1986) Analysis of deletions in DNA from patients with Becker and Duchenne muscular dystrophy. *Nature* **322**, 73–7.

Kunkel T., Niu Q.W., Chan Y.S. & Chua N.H. (1999) Inducible isopentenyl transferase as a high-efficiency marker for plant transformation. *Nature Biotechnol.* **17**, 916–19.

Kunkel T.A. (1985) Rapid and efficient site-specific mutagenesis without phenotypic selection. *Proc. Nat. Acad. Sci. USA* **82**, 488–92.

Kuo C.-L. & Campbell J.L. (1983) Cloning of *Saccharomyces cerevisiae* DNA replication genes: isolation of the CDC8 gene and two genes that compensate for the cdc8-1 mutation. *Mol. Cell. Biol.* **3**, 1730–7.

Kurata S., Tsukakoshi M., Kasuya T. & Ikawa Y. (1986) The laser method for efficient introduction of foreign DNA into cultured cells. *Exp. Cell Res.* **162**, 372–8.

Kurjan J. & Herskowitz I. (1982) Structure of a yeast pheromone (MF alpha): a putative alpha factor precursor contains four tandem copies of mature alpha factor. *Cell* **30**, 933–43.

Kurland C.G. (1987) Strategies for efficiency and accuracy in gene expression. 1. The major codon preference: a growth optimization strategy. *Trends Biochem. Sci.* **12**, 126–8.

Kurland C.G. & Dong H. (1996) Bacterial growth inhibition by overproduction of proteins. *Mol. Microbiol.* **21**, 1–4.

Kurtz D.T. & Nicodemus C.F. (1981) Cloning of α_{2u} globulin cDNA using a high efficiency technique for the cloning of trace messenger RNAs. *Gene* **13**, 145–52.

Labosky P.A., Barlow D.P. & Hogan B.L. (1994) Mouse embryonic germ (EG) cell lines: transmission through the germline and differences in the methylation imprint of insulin-like growth factor 2 receptor (Igf2r) gene compared with embryonic stem (ES) cell lines. *Development* **120**, 3197–204.

Labow M.A., Baim S.B., Shenk T. & Levine A. (1990) Conversion of the lac repressor into an allosterically regulated transcriptional activator for mammalian cells. *Mol. Cell Biol.* **10**, 3343–56.

Lacks S.A., Lopez P., Greenberg B. & Espinosa M. (1986) Identification and analysis of genes for tetracycline resistance and replication functions in the broad-host-range plasmid pLS1. *J. Mol. Biol.* **192**, 753–5.

Lacombe M.L., Ladant D., Mutszel R. & Veron M. (1987) Gene isolation by direct *in situ* cAMP binding. *Gene* **58**, 29–36.

Lacy E., Roberts S., Evans E.P., Burtenshaw M.D. & Constantini F.D. (1983) A foreign β-globin gene in transgenic mice: integration at abnormal chromosomal positions and expression in inappropriate tissues. *Cell* **34**, 343–58.

Lai E.M. & Kado C.I. (2000) The T-pilus of *Agrobacterium tumefaciens*. *Trends Microbiol.* **8**, 361–9.

Lalioti M.D. & Heath J.K. (2001) A new method for generating point mutations in bacterial artificial chromosomes by homologous recombination in *Escherichia coli*. *Nucl. Acids Res.* **29**, e14.

Lamb B.T. & Gerhart J.D. (1995) YAC transgenics and the study of genetics and human disease. *Curr. Opin. Genet. Devel.* **5**, 342–8.

Lambrigts D., Sachs D.H. & Cooper D.K.C. (1998) Discordant organ xenotransplantation in primates: world experience and current status. *Transplantation* **66**, 547–61.

Lamont P.J., Davis M.B. & Wood N.W. (1997) Identification and sizing of the GAA trinucleotide repeat expansion of Friedreich's ataxia in 56 patients – clinical and genetic correlates. *Brain* **120**, 673–80.

Land H., Grey M., Hanser H., Lindenmaier W. & Schutz G. (1981) 5′-Terminal sequences of eucaryotic mRNA can be cloned with a high efficiency. *Nucl. Acids Res.* **9**, 2251–66.

Landegran U. (1996) The challengers to PCR. a proliferation of chain reactions. *Curr. Opin. Biotechnol.* **7**, 95–7.

Landford R.E. (1988) Expression of simian virus 40 T antigen in insect cells using a baculovirus expression vector. *Virology* **167**, 72–81.

Lane C.D., Colman A., Mohun T. *et al.* (1980) The *Xenopus* oocyte as a surrogate secretory system: the specificity of protein export. *Eur. J. Biochem.* **111**, 225–35.

Lane D., Prentki P. & Chandler M. (1992) Use of gel retardation to analyze protein–nucleic acid interactions. *Microbiol. Rev.* **56**, 509–28.

Langley K.E. & Zabin I. (1976) Beta-galactosidase alpha-complementation: properties of the complemented enzyme and mechanism of the complementation reaction. *Biochemistry* **15**, 4866–75.

Langley K.E., Villarejo M.R., Fowler A.V., Zamenhof P.J. & Zabin I. (1975) Molecular basis of beta-galactosidase alpha-complementation. *Proc. Nat. Acad. Sci. USA* **72**, 1254–7.

Langridge W.H.R., Fitzgerald K.J., Koncz C., Schell J. & Szalay A.A. (1989) Dual promoter of *Agrobacterium tumefaciens* mannopine synthase genes is regulated by plant growth hormones. *Proc. Nat. Acad. Sci. USA* **86**, 3219–23.

Lapeyre B. & Amalric F. (1985) A powerful method for the preparation of cDNA libraries: isolation of cDNA encoding a 100-kDa nucleolar protein. *Gene* **37**, 215–20.

Larsson S., Hotchkiss G., Andang M. *et al.* (1994) Reduced β2-macroglobulin mRNA levels in transgenic mice expressing a designed hammerhead ribozyme. *Nucl. Acids Res.* **22**, 2242–8.

LaSalle G.L., Robert J.J., Berrard S. *et al.* (1993) An adenovirus vector for gene transfer into neurons and glia in the brain. *Science* **259**, 988–90.

Laski F.A., Rio D.C. & Rubin G.M. (1986) Tissue specificity of *Drosophila* P element transposition is regulated at the level of mRNA splicing. *Cell* **44**, 7–19.

Lasko M., Sauer B., Mosinger B. *et al.* (1992) Targeted oncogene activation by site-specific recombination in transgenic mice. *Proc. Nat. Acad. Sci. USA* **89**, 6232–6.

Lathe R. (1985) Synthetic oligonucleotide probes deduced from amino acid sequence data. *J. Mol. Biol.* **183**, 1–12.

Lau D., Kuzma G., Wei C.-M., Livingston D.J. & Hsiung N. (1987) A modified human tissue plasminogen activator with extended half-life *in vivo*. *Biotechnology* **5**, 953–8.

Laufs J., Wirtz U., Kammann M. *et al.* (1990) Wheat dwarf virus *Ac/Ds* vectors: expression and excision of transposable elements introduced into various cereals by a viral replicon. *Proc. Nat. Acad. Sci. USA* **87**, 7752–6.

Laursen C.M., Krzyzek R.A., Flick C.E., Anderson P.C. & Spencer T.M. (1994) Production of fertile transgenic maize by electroporation of suspension culture cells. *Plant Mol. Biol.* **24**, 51–61.

La Vallie E.R. & McCoy J.M. (1995) Gene fusion expression systems in *Escherichia coli*. *Curr. Opin. Biotechnol.* **5**, 501–6.

La Vallie E.R., Di Blasio E.A., Kovacic S. *et al.* (1993) A thioredoxin gene fusion expression system that circumvents inclusion body formation in the *E. coli* cytoplasm. *Bio/technology* **11**, 187–93.

Law M.-F., Byrne J. & Hawley P.M. (1983) A stable bovine papillomavirus hybrid plasmid that expresses a dominant selective trait. *Mol. Cell. Biol.* **3**, 2110–15.

Lawrence M.S., Ho D.Y., Dash R. & Sapolsky R.M. (1995) Herpes simplex virus vectors overexpressing the glucose transporter gene protect against seizure-induced neuron loss. *Proc. Nat. Acad. Sci. USA* **92**, 7247–51.

Lawyer F.C., Stoffel S., Saiki R. *et al.* (1989) Isolation, characterization, and expression in *E. coli* of the DNA polymerase from *Thermus aquaticus*. *J. Biol. Chem.* **264**, 6427–37.

Lay Thein S. & Wallace R.B. (1986) In *Human Genetic Diseases. A Practical Approach*, ed. Davies I.E., pp. 33–50. IRL Press, Oxford.

Lazarus R.A., Seymour J.L., Stafford R.K. *et al.* (1990) A biocatalytic approach to vitamin C production: metabolic pathway engineering of *Erwinia herbicola*. In *Biocatalysis*, ed. Abramowitz D., pp. 136–55. Van Nostrand Reinhold, New York.

Lazo G.R., Stein P.A. & Ludwig R.A. (1991) A DNA transformation competent *Arabidopsis* genomic library in *Agrobacterium*. *Biotechnology* **9**, 963–7.

Leder P., Tiemeier D. & Enquist L. (1977) EK2 derivatives of bacteriophage lambda useful in the cloning of DNA from higher organisms: the λgt WES system. *Science* **196**, 175–7.

Lederberg S. (1957) Suppression of the multiplication of heterologous bacteriophages in lysogenic bacteria. *Virology* **3**, 496–513.

Lederberg S. & Meselson M. (1964) Degradation of non-replicating bacteriophage DNA in non-accepting cells. *J. Mol. Biol.* **8**, 623–8.

Leduc N., Matthys-Rochon E., Rougier M. *et al.* (1996) Isolated maize zygotes mimic *in vivo* early development and express microinjected genes when cultured *in vitro*. *Devel. Biol.* **10**, 190–203.

Lee C.C., Wu X., Gibbs R.A. *et al.* (1988) Generation of cDNA probes directed by amino acid sequence: cloning of urate oxidase. *Science* **239**, 1288–91.

Lee F., Mulligan R., Berg P. & Ringold G. (1981) Glucocorticoids regulate expression of dihydrofolate reductase cDNA in mouse mammary tumour virus chimaeric plasmids. *Nature* **294**, 228–32.

Lee J.W., Choi H.S., Gyuris J., Brent R. & Moore D.D. (1995) Two classes of proteins dependent on either the presence or absence of thyroid hormone for interaction with the thyroid hormone receptor. *Mol. Endocrinol.* **9**, 243–54.

Lee R.C., Feinbaum R.L. & Ambros V. (1993) The *C. elegans* heterochromatic gene *lin-4* encodes small RNAs with antisense complementarity to *lin-14*. *Cell* **75**, 843–54.

Leemans J., Deblaere R., Willmitzer L. *et al.* (1982a) Genetic identification of functions of T_L–DNA transcripts in octopine crown galls. *EMBO J.* **1**, 147–52.

Leemans J., Langenakens J., De Greve H. *et al.* (1982b) Broad-host-range cloning vectors derived from the W-plasmid Sa. *Gene* **19**, 361–4.

Leenhouts K.J., Tolner B., Bron S. *et al.* (1991) Nucleotide sequence and characterization of the broad-host-range lactococcal plasmid pWV01. *Plasmid* **26**, 55–66.

Lefebvre D.D., Miki B.L. & Laliberte J.F. (1987) Mammalian metallothionein functions in plants. *Biotechnology* **5**, 1053–6.

Lehmann M., Pasamontes L., Lassen S.F. & Wyss M. (2000) The consensus concept for thermostability engineering of proteins. *Biochim. Biophys. Acta* **1543**, 408–15.

Lehoux D.E. & Levesque R.C. (2000) Detection of genes essential in specific niches by signature-tagged mutagenesis. *Curr. Opin. Biotechnol.* **11**, 434–9.

Leite A., Kemper E.L., Da Silva M.J. *et al.* (2000) Expression of correctly processed human growth hormone in seeds of transgenic tobacco plants. *Mol. Breeding* **6**, 47–53.

Le Loir Y., Gruss A., Ehrlich S.D. & Langella P. (1998) A nine-residue synthetic propeptide enhances secretion efficiency of heterologous proteins in *Lactococcus lactis*. *J. Bacteriol.* **180**, 1895–903.

Lennon G.G. & Lehrach H. (1991) Hybridization analyses of arrayed cDNA libraries. *Trends Genet.* **7**, 314–17.

Leno G.H. & Laskey R.A. (1991) The nuclear membrane determines the timing of replication in *Xenopus* egg extracts. *J. Cell Biol.* **112**, 557–66.

Lerner M.R. (1994) Tools for investigating functional interactions between ligands and G-protein-coupled receptors. *Trends Neurosci.* **17**, 142–6.

Lester S.C., LeVan S.K., Steglich C. & DeMars R. (1980) Expression of human genes for adenine phosphoribosyltransferase and hypoxanthine-guanine phosphoribosyltransferase after genetic transformation of mouse cells with purified human DNA. *Somat. Cell Genet.* **6**, 24159.

Levine M., Rubin G.M. & Tjian R. (1984) Human DNA sequences homologous to a protein coding region conserved between homeotic genes of *Drosophila*. *Cell* **38**, 667–73.

Lewanodski M. & Martin G.R. (1997) Cre-mediated chromosome loss in mice. *Nature Genet.* **17**, 223–5.

Lewis P.F. & Emmerman M. (1994) Passage through mitosis is required for oncoretroviruses but not for human immunodefficiency virus. *J. Virol.* **68**, 510–16.

Lewontin R.C. & Hartl D.L. (1991) Population genetics in forensic DNA typing. *Science* **254**, 1745–50.

Li H., Wang J., Wilhelmsson H. *et al.* (2000) Genetic modification of survival in tissue-specific knockout mice with mitochondrial cardiomyopathy. *Proc. Nat. Acad. Sci. USA* **97**, 3467–72.

Li H.W., Lucy A.P., Guo H.S. *et al.* (1999) Strong host resistance targeted against a viral suppressor of the plant gene silencing defense mechanism. *EMBO J.* **18**, 2683–91.

Li Q., Harju S. & Peterson K.R. (1999) Locus control regions coming of age at a decade plus. *Trends Genet.* **15**, 403–8.

Li Z.W., Stark G., Gotz J. *et al.* (1996) Generation of mice with a 200-kb amyloid precursor protein gene deletion by Cre-recombinase-mediated site-specific recombination in embryonic stem cells. *Proc. Nat. Acad. Sci. USA* **93**, 6158–62.

Liang P. & Pardee A.B. (1992) Differential display analysis of eukaryotic messenger RNA by means of the polymerase chain reaction. *Science* **257**, 967–71.

Lian P., Averboukh L., Keyomarsi K., Saeger R. & Pardee A.B. (1992) Differential display and cloning of messenger RNAs from human breast cancer verses mammary epithelial cells. *Cancer Res.* **52**, 6966–8.

Liberles S.D., Diver S.T., Austin D.J. & Schreiber S.L. (1997) Inducible gene expression and protein translocation using nontoxic ligands identified by a mammalian three-hybrid system. *Proc. Nat. Acad. Sci. USA* **94**, 7825–30.

Licitra E.J. & Liu J.O. (1996) A three-hybrid system for detecting small ligand–protein receptor interactions. *Proc. Nat. Acad. Sci. USA* **93**, 12817–21.

Lien L.L., Lee Y. & Orkin S.H. (1997) Regulation of the myeloid-cell-expressed human gp91-*phox* gene as studied by transfer of yeast artificial chromosome clones into embryonic stem cells: suppression of a variegated cellular pattern of expression requires a full complement of distant *cis* elements. *Mol. Cell. Biol.* **17**, 2279–90.

Lilie H., Schwarz E. & Rudolph R. (1998) Advances in refolding of proteins produced in *E. coli. Curr. Opin. Biotechnol.* **9**, 497–501.

Liljestrom P. & Garoff H. (1993) A new generation of animal cell expression vectors based in the Semliki Forest virus replicon. *Biotechnology* **9**, 1356–61.

Lin Q., Chen Z., Antoniw J. & White R. (1991) Isolation and characterization of a cDNA clone encoding the antiviral protein from *Phytolacca americana. Plant Mol. Biol.* **17**, 609–14.

Lin W., Anuratha C.S., Datta K. *et al.* (1995) Genetic engineering of rice for resistance to sheath blight. *Biotechnology* **3**, 686–91.

Lindbo J.A. & Dougherty W.G. (1992) Untranslatable transcripts of the tobacco etch virus coat protein gene sequence can interfere with tobacco etch virus replication in transgenic plants and protoplasts. *Virology* **189**, 725–33.

Lindsey K., Wei W., Clarke M.C. *et al.* (1993) Tagging genomic sequences that direct transgene expression by activation of a promoter trap in plants. *Transgenic Res.* **2**, 33–47.

Ling M.M. & Robinson B.H. (1998) Approaches to DNA mutagenesis: an overview. *Anal. Biochem.* **254**, 157–78.

Lipshutz R.J., Taverner F., Hennessy K., Hartzell G. & Davis R. (1994) DNA sequence confidence estimation. *Genomics* **19**, 417–24.

Lipshutz R.J., Fodor S.P.A., Gingeras T.R. & Lockhart D.J. (1999) High density synthetic oligonucleotide arrays. *Nature Genet. Suppl.* **21**, 20–4.

Lis J.T., Simon J.A. & Sutton C.A. (1983) New heat shock puffs and β-galactosidase activity resulting from transformation of *Drosophila* with an hsp70-*lacZ* hybrid gene. *Cell* **35**, 403–10.

Lisitsyn N., Lisitsyn N. & Wigler M. (1993) Cloning the difference between two complex genomes. *Science* **259**, 946–51.

Lisser S. & Margalit, H. (1993) Compilation of *E. coli* mRNA promoter sequences. *Nucl. Acids Res.* **21**, 1507–16.

Littlewood T.D., Hancock D.C., Danielian P.S., Parker M.G. & Evan G.I. (1995) A modified oestrogen receptor ligand-binding domain as an improved switch for the regulation of heterologous proteins. *Nucl. Acids Res.* **23**, 1686–90.

Liu D., Raghothama K.G., Hasegawa P.M. & Bressan R.A. (1994) Osmotin overexpression in potato delays development of disease symptoms. *Proc. Nat. Acad. Sci. USA* **91**, 1888–92.

Liu H. & Rashidbaigi A. (1990) Comparison of various competent cell preparation methods for high efficiency DNA transformation. *Biotechniques* **8**, 21–5.

Liu Y.-G., Shirano Y., Fukaki H. *et al.* (1999) Complementation of plant mutants with large genomic DNA fragments by a transformation-competent artificial chromosome vector accelerates positional cloning. *Proc. Nat. Acad. Sci. USA* **96**, 6535–40.

Livak K.J., Flood S.J., Marmaro J., Giusti W. & Deetz K. (1995) Oligonucleotides with fluorescent dyes at opposite ends provide a quenched probe system useful for detecting PCR product and nucleic acid hybridization. *PCR Methods Appl.* **4**, 357–62.

Lo D., Pursel V., Linton P.J. *et al.* (1991) Expression of mouse IgA by transgenic mice, pigs and sheep. *Eur. J. Immunol.* **21**, 1001–6.

Lo D.C., McAllister A.K. & Katz L.C. (1994) Neuronal transfection in brain slices using particle-mediated gene transfer. *Neuron* **13**, 1263–8.

Loake G.J., Ashby A.M. & Shaw C.H. (1988) Attraction of *Agrobacterium tumefaciens* C58C1 towards sugars involves a highly sensitive chemotaxis system. *J. Gen. Microbiol.* **134**, 1427–32.

Lobban P.E. & Kaiser A.D. (1973) Enzymatic end-to-end joining of DNA molecules. *J. Mol. Biol.* **78**, 453–71.

Locket T.J., Lewy D., Holmes P., Medveezky K. & Saint R. (1992) The rough (ro+) gene is a dominant marker in germ line transformation of *Drosophila melanogaster. Gene* **114**, 187–193.

Lockhart D.J., Dong H., Byrne M.C. *et al.* (1996) Expression monitoring by hybridization to high-density oligonucleotide arrays. *Nature Biotechnol.* **14**, 1675–80.

Loenen W.A.M. & Brammar W.J. (1980) A bacteriophage lambda vector for cloning large DNA fragments made with several restriction enzymes. *Gene* **10**, 249–59.

Logan J.S. (2000) Prospects for xenotransplantation. *Curr. Opin. Immunol.* **12**, 563–8.

Lohnes D., Kastner A., Dierich M. *et al.* (1993) Function of retinoic acid receptor gamma in the mouse. *Cell* **73**, 643–58.

Lokman B.C., Heerikshuizen M., Van den Broek A. *et al.* (1997) Regulation of the *Lactobacillus pentosus* xylAB operon. *J. Bacteriol.* **179**, 5391–7.

Lomonossoff G.P. & Hamilton W.D.O. (1999) Cowpea mosaic virus-based vaccines. *Curr. Top. Microbiol. Immunol.* **240**, 177–89.

Lopata M.A., Cleveland D.W. & Sollner-Webb B. (1984) High level expression of a chloramphenicol acetyltransferase gene by DEAE-dextran mediated DNA transfection coupled with a dimethyl sulfoxide or glycerol shock treatment. *Nucl. Acids Res.* **12**, 5707–17.

Lorenz M.G. & Wackernagel W. (1994) Bacterial gene transfer by natural genetic transformation in the environment. *Microbiol. Rev.* **58**, 563–602.

Lory S. (1998) Secretion of proteins and assembly of bacterial surface organelles: shared pathways of extracellular protein targeting. *Curr. Opin. Microbiol.* **1**, 27–35.

Lorz H., Baaker B. & Schell J. (1985) Gene transfer to cereal cells mediated by protoplast transformation. *Mol. Gen. Genet.* **199**, 178–82.

Love J., Gribbin C., Mather C. & Sang H. (1994) Transgenic birds by DNA microinjection. *Biotechnology* **12**, 60–3.

Lowman H.B., Bass S.H., Simpson N. & Wells J.A. (1991) Selecting high-affinity binding proteins by monovalent phage display. *Biochemistry* **30**, 10832–8.

Lte A., Kemper E.L., Da Silva M.J. *et al.* (2000) Expression of correctly processed human growth hormone in seeds of transgenic tobacco plants. *Mol. Breeding* **6**, 47–53.

Lu S., Lyngholm L., Yang G. *et al.* (1994) Tagged mutations at the Tox 1 locus of *Cochliobolus heterostrophus* by restriction enzyme-mediated integration. *Proc. Nat. Acad. Sci. USA* **91**, 12649–53.

Lu Z., DiBlasio-Smith E.A., Grant K.L. *et al.* (1996) Histidine patch thioredoxins: mutant forms of thioredoxins with metal chelating affinity that provide for convenient purifications of thioredoxin fusion proteins. *J. Biol. Chem.* **271**, 5059–65.

Lubbert H., Hoffman B.J., Snutch T.P. *et al.* (1987) cDNA cloning of a serotonin 5-HT1C receptor by electrophysiological assays of messenger RNA-injected *Xenopus* oocytes. *Proc. Nat. Acad. Sci. USA* **84**, 4332–6.

Lucca P., Hurrell R. & Potrykus I. (2001) Genetic engineering approaches to improve the bioavailability and the level of iron in rice grains. *Theor. Appl. Genet.* **102**, 392–7.

Lucklow V.A. & Summers M.D. (1988) Signals important for high level expression of foreign genes in *Autographa californica* nuclear polyhedrosis virus expression vectors. *Virology* **167**, 56–71.

Ludecke H.J., Senger G., Claussen U. & Horsthemke B. (1989) Cloning defined regions of the human genome by microdissection of banded chromosomes and enzymatic amplification. *Nature* **338**, 348–50.

Ludecke H.J., Senger G., Claussen U. & Horsthemke B. (1990) Construction and characterization of band-specific DNA libraries. *Hum. Genet.* **84**, 512–16.

Lundberg K.S., Shoemaker D.D., Adams M.W. *et al.* (1991) High fidelity amplification using a thermostable DNA polymerase isolated from *Pyrococcus furiosus*. *Gene* **108**, 1–4.

Lundstrom K. (1997) Alphaviruses as expression vectors. *Curr. Opin. Biotechnol.* **8**, 578–82.

Lupton S. & Levine A.J. (1985) Mapping genetic elements of Epstein–Barr virus that facilitate extrachromosomal persistence of Epstein–Barr virus-derived plasmids in human cells. *Mol. Cell. Biol.* **5**, 2533–42.

Luria S.E. (1953) Host-induced modifications of viruses. *Cold Spring Harbor Symp. Quant. Biol.* **18**, 237–44.

Lusky M. & Botchan M. (1981) Inhibitory effect of specific pBR322 DNA sequences upon SV40 replication in simian cells. *Nature* **293**, 79–81.

Luthmann H. & Magnusson G. (1983) High efficiency polyoma DNA transfection of chloroquine-treated cells. *Nucl. Acids Res.* **11**, 1295–308.

Lydiate D.J., Malpartida F. & Hopwood D.A. (1985) The *Streptomyces* plasmid SCP2*: its functional analysis and development into useful cloning vectors. *Gene* **35**, 223–35.

Ma H. Kunes S., Schatz P.J. & Botstein D. (1987) Plasmid construction by homologous recombination in yeast. *Gene* **58**, 201–16.

Ma J., Yanofsky M.F., Klee H.J., Bowman B.L. & Meyerowitz, E.M. (1992) Vectors for plant transformation and cosmid libraries. *Gene* **117**, 161–7.

Ma J.K., Hiatt A., Hein M. *et al.* (1995) Generation and assembly of secretory antibodies in plants. *Science* **268**, 716–19.

Ma J.K., Hikmat B.Y., Wycoff K. *et al.* (1998) Characterization of a recombinant plant monoclonal secretory antibody and preventive immunotherapy in humans. *Nature Med.* **4**, 601–6.

Ma S.W., Zhao D.L., Yin Z.Q. *et al.* (1997) Transgenic plants expressing autoantigens fed to mice to induce oral immune tolerance. *Nature Med.* **3**, 793–6.

McBride K.E. & Summerfelt K.R. (1990) Improved binary vectors for *Agrobacterium*-mediated plant transformation. *Plant Mol. Biol.* **14**, 269–76.

McCabe D. & Christou P. (1993) Direct DNA transfer using electric discharge particle acceleration (Accell® technology). *Plant Cell Tissue Organ Cult.* **33**, 227–36.

McCabe D.E., Swain W.F., Martinell B.J. & Christou P. (1988) Stable transformation of soybean (*Glycine max*) by particle acceleration. *Biotechnology* **6**, 923–6.

McCarthy J.J. & Hilfiker R. (2000) The use of single-nucleotide polymorphism maps in pharmacogenomics. *Nature Biotechnol.* **18**, 505–8.

McClelland M., Mathieu-Daude F. & Welsh J. (1995) RNA fingerprinting and differential display using arbitrarily-primed PCR. *Trends Genet.* **11**, 242–6.

McCormick A.A., Kumagai M.H., Hanley K. *et al.* (1999) Rapid production of specific vaccines for lymphoma by expression of the tumor-derived single-chain Fv epitopes in tobacco plants. *Proc. Nat. Acad. Sci. USA* **96**, 703–8.

McCreath K.J., Howcroft J., Campbell K.H.S. *et al.* (2000) Production of gene targeted sheep by nuclear transfer from cultured somatic cells. *Nature* **405**, 1066–9.

McCutchan J.H. & Pango J.S. (1968) Enhancement of the infectivity of simian virus 40 deoxyribonucleic acid with diethylaminoethyl dextran. *J. Nat. Cancer Inst.* **41**, 351–7.

McDaniel R., Thamchaipenet A., Gustafsson C. *et al.* (1999) Multiple genetic modifications of the erythromycin polyketide synthase to produce a library of novel 'unnatural' natural products. *Proc. Nat. Acad. Sci. USA* **96**, 1846–51.

McDowell D. (1999) PCR: factors affecting reliability and validity. In *Analytical Molecular Biology: Quality and Validation*, eds Saunders G.C. & Parkes H.C., pp. 58–80. Royal Society of Chemistry, London.

McEachern M.J., Bott M.A., Tooker P.A. & Helinski D.R. (1989) Negative control of plasmid R6K replication: possible role of intermolecular coupling of replication origins. *Proc. Nat. Acad. Sci. USA* **86**, 7942–6.

McElroy D. & Brettel R.I.S. (1994) Foreign gene expression in transgenic cereals. *Trends Biotechnol.* **12**, 62–8.

McElroy D., Blowers A.D., Jenes B. & Wu R. (1991) Construction of expression vectors based on the rice actin-1 (Act-1), 5′ region for use in monocot transformation. *Mol. Gen. Genet.* **231**, 150–60.

McElroy D., Chamberlain D.A., Moon E. & Wilson K.J. (1995) Development of *gusA* reporter gene constructs for cereal transformation – availability of plant transformation vectors from the CAMBIA molecular genetic resource center. *Mol. Breeding* **1**, 27–37.

McGarvey P.B., Hammond J., Dienelt M.M. *et al.* (1995) Expression of the rabies virus glycoprotein in transgenic tomatoes. *Biotechnology* **13**, 1484–7.

Macgregor P.F., Abate C. & Curran T. (1990) Direct cloning of leucine zipper proteins: June binds cooperatively to the CRE with CRE-BP1. *Oncogene* **5**, 451–8.

McKnight R.A., Shamay A., Sankaran L., Wall R.J. & Hennighausen L. (1992) Matrix-attachment regions can impart position-independent regulation of a tissue-specific gene in transgenic mice. *Proc. Nat. Acad. Sci. USA* **89**, 6943–7.

McLain L., Porta C., Lomonossoff G.P., Durrani Z. & Dimmock, N.J. (1995) Human immunodeficiency virus type I neutralizing antibodies raised to a gp4l peptide expressed on the surface of a plant virus. *AIDS Res. Hum. Retroviruses* **11**, 327–34.

McLain L., Durrani Z., Wisniewski L.A. *et al.* (1996a) Stimulation of neutralising antibodies to human immunodeficiency virus type I in three strains of mice immunized with a 22 amino acid peptide peptide of gp4l expressed on the surface of a plant virus. *Vaccine* **14**, 799–810.

McLain L., Durrani Z., Dimmock N.J. *et al.* (1996b) A plant virus HIV-1 chimaera stimulates antibody that neutralises HIV-I. In *Vaccines 96*, eds Brown F., Burton D.R., Collier J., Mekalanos J. & Norrby E. Cold Spring Harbor Laboratory Press, Cold Spring Harbor, New York.

McLaughlin J.A., Davies L. & Seamark R.F. (1990) *In vitro* embryo culture in the production of identical Merino lambs by nuclear transplantation. *Reprod. Fert. Devel.* **2**, 619–22.

McLaughlin S.K., Collis P., Hermonat P.L. & Muzyczka N. (1988) Adeno-associated virus general transduction vectors: analysis of proviral structures. *J. Virol.* **62**, 1963–73.

Macleod K.F. & Jacks T. (1999) Insights into cancer from transgenic mouse models. *J. Pathol.* **187**, 43–60.

McMahon A.P. & Bradley A. (1990) The *Wnt-1* (*int-1*) proto-oncogene is required for development for a large region of the mouse brain. *Cell* **62**, 1073–85.

McPherson D.T. (1988) Codon preference reflects mistranslational constructs: a proposal. *Nucl. Acids Res.* **16**, 4111–20.

Madisen L., Krumm A., Hebbes T.R. & Groudine M. (1998) The immunoglobulin heavy chain locus control region increases histone acetylation along linked c-*myc* genes. *Mol. Cell Biol.* **18**, 6281–92.

Maeda S., Kawai T., Obinata M. *et al.* (1985) Production of human α-interferon in silkworm using a baculovirus vector. *Nature* **315**, 592–4.

Maes T., De Keukeleire P. & Gerats T. (1999) Plant tagnology. *Trends Plant Sci.* **4**, 90–6.

Magari S.R., Rivera V.M., Luliucci J.D., Gilman M. & Cerasoil F. Jr (1997) Pharmacologic control of a humanized gene therapy system implanted into nude mice. *J. Clin. Invest.* **100**, 2865–72.

Magnuson N.S., Linzmaier P.M., Reeves R. *et al.* (1998) Secretion of biologically active human interleukin-2 and interleukin-4 from genetically modified tobacco cells in suspension culture. *Protein Expr. Purif.* **13**, 45–52.

Magram J., Chada K. & Costantini F. (1985) Developmental regulation of a cloned adult β-globin gene in transgenic mice. *Nature* **315**, 338–40.

Maguin E., Prevost H., Ehrlich S.D. & Gruss A. (1996) Efficient insertional mutagenesis in lactococci and other Gram-positive bacteria. *J. Bacteriol.* **178**, 931–5.

Mahan M.J., Slauch J.M. & Mekalanos J.J. (1993) Selection of bacterial virulence genes that are specifically induced in host tissues. *Science* **259**, 686–8.

Maliga P. (1993) Towards plastid transformation in flowering plants. *Trends Biotechnol.* **11**, 101–7.

Malpartida R. & Hopwood D.A. (1984) Molecular cloning of the whole biosynthetic pathway of a *Streptomyces* antibiotic and its expression in a heterologous host. *Nature* **309**, 462–4.

Mancini M., Hadchouel M., Davis H.L. *et al.* (1996) DNA-mediated immunization in a transgenic mouse model of the hepatitis B surface antigen chronic carrier state. *Proc. Nat. Acad. Sci. USA* **93**, 12496–501.

Mandel M. & Higa A. (1970) Calcium-dependent bacteriophage DNA infection. *J. Mol. Biol.* **53**, 159–62.

Maniatis T., Kee S.G., Efstratiadis A. & Kafatos F.C. (1976) Amplification and characterisation of a β-globin gene synthesized *in vitro*. *Cell* **8**, 163–82.

Maniatis T., Hardison R.C., Lacy E. *et al.* (1978) The isolation of structural genes from libraries of eucaryotic DNA. *Cell* **15**, 687–701.

Mann R., Mulligan R.C. & Blatimore D. (1983) Construction of a retrovirus packaging mutant and its use to produce helper-free defective retrovirus. *Cell* **33**, 153–9.

Mansour S.L., Thomas K.R. & Capecchi M.R. (1988) Disruption of the proto-oncogene int-2 in mouse embryo-derived stem cells: a general strategy for targeting mutations to non-selectable genes. *Nature* **336**, 348–52.

Mansour S.L., Goddard J.M. & Capecchi M.R. (1993) Mice homozygous for a targeted disruption of the proto-oncogene *int-2* have developmental defects in the tail and inner ear. *Development* **117**, 13–28.

Mantei N., Boll W. & Weissman C. (1979) Rabbit beta-globin mRNA production in mouse L cells transformed with cloned rabbit beta-globin chromosomal DNA. *Nature* **281**, 40–6.

Marconi P., Krisky D., Oligino T. *et al.* (1996) Replication-defective herpes simplex virus vectors for gene transfer *in vivo. Proc. Nat. Acad. Sci. USA* **93**, 11319–20.

Margolskee R.F. (1992) Epstein–Barr virus-based expression vectors. *Curr. Top. Microbiol. Immunol.* **185**, 67–95.

Margolskee R.F., Kavathas P. & Berg P. (1988) Epstein–Barr virus shuttle vector for stable episomal replication of cDNA expression libraries in humans. *Mol. Cell. Biol.* **8**, 2337–47.

Marini N., Hiiyanna K.T. & Benbow R.M. (1989) Differential replication of circular DNA molecules co-injected into early *Xenopus laevis* embryos. *Nucl. Acids Res.* **17**, 5793–808.

Marinus M.G., Carraway M., Frey A.Z., Brown L. & Arraj J.A. (1983) Insertion mutations in the *dam* gene of *Escherichia coli* K-12. *Mol. Gen. Genet.* **192**, 288–9.

Markowitz D., Goff S. & Bank A. (1988) A safe packaging line for gene transfer: separating viral genes on two different plasmids. *J. Virol.* **62**, 1120–4.

Marshall E. (1999) Gene therapy death prompts review of adenovirus vector. *Science* **286**, 2244–5.

Marshall J., Molloy R., Moss G.W., Howe J.R. & Hughes T.E. (1995) The jellyfish green fluorescent protein: a new tool for studying ion channel expression and function. *Neuron* **14**, 211–15.

Marsh-Armstrong N., Huang H., Berry D.L. & Brown D.D. (1999) Germ line transmission of transgenes in *Xenopus laevis. Proc. Nat. Acad. Sci. USA* **96**, 14389–93.

Martin G.R. (1981) Isolation of a pluripotent cell line from early mouse embryos cultured in medium conditioned by teratocarcinoma stem cells. *Proc. Nat. Acad. Sci. USA* **78**, 7634–8.

Martinez A., Sparks C., Hart C.A., Thompson J. & Jepson I. (1999) Ecdysone agonist inducible transcription in transgenic tobacco plants. *Plant J.* **19**, 97–106.

Martzen M.R., McCraith S.M., Spinelli S.L. *et al.* (1999) A biochemical genomics approach for identifying genes by the activity of their products. *Science* **286**, 1153–5.

Maruyama K. & Sugano S. (1994) Oligo-capping: a simple method to replace the cap structure of eukaryotic mRNAs with oligoribonucleotides. *Gene* **138**, 171–4.

Mascarenhas D., Mettler I.J., Pierce D.A. & Lowe H.W. (1990) Intron-mediated enhancement of heterologous gene expression in maize. *Plant Mol. Biol.* **15**, 913–20.

Mason D.A., Moore J.D., Green S.A. & Liggett S.S. (1999) A gain-of-function polymorphism in a G-protein coupling domain of the human beta1-adrenergic receptor. *J. Biol. Chem.* **274**, 12670–4.

Mason H.S. & Arntzen C.J. (1995) Transgenic plants as vaccine production systems. *Trends Biotechnol.* **13**, 388–92.

Mason H.S., Lam D.M. & Arntzen C.J. (1992) Expression of hepatitis B surface antigen in transgenic plants. *Proc. Nat. Acad. Sci. USA* **89**, 11745–9.

Mason H.S., Ball J.M., Shi J.J. *et al.* (1996) Expression of Norwalk virus capsid protein in transgenic tobacco and potato and its oral immunogenicity in mice. *Proc. Nat. Acad. Sci. USA* **93**, 5335–40.

Massoud M., Bischoff R., Dalemans W. *et al.* (1991) Expression of active recombinant human alpha l-antitrypsin in transgenic rabbits. *J. Biotechnol.* **18**, 193–204.

Massoud M., Attal J., Thepot D. *et al.* (1996) The deleterious effects of human erythropoietin gene driven by the rabbit whey acidic protein gene promoter in transgenic rabbits. *Reprod. Nutr. Devel.* **36**, 555–63.

Masu Y., Nakayama K., Tamaki H. *et al.* (1987) cDNA cloning of bovine substance-K receptor through oocyte expression system. *Nature* **329**, 836–8.

Matsumoto K., Yoshimatsu T. & Oshima Y. (1983) Recessive mutations conferring resistance to carbon catabolite repression of galactokinase synthesis in *Saccharomyces cerevisiae. J. Bacteriol.* **153**, 1405–14.

Matsumoto S., Ikura K., Ueda M. & Sasaki R. (1995) Characterisation of a human glycoprotein (erythropoetin) produced in cultured tobacco cells. *Plant Mol. Biol.* **27**, 1163–72.

Mattila P., Korpela J., Tenkanen T. & Pitkanen K. (1991) Fidelity of DNA synthesis by the *Thermococcus litoralis* DNA polymerase – an extremely heat stable enzyme with proofreading activity. *Nucl. Acids Res.* **18**, 4967–73.

Matz M.V., Fradkov A.F., Labas Y.A. *et al.* (1999) Fluorescent proteins from nonbioluminescent *Anthozoa* species. *Nature Biotechnol.* **17**, 969–73.

Matzeit V., Schaefer S., Kammann M. *et al.* (1991) Wheat dwarf virus vectors replicate and express foreign genes in cells of monocotyledonous plants. *Plant Cell* **3**, 247–58.

Matzke A.J.M. & Chilton M.-D. (1981) Site-specific insertion of genes into T-DNA of the *Agrobacterium* tumour-inducing plasmid: an approach to genetic engineering of higher plant cells. *J. Mol. Appl. Genet.* **1**, 39–49.

Matzke A.J.M. & Matzke M.A. (1998) Position effects and epigenetic silencing of plant transgenes. *Curr. Opin. Plant Biol.* **1**, 142–8.

Matzuk M.M., Finegold M.J., Su J.G., Hsueh A.J. & Bradley A. (1992) Alpha inhibin is a tumour-suppressor gene with gonadal specificity in mice. *Nature* **360**, 313–19.

Maxam A.M. & Gilbert W. (1977) A new method for sequencing DNA. *Proc. Nat. Acad. Sci. USA* **74**, 560–4.

Mazodier P. & Davies J. (1991) Gene transfer between distantly related bacteria. *Ann. Rev. Genet.* **25**, 147–71.

Mazodier P., Petter R. & Thompson C. (1989) Intergeneric conjugation between *Escherichia coli and Streptomyces* species. *J. Bacteriol.* **171**, 3583–5.

McMahon A.P. & Bradley A. (1990) The *Wnt-1* (*int-1*) proto-oncogene is required for development for a large region of the mouse brain. *Cell* **62**, 1073–85.

McPherson D.T. (1988) Codon preference reflects mistranslational constructs: a proposal. *Nucl. Acids Res.* **16**, 4111–20.

Mead D.A., Pey N.K., Herrnstadt C., Marcil R.A. & Smith L.M. (1991) A universal method for the direct cloning of PCR amplified nucleic acid. *Biotechnology* **9**, 657–63.

Meade H. (1999) Taking ATIII from goats through clinical trials. In *Transgenic Animal Research Conference, Tahoe City USA*, p. 33.

Meade H., Gates L., Lacy E. & Lonberg N. (1990) Bovine alpha$_{S1}$-casein gene sequences direct high level expression of active human urokinase in mouse milk. *Biotechnology* **8**, 443–6.

Mecsas J. & Sugden B. (1987) Replication of plasmids derived from bovine papilloma virus type 1 and Epstein–Barr virus in cells in culture. *Ann. Rev. Cell Biol.* **3**, 87–108.

Medberry S.L., Dale E.C., Qin M. & Ow D.W. (1995) Intrachromosomal rearrangements generated by site-specific recombination. *Nucl. Acids Res.* **23**, 485–90.

Melchers L.S. & Stuiver M.H. (2000) Novel genes for disease-resistance breeding. *Curr. Opin. Plant Biol.* **3**, 147–52.

Mellon P., Parker V., Gluzman Y. & Maniatis T. (1981) Identification of DNA sequences required for transcription of the human α-globin gene in a new SV40 host–vector system. *Cell* **27**, 279–88.

Mellor J., Malim M., Gull K. *et al.* (1985) Reverse transcriptase activity and Ty RNA are associated with virus-like particles in yeast. *Nature* **318**, 583–6.

Mellor J., Dobson M., Kingsman A.J. & Kingsman S.M. (1987) A transcriptional activator is located in the coding region of the yeast PGK gene. *Nucl. Acids Res.* **15**, 6243–59.

Melton D.A. (1987) Translation of messenger RNA in injected frog oocytes. *Methods Enzymol.* **152**, 288–96.

Memelink J., Menke F.L.H., Van der Fits L. & Kijne J.W. (2000) Transcriptional regulators to modify secondary metabolism. In *Metabolic Engineering of Plant Secondary Metabolism*, eds Verpoorte R. & Alfermann A.W., pp. 111–25. Kluwer Academic Press, The Netherlands.

Mendez M.J., Green L.L., Corvalan J.R. *et al.* (1997) Functional transplant of megabase human immunoglobulin loci recapitulates human antibody response in mice. *Nature Genet.* **15**, 146–56.

Merchlinsky M., Eckert D., Smith E. & Zauderer M. (1997) Construction and characterization of vaccinia direct ligation vectors. *Virology* **238**, 444–51.

Mermod N., Ramos J.L., Lehrbach P.R. & Timmis K.N. (1986) Vector for regulated expression of cloned genes in a wide range of Gram-negative bacteria. *J. Bacteriol.* **167**, 447–54.

Meselson M. & Yuan R. (1968) DNA restriction enzyme from *E. coli. Nature* **217**, 1110–14.

Messing J., Gronenborn B., Muller-Hill B. & Hofschneider P.H. (1977) Filamentous coliphage M13 as a cloning vehicle: insertion of a *Hind*II fragment of the *lac* regulatory region in M13 replicative form *in vitro. Proc. Nat. Acad. Sci. USA* **74**, 3642–6.

Messing J., Crea R. & Seeburg P.H. (1981) A system for shotgun DNA sequencing. *Nucl. Acids Res.* **9**, 309–21.

Mett V.L., Lochhead L.B. & Reynolds P.H.S. (1993) Copper controllable gene expression system for whole plants. *Proc. Nat. Acad. Sci. USA* **90**, 4567–71.

Metzger D. & Feil R. (1999) Engineering the mouse genome by site-specific recombination. *Curr. Opin. Biotechnol.* **10**, 470–6.

Metzger D., White J.H. & Chambon P. (1988) The human estrogen receptor functions in yeast. *Nature* **334**, 31–6.

Mewes H.W., Albermann K., Bahr M. *et al.* (1997) Overview of the yeast genome. *Nature* 387 (Suppl. 7–65).

Meyer P., Heidmann I., Forkmann G. & Saedler H. (1987) A new petunia flower colour generated by transformation of a mutant with a maize gene. *Nature* **330**, 677–8.

Meyer P., Heidmann I. & Niedenhof I. (1992) The use of African cassava mosaic virus as a vector system for plants. *Gene* **110**, 213–17.

Michel B., Palla E., Niaudet B. & Ehrlich S.D. (1980) DNA cloning in *Bacillus subtilis*. III. Efficiency of random-segment cloning and insertional inactivation vectors. *Gene* **12**, 147–54.

Michel B., Niaudet B. & Ehrlich S.D. (1982) Intramolecular recombination during plasmid transformation of *Bacillus subtilis* competent cells. *EMBO J.* **1**, 1565–71.

Michelson A.M. & Orkin S.H. (1982) Characterization of the homoploymer tailing reaction catalyzed by terminal deoxynucleotidyl transferase: implications for the cloning of cDNA. *J. Biol. Chem.* **256**, 1473–82.

Mierendorf R.C., Percy C. & Young R.A. (1987) Gene isolation by screening λgt11 libraries with antibodies. *Methods Enzymol.* **152**, 458–69.

Miller A.D. (1992a) Human gene therapy comes of age. *Nature* **357**, 455–60.

Miller A.D. (1992b) Retroviral vectors. *Curr. Top. Microbiol. Immunol.* **185**, 1–24.

Miller A.D., Jolly D.J., Friedmann T. & Verma I.M. (1983) A transmissible retrovirus expressing human hypoxanthine phosphoriboryltransferase (HPRT): gene transfer into cells obtained from humans deficient in HPRT. *Proc. Nat. Acad. Sci. USA* **80**, 4709–13.

Miller C.A., Beaucage S.L. & Cohen S.N. (1990) Role of DNA superhelicity in partitioning of the pSC101 plasmid. *Cell* **62**, 127–33.

Miller D.G., Adam M.A. & Miller A.D. (1990) Gene transfer by retrovirus vectors occurs only in cells that are actively replicating at the time of infection. *Mol. Cell Biol.* **10**, 4239–42.

Miller G. (1985) Epstein–Barr virus. In *Virology*, ed. Fields B.N., pp. 563–89. Raven Press, New York.

Miller L.K., Miller D.W. & Adang M.J. (1983) An insect virus for genetic engineering: developing baculovirus polyhedrin substitution vectors. In *Genetic Engineering in Eukaryotes*, eds Lurgin P.F. & Kleinhofs A. Plenum Press, New York.

Milman G. & Herzberg M. (1981) Efficient DNA transfection and rapid assay for thymidine kinase activity and viral antigenic determinants. *Somat. Cell Genet.* **7**, 161–70.

Milot E., Strouboulis J., Trimborn T. *et al.* (1996) Heterochromatin effects on the frequency and duration of LCR-mediated gene transcription. *Cell* **87**, 105–14.

Minshull J. & Stemmer W. (1999) Protein evolution by molecular breeding. *Curr. Opin. Chem. Biol.* **3**, 284–90.

Mirkovitch J., Mirault M.E. & Laemmli U.K. (1984) Organization of the higher-order chromatin loop: specific DNA attachment sites on nuclear scaffold. *Cell* **39**, 223–32.

Miroux B. & Walker J.E. (1996) Over-production of protein in *Escherichia coli*: mutant hosts that allow synthesis of some membrane proteins and globular proteins at high levels. *J. Mol. Biol.* **260**, 289–98.

Mlynarova L., Loonen A., Heldens J. *et al.* (1994) Reduced position effect in mature transgenic plants conferred by the chicken lysozyme matrix-attachment region. *Plant Cell* **6**, 417–26.

Moerman D.G., Benian G.M. & Waterston R.H. (1986) Molecular cloning of the muscle gene unc-22 in *Caenorhabditis elegans* by Tc1 transposon tagging. *Proc. Nat. Acad. Sci. USA* **83**, 2579–83.

Monahan P.E. & Samulski R.J. (2000) Adeno-associated virus vectors for gene therapy: more pros than cons? *Mol. Med. Today* **6**, 433–40.

Montaya A.L., Chilton M.-D., Gordon M.P., Sciaky D. & Nester E.W. (1977) Octopine and nopaline metabolism in *Agrobacterium tumefaciens* and crown gall tumor cells: role of plasmid genes. *J. Bacteriol.* **129**, 101–7.

Moolten F.L. (1986) Tumour chemosensitivity conferred by inserted herpes thymidine kinase genes: paradigm for a prospective cancer control strategy. *Cancer Res.* **46**, 5276–81.

Moore R.C., Redhead N.J., Selfridge J. *et al.* (1995) Double replacement gene targeting for the production of a series of mouse strains with different prion protein gene alterations. *Biotechnology* **13**, 999–104.

Morgan D.O. & Roth R.A. (1988) Analysis of intracellular protein function by antibody injection. *Immunol. Today* **9**, 84–8.

Morsey M.A. & Caskey C.T. (1999) Expanded-capacity adenoviral vectors – the helper-dependent vectors. *Mol. Med. Today* **5**, 18–24.

Moss B. (1996) Genetically engineered poxviruses for recombinant gene expression, vaccination and safety. *Proc. Nat. Acad. Sci. USA* **93**, 11341–8.

Mottes M., Grandi G., Sgaramella V. *et al.* (1979) Different specific activities of the monomeric and oligomeric forms of plasmid DNA in transformation of *B. subtilis* and *E. coli. Mol. Gen. Genet.* **174**, 281–6.

Mountford P.S. & Smith A.G. (1995) Internal ribosome entry sites and dicistronic RNAs in mammalian transgenesis. *Trends Genet.* **11**, 179–84.

Moxham C.M., Hod Y. & Malbon C. (1993) Induction of G_{ia2}-specific antisense RNA *in vivo* inhibits neonatal growth. *Science* **260**, 991–5.

Muesing M., Tamm J., Shepard H.M. & Polisky B. (1981) A single base pair alteration is responsible for the DNA overproduction phenotype of a plasmid copy-number mutant. *Cell* **24**, 235–42.

Muller S., Sanda I.S., Kamp-Hansen P. & Dalboge H. (1998) Comparison of expression systems in the yeasts *Saccharomyces cerevisiae, Hansenula polymorpha, Kluyveromyces lactis, Schizosaccharomyces pombe* and *Yarrowia lipolytica*: cloning of two novel promoters from *Y. lipolytica. Yeast* **14**, 1267–83.

Muller U. (1999) Ten years of gene targeting: targeted mouse mutants, from vector design to phenotype analysis. *Mech. Devel.* **82**, 3–21.

Muller W.J., Naujokas M.A. & Hassel J.A. (1984) Isolation of large T antigen-producing mouse cell lines capable of supporting replication of polyomavirus-plasmid recombinants. *Mol. Cell. Biol.* **4**, 2406–12.

Muller-Rober B., Sonnewald U. & Willmitzer L. (1992) Inhibition of the ADP-glucose pyrophosphorylase in transgenic potatoes leads to sugar-storing tubers and influences tuber formation and expression of tuber storage proteins. *EMBO J.* **11**, 1229–38.

Mulligan R.C. & Berg P. (1980) Expression of a bacterial gene in mammalian cells. *Science* **209**, 1422–7.

Mulligan R.C. & Berg P. (1981a) Factors governing the expression of a bacterial gene in mammalian cells. *Mol. Cell. Biol.* **1**, 449–59.

Mulligan R.C. & Berg P. (1981b) Selection for animal cells that express the *Escherichia coli* gene coding for xanthine-guanine phosphoribosyl-transferase. *Proc. Nat. Acad. Sci. USA* **78**, 2072–6.

Mullis K.B. (1990) The unusual origin of the polymerase chain reaction. *Sci. Am.* **262**, 56–65.

Mullis K.B. & Faloora F. (1987) Specific synthesis of DNA *in vitro* via a polymerase catalyzed chain reaction. *Methods Enzymol.* **155**, 335–50.

Munir M., Rossiter B. & Caskey C. (1990) Antisense RNA production in transgenic mice. *Somat. Cell Mol. Genet.* **16**, 383–94.

Muotri A.R., Da Veiga Pereira L., Dos Reis Vasques L. & Menck C.F.M. (1999) Ribozymes and the anti-gene therapy: how a catalytic RNA can be used to inhibit gene function. *Gene* **237**, 303–10.

Murdock D., Ensley B.D., Serdar C. & Thalen M. (1993) Construction of metabolic operons catalyzing the *de novo* biosynthesis of indigo in *Escherichia coli*. *Biotechnology* **11**, 381–6.

Murphy D.J. (1999) Production of novel oils in plants. *Curr. Opin. Biotechnol.* **10**, 175–80.

Murray K. & Murray N.E. (1975) Phage lambda receptor chromosomes for DNA fragments made with restriction endonuclease III of *Haemophilus influenzae* and restriction endonuclease I of *Escherichia coli*. *J. Mol. Biol.* **98**, 551–64.

Murray M.J., Kaufman R.J., Latt S.A. & Weinberg R.A. (1983) Construction and use of a dominant, selectable marker: a Harvey sarcoma virus-dihydrofolate reductase chimera. *Mol. Cell. Biol.* **3**, 32–43.

Murray N.E. (1983) Phage lambda and molecular cloning. In *The Bacteriophage Lambda*, eds Hendrix R.W., Roberts J.W., Stahl F.W. & Weisberg R.A., Lambda II (Monograph No. 13), Vol. 2. Cold Spring Harbor Laboratory, Cold Spring Harbor, New York.

Murray N.E. & Kelley W.S. (1979) Characterization of λ*pol*A transducing phages: effective expression of the *E. coli pol* A gene. *Mol. Gen. Genet.* **175**, 77–87.

Murray N.E., Manduca de Ritis P. & Foster L.A. (1973a) DNA targets for the *Escherichia coli* K restriction system analysed genetically in recombinants between phages phi-80 and Lambda. *Mol. Gen. Genet.* **120**, 261–81.

Murray N.E., Batten P.L. & Murray K. (1973b) Restriction of bacteriophage lambda by *Escherichia coli* K. *J. Mol. Biol.* **81**, 395–407.

Mutzel R., Baeuerle A., Jung S. & Dammann H. (1990) Prophage lambda libraries for isolating cDNA clones by functional screening. *Gene* **96**, 205–11.

Muyrers J.P.P., Zhang Y., Benes V. *et al.* (2000) Point mutation of bacterial artificial chromosomes by ET recombination. *EMBO Rep.* **1**, 239–43.

Muzyczka N. (1992) Use of adeno-associated virus as a general transduction vector for mammalian cells. *Curr. Top. Microbiol. Immunol.* **158**, 97–129.

Myagkikh M., Alpanah S., Markham P.D. *et al.* (1996) Multiple immunizations with attenuated poxvirus HIV type 2 recombinants and subunit boosts required for protection of rhesus macaques. *AIDS Res. Hum. Retroviruses* **12**, 985–92.

Myers R.M. & Tjian R. (1980) Construction and analysis of simian virus 40 origins defective in tumor antigen binding and DNA replication. *Proc. Nat. Acad. Sci. USA* **77**, 6491–5.

Nagaraju J., Kanda T., Yukuhiro K. *et al.* (1996) Attempt at transgenesis of the silkworm (*Bombyx mon* L.) by egg-injection of foreign DNA. *Appl. Entomol. Zool.* **31**, 487–96.

Nagatani N., Honda H., Shimada T. & Kobayashi T. (1997) DNA delivery into rice cells and transformation using silicon carbide whiskers. *Biotechnol. Techniques* **11**, 471–3.

Nakai H., Storm T.A. & Kay M.A. (2000) Increasing the size of rAAV-mediated expression cassettes *in vivo* by intermolecular joining of two complementary vectors. *Nature Biotechnol.* **18**, 527–32.

Naldini L. (1998) Lentiviruses as gene transfer agents for delivery to non-dividing cells. *Curr. Opin. Biotechnol.* **9**, 457–63.

Naldini L., Blomer U., Gallay O. *et al.* (1996) *In vivo* gene delivery and stable transfection of nondividing cells by a lentiviral vector. *Science* **272**, 263–7.

Napoli C., Lemieux C.H. & Jorgensen R. (1990) Introduction of a chimeric chalcone synthese gene into petunia results in reversible co-suppression of homologous genes *in trans*. *Plant Cell* **2**, 279–89.

Nasmyth K.A. & Reed S.I. (1980) Isolation of genes by complementation in yeast: molecular cloning of a cell-cycle gene. *Proc. Nat. Acad. Sci. USA* **77**, 2119–23.

Naviaux R.K. & Verma I.M. (1992) Retroviral vectors for persistent expression *in vivo*. *Curr. Biol.* **3**, 540–7.

Naylor L.H. (1999) Reporter gene technology – the future looks bright. *Biochem. Pharmacol.* **58**, 749–57.

Nedivi E., Hevroni D., Nato D., Israeli D. & Citri Y. (1993) Numerous candidate plasticity-related genes revealed by differential cDNA cloning. *Nature* **363**, 718–22.

Negrotto D., Jolley M., Beer S., Wenck A.R. & Hansen G. (2000) The use of phosphomannose isomerase as a selectable marker to recover transgenic maize plants (*Zea mays* L.) via *Agrobacterium* transformation. *Plant Cell Rep.* **19**, 798–803.

Negrutiu I., Shillito R., Potrykus I., Biasini G. & Sala F. (1987) Hybrid genes in the analysis of transformation conditions. I. Setting up a simple method for direct gene transfer in plant protoplasts. *Plant Mol. Biol.* **8**, 363–73.

Nellen W. & Lichtenstein C. (1993) What makes an mRNA anti-sense-itive? *Trends Biochem. Sci.* **18**, 419–23.

Ness J.E., Welch M., Giver L. *et al.* (1999) DNA shuffling of subgenomic sequences of subtilisin. *Nature Biotechnol.* **17**, 893–6.

Ness J.E., Del Cardayre S.B., Minshull J. & Stemmer W.P. (2000) Molecular breeding: the natural approach to protein design. *Adv. Protein Chem.* **55**, 261–92.

Newman J.R. & Fuqua C. (1999) Broad-host-range expression vectors that carry the ʟ-arabinose-inducible *Escherichia coli araBAD* promoter and the *araC* regulator. *Gene* **227**, 197–203.

Ng R. & Abelson J. (1980) Isolation and sequence of the gene for actin in *Saccharomyces cerevisiae*. *Proc. Nat. Acad. Sci. USA* **77**, 3912–16.

Niaudet B., Goze A. & Ehrlich S.D. (1982) Insertional mutagenesis in *Bacillus subtilis*: mechanism and use in gene cloning. *Gene* **19**, 277–84.

Nida D.L., Kolacz K.H., Buehler R.E. *et al.* (1996) Glyphosate-tolerant cotton: genetic characterization and protein expression. *J. Agric. Food Chem.* **44**, 1960–6.

Nielsen L.B., Kahn D., Duell T. *et al.* (1998) Apolipoprotein B gene expression in a series of human apolipoprotein B transgenic mice generated with recA-assisted restriction endonuclease cleavage modified bacterial artificial chromosomes: an intestine-specific enhancer element is located between 54 and 62 kilobases 5' to the structure gene. *J. Biol. Chem.* **273**, 21800–7.

Nir I., Kedzierski W., Chen S. & Travis G.H. (2000) Expression of Bcl-2 protects against photoreceptor degeneration in retinal degeneration slow (rds) mice. *J. Neurosci.* **20**, 2150–4.

Nitsche E.M., Moquin A., Adams P.S. *et al.* (1996) Differential display RT-PCR of total RNA from human foreskin fibroblasts for investigation of androgen-dependent gene expression. *Am. J. Med. Genet.* **63**, 231–8.

No D., Yao T.-P. & Evans R.M. (1996) Ecdysone-inducible gene expression in mammalian cells and transgenic mice. *Proc. Nat. Acad. Sci. USA* **93**, 3346–51.

Noolandi J., Forsyth J. & Shi A-C. (1993) Sequencing using pulsed field and image reconstruction. *Proc. Nat. Acad. Sci.* **90**, 10101–4.

Noma Y., Sideras P., Natto T. *et al.* (1986) Cloning of cDNA encoding the murine IgG1 induction factor by a novel strategy using SP6 promoter. *Nature* **319**, 640–6.

Norgren R.B. Jr & Lehman M.N. (1998) Herpes simplex virus as a transneuronal tracer. *Neurosci. Biobehav. Rev.* **22**, 695–708.

Norman T.C., Smith D.L., Sorger P.K. *et al.* (1999) Genetic selection of peptide inhibitors of biological pathways. *Science* **285**, 591–5.

Norrander J., Kempe T. & Messing J. (1983) Construction of improved M13 vectors using oligodeoxynucleotide-directed mutagenesis. *Gene* **27**, 101–6.

Norton P.A. & Coffin J.M. (1985) Bacterial β-galactosidase as a marker of Rous sarcoma virus gene expression and replication. *Mol. Cell Biol.* **5**, 281–90.

Nugent M.E., Primrose S.B. & Tacon W.C.A. (1983) The stability of recombinant DNA. *Devel. Ind. Microbiol.* **24**, 271–85.

Nunberg J.H., Wright D.K., Cole G.E. *et al.* (1989) Identification of the thymidine kinase gene of feline herpesvirus: use of degenerate oligonucleotides in the polymerase chain reaction to isolate herpesvirus gene homologs. *J. Virol.* **63**, 3240–9.

Nur I., Szyf M., Razin A. *et al.* (1985) Procaryotic and eucaryotic traits of DNA methylation in spiroplasmas (mycoplasmas). *J. Bacteriol.* **164**, 19–24.

Oard J.H., Linscombe S.D., Braverman M.P. *et al.* (1996) Development, field evaluation, and agronomic performance of transgenic herbicide resistant rice. *Mol. Breeding* **2**, 359–68.

Ochman H., Gerber S.A. & Hartl D.L. (1988) Genetic applications of an inverse polymerase chain reaction. *Genetics* **120**, 621–5.

Ockert D., Schmitz M., Hampl M. & Rieber E.P. (1999) Advances in cancer immunotherapy. *Immunol. Today* **20**, 63–5.

Odell J.T., Hoopes J.L. & Vermerris W. (1994) Seed-specific gene activation mediated by the Cre/lox site specific recombination system. *Plant Physiol.* **106**, 447–58.

Oeller P.W., Miw-Wong L., Taylor L.P., Pike D.A. & Theologis A. (1991) Reversible inhibition of tomato fruit senescence by antisense RNA. *Science* **254**, 437–9.

Offield M.F., Hirsch N. & Grainger R.M. (2000) The development of *Xenopus tropicalis* transgenic lines and their use in studying lens development timing in living embryos. *Development* **127**, 1789–97.

Ogonah O.W., Freedman R.B., Jenkins N., Patel K. & Rooney B.C. (1996) Isolation and characterization of an insect cell line able to perform complex N-linked glycosylation on recombinant proteins. *Nature Biotechnol.* **14**, 197–202.

O'Gorman S., Fox D.T. & Wahl G.M. (1991) Recombinase mediated gene activation and site-specific integration in mammalian cells. *Science* **251**, 1351–5.

O'Gorman S., Dagenais N.A., Qian M. & Marchuk Y. (1997) Protamine-Cre recombinase transgenes efficiently recombine target sequences in the male germ line of mice, but not in embryonic stem cells. *Proc. Nat. Acad. Sci. USA* **94**, 14602–7.

Oh S.H. & Chater K.F. (1997) Denaturation of circular or linear DNA facilitates targeted integrative transformation of *Streptomyces coelicolor* A3 (2): possible relevance to other organisms. *J. Bacteriol.* **179**, 122–7.

O'Hare K. & Rubin G.M. (1983) Structures of P transposable elements and their sites of insertion and excision in the *Drosophila melanogaster* genome. *Cell* **34**, 25–35.

O'Hare K., Benoist C. & Breathnach R. (1981) Transformation of mouse fibroblasts to methotrexate resistance. *Proc. Nat. Acad. Sci. USA* **78**, 1527–31.

Okamato K. & Beach D. (1994) Cyclin G is a transcription target of the p53 tumor suppressor protein. *EMBO J.* **13**, 4816–22.

O'Kane C.J. & Gehring W.J. (1987) Detection *in situ* of genetic regulatory elements in *Drosophila*. *Proc. Nat. Acad. Sci. USA* **84**, 9123–7.

O'Kane C.J. & Moffat K.G. (1992) Selective cell ablation and genetic surgery. *Curr. Opin. Genet. Devel.* **2**, 602–7.

Okayama H. & Berg P. (1982) High-efficiency cloning of full-length cDNA. *Mol. Cell. Biol.* **2**, 161–70.

Okubo K. & Matsubara K. (1997) Complementary DNA sequence (EST) collections and the expression information of the human genome. *FEBS Lett.* **403**, 225–9.

Okubo K., Hori N., Matoba R. *et al.* (1992) Large scale cDNA sequencing for analysis of quantitative and qualitative aspects of gene expression. *Nature Genet.* **2**, 173–9.

Old J.M., Ward R.H.T., Petrov M. *et al.* (1982) First trimester diagnosis for haemoglobinopathies: a report of 3 cases. *Lancet* **ii**, 1413–16.

Old R., Murray K. & Roizes G. (1975) Recognition sequence of restriction endonuclease III from *Haemophilus influenzae*. *J. Mol. Biol.* **92**, 331–9.

Oliver S.G. (1996) From DNA sequence to biological function. *Nature* **379**, 597–600.

Oliver S.G. & 146 others (1992) The complete DNA sequence of yeast chromosome III. *Nature* **357**, 38–46.

Olivera B.M., Hall Z.W. & Lehman I.R. (1968) Enzymatic joining of polynucleotides. V. A DNA adenylate intermediate in the polynucleotide joining reaction. *Proc. Nat. Acad. Sci. USA* **61**, 237–44.

Olsen M.J., Stephens D., Griffiths D. *et al.* (2000) Function-based isolation of novel enzymes from a large library. *Nature Biotechnol.* **18**, 1071–4.

Olson M.V. (1981) Applications of molecular cloning to *Saccharomyces*. In *Genetic Engineering*, eds Setlow J.K. & Hollaender A. Plenum Press, New York.

Olszewska E. & Jones K. (1988) Vacuum blotting enhances nucleic acid transfer. *Trends Genet.* **4**, 92–4.

Omirulleh S., Abraham M., Golovkin M. *et al.* (1993) Activity of a chimeric promoter with doubled CaMV 35S enhancer element in protoplast-derived cells and transgenic plants in maize. *Plant Mol. Biol.* **21**, 415–28.

Ooms G., Hooykaas P.J.J., Moolenaar G. & Schilperoort R.A. (1981) Crown gall plant tumours of abnormal morphology induced by *Agrobacterim tumefaciens* carrying mutated octopine Ti plasmids: analysis of T-DNA functions. *Gene* **14**, 33–50.

O'Reilley D.R., Miller L.K. & Luckow V.A. (1992) *Baculovirus Expression Vectors: A Laboratory Manual.* W.H. Freeman, San Francisco.

Orkin S.H., Little P.F.R., Kazazian H.H. & Boehm C. (1982) Improved detection of the sickle mutation by DNA analysis. *N. Engl. J. Med.* **307**, 32–6.

Orkin S.H., Daddona P.E., Shewach D.S. *et al.* (1983) Molecular cloning of human adenosine deaminase gene sequences. *J. Biol. Chem.* **258**, 2753–6.

Orrantia E. & Chang P.L. (1990) Intracellular distribution of DNA internalized through calcium phosphate precipitation. *Exp. Cell Res.* **190**, 170–4.

Orr-Weaver T.L., Szostak J.W. & Rothstein R.L. (1981) Yeast transformation: a model system for the study of recombination. *Proc. Nat. Acad. Sci. USA* **78**, 6354–8.

Osborne M.A., Zenner G., Lubinus M. *et al.* (1996) The inositol 5′-phosphatase SHIP binds to immunoreceptor signaling motifs and responds to high affinity IgE receptor aggregation. *J. Biol. Chem.* **271**, 29271–8.

Osbourne B.I., Wirtz U. & Baker B. (1995) A system for insertional mutagenesis and chromosomal rearrangement using the *Ds* transposon and Cre-*lox*. *Plant J.* **7**, 687–701.

Osman N., McKenzie I.F.C., Ostenreid K. *et al.* (1997) Combined transgenic expression of α-galactosidase and α1,2-fucosyltransferase leads to optimal reduction in the major xenoepitope Galα (1,3) Gal. *Proc. Nat. Acad. Sci. USA* **94**, 14677–82.

Ostrowski M.C., Richard-Foy H., Wolford R.G., Berard D.S. & Hager G.L. (1983) Glucocorticoid regulation of transcription at an amplified episomal promoter. *Mol. Cell. Biol.* **3**, 2045–57.

O'Sullivan D.J., Walker S.A., West S.G. & Klaenhammer T.R. (1996) Development of an expression strategy using a lytic phage to trigger explosive plasmid amplification and gene expression. *Bio/Technology* **14**, 82–7.

Ow D.W. (1996) Recombinase-directed chromosome engineering in plants. *Curr. Opin. Biotechnol.* **7**, 181–6.

Ow D.W., Wood K.V., DeLuca M. *et al.* (1986) Transient and stable expression of the firefly luciferase gene in plant cells and transgenic plants. *Science* **234**, 856–9.

Padgett H.S., Epel B.L., Kahn T.W. *et al.* (1996) Distribution of tombavirus movement in infected cells and implications for cell-to-cell spread of infection. *Plant J.* **10**, 1079–88.

Padgette S.R., Kolacz K.H., Delannay X. *et al.* (1996) Development, identification, and characterization of a glyphosate-tolerant soybean line. *Crop Sci.* **35**, 1451–61.

Pain B., Chenevier P. & Samarut J. (1999) Chicken embryonic stem cells and transgenic strategies. *Cells Tissues Organs* **165**, 212–19.

Palauqui J.C., Elmayan T., Pollien J.M. & Vaucheret H. (1997) Systemic acquired silencing: transgene-specific post-transcriptional silencing is transmitted by grafting from silenced stocks to non-silenced scions. *EMBO J.* **16**, 4738–45.

Pal-Bhadra M., Bhadra U. & Birchler J.A. (1997) Cosuppression in *Drosophila*: gene silencing of alcohol dehydrogenase by *white-Adh* transgenes is polycomb-dependent. *Cell* **90**, 479–90.

Palmer K.E. & Rybicki E.P. (1997) The use of geminiviruses in biotechnology and plant molecular biology, with particular focus on Mastreviruses. *Plant Sci.* **129**, 115–30.

Palmiter R.D. & Brinster R.L. (1986) Germ-line transformation of mice. *Ann. Rev. Genet.* **20**, 465–99.

Palmiter R.D., Brinster R.L., Hammer R.E. *et al.* (1982a) Dramatic growth of mice that develop from eggs microinjected with metallothionein–growth hormone fusion genes. *Nature* **300**, 611–15.

Palmiter R.D., Chen H.Y. & Brinster R.L. (1982b) Differential regulation of metallothionein–thymidine kinase fusion genes in transgenic mice and their offspring. *Cell* **29**, 701–10.

Palzkill T.G. & Newlon C.S. (1988) A yeast replication origin consists of multiple copies of a small conserved sequence. *Cell* **53**, 441–50.

Pansegrau W., Lanka E., Barth P. *et al.* (1994) Complete nucleotide sequence of Birmingham IncP alpha plasmids: compilation and comparative analysis. *J. Mol. Biol.* **239**, 623–63.

Pantoliano M.W., Whitlow M., Wood J.E. *et al.* (1988) The engineering of binding affinity at metal ion binding sites for the stability of proteins: subtilisin as a test case. *Biochemistry* **27**(22), 8311–17.

Pantoliano M.W., Whitlow M., Wood J.F. *et al.* (1989) Large increases in general stability for subtilisin BPN′ through incremental changes in the free energy of unfolding. *Biochemistry* **28**(18), 7205–13.

Paoletti E. (1996) Applications of pox virus vectors to vaccination: an update. *Proc. Nat. Acad. Sci. USA* **93**, 11349–53.

Parekh R.N., Shaw M.R. & Wittrup K.D. (1996) An integrating vector for tunable, high copy, stable integration into the dispersed Ty sites of *Saccharomyces cerevisiae*. *Biotechnol. Progress* **12**, 16–21.

Parinov S., Sevugan M., Ye D. *et al.* (1999) Analysis of flanking sequences from *dissociation* insertion lines: a database for reverse genetics in *Arabidopsis*. *Plant Cell* **11**, 2263–70.

Park S.H., Pinson S.R.M. & Smith R.H. (1996) T-DNA integration into genomic DNA of rice following *Agrobacterium* inoculation of isolated shoot apices. *Plant Mol. Biol.* **32**, 1135–48.

Parke D., Ornston L.N. & Nester E.W. (1987) Chemotaxis to plant phenolic inducers of virulence genes is constitutively expressed in the absence of the Ti plasmid in *Agrobacterium tumefaciens*. *J. Bacteriol.* **169**, 5336–8.

Parmley S.E. & Smith P.G. (1988) Antibody-selectable filamentous fd phage vectors: affinity purification of target genes. *Gene* **73**, 305–18.

Parolin C., Dorfman T., Palu G., Gottlinger H. & Sodroski J. (1994) Analysis in human immunodeficiency virus type 1 vectors of *cis*-acting sequences that affect gene transfer into human lymphocytes. *J. Virol.* **68**, 3888–95.

Parrish S., Fleenor J., Xu S., Mello C. & Fire A. (2000) Functional anatomy of a dsRNA trigger: differential requirement for the two trigger strands in RNA interference. *Mol. Cell* **6**, 1077–87.

Parsell D.A. & Lindquist S. (1983) The function of heat-shock protein in stress tolerance – degradation and reactivation of damaged proteins. *Ann. Rev. Genet.* **27**, 437–96.

Patience C., Patton G.S. & Takeuchi Y. (1998) No evidence of pig DNA or retroviral infection in patients with short-term extracorporeal connection to pig kidneys. *Lancet* **352**, 699–701.

Patton J.S., Gomes X.V. & Geyer P.K. (1992) Position-independent germline transformation in *Drosophila* using a cuticle pigmentation gene as a selectable marker. *Nucl. Acids Res.* **20**, 5859–60.

Patzer E.J., Nakamura G.R., Hershberg R.D. *et al.* (1986) Cell culture derived recombinant HBsAg is highly immunogenic and protects chimpanzees from infection with hepatitis B virus. *Biotechnology* **4**, 630–6.

Paulus W., Baur I., Boyce F.M., Breakefield X.O. & Reeves S.A. (1996) Self contained, tetracycline regulated retroviral vector system for gene delivery to mammalian cells. *J. Virol.* **70**, 62–7.

Pearse M.J., Cowan P., Van Dendren B.J. & D'Apice A. (1997) The role of complement activation in xenograft rejection. *Curr. Opin. Organ Transpl.* **2**, 103–8.

Peden K.W.C. (1983) Revised sequence of the tetracycline-resistance gene of pBR322. *Gene* **22**, 277–80.

Peebles C.L., Ogden R.C., Knapp G. & Abelson J. (1979) Splicing of yeast transfer RNA precursors: a two-stage reaction. *Cell.* **18**, 27–36.

Peeters B.P.H., De Boer J.H., Bron S. & Venema G. (1988) Structural plasmid instability in *Bacillus subtilis*: effect of direct and inverted repeats. *Mol. Gen. Gent.* **212**, 450–8.

Peijnenburg A.A.C.M., Bron S. & Venema G. (1987) Structural plasmid instability in recombination- and repair-deficient strains of *Bacillus subtilis*. *Plasmid* **17**, 167–70.

Pennock G.D., Shoemaker C. & Miller L.K. (1984) Strong and regulated expression of *E. coli* β-galactosidase in insect cells using a baculovirus vector. *Mol. Cell. Biol.* **4**, 399–406.

Peralta E.G., Hellmiss R. & Ream W. (1986) *Overdrive*, a T-DNA transmission enhancer on the *A. tumefaciens* tumour-inducing plasmid. *EMBO J.* **5**, 1137–42.

Perkus M.E., Tartaglia J. & Paoletti E. (1995) Poxvirus-based vaccine candidates for cancer, AIDS, and other infectious diseases. *J. Leukocyte Biol.* **58**, 1–13.

Perlak F.J., Deaton R.W., Armstrong T.A. *et al.* (1990) Insect resistant cotton plants. *Biotechnology* **8**, 939–43.

Perlak F.J., Fuchs R.L., Dean D.A., McPherson S.L. & Fischhoff D.A. (1991) Modification of the coding sequence enhances plant expression of insect control protein genes. *Proc. Nat. Acad. Sci. USA* **88**, 3324–8.

Perlak F.J., Stone T.B., Muskopf Y.M. *et al.* (1993) Genetically improved potatoes: protection from damage by Colorado potato beetles. *Plant Mol. Biol.* **22**, 313–21.

Perry A.C.F., Wakayama T., Kishikawa H. *et al.* (1999) Mammalian transgenesis by intracytoplasmic sperm injection. *Science* **284**, 1180–3.

Perry L.J. & Wetzel R. (1984) Disulfide bond engineered into T4 lysozyme: stabilization of the protein toward thermal inactivation. *Science* **226**, 555–7.

Perucho M., Hanahan D., Lipsich L. & Wigler M. (1980a) Isolation of the chicken thymidine kinase gene by plasmid rescue. *Nature* **285**, 207–10.

Perucho M., Hanahan D. & Wigler M. (1980b) Genetic and physical linkage of exogenous sequences in transformed cells. *Cell* **22**, 309–17.

Peschel A., Ottenwalder B. & Gotz F. (1996) Inducible production and cellular location of the epidermin biosynthetic enzyme EpiB and improved staphylococcal expression system. *FEMS Microbiol. Lett.* **137**, 279–84.

Petersen B.E., Bowen W.C., Patrene K.D. *et al.* (1999) Bone marrow as a potential source of hepatic oval cells. *Science* **284**, 1168–70.

Petrusyte M., Bitinaite J., Menkevicius S. *et al.* (1988) Restriction endonucleases of a new type. *Gene* **74**, 89–91.

Petters R.M. & Sommer J.R. (2000) Transgenic animals and models of human disease. *Transgenic Res.* **9**, 347–51.

Pham C.T.N., McIvor D.M., Hug B.A., Heusel J.W. & Ley T.J. (1996) Long range disruption of gene expression by a selectable marker cassette. *Proc. Nat. Acad. Sci. USA* **93**, 13090–5.

Picard D. (1994) Regulation of protein function through expression of chimaeric proteins. *Curr. Opin. Biotechnol.* **5**, 511–15.

Pichel J.G., Lakso M. & Westphal H. (1993) Timing of SV40 oncogene activation by site-specific recombination determines subsequent tumour progression during murine lens development. *Oncogene* **8**, 3333–42.

Pierce J.C., Sauer B. & Sternberg N. (1992) A positive selection vector for cloning high molecular weight DNA by the bacteriophage P1 system: improved cloning efficiency. *Proc. Nat. Acad. Sci. USA* **89**, 2056–60.

Pietrokovski S. (1998) Modular organization of inteins and C-terminal autocatalytic domains. *Protein Sci.* **7**, 64–71.

Pigac J. & Schrempf H. (1995) A simple and rapid method of transformation of *Streptomyces rimosus* R6 and other streptomycetes by electroporation. *Appl. Environ. Microbiol.* **61**, 352–6.

Pikaart M.J., Feng J. & Villepointeau B. (1998) The polyomavirus enhancer activates chromatin accessibility on integration into the *HPRT* gene. *Mol. Cell. Biol.* **12**, 5785–92.

Pirrotta V., Hadfield C. & Pretorius G.H.J. (1983) Microdissection and cloning of the *white* locus and the 3B1–3C2 region of the *Drosophila* X chromosome. *EMBO J.* **2**, 927–34.

Pittius C.W., Hennighausen L., Lee E. *et al.* (1988) A milk protein gene promoter directs expression of human tissue plasminogen activator cDNA to the mammary gland in transgenic mice. *Proc. Nat. Acad. Sci. USA* **85**, 5874–8.

Pizza M. & 35 others (2000) Identification of vaccine candidates against serogroup B meningococcus by whole-genome sequencing. *Science* **287**, 1767–8.

Plasterk R.H.A. & Ketting R.F. (2000) The silence of the genes. *Curr. Opin. Genet. Devel.* **10**, 562–7.

Poirier J. & many others (1995) Apolipoprotein E4 allele as a predictor of cholinergic deficit and treatment outcome in Alzheimer disease. *Proc. Nat. Acad. Sci. USA* **92**, 12260–4.

Polejaeva I.A., Chen S.-H., Vaught T.D. *et al.* (2000) Cloned pigs produced by nuclear transfer from adult somatic cells. *Nature* **407**, 86–90.

Pollock D. & Gaken J. (1995) Antisense oligonucleotides: a survey of recent literature, possible mechanisms of action and therapeutic progress. In *Functional Analysis of the Human Genome*, eds Farzaneh F. & Cooper D.N., pp. 241–65. BIOS Scientific Publishers, Oxford.

Polo J.M., Belli B.A., Driver D.A. *et al.* (1999) Stable alphavirus packaging cell lines of Sindbis virus and Semliki Forest virus-derived vectors. *Proc. Nat. Acad. Sci. USA* **96**, 4598–603.

Poquet I., Ehrlich S.D. & Gruss A. (1998) An export-specific reporter designed for Gram-positive bacteria: application to *Lactococcus lactis*. *J. Bacteriol.* **180**, 1904–12.

Porceddu A., Falorni A., Ferradini N. *et al.* (1999) Transgenic plants expressing human glutamic acid decarboxylase (GAD65), a major autoantigen in insulin-dependent diabetes mellitus. *Mol. Breeding* **5**, 553–60.

Porta C. & Lomonossoff G.P. (1996) Use of viral replicons for the expression of genes in plants. *Mol. Biotechnol.* **5**, 209–21.

Porta C. & Lomonossoff G.P. (2001) Viruses as vectors for the expression of foreign sequences in plants. *Biotechnol. Genet. Eng. Rev.* **19** (in press).

Porta C., Spall V.E., Loveland J. *et al.* (1994) Development of cowpea mosaic virus as a high yielding system for the presentation of foreign peptides. *Virology* **202**, 949–55.

Potrykus I., Paszkowski J., Saul M.W., Petruska J. & Shillito R.D. (1985a) Molecular and general genetics of a hybrid foreign gene introduced into tobacco by direct gene transfer. *Mol. Gen. Genet.* **199**, 169–77.

Potrykus I., Saul M.W., Petruska J., Paszkowski J. & Shillito R. (1985b) Direct gene transfer to cells of a graminaceous monocot. *Mol. Gen. Genet.* **199**, 183–8.

Potter H. (1988) Electroporation in biology: methods, applications and instrumentation. *Anal. Biochem.* **174**, 361–73.

Potter H., Weir L. & Leder P. (1984) Enhancer-dependent expression of human K immunoglobulin genes introduced into mouse pre-B lymphocytes by electroporation. *Proc. Nat. Acad. Sci. USA* **81**, 7161–5.

Poustka A. & Lehrach H. (1986) Jumping libraries and linking libraries: the next generation of molecular tools in mammalian genetics. *Trends Genet.* **2**, 174–9.

Powell S.K. & 8 others (2000) Breeding of retroviruses by DNA shuffling for improved stability and processing yields. *Nature Biotechnol.* **18**, 1279–82.

Powell-Abel P., Nelson R.S., De B. *et al.* (1986) Delay of disease development in transgenic plants that express the tobacco mosaic virus coat protein gene. *Science* **232**, 738–43.

Poyart C. & Trieu-Cuot P. (1997) A broad-host-range mobilizable shuttle vector for the construction of transcriptional fusions to beta-galactosidase in Gram-positive bacteria. *FEMS Microbiol. Lett.* **156**, 193–8.

Poznansky M., Lever A., Bergeron L., Haseltine W. & Sodroski J. (1991) Gene transfer into human lymphocytes by a defective human immunodeficiency virus type 1 vector. *J. Virol.* **65**, 532–6.

Prelle K., Vassiliev I.M., Vassilieva S.G., Wolf E. & Wobus A.M. (1999) Establishment of pluripotent cell lines from vertebrate species – present status and future prospects. *Cells Tissues Organs* **165**, 220–36.

Pridmore R.D. (1987) New and versatile cloning vectors with kanamycin resistance marker gene. *Gene* **56**, 309–12.

Primrose S.B. & Ehrlich S.D. (1981) Isolation of plasmid deletion mutants and a study of their instability. *Plasmid* **6**, 193–201.

Primrose S.B. & Twyman R.M. (2002) *Principles of Genome Analysis*, 3rd edn. Blackwell Science, Oxford.

Primrose S.B., Derbyshire P., Jones I.M., Nugent M.E. & Tacon W.C.A. (1983) Hereditary instability of recombinant DNA molecules. In *Bioactive Microbial Products 2: Development and Production*, eds Nisbet L.J. & Winstanley D.J., pp. 63–77. Academic Press, London.

Probst F.J., Fridell R.A., Raphael Y. *et al.* (1998) Correction of deafness in shaker-2 mice by an unconventional myosin in a BAC transgene. *Science* **280**, 1444–7.

Projan S.J. & Archer G.L. (1989) Mobilization of the relaxable *Staphylococcus aureus* plasmid pC221 by the conjugative plasmid pGO1 involves three pC221 loci. *J. Bacteriol.* **171**, 1841–5.

Pruchnic R., Cao B., Peterson Z.Q. *et al.* (2000) The use of adeno-associated virus to circumvent the maturation-dependent viral transduction of myofibres. *Hum. Gene Ther.* **11**, 521–36.

Prunkard D., Cottingham I., Garner I. *et al.* (1996) High-level expression of recombinant human fibrinogen in the milk of transgenic mice. *Nature Biotechnol.* **14**, 867–71.

Ptashne M. (1967a) Isolation of the λ phage repressor. *Proc. Nat. Acad. Sci. USA* **57**, 306–13.

Ptashne M. (1967b) Specific binding of the λ phage repressor to λDNA. *Nature* **214**, 232–4.

Ptashne M. (1992) *A Genetic Switch*, 2nd edn. Blackwell Science Ltd, Oxford.

Pugachev K.V., Mason P.W., Shope R.E. & Frey T.K. (1995) Double-subgenomic Sindbis virus recombinants expressing immunogenic proteins of Japanese encephalitis virus induce significant protection in mice against lethal JEV infection. *Virology* **212**, 587–94.

Putney S.D., Herlihy W.C. & Schimmel P. (1983) A new troponin T and cDNA clones for 13 different muscle proteins, found by shotgun sequencing. *Nature* **302**, 718–21.

Puyet A., Sandoval M., López P. *et al.* (1987) A simple medium for rapid regeneration of *Bacillus subtilis* protoplasts transformed with plasmid DNA. *FEBS Microbiol. Lett.* **40**, 1–5.

Qian Z. & Wilusz J. (1993) Cloning cDNA encoding an RNA-binding protein by screening expression libraries using a northwestern strategy. *Ann. Biochem.* **212**, 547–54.

Qin M., Bayley C., Stockton T. & Ow D.W. (1994) Cre recombinase-mediated site-specific recombination between plant chromosomes. *Proc. Nat. Acad. Sci. USA* **91**, 1706–10.

Quon D. Wang Y., Catalano W.R. *et al.* (1991) Formation of β-amyloid protein deposits in the brains of transgenic mice. *Nature* **352**, 239–41.

Rabinowitz J.E. & Samulski J. (1998) Adeno-associated virus expression systems for gene transfer. *Curr. Opin. Biotechnol.* **9**, 470–5.

Radloff R., Bauer W. & Vinograd J. (1967) A dye-buoyant-density method for the detection and isolation of closed circular duplex DNA: the closed circular DNA in HeLa cells. *Proc. Nat. Acad. Sci. USA* **57**, 1514–21.

Rahman M.A., Mak R., Ayad H., Smith A. & Maclean N. (1998) Expression of a novel piscine growth hormone gene results in growth enhancement intransgenic tilapia (*Orechromis niloticus*). *Transgenic Res.* **7**, 357–69.

Raineri D.M., Bottino P., Gordon M.P. & Nester E.W. (1990) *Agrobacterium*-mediated transformation of rice (*Oryza sativa* L.). *Biotechnology* **9**, 33–8.

Rainey P.B. (1999) Adaptation of *Pseudomonas fluorescens* to the plant rhizosphere. *Environ. Microbiol.* **1**, 243–57.

Ramirez-Solis R., Lui P. & Bradley A. (1995) Chromosome engineering in mice. *Nature* **378**, 720–4.

Rashid H., Yokoi S., Toriyama K. & Hinata K. (1996) Transgenic plant production mediated by *Agrobacterium* in Indica rice. *Plant Cell Rep.* **15**, 727–30.

Rathus C., Bower R. & Birch R.G. (1993) Effects of promoter, intron and enhancer elements on transient gene expression in sugarcane and carrot protoplasts. *Plant Mol. Biol.* **23**, 616–18.

Ratzkin B. & Carbon J. (1977) Functional expression of cloned yeast DNA in *Escherichia coli*. *Proc. Nat. Acad. Sci. USA* **74**, 487–91.

Reddy P.H., Williams M. & Tagle D.A. (1999) Recent advances in the understanding the pathogenesis of Huntington's disease. *Trends Neurosci.* **22**, 248–55.

Reddy S., Rayburn H., Von Melchner H. & Ruley H.E. (1992) Fluorescence-activated sorting of totipotent embryonic stem cells expressing developmentally regulated lacZ fusion genes. *Proc. Nat. Acad. Sci. USA* **89**, 6721–5.

Reeves R.H., Baxter L.L. & Richtsmeier J.T. (2001) Too much of a good thing: mechanisms of gene action in Down syndrome. *Trends Genet.* **17**, 83–8.

Reidhaar-Olson J.F. & Sauer R.T. (1988) Combinatorial cassette mutagenesis as a probe of the informational content of protein sequences. *Science* **241**, 53–7.

Reisman D., Yates J. & Sugden B. (1985) A putative origin of replication of plasmids derived from Epstein–Barr virus composed of two *cis*-acting components. *Mol. Cell. Biol.* **5**, 1822–32.

Ren Z. & Black L.W. (1998) Phage T4 SOC and HOC display of biologically active, full length proteins on the viral capsid. *Gene* **215**, 439–44.

Renart J. & Sandoval I.V. (1984) Western blots. *Methods Enzymol.* **104**, 455–60.

Renault P., Corthier G., Goupil N., Delorme C. & Ehrlich S.D. (1996) Plasmid vectors for Gram-positive bacteria switching from high to low copy number. *Gene* **183**, 175–82.

Resnick J.L., Bixler L.S., Cheng L. & Donovan P.J. (1992) Long-term proliferation of mouse primordial germ cells in culture. *Nature* **359**, 550–1.

Reynolds P.N., Feng M. & Curiel D.T. (1999) Chimeric viral vectors – the best of both worlds. *Mol. Med. Today* **5**, 25–31.

Richards J.E., Gilliam T.C., Cole J.L. *et al.* (1988) Chromosome jumping from D4S10 (G8) toward the Huntington disease gene. *Proc. Nat. Acad. Sci. USA* **85**, 6437–41.

Richardson J.H. & Marasco W.A. (1995) Intracellular antibodies: development and therapeutic potential. *Trends Biotechnol.* **13**, 306–10.

Richardson J.H., Sodroski J.G., Waldmann T.A. & Marasco W.A. (1995) Phenotypic knockout of the high-affinity human interleukin 2 receptor by intracellular single-chain antibodies against the alpha subunit of the receptor. *Proc. Nat. Acad. Sci. USA* **92**, 3137–41.

Richardson J.H., Hofmann W., Sodroski J.G. & Marasco W.A. (1998) Intrabody-mediated knockout of the high affinity IL-2 recdeptor in primary human T cells using a bicistronic lentivirus vector. *Gene Ther.* **5**, 635–44.

Rijkers T., Peetz A. & Ruther U. (1994) Insertional mutagenesis in transgenic mice. *Transgenic Res.* **3**, 203–15.

Rinas U., Tsai L.B., Lyons D. *et al.* (1992) Cysteine to serine substitutions in basic fibroblast growth factor: effect on inclusion body formation and proteolytic susceptibility during *in vitro* refolding. *Biotechnology* **10**, 435–40.

Ringquist S., Shinedling S., Barrick D. *et al.* (1992) Translation initiation in *Escherichia coli*: sequences within the ribosome-binding site. *Mol. Microbiol.* **6**, 1219–29.

Rio D.C., Clark S.G. & Tjian R. (1985) A mammalian host–vector system that regulates expression and amplification of transfected genes by temperature induction. *Science* **227**, 23–8.

Rivera V.M. *et al.* (1996) A humanized system for pharmacologic control of gene expression. *Nature Med.* **2**, 1028–32.

Robbins P.D., Tahara H. & Ghivizzani S.C. (1998) Viral vectors for gene therapy. *Trends Biotechnol.* **16**, 35–40.

Roberts J.M. & Axel R. (1982) In *Gene Amplification*, ed. Schimke R., p. 251. Cold Spring Harbor, New York.

Roberts L. (1992) NIH gene patents: round two. *Science* **255**, 912–13.

Roberts M.W. & Rabinowitz J.C. (1989) The effect of *Escherichia coli* ribosomal protein S1 on the translational specificity of bacterial ribosomes. *J. Biol. Chem.* **264**, 2228–35.

Robertson E., Bradley A., Kuehn M. & Evans M. (1986) Germ line transformation of genes introduced into cultured pluripotential cells by retroviral vector. *Nature* **323**, 445–8.

Robinson K., Gilbert W. & Church G.M. (1994) Large scale bacterial gene discovery by similarity search. *Nature Genet.* **7**, 205–14.

Rodriguez R.L., West R.W., Heyneker H.L., Bolivar P. & Boyer H.W. (1979) Characterizing wild-type and mutant promoters of the tetracycline resistance gene in pBR313. *Nucl. Acids Res.* **6**, 3267–87.

Roe T., Reynolds T.C., Yu G. & Brown P.O. (1993) Integration of murine leukemia virus DNA depends on mitosis. *EMBO J.* **12**, 2099–108.

Rogers J., Goedert M. & Wilson P.M. (1988) An extra sequence in the lambda EMBL3 polylinker. *Nucleic Acid Res.* **16**, 1633.

Roman M., Martin-Orozco E., Goodman J.S. *et al.* (1997) Immunostimulatory DNA sequences function as T helper-1-promoting adjuvants. *Nature Med.* **3**, 849–54.

Romanos M.A., Scorer C.A. & Clare J.J. (1992) Foreign gene expression in yeast: a review. *Yeast* **8**, 423–88.

Rommens J.M., Iannuzzi M.C., Kerem B.-S. *et al.* (1989) Identification of the cystic fibrosis gene: chromosome walking and jumping. *Science* **245**, 1059–65.

Rondon I.J. & Marasco W.A. (1997) Intracellular antibodies (intrabodies) for gene therapy of infectious diseases. *Ann. Rev. Microbiol.* **51**, 257–83.

Rose M.D., Novick P., Thomas J.H., Botstein D. & Fink G.R. (1987) A *Saccharomyces cerevisiae* genomic plasmid bank based on a centromere-containing shuttle vector. *Gene* **60**, 237–43.

Rosenberg S. plus 14 others (1990) Gene transfer into humans – immunotherapy of patients with advanced melonoma, using tumor-infiltrating lymphocytes modified by retroviral gene transduction. *N. Engl. J. Med.* **323**, 570–8.

Rossi J.J. (1995) Controlled, targeted, intracellular expression of ribozymes: progress and problems. *Trends Biotechnol.* **13**, 301–6.

Ross-Macdonald P., Sheehan A., Roeder G.S. & Snyder M. (1997) A multipurpose transposon system for analyzing protein production, localization, and function in *Saccharomyces cerevisiae*. *Proc. Nat. Acad. Sci. USA* **94**, 190–5.

Ross-Macdonald P. & 17 others (1999) Large-scale analysis of the yeast genome by transposon tagging and gene disruption. *Nature* **402**, 413–18.

Rothberg B.E.G. (2001) Mapping a role for SNPs in drug development. *Nature Biotechnol.* **19**, 209–11.

Rothman J.E. & Orci L. (1992) Molecular dissection of the secretory pathway. *Nature* **355**, 409–15.

Rousset M., Casalot L., Rapp-Giles B.J. *et al.* (1998) New shuttle vectors for the introduction of cloned DNA in *Desulfovibrio*. *Plasmid* **39**, 114–22.

Roychoudhury R., Jay E. & Wu R. (1976) Terminal labelling and addition of homopolymer tracts to duplex DNA fragments by terminal deoxynucleotidyl transferase. *Nucl. Acids Res.* **3**, 863–77.

Rubin G.M. & Spradling A.C. (1982) Genetic transformation of *Drosophila* with transposable element vectors. *Science* **218**, 348–53.

Rubin G.M. & Spradling A. (1983) Vectors for P element-mediated gene transfer in *Drosophila*. *Nucl. Acids Res.* **11**, 6341–51.

Rubin G.M., Kidwell M.G. & Bingham P.M. (1982) The molecular basis of P–M hybrid dysgenesis: the nature of induced mutations. *Cell* **29**, 987–94.

Rubinstein J.L., Brice A.E., Ciaranello R.D. *et al.* (1990) Subtractive hybridization system using single stranded phagemids with directional inserts. *Nucl. Acids Res.* **18**, 4833–42.

Rudnicki M.A., Braun B., Hinuma S. & Jaenisch R. (1992) Inactivation of *myoD* in mice leads to up-regulation of the myogenic HLH gene *myf*-5 and results in apparently normal muscle development. *Cell* **71**, 383–90.

Rudolph H. & Hinnen A. (1987) The yeast PHO5 promoter: phosphate control elements and sequences mediating mRNA start-site selection. *Proc. Nat. Acad. Sci. USA* **84**, 1340–4.

Rudolph N.S. (1999) Biopharmaceutical production in transgenic livestock. *Trends Biotechnol.* **17**, 367–74.

Rusconi S. & Schaffner W. (1981) Transformation of frog embryos with a rabbit β-globin gene. *Proc. Nat. Acad. Sci. USA* **78**, 5051–5.

Russel S.H., Hoopes J.L. & Odell J.T. (1992) Directed excision of a transgene from the plant genome. *Mol. Gen. Genet.* **234**, 49–59.

Russell D.W. & Zinder N.D. (1987) Hemimethylation prevents DNA replication in *E. coli. Cell* **50**, 1071–9.

Ruysscharet J.-M., El Ouahabi A., Willeaume V. *et al.* (1994) A novel cationic ampiphile for transfection of mammalian cells. *Biochem. Biophys. Res. Commun.* **203**, 1622–8.

Ruzzo A., Andreoni F. & Magnani M. (1998) Structure of the human hexokinase type I gene and nucleotide sequence of the 5′ flanking region. *Biochem. J.* **331**, 607–13.

Sabbioni M., Negrini P., Rimessi R., Manservigi R. & Barbatibrodano G. (1995) A BK virus episomal vector for constitutive high expression of exogenous cDNAs in human cells. *Arch. Virol.* **140**, 335–9.

Sadowski P.D. (1993) Site-specific genetic recombination: hops, flips and flops. *FASEB J.* **7**, 760–7.

Saez E., No D., West A. & Evans R.M. (1997) Inducible gene expression in mammalian cells and transgenic mice. *Curr. Opin. Biotechnol.* **8**, 608–16.

Sager R., Anisowicz A., Neveu M., Lian P. & Sotiropoulou G. (1993) Identification by differential display of alpha 6 integrin as a candidate tumor suppressor gene. *FASEB J.* **7**, 964–70.

Sagerstrom C.G., Sun B.I. & Sive H.L. (1997) Subtractive cloning: past, present and future. *Ann. Rev. Biochem.* **66**, 751–83.

Saiki R., Gelfond D.H., Stoffel S. *et al.* (1988) Primer-directed enzymatic amplification of DNA with a thermostable DNA polymerase. *Science* **239**, 487–91.

Saint C.P., Alexander S. & McClure N.C. (1995) pTIM3, a plasmid delivery vector for a transposon-based inducible marker gene system in Gram-negative bacteria. *Plasmid* **34**, 165–74.

St Ogne, L., Furth P.A. & Gruss P. (1996) Temporal control of the Cre recombinase in transgenic mice by a tetracycline responsive promoter. *Nucl. Acids Res.* **24**, 3875–7.

Salyers A.A., Shoemaker N.B., Stevens A.M. & Li L.-Y. (1995) Conjugative transposons. an unusual and diverse set of integrated gene transfer elements. *Microbiol. Rev.* **59**, 579–90.

Sambrook J., Rodgers L., White J. & Getling M.J. (1985) Lines of BPV-transformed murine cells that constitutively express influenza virus heamagglutinin. *EMBO J.* **4**, 91–103.

Samulski R.J., Chang L.-S. & Shenk T. (1989) Helper-free stocks of recombinant adeno-associated viruses: normal integration does not require viral gene expression. *J. Virol.* **63**, 3822–8.

Sanchez O., Navarro R.E. & Aguirre J. (1998) Increased transformation frequency and tagging of developmental genes in *Aspergillus nidulans* by restriction enzyme-mediated integration (REMI). *Mol. Gen. Genet.* **258**, 89–94.

Sandrin M.S. & McKenzie I.F.C. (1999) Recent advances in xenotransplantation. *Curr. Opin. Immunol.* **11**, 527–31.

Sandrin M.S., Vaughan H.A. & McKenzie I.F.C. (1994) Identification of Gal (α1,3) Gal as the major epitope for pig-to-human vascularised xenografts. *Transpl. Rev.* **8**, 134–49.

Sandrin M.S., Fodor W.L., Mouhtouris E. *et al.* (1995) Enzymatic remodeling of the carbohydrate surface of a xenogenic cell substantially reduces human antibody binding and complement-mediated cytolysis. *Nature Med.* **1**, 1261–7.

Sanford J.C., Klein T.M., Wolf E.D. & Allen N. (1987) Delivery of substances into cells and tissues using a particle bombardment process. *J. Particle Sci. Techniques* **6**, 559–63.

Sanford J.C., Devit M.J., Russell J.A. *et al.* (1991) An improved, helium-driven biolistic device. *Technique* **3**, 3–16.

Sanford J.C., Smith F.D. & Russel J.A. (1993) Optimizing the biolistic process for different biological applications. *Methods Enzymol.* **217**, 483–509.

Sanger F., Air G.M., Barrell B.G. *et al.* (1977a) Nucleotide sequence of bacteriophage ΦX174DNA. *Nature* **265**, 687–95.

Sanger F., Nicklen S. & Coulson A.R. (1977b) DNA sequencing with chain terminating inhibitors. *Proc. Nat. Acad. Sci. USA* **74**, 5463–7.

Sanger F., Coulson A.R., Hong G.-F., Hill D.F. & Petersen G.B. (1982) Nucleotide sequence of bacteriophage lambda DNA. *J. Mol. Biol.* **162**, 729–73.

Santini C., Brennan D., Mennuni C. *et al.* (1998) Efficient display of an HCV cDNA expression library as C-terminal fusion to the capsid protein D of bacteriophage lambda. *J. Mol. Biol.* **282**, 125–35.

Sarkar G. & Sommer S.S. (1990) The 'megaprimer' method of site-directed mutagenesis. *Biotechniques* **8**, 404–7.

Sarver N., Gruss P., Law M.-F., Khoury G. & Howley P.M. (1981a) Bovine papilloma virus deoxyribonucleic acid: a novel eucaryotic cloning vector. *Mol. Cell. Biol.* **1**, 486–96.

Sarver N., Gruss P., Law M.-F., Khoury G. & Howley P.M. (1981b) Rat insulin gene covalently linked to bovine papilloma virus DNA is expressed in transformed mouse cells. In *Development Biology Using Purified Genes*, eds Brown D. & Fox C.R., ICN–UCLA Symposia on Molecular and Cellular Biology, Vol. 23. Academic Press, New York.

Sauer B. (1994) Site-specific recombination: developments and applications. *Curr. Opin. Biotechnol.* **5**, 521–7.

Sauer B. & Henderson N. (1990) Targeted insertion of exogenous DNA into the eukaryotic genome by the Cre recombinase. *New Biol.* **2**, 441–9.

Savage M.O. & Fallon J.F. (1995) *fgf2* messenger RNA and its antisense message are expressed in a developmentally specific manner in the chick limb bud and mesonephros. *Devel. Dynamics* **200**, 343–53.

Sawa S., Watanabe K., Goto K. *et al.* (1999) *FILAMEN-TOUS FLOWER*, a meristem and organ identity gene of *Arabidopsis*, encodes a protein with zinc finger and HMG-related domains. *Genes Devel.* **13**, 1079–88.

Sayers J.R. & Eckstein F. (1991) A single-strand specific endonuclease activity copurifies with overexpressed T5 D15 exonuclease. *Nucl. Acids Res.* **19**, 4127–32.

Scalenghe F., Turco E., Edstrom J.E., Pirotta V. & Melli M. (1981) Micro-dissection and cloning of DNA from a specific region of *Drosophila melanogaster* polytene chromosomes. *Chromosoma* **82**, 205–16.

Schaefer B.C. (1995) Revolutions in RACE: new strategies for polymerase chain reaction cloning of full length cDNA ends. *Ann. Biochem.* **277**, 255–73.

Schaefer-Ridder M., Wang Y. & Hofschneider P.H. (1982) Liposomes as gene carriers – efficient transformation of mouse L-cells by thymidine kinase gene. *Science* **215**, 166–8.

Schafer W., Gorz A. & Kahl G. (1987) T-DNA integration and expression in a monocot crop plant after induction of *Agrobacterium. Nature* **327**, 529–31.

Schaffner W. (1980) Direct transfer of cloned genes from bacteria to mammalian cells. *Proc. Nat. Acad. Sci. USA* **77**, 2163–9.

Scharf S.J., Horn G.T. & Erlich H.A. (1986) Direct cloning and sequence analysis of enzymatically amplified genomic sequences. *Science* **233**, 1076–8.

Scheidereit C., Greisse S., Westphal H.M. & Beato M. (1983) The glucocorticoid receptor binds to defined nucleotide sequences near the promoter of mouse mammary tumour virus. *Nature* **304**, 749–52.

Schein C.H. (1991) Optimizing protein folding to the native state in bacteria. *Curr. Opin. Biotechnol.* **2**, 746–50.

Schein C.H. & Noteborn M.H.M. (1988) Formation of soluble recombinant proteins in *Escherichia coli* is favored by lower growth temperature. *Biotechnology* **6**, 291–4.

Scheller R.H., Dickerson R.E., Boyer H.W., Riggs A.D. & Itakura K. (1977) Chemical synthesis of restriction enzyme recognition sites useful for cloning. *Science* **196**, 177–80.

Schena M. & Yamamoto K.R. (1988) Mammalian glucocorticoid receptor derivatives enhance transcription in yeast. *Science* **241**, 965–7.

Schena M., Shalon D., Davis R.W. & Brown P.O. (1995) Quantitative monitoring of gene expression patterns with a complementary DNA microarray. *Science* **270**, 467–70.

Schena M., Heller R.A., Theriault T.P. *et al.* (1998) Microarrays: biotechnology's discovery platform for functional genomics. *Trends Biotechnol.* **16**, 301–6.

Scherzinger E., Bagdasarian M.M., Scholz P. *et al.* (1984) Replication of the broad host range plasmid RSF1010: requirement for three plasmid-encoded proteins. *Proc. Nat. Acad. Sci. USA* **81**, 654–8.

Scheule R.K. & Cheng S.H. (1996) Liposome delivery systems. In *Gene Therapy*, eds Lemoine N.R. & Cooper D.N., pp. 93–112. BIOS Scientific Publishers, Oxford, UK.

Scheidner G., Morral N., Parks R.J. *et al.* (1998) Genomic DNA transfer with high capacity adenovirus vector results in improved *in vivo* gene expression and decreased toxicity. *Nature Genet.* **18**, 180–3.

Schiestl R.H. & Petes T.D. (1991) Integration of DNA fragments by illegitimate recombination in *Saccharomyces cerevisiae. Proc. Nat. Acad. Sci. USA* **88**, 7585–9.

Schillberg J., Zimmermann J., Findlay K. & Fischer R. (2001) Plasma membrane display of anti-viral single chain Fv fragments confers resistance to tobacco mosaic virus. *Mol. Breeding* in press.

Schimke R.T., Kaufman R.J., Alt F.W. & Kellems R.F. (1978) Gene amplification and drug resistance in cultured murine cells. *Science* **202**, 1051–5.

Schlesinger S. & Dubensky T.W. Jr (1999) Alphavirus vectors for gene expression and vaccines. *Curr. Opin. Biotechnol.* **10**, 434–9.

Schnieke A.E., Kind A.J., Ritchie W.A. *et al.* (1997) Human factor IX transgenic sheep produced by transfer of nuclei from transfected fetal fibroblasts. *Science* **278**, 2130–3.

Schocher R.J., Shillito R.D., Saul M.W., Paszkowski S.J. & Potrykus I. (1986) Co-transformation of unlinked foreign genes into plants by direct gene transfer. *Biotechnology* **4**, 1093–6.

Scholthof H.B., Scholthof K.B.G. & Jackson A.O. (1996) Plant virus gene vectors for transient expression of foreign proteins in plants. *Ann. Rev. Phytopathol.* **34**, 299–323.

Scholz P., Haring V., Wittman-Leibold B. *et al.* (1989) Complete nucleotide sequence and gene organization of the broad-host-range plasmid RSF1010. *Gene* **75**, 271–88.

Schreiber S.L. (1991) Chemistry and biology of the immunophillins and their immunosuppressive ligands. *Science* **251**, 283–7.

Schuler T.H., Poppy G.M., Kerry B.R. & Denholm I. (1998) Insect resistant transgenic plants. *Trends Biotechnol.* **16**, 168–75.

Schultze N., Burki Y., Lang Y., Certa U. & Bluethmann H. (1996) Efficient control of gene expression by single step integration of the tetracycline system in transgenic mice. *Nature Biotechnol.* **14**, 499–503.

Schultz L.D., Hofmann K.J., Mylin L.M. *et al.* (1987) Regulated over-production of the *GAL4* gene product greatly increases expression from galactose-inducible promoters on multi-copy expression vectors in yeast. *Gene* **61**, 123–33.

Schwartz D.C. & Cantor C.R. (1984) Separation of yeast chromosomal-sized DNAs by pulsed field gradient gel electrophoresis. *Cell* **37**, 67–75.

Schwartz D.C. & Koval M. (1989) Conformational dynamics of individual DNA molecules during gel electrophoresis. *Nature* **338**, 520–2.

Schwartzberg P.L., Goff S.P. & Robertson E.J. (1989) Germ line transmission of a c-*abl* mutation produced by targeted gene disruption in ES cells. *Science* **246**, 799–803.

Schwecke T. & 15 others. (1995) The biosynthetic gene cluster for the polyketide immunosuppressant rapamycin. *Proc. Nat. Acad. Sci. USA* **92**, 7839–43.

Schwenk F., Kuhn R., Angrand P.O., Rajewsky K. & Stewart A.F. (1998) Temporally and spatially regulated somatic mutagenesis in mice. *Nucl. Acids Res.* **26**, 1427–32.

Sclimenti C.R. & Calos M.P. (1998) Epstein–Barr virus vectors for gene expression and transfer. *Curr. Opin. Biotechnol.* **9**, 476–9.

Scott J.K. & Smith G.P. (1990) Searching for peptide ligands with an epitope library. *Science* **249**, 386–90.

Seelke R., Kline B., Aleff R., Porter R.D. & Shields M.S. (1987) Mutations in the *recD* gene of *Escherichia coli* that raise the copy number of certain plasmids. *J. Bacteriol.* **169**, 4841–4.

Selker E.U. (1997) Epigenetic phenomena in filamentous fungi: useful paradigms or repeat-induced confusion? *Trends Genet.* **13**, 296–301.

Selker E.U. (1999) Gene silencing: repeats that count. *Cell* **97**, 157–60.

SenGupta D.J., Wickens M. & Fields S. (1999) Identification of RNAs that bind to a specific protein using the yeast three hybrid system. *RNA* **5**, 596–601.

Serrano L., Day A.G. & Fersht A.R. (1993) Step-wise mutation of barnase to binase: a procedure for engineering increased stability of proteins and an experimental analysis of the evolution of protein stability. *J. Mol. Biol.* **233**, 305–12.

Sgaramella V. (1972) Enzymatic oligomerization of bacteriophage P22 DNA and of linear simian virus 40 DNA. *Proc. Nat. Acad. Sci. USA* **69**, 3389–93.

Shah D.M., Horsch R.B., Klee H.J. *et al.* (1986) Engineering herbicide tolerance in transgenic plants. *Science* **233**, 478–81.

Shalon D., Smith S.J. & Brown P.O. (1996) A DNA microarray system for analysing complex DNA samples using two-colour fluorescent probe hybridization. *Genome Res.* **6**, 639–45.

Sharp P.M. & Bulmer M. (1988) Selective differences among translation termination codons. *Gene* **63**, 141–5.

Shaw C.H., Watson M.D., Carter G.H. & Shaw C.H. (1984) The right hand copy of the nopaline Ti-plasmid 25 bp repeat is required for tumour formation. *Nucl. Acids Res.* **12**, 6031–41.

Shaw G. & Kamen R. (1986) A conserved AU sequence from the 3′-untranslated region of GM-CSF messenger RNA mediates selective messenger RNA degradation. *Cell* **46**, 659–67.

Sheehy R.E., Kramer M. & Hiatt W.R. (1988) Reducion of polygalacturonase in tomato fruit by antisense RNA. *Proc. Nat. Acad. Sci. USA* **85**, 8805–9.

Sheen J., Hwang S., Niwan Y., Kobayashi H. & Galbraith D.W. (1995) Green fluorescent protein as a new vital marker in plant cells. *Plant J.* **8**, 777–84.

Sheng O.J. & Citovsky V. (1996) *Agrobacterium* plant cell DNA transport: have virulence proteins will travel. *Plant Cell* **8**, 1699–710.

Sheng Y., Mancino V. & Birren B. (1995) Transformation of *Escherichia coli* with large DNA molecules by electroporation. *Nucl. Acids Res.* **23**, 1990–6.

Shepherd N.S. & Smoller D. (1994) The O1 vector system for the preparation and screening of genomic libraries. *Genet. Eng.* **16**, 213–28.

Shepherd N.S., Pfronger B.D., Coulby J.N. *et al.* (1994) Preparation and screening of an arrayed human genomic library generated with the P1 cloning system. *Proc. Nat. Acad. Sci. USA* **91**, 2629–33.

Shi Y., Wang M.B., Powell K.S. *et al.* (1994) Use of the rice sucrose synthase-1 promoter to direct phloem-specific expression of beta-glucuronidase and snowdrop lectin genes in transgenic tobacco plants. *J. Exp. Botany* **45**, 623–31.

Shibata D. & Lui Y.-G. (2000) *Agrobacterium*-mediated plant transformation with large DNA fragments. *Trends Plant Sci.* **5**, 354–7.

Shida H. (1986) Nucleotide sequence of the vaccinia virus hemagglutinin gene. *Virology* **150**, 451–62.

Shillito R.D., Saul M.W., Pazkowski J., Muller M. & Potrykus I. (1985) High efficiency direct gene transfer to plants. *Biotechnology* **3**, 1099–103.

Shimada N., Toyoda-Yamamoto A., Nagamine J. *et al.* (1990) Control of expression of *Agrobacterium vir* genes by synergistic actions of phenolic signal molecules and monosaccharides. *Proc. Nat. Acad. Sci. USA* **87**, 6684–8.

Shimada T., Fujii H., Mitsuya A. & Nienhuis W. (1991) Targeted and highly efficient gene transfer into CD4+ cells by a recombinant human immunodeficiency virus retroviral vector. *J. Clin. Invest.* **88**, 1043–7.

Shimamoto K., Terada R., Izawa T. & Fujimoto H. (1988) Fertile transgenic rice plants regenerated from transformed protoplasts. *Nature* **338**, 274–6.

Shimotohno K. & Temin H.M. (1982) Loss of intervening sequences in genomic mouse α-globin DNA inserted in an infectious retrovirus vector. *Nature* **299**, 265–8.

Shizuya H., Birren B., Kim U.-J. *et al.* (1992) Cloning and stable maintenance of 300-kilobase-pair fragments of human DNA in *Escherichia coli* using an F-factor-based vector. *Proc. Nat. Acad. Sci. USA* **89**, 8794–7.

Shoemaker D.D., Lashkari D.A., Morris D., Mittmann M. & Davis R.W. (1996) Quantitative phenotypic analysis of yeast deletion mutants using a highly parallel molecular bar-coding strategy. *Nature Genet.* **14**, 450–6.

Short J.M., Fernandez J.M., Sorge J.A. & Huse W.D. (1988) λZAP: a bacteriophage lambda expression vector with *in vivo* excision properties. *Nucl. Acids Res.* **16**, 7583–600.

Shuman S. (1994) Novel approach to molecular cloning and polynucleotide synthesis using vaccinia DNA topoisomerase. *J. Biol. Chem.* **269**, 32678–84.

Shusta E.V., Raines R.T., Pluckthun A. & Wittrup K.D. (1998) Increasing the secretory capacity of *Saccharomyces cerevisiae* for production of single-chain antibody fragments. *Nature Biotechnol.* **16**, 773–7.

Shusta E.V., Van Antwerp J. & Wittrup K.D. (1999) Biosynthetic polypeptide libraries. *Curr. Opin. Biotechnol.* **10**, 117–22.

Sidhu S.S. (2000) Phage display in pharmaceutical biotechnology. *Curr. Opin. Biotechnol.* **11**, 610–16.

Sidhu S.S., Lowman H.B., Cunningham B.C. & Wells J.A. (2000) Phage display for selection of novel binding peptides. *Methods Enzymol.* **328**, 333–63.

Siegele D.A. & Hu J.C. (1997) Gene expression from plasmids containing the *ara*BAD promoter at subsaturating inducer concentrations represents mixed populations. *Proc. Nat. Acad. Sci. USA* **94**, 8168–72.

Signs M.W. & Flores H.E. (1990) The biosynthetic potential of plant roots. *Bio-Essays* **12**, 282–5.

Sijmons P.C., Dekker B.M.M., Schrammeijer B. *et al.* (1990) Production of correctly processed human serum albumin in transgenic plants. *Biotechnology* **8**, 217–21.

Sikorski R. & Peters R. (1997) Transgenics on the Internet. *Nature Biotechnol.* **15**, 289.

Sikorski R.S., Michaud W., Levin H.L., Boeke J.D. & Hieter P. (1990) Trans-kingdom promiscuity. *Nature* **345**, 581–2.

Simonato M., Manservigi R., Marconi P. & Glorioso J. (2000) Gene transfer into neurones for the molecular analysis of behaviour: focus on herpes simplex vectors. *Trends Neurosci.* **23**, 183–90.

Simons J.P., McClenaghan M. & Clark A.J. (1987) Alteration of the quality of milk by expression of sheep beta-lactoglobulin in transgenic mice. *Nature* **328**, 530–2.

Simons R. & Kleckner N. (1988) Biological regulation by antisense RNA in prokaryotes. *Ann. Rev. Genet.* **22**, 567–600.

Simonsen C.C. & Levinson A.D. (1983) Isolation and expression of an altered mouse dihydrofolate reductase cDNA. *Proc. Nat. Acad. Sci. USA* **80**, 2495–9.

Simpson K., McGuigan A. & Huxley C. (1996) Stable episomal maintenance of yeast artificial chromosomes in human cells. *Mol. Cell Biol.* **16**, 5117–26.

Singer-Sam J., Simmer R.L., Keith D.H. *et al.* (1983) Isolation of a cDNA clone for human X-linked 3-phosphoglycerate kinase by use of a mixture of synthetic oligodeoxyribonucleotides as a detection probe. *Proc. Nat. Acad. Sci. USA* **80**, 802–6.

Singh H. (1993) Specific recognition site probes for isolating genes encoding DNA-binding proteins. *Methods Enzymol.* **218**, 551–67.

Singh H., LeBowitz J.H., Bladwin A.S.J. & Sharp P.A. (1988) Molecular cloning of enhancer-binding protein: isolation by screening of an expression library with a recognition site DNA. *Cell* **52**, 415–23.

Sinn E., Muller W., Pattengale P. *et al.* (1987) Coexpression of MMTV/v-Ha-*ras* and MMTV/c-*myc* genes in transgenic mice: synergistic action of oncogenes *in vivo.* *Cell* **49**, 465–75.

Sinn P.L. & Sigmund C.D. (2000) Transgenic models as tools for studying the regulation of human renin expression. *Regul. Pept.* **86**, 77–82.

Sive H.L. & St John T. (1988) A simple subtractive hybridization technique employing photoactivable biotin and phenol extracton. *Nucl. Acids Res.* **16**, 10937.

Sizemore C., Wieland B., Gotz F. & Hillen W. (1991) Regulation of the *Staphylococcus xylosus* xylose utilization genes at the molecular level. *J. Bacteriol.* **174**, 3042–8.

Sjoberg E.M., Suomalainen M. & Garoff H. (1994) A significantly improved Semliki Forest virus expression system based on translation enhancer segments from the viral capsid gene. *Biotechnology* **12**, 1127–31.

Skarnes W.C., Auerbach B.A. & Joyner A.L. (1992) A gene trap approach in mouse embryonic stem cells: the *lacZ* reporter is activated by splicing, reflects endogenous gene expression, and is mutagenic in mice. *Genes Devel.* **6**, 903–18.

Skarnes W.C., Moss J.E., Hurtley S.M. & Beddington R.S. (1995) Capturing genes encoding membrane and secreted proteins important for mouse development. *Proc. Nat. Acad. Sci. USA* **92**, 6592–6.

Sklar M.D., Thompson E., Welsh M.J. *et al.* (1991) Depletion of c-*myc* with specific antisense sequences reverses the transformed phenotype in ras oncogene-transformed NIH 3T3 cells. *Mol. Cell Biol.* **11**, 3699–710.

Skow L.C., Burkhart B.A., Johnson F.M., Popp P.A. & Popp D.M. (1983) A mouse model for beta-thalassemia. *Cell* **34**, 1043–52.

Slilaty S.N. & Lebel S. (1998) Accurate insertional inactivation of lacZalpha: construction of pTrueBlue and M13 TrueBlue cloning vectors. *Gene* **213**, 83–91.

Smerdou C. & Liljestrom P. (1999) Two helper RNA system for production of recombinant Semliki Forest virus particles. *J. Virol.* **73**, 1092–8.

Smith A.J., De Sousa M.A., Kwabi-Addo B. *et al.* (1995) A site-directed chromosomal translocation induced in embryonic stem cells by Cre-*loxP* recombination. *Nature Genet.* **9**, 376–85.

Smith C., Watson C., Ray J. *et al.* (1988) Antisense RNA inhibition of polygalacturonase gene expression in transgenic tomatoes. *Nature* **334**, 724–6.

Smith C.J.S., Watson C.F., Bird C.R. *et al.* (1990) Expression of a truncated tomato polygalacturonase gene inhibits expression of the endogenous gene in transgenic plants. *Mol. Gen. Genet.* **224**, 477–81.

Smith D.J., Stevens M.E., Sudanagunta S.P. *et al.* (1997) Functional screening of 2 Mb of human chromosome 21q22.2 in transgenic mice implicates minibrain in learning defects associated with Down Syndrome. *Nature Genet.* **16**, 28–36.

Smith E.F. & Townsend C.O. (1907) A plant-tumor of bacterial origin. *Science* **25**, 671–3.

Smith G.E., Summers M.D. & Fraser M.J. (1983) Production of human β-interferon in insect cells infected with a baculovirus expression vetor. *Mol. Cell. Biol.* **3**, 2156–65.

Smith G.L., Mackett M. & Moss B. (1983a) Infectious vaccinia virus recombinants that express hepatitis B virus surface antigen. *Nature* **302**, 490–5.

Smith H., Bron S., Ven Ee J. & Venema G. (1987) Construction and use of signal sequence selection vectors in *Escherichia coli* and *Bacillus subtilis. J. Bacteriol.* **169**, 3321–8.

Smith H.B., Larimer F.W. & Hartman F.C. (1990) An engineered change in substrate specificity of ribulosebisphosphate carboxylase/oxygenase. *J. Biol. Chem.* **265**, 1243–5.

Smith H.O. & Nathans D. (1973) A suggested nomenclature for bacterial host modification and restriction systems and their enzymes. *J. Mol. Biol.* **81**, 419–23.

Smith H.O. & Wilcox K.W. (1970) A restriction enzyme from *Hemophilus influenzae*. I. Purification and general properties. *J. Mol. Biol.* **51**, 379–91.

Smith K., Johnson K., Bryan T. *et al.* (1993) The *APC* gene product in normal and tumor cells. *Proc. Nat. Acad. Sci. USA* **90**, 2846–50.

Smith L.C. & Wilmut I. (1989) Influence of nuclear and cytoplasmic activity on the development *in vivo* of sheep embryos after nuclear transplantation. *Biol. Reprod.* **40**, 1027–35.

Smith P.A., Tripp B.C., DiBlasio-Smith E.A. *et al.* (1998) A plasmid expression system for quantitative *in vivo* biotinylation of thioredoxin fusion proteins in *Escherichia coli. Nucl. Acids Res.* **26**, 1414–20.

Smith S.B., Aldridge P.K. & Callis J.B. (1989) Observation of individual DNA molecules undergoing gel electrophoreis. *Science* **243**, 203–6.

Smithies O. (1993) Animal models of human genetic diseases. *Trends Genet.* **9**, 112–16.

Smithies O., Gregg R.G., Boggs S.S., Koralewski M.A. & Kucherlapati R. (1985) Insertion of DNA sequences into the human β-globin locus by homologous recombination. *Nature* **317**, 230–4.

Snouwaert J.N., Brigman K.K., Latour A.M. *et al.* (1992) An animal-model for cystic-fibrosis made by gene targeting. *Science* **257**, 1083–8.

Snyder R.O., Miao C.H., Patijn G.A. *et al.* (1997) Persistent and therapeutic concentrations of human factor IX in mice after hepatic gene transfer of recombinant AAV vectors. *Nature Genet.* **16**, 270–6.

Somers D.A., Rines H.W., Gu W., Kaeppler H.F. & Bushnell W.R. (1992) Fertile, transgenic oat plants. *Biotechnology* **10**, 1589–94.

Sorge J.A. (1988) Bacteriophage lambda cloning vectors. In *Vectors: A Survey of Molecular Cloning Vectors and Their Uses*, eds Rodriguez R.L. & Denhardt D.T., pp. 43–60. Butterworth Press, Boston, Massachusetts.

Soriano P. (1995) Gene targeting in ES cells. *Ann. Rev. Neurosci.* **18**, 1–18.

Sosio M., Giusino F., Cappellano C. *et al.* (2000) Artificial chromosomes for antibiotic-producing actinomycetes. *Nature Biotechnol.* **18**, 343–5.

Southern E.M. (1996a) DNA chips: analysing sequence by hybridization to oligonucleotides on a large scale. *Trends Genet.* **12**, 110–15.

Southern E.M. (1996b) High-density gridding: techniques and applications *Curr. Opin. Biotechnol.* **7**, 85–8.

Southern E.M. (1975) Detection of specific sequences among DNA fragments separated by gel electrophoresis. *J. Mol. Biol.* **98**, 503–17.

Southern E.M. (1979a) Measurement of DNA length by gel electrophoresis. *Anal. Biochem.* **100**, 319–23.

Southern E.M. (1979b) Gel electrophoresis of restriction fragments. *Methods Enzymol.* **68**, 152–76.

Southern P.J. & Berg P. (1982) Transformation of mammalian cells to antibiotic resistance with a bacterial gene under the control of the SV40 early region promoter. *J. Mol. Appl. Genet.* **1**, 327–41.

Speek M., Raff J.W., Harrison Lavoie K., Little P.F.R. & Glover D.M. (1988) Smart2, a cosmid vector with a phage lambda origin for both systematic chomosome walking and P-element mediated gene transfer in *Drosophila. Gene* **64**, 173–7.

Spellman P.T., Sherlock G., Zhang M.Q. *et al.* (1998) Comprehensive identification of cell cycle-regulated gene of the yeast *Saccharomyces cerevisiae* by microarray hybridization. *Mol. Biol. Cell* **9**, 3273–3297.

Spiker S. & Thompson W.F. (1996) Nuclear matrix attachment regions and transgene expression in plants. *Plant Physiol.* **110**, 15–21.

Spiller B., Gershenson A., Arnold F.H. & Stevens R.C. (1999) A structural view of evolutionary divergence. *Proc. Nat. Acad. Sci. USA* **96**, 12305–10.

Spradling A.C. & Rubin G.M. (1982) Transposition of cloned P elements into *Drosophila* germ line chromosomes. *Science* **218**, 341–7.

Sprague K.V., Faulds D.H. & Smith G.R. (1978) A single basepair change creates a *chi* recombinational hotspot in bacteriophage λ. *Proc. Nat. Acad. Sci. USA* **75**, 6182–6.

Spreng S., Dietrich G. & Niewiesk S. *et al.* (2000) Novel bacterial systems for the delivery of recombinant protein or DNA. *FEMS Immunol. Med. Microbiol.* **27**, 299–304.

Stacey G.N., Bolton B.J. & Doyle E. (1992) DNA fingerprinting transforms the art of cell authentication. *Nature* **357**, 261–2.

Stachel S.E., Messens E., Van Montagu M. & Zambryski P. (1985) Identification of the signal molecules produced by wounded plant cells that activate T-DNA transfer in *Agrobacterium tumefaciens. Nature* **318**, 624–9.

Stagljar I., Bourquin J.P. & Schaffner W. (1996) Use of the two-hybrid system and random sonicated DNA to identify the interaction domain of a protein. *Biotechniques* **21**, 430–2.

Stallcup M.R., Sharrock W.J. & Rabinowitz J.C. (1974) Ribosome and messenger specificity in protein synthesis by bacteria. *Biochem. Biophys. Res. Commun.* **58**, 92–8.

Stanley J. (1983) Infectivity of the cloned geminivirus genome requires sequences from both DNAs. *Nature* **305**, 643–5.

Stark D.M., Timmermann K.P., Barry G.F., Preiss J. & Kishore G.M. (1992) Regulation of the amount of starch in plant tissues by ADP glucose phosphorylase. *Science* **258**, 287–92.

Staub J. & Maliga P. (1992a) Long regions of homologous DNA are incorporated into the tobacco plastid genome by transformation. *Plant Cell* **4**, 39–45.

Staub J. & Maliga P. (1992b) High-frequency plastid transformation in tobacco by selection for a chimeric *aadA* gene. *Proc. Nat. Acad. Sci. USA* **90**, 913–17.

Staub J., Garcia B., Graves J. *et al.* (2000) High-yield production of a human therapeutic protein in tobacco chloroplasts. *Nature Biotechnol.* **18**, 333–8.

Staudt L.M., Clerc R.G., Singh H. *et al.* (1988) Cloning of a lymphoid-specific cDNA encoding a protein binding the regulatory octamer DNA motif. *Science* **241**, 577–9.

Stavropoulos T.A. & Strathdee C.A. (1998) An enhanced packaging system for helper-dependent herpes simplex virus vectors. *J. Virol.* **72**, 7137–43.

Stearns T., Ma H. & Botstein D. (1990) Manipulating the yeast genome using plasmid vectors. *Methods Enzymol.* **185**, 280–97.

Steck T.R. (1997) Ti plasmid type affects T-DNA processing in *Agrobacterium tumefaciens. FEMS Microbiol. Lett.* **147**, 121–5.

Steinbuchel A. & Fuchtenbusch B. (1998) Bacterial and other biological systems for polyester production. *Trends Biotechnol.* **16**, 419–27.

Steller H. & Pirrotta V. (1985) A transposable P vector that confers selectable G418 resistance of *Drosophila* larvae. *EMBO J.* **4**, 167–71.

Stenger D.C., Revington G.N., Stevenson M.C. & Bisaro D.M. (1991) Replicational release of geminivirus genomes from tandemly repeated copies – evidence for rolling-circle replication of a plant viral-DNA. *Proc. Nat. Acad. Sci. USA* **88**, 8029–33.

Sternberg N. (1990) Bacteriophage P1 cloning system for the isolation, amplification and recovery of DNA fragments as large as 100 kilobase pairs. *Proc. Nat. Acad. Sci. USA* **87**, 103–7.

Sternberg N. (1994) The P1 cloning system – past and future. *Mammalian Genome* **5**, 397–404.

Stief A., Winter D.M., Stratling W.H. & Sippel A.E. (1989) A nuclear DNA attachment element mediates elevated and position-independent gene activity. *Nature* **341**, 343–5.

Stinchcomb D.T., Struhl K. & Davis R.W. (1979) Isolation and characterization of a yeast chromosomal replicator. *Nature* **282**, 39–43.

Stoesser G., Sterk P., Tuli M.A., Stoehr P.J. & Cameron G.N. (1997) The EMBL nucleotide sequence database. *Nucl. Acids Res.* **25**, 7–13.

Storb U., O'Brien R.L., McMullen M.D., Gollahon K.A. & Brinster R.L. (1984) High expression of cloned immunoglobulin kappa gene in transgenic mice is restricted to B-lymphocytes. *Nature* **310**, 238–48.

Stoss O., Mogk A. & Schumann W. (1997) Integrative vector for constructing single-copy translational fusions between regulatory regions of *Bacillus subtilis* and the *bgsB* reporter gene encoding a heat-stable beta-galactosidase. *FEMS Microbiol. Lett.* **150**, 49–54.

Stover C.K. & 13 others (1991) New use of BCG for recombinant vaccines. *Nature* **351**, 456–60.

Strathdee C.A., Gavish H., Shannon W.R. & Buchwald M. (1992) Cloning of cDNAs for Fanconi's anaemia by functional complementation. *Nature* **356**, 763–7.

Struhl K. (1983) The new yeast genetics. *Nature* **305**, 391–7.

Struhl K., Stinchcomb D.T., Scherer S. & Davis R.W. (1979) High-frequency transformation of yeast: autonomous replication of hybrid DNA molecules. *Proc. Nat. Acad. Sci. USA* **76**, 1035–9.

Studier F.W. (1991) Use of bacteriophage T7 lysozyme to improve an inducible T7 expression system. *J. Mol. Biol.* **219**, 37–44.

Studier F.W., Rosenberg A.H., Dunn J.J. & Dubendorff J.W. (1990) Use of T7 RNA polymerase to direct expression of cloned genes. *Methods Enzymol.* **185**, 60–89.

Su H., Chang J.C., Xu S.M. & Kan Y.W. (1996) Selective killing of AFP-positive hepatocellular carcinoma cells by adeno-associated virus transfer of the herpes simplex virus thymidine kinase gene. *Hum. Gene Ther.* **7**, 463–70.

Subramani S., Mulligan R. & Berg P. (1981) Expression of mouse dihydrofolate reductase complementary deoxyribonucleic acid in simian virus 40 vectors. *Mol. Cell. Biol.* **1**, 854–64.

Sugden B., Marsh K. & Yates J. (1985) A vector that replicates as a plasmid and can be efficiently selected in B-lymphoblasts transformed by Epstein–Barr virus. *Mol. Cell. Biol.* **5**, 410–13.

Suggs S.V., Wallace R.B., Hirose T., Kawashima E.H. & Itakura K. (1981) Use of synthetic oligonucleotides as hybridization probes. III. Isolation of cloned cDNA sequences for human beta-2-microglobulin. *Proc. Nat. Acad. Sci. USA* **78**, 6613–17.

Sugimoto Y., Aksentijevich I., Gottesman M.M. & Pastan I. (1994) Efficient expression of drug-selectable genes in retroviral vectors under control of an internal ribosome entry site. *Biotechnology* **12**, 694–8.

Sukchawalit R., Vattanaviboon P., Sallabhan R. & Mongkolsuk S. (1999) Construction and characterization of regulated L-arabinose-inducible broad host range expression vectors in *Xanthomonas. FEMS Microbiol. Lett.* **181**, 217–23.

Sumikawa K., Houghton M., Emtage J., Richards B. & Barnard E. (1981) Active multi-subunit ACh receptor assembled by translation of heterologous mRNA in *Xenopus* oocytes. *Nature* **292**, 862.

Summers D. (1998) Timing, self-control and a sense of direction are the secrets of multicopy plasmid stability. *Mol. Microbiol.* **29**, 1137–45.

Summers D.K. & Sherratt D.J. (1984) Multimerization of high copy number plasmids causes instability: Col E1 encodes a determinant essential for plasmid monomerization and stability. *Cell* **36**, 1097–103.

Sun L., Li J. & Xiao X. (2000) Overcoming AAV vector size limitation through viral DNA heterodimerization. *Nature Med.* **6**, 599–602.

Sun T.-Q., Fenstermacher D.A. & Vos J.-M.H. (1994) Human artificial episomal chromosomes for cloning large DNA fragments in human cells. *Nature Genet.* **8**, 33–41.

Sundaresan V. (1996) Horizontal spread of transposon mutagenesis: new uses for old elements. *Trends Plant Sci.* **6**, 184–90.

Sundaresan V., Springer P., Volpe T. *et al.* (1995) Patterns of gene action in plant development revealed by enhancer trap and gene trap transposable elements. *Genes Devel.* **9**, 1797–810.

Sussman D.J. & Milman G. (1984) Short-term, high-efficiency expression of transfected DNA. *Mol. Cell. Biol.* **4**, 1641–3.

Sutcliffe J.G. (1979) Complete nucleotide sequence of the *Escherichia coli* plasmid pBR322. *Cold Spring Harbor Symp. Quant. Biol.* **43**(1), 77–90.

Sutter G. & Moss B. (1992) Non-replicating vaccinia vector efficiently expresses recombinant genes. *Proc. Nat. Acad. Sci. USA* **89**, 10847–51.

Suzuki Y., Yoshimoto-Nakagawa K., Maruyama K., Suyama A. & Sugano S. (1997) Construction and characterization of full length-enriched and a 5′-end-enriched cDNA library. *Gene* **200**, 149–56.

Suzuki Y., Ishihara D., Sasaki M. *et al.* (2000) Statistical analysis of the 5′-untranslated region of human mRNA using oligo capped cDNA libraries. *Genomics* **64**, 286–97.

Svab Z., Harper E.C., Jones J.D.G. & Maliga P. (1990a) Aminoglycoside 3′-adenyltransferase confers resistance to spectinomycin and streptomycin in *Nicotiana tabacum* plants. *Plant Mol. Biol.* **14**, 197–205.

Svab Z., Hajdukiewcz P. & Maliga P. (1990b) Stable transformation of plastids in higher plants. *Proc. Nat. Acad. Sci. USA* **87**, 8526–30.

Swick A.G., Janicot M., Chenevalkastelic T., McLenithan J.C. & Lane M.D. (1992) Promoter cDNA-directed heterologous protein expression in *Xenopus laevis* oocytes. *Proc. Nat. Acad. Sci. USA* **89**, 1812–16.

Swift G.H., Hammer R.E., MacDonald R.J. & Brinster R.L. (1984) Tissue specific expression of the rat pancreatic elastase 1 gene in transgenic mice. *Cell* **38**, 639–46.

Sykes B. & Irven C. (2000) Surnames and the Y chromosome. *Am. J. Hum. Genet.* **66**, 1417–19.

Szczylik C., Skorski T., Nicolaides N.C. *et al.* (1991) Selective inhibition of leukaemia cell proliferation by BCR-ABL antisense oligonucleotides. *Science* **253**, 262–5.

Szostak J.W. & Blackburn E.H. (1982) Cloning yeast telomeres on linear plasmid vectors. *Cell* **29**, 245–55.

Szybalska E.H. & Szybalski (1962) Genetics of human cell lines IV. DNA-mediated heritable transformation of a biochemical trait. *Proc. Nat. Acad. Sci. USA* **48**, 2026–31.

Tabara H., Grishok A. & Mello C.C. (1998) RNAi in *C. elegans*: soaking in the genome sequence. *Science* **282**, 430–1.

Tabor S. & Richardson C.C. (1995) A single residue in DNA polymerases of the *Escherichia coli* DNA polymerase I family is critical for distinguishing between deoxy- and dideoxyribonucleotides. *Proc. Nat. Acad. Sci.* **92**, 6339–43.

Tackaberry E.S., Dudani A.K., Prior F. *et al.* (1999) Development of biopharmaceuticals in plant expression systems: cloning, expression and immunological reactivity of human cytomegalovirus glycoprotein B (UL55) in seeds of transgenic tobacco. *Vaccine* **17**, 3020–9.

Tacket C.O., Reid R.H., Boedeker E.C. *et al.* (1994) Enteral immunisation and challenge of volunteers given enterotoxigenic *E. coli* CFA/II encapsulated in biodegradable microspheres. *Vaccine* **12**, 1270–4.

Tait R.C., Lundquist R.C. & Kado C.I. (1982) Genetic map of the crown gall suppressive *Inc*W plasmid pSa. *Mol. Gen. Genet.* **186**, 10–15.

Tait R.C., Close T.J., Lundquist R.C. *et al.* (1983) Construction and characterization of a versatile broad host range DNA cloning system for Gram-negative bacteria. *Biotechnology* **1**, 269–75.

Takagi H., Morinaga Y., Tsuchiya M., Ikemura H. & Inouye M. (1988) Control of folding of proteins secreted by a high expression secretion vector, pIN-111-ompA: 16-fold increase in production of active subtilisin E in *Escherichia coli*. *Biotechnology* **6**, 948–50.

Takamatsu N., Ishikawa M., Meshi T. & Okada Y. (1987) Expression of bacterial chloramphenicol acetyltransferase gene in tobacco plants mediated by TMV-RNA. *EMBO J.* **6**, 307–11.

Takeuchi T., Yamazaki Y., Katoh-Fukui Y. *et al.* (1995) Gene trap capture of a novel mouse gene, jumonji, required for neural tube formation. *Genes Devel.* **9**, 1211–22.

Takeuchi Y., Dotson M. & Keen N.T. (1992) Plant transformation: a simple particle bombardment device based on flowing helium. *Plant Mol. Biol.* **18**, 835–9.

Takumi T. (1997) Use of PCR for cDNA library screening. *Methods Mol. Biol.* **67**, 339–44.

Takumi T. & Lodish H.F. (1994) Rapid cDNA cloning by PCR screening. *Biotechniques* **17**, 443–4.

Tamura T., Thibert C., Royer C. *et al.* (1999) Germline transformation of the silkworm *Bombyx mori* L. using a piggyBac transposon-derived vector. *Nature Biotechnol.* **18**, 81–4.

Tao H., Bausch C., Richmond C., Blattner F.R. & Conway T. (1999) Functional genomics: expression analysis of *Escherichia coli* growing on minimal and rich media. *J. Bacteriol.* **181**, 6425–40.

Tartaglia J. & Paoletti E. (1988) Recombinant vaccinia virus vaccines. *Trends Biotechnol.* **6**, 43–6.

Tartaglia J., Perkus M.E., Taylor J. *et al.* (1992) NYVAC: a highly attenuated strain of vaccinia virus. *Virology* **188**, 217–32.

Tate P., Lee M., Tweedie S., Skarnes W.C. & Bickmore W.A. (1998) Capturing novel mouse genes encoding chromosomal and other nuclear proteins. *J. Cell Sci.* **111**, 2575–85.

Tateno Y. & Gojobori T. (1997) DNA databank of Japan in the age of information biology. *Nucl. Acids Res.* **25**, 14–17.

Tatusov R.L., Mushegian A.R., Bork P. *et al.* (1996) Metabolism and evolution of Haemophilus influenzae deduced from a whole-genome comparison with *Escherichia coli*. *Curr. Biol.* **6**, 279–91.

Tavernarakis N., Wang S.L., Dorokov M., Ryazanov A. & Driscoll M. (2000) Heritable and inducible genetic interference by double-stranded RNA encoded by transgenes. *Nature Genet.* **24**, 180–3.

Tavladoraki P., Benvenuto E., Trinca S. *et al.* (1993) Transgenic plants expressing a functional single-chain Fv antibody are specifically protected from virus attack. *Nature* **366**, 469–72.

Taylor G.R. & Logan W.P. (1995) The polymerase chain reaction: new variations on an old theme. *Curr. Opin. Biotechnol.* **6**, 24–9.

Teeri T.H., Herrera-Estrella L., Depicker A., Van Montagu M. & Palva T. (1986) Identification of plant promoters by T-DNA-mediated transcriptional fusions to the *nptII* gene. *EMBO J.* **5**, 1755–60.

Tepfer D. (1984) Transformation of several species of higher plants by *Agrobacterium rhizogenes*: sexual transmission of the transformed genotype and phenotype. *Cell* **37**, 959–67.

Terashima M., Murai Y., Kawamura M. *et al.* (1999) Production of functional human alpha l-antitrypsin by plant cell culture. *Appl. Microbial Biotechnol.* **52**, 516–23.

Te Riele H., Maandag E.R. & Berns A. (1992) Highly efficiency gene targeting in embryonic stem cells through homologous recombination with isogenic DNA constructs. *Proc. Nat. Acad. Sci. USA* **89**, 5128–32.

Terskikh A., Fradkov A., Ermakova G. *et al.* (2000) 'Fluorescent timer': protein that changes colour with time. *Science* **290**, 1585–8.

Thomas J.G., Ayling A. & Baneyx F. (1997) Molecular chaperones, folding catalysts and the recovery of active recombinant proteins from *E. coli*: to fold or to refold. *Appl. Biochem. Biotechnol.* **66**, 197–238.

Thomas K.R. & Capecchi M.R. (1987) Site-directed mutagenesis by gene targeting in mouse embryo-derived stem cells. *Cell* **51**, 503–12.

Thomas K.R., Folger K.R. & Cappechi M.R. (1986) High frequency targeting of genes to specific sites in the mammalian genome. *Cell* **44**, 419–28.

Thomas M. & Davis R.W. (1975) Studies on the cleavage of bacteriophage lambda DNA with EcoRI restriction endonuclease. *J. Mol. Biol.* **91**, 315–28.

Thomas M., Cameron J.R. & Davis R.W. (1974) Viable molecular hybrids of bacteriophage lambda and eukaryotic DNA. *Proc. Nat. Acad. Sci. USA* **71**, 4579–83.

Thomas P.S. (1980) Hybridization of denatured RNA and small DNA fragments transferred to nitrocellulose. *Proc. Nat. Acad. Sci. USA* **77**, 5201–5.

Thompson E.M., Nagata S. & Tsuji F.I. (1990) *Vargula hilgendorfii* luciferase: a secreted reporter enzyme for monitoring gene expression in mammalian cells. *Gene* **96**, 257–62.

Thompson J.A., Drayton P.R., Frame B.R., Wang K. & Dunwell J.M. (1995) Maize transformation utilizing silicon carbide whiskers – a review. *Euphytica* **85**, 75–80.

Thorey I.S., Muth K., Russ A.P. *et al.* (1998) Selective disruption of genes transiently induced in differentiating mouse embryonic stem cells by using gene trap mutagenesis and site-specific recombination. *Mol. Cell Biol.* **18**, 3081–8.

Thorsted P.B. & 11 others (1998) Complete sequence of the IncPbeta plasmid R751: implications for evolution and organisation of the IncP backbone. *J. Mol. Biol.* **282**, 969–90.

Tickle C. & Eichele G. (1994) Vertebrate limb development. *Ann. Rev. Cell Biol.* **10**, 121–52.

Timmermans M.C.P., Das O.P. & Messing J. (1992) *Trans* replication and high copy numbers of wheat dwarf virus vectors in maize cells. *Nucl. Acids Res.* **20**, 4047–54.

Timmermans M.C.P., Das O.P. & Messing J. (1994) Geminiviruses and their uses as extrachromosomal replicons. *Ann. Rev. Plant Physiol. Plant Mol. Biol.* **45**, 79–112.

Timmons L. & Fire A. (1998) Specific interference by ingested dsRNA. *Nature* **395**, 854.

Tingay S., McElroy D., Kalla R. *et al.* (1997) *Agrobacterium tumefaciens*-mediated barley transformation. *Plant J.* **11**, 1369–76.

Tinland B. (1996) The integration of T-DNA into plant genomes. *Trends Plant Sci.* **1**, 179–84.

Tinland B., Koukolikova-Nicola Z., Hall M.N. & Hohn B. (1992) The T-DNA-linked *vir*D2 protein contains two distinct functional nuclear localization signals. *Proc. Nat. Acad. Sci. USA* **89**, 7442–6.

Tissier A.F., Marillonnet S., Klimyuk V. *et al.* (1999) Multiple independent defective *suppressor-mutator* transposon insertions in *Arabidopsis*: a tool for functional genomics. *Plant Cell* **11**, 1841–52.

Tjalsma H., Bolhuis A., Jongbloed J.D.H., Bron S. & Van Dijl J.M. (2000) Signal peptide-dependent protein transport in *Bacillus subtilis*: a genome-based survey of the secretome. *Microbiol. Mol. Biol. Rev.* **64**, 515–47.

Tomashek J.J., Sonnenburg J.L., Artimovich J.M. & Klionsky D.J. (1996) Resolution of subunit interactions and cytoplasmic subcomplexes of the yeast vacuolar proton-translocating ATPase. *J. Biol. Chem.* **271**, 10397–404.

Tomes D.T., Wessinger A.K., Ross M. *et al.* (1990) Transgenic tobacco plants and their progeny derived from microprojectile bombardment of tobacco leaves. *Plant Mol. Biol.* **14**, 261–8.

Tomizawa J.-I. & Itoh T. (1981) Plasmid ColE1 incompatibility determined by interaction of RNA I with primer transcript. *Proc. Nat. Acad. Sci. USA* **78**, 6096–100.

Tomizawa J.-I. & Itoh T. (1982) The importance of RNA secondary structure in ColE1 primer formation. *Cell* **31**, 575–83.

Tomizuka K., Yoshida H., Uejima H. *et al.* (1997) Functional expression and germline transmission of a human chromosome fragment in chimaeric mice. *Nature Genet.* **16**, 133–43.

Toole J.J., Knopf J.L., Wozney J.M. *et al.* (1984) Molecular cloning of a cDNA encoding human antihaemophilic factor. *Nature* **321**, 342–7.

Topping J.F., Wei W. & Lindsey K. (1991) Functional tagging of regulatory elements in the plant genome. *Development* **112**, 1009–19.

Torbert K., Rines H.W. & Somers D.A. (1995) Use of paromomycin as a selective agent for oat transformation. *PCR* **14**, 635–40.

Touraev A., Stoger E., Voronin V. & Heberle-Bors E. (1997) Plant male germ line transformation. *Plant J.* **12**, 949–56.

Towbin H., Staehelin T. & Gordon J. (1979) Electrophoretic transfer of proteins from polyacrylamide gels to nitrocellulose sheets: procedure and some applications. *Proc. Nat. Acad. Sci. USA* **76**, 4350–4.

Townes T.M. *et al.* (1985) Erythroid-specific expression of human β-globin genes in transgenic mice. *EMBO J.* **4**, 1715–23.

Townsend R., Watts J. & Stanley J. (1986) Synthesis of viral-DNA forms in *Nicotiana-plumbaginifolia* protoplasts inoculated with cassava latent virus (CLV) – evidence for the independent replication of one-component of the CLV genome. *Nucl. Acids Res.* **14**, 1253–65.

Traboni C., Cortese R., Cilibert G. & Cesarini G. (1983) A general method to select M13 clones carrying base pair substitution mutants constructed *in vitro*. *Nucl. Acids Res.* **11**, 4229–39.

Tricoli D.M., Carney K.J., Russell P.F. *et al.* (1995) Field evaluation of transgenic squash containing single or multiple virus coat protein gene constructs for resistance to cucumber mosaic virus, watermelon mosaic virus 2, and zucchini yellow mosaic virus. *Biotechnology* **13**, 1458–65.

Trieu-Cuot P., Carlier C., Martin P. & Courvalin P. (1987) Plasmid transfer by conjugation from *Escherichia coli* to Gram positive bacteria. *FEMS Microbiol. Lett.* **48**, 289–94.

Trieu-Cuot P., Carlier C. & Courvalin P. (1988) Conjugative plasmid transfer from *Enterococcus faecalis* to *Escherichia coli*. *J. Bacteriol.* **170**, 4388–91.

Troester H., Bub S., Hunziker A. & Trendelenburg M.F. (2000) Stability of DNA repeats in *Escherichia coli dam* mutant strains indicates a Dam methylation-dependent DNA deletion process. *Gene* **258**(1–2), 95–108.

Trudel M., Saadane N., Garel M.C. *et al.* (1991) Towards a transgenic mouse model of sickle cell disease: hemoglobin SAD. *EMBO J.* **10**, 3157–65.

Tseng W.-C. & Huang L. (1998) Liposome-based gene therapy. *Pharmacol. Sci. Technol. Today* **1**, 206–13.

Tsien R.Y. (1998) The green fluorescent protein. *Ann. Rev. Biochem.* **67**, 509–44.

Tsien R.Y. & Miyawaki A. (1998) Seeing the machinery of live cells. *Science* **280**, 1954–5.

Tuite M.F., Dobson M.J., Roberts N.A. *et al.* (1982) Regulated high efficiency expression of human interferon-alpha in *Saccharomyces cerevisiae*. *EMBO J.* **1**, 603–8.

Tumbula D.L. & Whitman W.B. (1999) Genetics of *Methanococcus*: possibilities for functional genomics in Archaea. *Mol. Microbiol.* **33**, 1–7.

Twigg A.J. & Sherratt D. (1980) *Trans*-complementable copy-number mutants of plasmid ColE1. *Nature* **283**, 216–18.

Tyurin M., Starodubtseva L., Kudryavtseva H., Voeykova T. & Livshits V. (1995) Electrotransformation of germinating spores of *Streptomyces* spp. *Biotechnol. Techniques* **9**, 737–40.

Tzfira T., Rhee Y., Chen M.H., Kunik T. & Citovsky V. (2000) Nucleic acid transport in plant–microbe interactions: the molecules that walk through the walls. *Ann. Rev. Microbiol.* **54**, 187–219.

Uchimiya H., Iwata M., Nojiri C. *et al.* (1993) Bialaphos treatment of transgenic rice plants expressing a *bar* gene prevents infection by the sheath blight pathogen (*Rhizoctonia solani*). *Biotechnology* **11**, 835–6.

Uetz P. & Hughes R.E. (2000) Systematic and large-scale two-hybrid screens. *Curr. Opin. Microbiol.* **3**, 303–8.

Ulaeto D. & Hruby D.E. (1994) Uses of vaccinia virus in vaccine delivery. *Curr. Opin. Biotechnol.* **5**, 501–4.

Ullrich A., Bell J.R., Chen E.Y. *et al.* (1985) Human insulin receptor and its relationship to the tyrosine kinase family of oncogenes. *Nature* **313**, 756–61.

Ulmer J.B., Donnelly J.J., Parker S.E. *et al.* (1993) Heterologous protection against influenza by injection of DNA encoding a viral protein. *Science* **259**, 1745–9.

Urlaub G., Kas E., Carothers A.M. & Chasin L.A. (1983) Deletion of the diploid dihydrofolate reductase locus from cultured mammalian cells. *Cell* **33**, 405–12.

Usdin K. & Grabczyk E. (2000) DNA repeat expansions and human disease. *Cell Mol. Life Sci.* **57**, 914–31.

Vaeck M., Reynaerts A., Hofte H. *et al.* (1987) Transgenic plants protected from insect attack. *Nature* **328**, 33–7.

Vagner V., Dervyn E. & Ehrlich S.D. (1998) A vector for systematic gene inactivation in *Bacillus subtilis*. *Microbiology* **144**, 3097–104.

Vain P., Finer K.R., Engler D.E., Pratt R.C. & Finer J.J. (1996) Intron-mediated enhancement of gene expression in maize (*Zea mays* L) and bluegrass (*Poa pratensis* L). *Plant Cell Rep.* **15**, 489–94.

Valancius V. & Smithies O. (1991) Testing an 'in–out' targeting procedure for making subtle genomic modifications in mouse embryonic stem cells. *Mol. Cell. Biol.* **11**, 1402–8.

Valdivia R.H. & Falkow S. (1997) Fluorescence-based isolation of bacterial genes expressed within host cells. *Science* **277**, 2007–11.

Valenzuela P., Medina A., Rutter W.J., Ammerer G. & Hall B.D. (1982) Synthesis and assembly of hepatitis B virus surface antigen particles in yeast. *Nature* **298**, 347–50.

Valle G., Jones E.A. & Colman A. (1982) Anti-ovalbumin monoclonal antibodies interact with their antigen in internal membranes of *Xenopus* oocytes. *Nature* **300**, 71–4.

Van den Elzen P.J.M., Townsend J., Lee K.Y. & Bedbrook J.R. (1985) A chimeric hygromycin resistance gene as a selectable marker in plant cells. *Plant Mol. Biol.* **5**, 299–302.

Van der Geest A.H.M. & Hall T.C. (1997) The β-phaseolin 5′ matrix attachment region acts as an enhancer facilitator. *Plant Mol. Biol.* **33**, 553–7.

Van der Krol A.R., Lenting P.E., Veenstra J. *et al.* (1988) An antisense chalcone synthase gene in transgenic plants inhibits flower pigmentation. *Nature* **333**, 866–9.

Van Deursen J., Fornerod M., Van Rees B. & Grosveld G. (1995) Cre-mediated site-specific translocation between nonhomologous mouse chromosomes. *Proc. Nat. Acad. Sci. USA* **92**, 7376–80.

Van Dyk T.K., Gatenby A.A. & LaRossa R.A. (1989) Demonstration by genetic suppression of interaction of GroE products with many proteins. *Nature* **342**, 451–3.

Vanhove B., Charreau B., Cassard A., Pourcel C. & Soulillou J.P. (1998) Intracellular expression in pig cells of anti-α1,3 galactosyl-transferase single-chain FV antibodies reduces Gal α1,3 Gal expression and inhibits cytotoxicity mediated by anti-Gal xenoantibodies. *Transplantation* **66**, 1477–85.

Van Larbeke N., Engler G., Holsters M. *et al.* (1974) Large plasmid in *Agrobacterium tumefaciens* essential for crown gall-inducing ability. *Nature* **252**, 169–70.

Van Sluys M.A., Tempe J. & Fedoroff N. (1987) Studies on the introduction and mobility of the maize *Activator* element in *Arabidopsis thaliana* and *Daucus carota*. *EMBO J.* **6**, 3881–9.

Vara J.A., Portela A., Ortin J. & Jiminez A. (1986) Expression in mammalian cells of a gene from *Streptomyces alboniger* conferring puromycin resistance. *Nucl. Acids Res.* **14**, 4617–24.

Vasil V., Castillo A., Fromm M. & Vasil I. (1992) Herbicide resistant fertile transgenic wheat plants obtained by microprojectile bombardment of regenerable embryogenic callus. *Biotechnology* **10**, 667–74.

Vegeto E., Allan G.F., Schrader W.T. *et al.* (1992) The mechanism of RU486 antagonism is dependent on the conformation of the carboxy-terminal tail of the human progesterone receptor. *Cell* **69**, 703–13.

Velculescu V.E., Zhang L., Vogelstein B. & Kinzler K.W. (1995) Serial analysis of gene expression. *Science* **270**, 484–7.

Vellanoweth R.L. (1993) Translation and its regulation: *Bacillus subtilis* and other Gram-positive bacteria. In *Biochemistry, Physiology and Molecular Genetics*, eds Sonenshein A.L., Hoch J.A., Losick B.R. ASM. Washington DC.

Venter J.C., Adams M.D., Myers E.W. *et al.* (2001) The sequence of the human genome. *Science* **291**, 1304–51.

Venter J.C., Smith H.O. & Hood L. (1996) A new strategy for genome sequencing. *Nature* **381**, 364–6.

Verch T., Yusibov V. & Koprowski H. (1998) Expression and assembly of a full-length monoclonal antibody in plants using a plant virus vector. *J. Immunol. Methods* **220**, 69–75.

Verpoorte R. (1998) Exploration of nature's chemodiversity: the role of secondary metabolites as leads for drug development. *Drug Devel. Today* **3**, 232–8.

Verpoorte R., Van der Heijden R. & Memelink J. (2000) Engineering the plant cell factory for secondary metabolite production. *Transgenic Res.* **9**, 323–43.

Vick L., Li Y. & Simkiss K. (1993) Transgenic birds from transformed primordial germ cells. *Proc. R. Soc. London B Biol. Sci.* **251**, 179–82.

Vidal M., Brachmann R.K., Fattaey A., Harlow E. & Boeke J.D. (1996a) Reverse two-hybrid and one-hybrid systems to detect dissociation of protein–protein and DNA–protein interaction. *Proc. Nat. Acad. Sci. USA* **93**, 10315–20.

Vidal M., Braun P., Chen E., Boeke J.D. & Harlow E. (1996b) Genetic characterization of a mammalian protein–protein interaction domain by using a yeast reverse two-hybrid system. *Proc. Nat. Acad. Sci. USA* **93**, 10321–26.

Vidan S. & Snyder M. (2001) Large-scale mutagenesis: yeast genetics in the genome era. *Curr. Opin. Biotechnol.* **12**, 28–34.

Vieira J. & Messing J. (1982) The pUC plasmids, an M13mp7-derived system for insertion mutagenesis and sequencing with synthetic universal primers. *Gene* **19**, 259–68.

Vieira J. & Messing J. (1987) Production of single-stranded plasmid DNA. *Methods Enzymol.* **153**, 3–11.

Vielkind J.R. (1992) Medaka and zebrafish: ideal as transient and stable transgenic systems. In: *Transgenic Fish*, eds Hew C.L. & Fletcher G.L., pp. 72–91. World Scientific Press.

Vinson C.R., LaMarco K.L., Johnson P.F., Landschulz W.H. & McKnight S.L. (1988) *In situ* detection of sequence-specific binding activity specified by a recombinant bacteriophage. *Genes Devel.* **2**, 801–6.

Visser R.G.F., Somhorst I., Kuipers G.F.J. *et al.* (1991) Inhibition of the expression of the gene for granule-bound synthase in potato by antisense constructs. *Mol. Gen. Genet.* **225**, 289–96.

Vize P.D. & Melton D.A. (1991) Assays for gene function in developing *Xenopus* embryos. *Methods Cell Biol.* **36**, 367–87.

Voelker T.A., Worrell A.C., Anderson L. *et al.* (1992) Fatty acid biosynthesis redirected to medium chains in transgenic oilseed plants. *Science* **257**, 72–4.

Voinnet O., Vain P., Angell S. & Baulcombe S.C. (1998) Systemic spread of sequence-specific transgene RNA degradation in plants is initiated by localized introduction of ectopic promoterless DNA. *Cell* **95**, 177–87.

Von Melchner H., DeGregori J.V., Rayburn H. *et al.* (1992) Selective disruption of genes expressed in totipotent embryonal stem cells. *Genes Devel.* **6**, 919–27.

Vos J.-M.H., Westphal E.-M. & Banerjee S. (1996) Infectious herpes vectors for gene therapy. In *Gene Therapy*, eds Lemoine N.R. & Cooper D.N., pp. 127–53. BIOS Scientific Publishers, Oxford, UK.

Voskuil M.I. & Chambliss G.H. (1998) The −16 region of *Bacillus subtilis* and other Gram-positive bacterial promoters. *Nucl. Acids Res.* **26**, 3584–90.

Voss A., Niersbach M., Han R. *et al.* (1995) Reduced virus infectivity in *N. tabacum* secreting a TMV-specific full size antibody. *Mol. Breeding* **1**, 39–50.

Wacker I., Kaether C., Kromer A. *et al.* (1997) Microtubule-dependent transport of secretory vesicles visualized in real time with a GFP-tagged secretory protein. *J. Cell Sci.* **110**, 1453–63.

Wagner K.U., Wall R.J., St-Ogne L. *et al.* (1997) Cre-mediated gene deletion in the mammary gland. *Nucl. Acids Res.* **25**, 4323–30.

Wagner R., Liedtke S., Kretzschmar E. *et al.* (1996) Elongation of the N-glycans of fowl plague virus hemagglutinin expressed in *Spodoptera frugiperda* (Sf9) cells by coexpression of human beta 1,2-N-acetylglucosaminyltransferase I. *Glycobiology* **6**, 165–75.

Wahl G.M., De Saint Vincent B.R. & DeRose M.L. (1984) Effect of chromosomal position on amplification of transfected genes in animal cells. *Nature* **307**, 516–20.

Wahl G.M., Lewis K.A., Ruiz J.C. *et al.* (1987) Cosmid vectors for rapid genomic walking, restriction mapping, and gene transfer. *Proc. Nat. Acad. Sci. USA* **84**, 2160–4.

Wakayama T., Perry A.C., Zuccotti M., Johnson K.R. & Yanagimachi R. (1998) Full term development of mice from enucleated oocytes injected with cumulus cell nuclei. *Nature* **394**, 369–74.

Walbot V. (2000) Saturation mutagenesis using maize transposons. *Curr. Opin. Plant Biol.* **3**, 103–7.

Walker M.D., Karlsson O., Edlund T., Barnett J. & Rutter W.J. (1986) Sequences controlling cell-specific expression of the rat insulin-1 gene. *J. Cell Biochem.* **73**, Suppl. 10.

Wall R.J. (1999) Biotechnology for the production of modified and innovative animal products: transgenic livestock bioreactors. *Livestock Prod. Sci.* **59**, 243–55.

Wallace R.B., Johnson P.F., Tanaka S. *et al.* (1980) Directed deletion of a yeast transfer RNA intervening sequence. *Science* **209**, 1396–400.

Wallace R.B., Schold M., Johnson M.J., Dembek P. & Itakura K. (1981) Oligonucleotide directed mutagenesis of the human β-globin gene: a general method for producing specific point mutations in cloned DNA. *Nucl. Acids Res.* **9**, 3647–56.

Walmsey A.M. & Arntzen C.J. (2000) Plants for delivery of edible vaccines. *Curr. Opin. Biotechnol.* **11**, 126–9.

Wan Y. & Lemaux P. (1994) Generation of large numbers of independently transformed fertile barley plants. *Plant Physiol.* **104**, 37–48.

Wang B., Ugen K.E., Srikantan V. *et al.* (1993) Gene inoculation generates immune responses against human immunodeficiency virus type 1. *Proc. Nat. Acad. Sci. USA* **90**, 4156–60.

Wang G.-L., Holsten T.E., Song W.-Y., Wang H.-P. & Ronald P.C. (1995a) Construction of a rice bacterial artificial chromosome library and identification of clones linked to the Xa-21 disease resistance locus. *Plant J.* **7**, 525–33.

Wang L., Ogburn C.E., Ware C.B. *et al.* (2000) Cellular Werner phenotypes in mice expressing a putative dominant-negative human WRN gene. *Genetics* **154**, 357–62.

Wang S. & Hazelrigg T. (1994) Implications for *bcd* mRNA localization from spatial distribution of Exu protein in *Drosophila* oogenesis. *Nature* **369**, 400–3.

Wang Y., De Mayo F.J., Tsai S.Y. & O'Malley B.W. (1997a) Ligand-inducible and liver-specific target gene expression in transgenic mice. *Nature Biotechnol.* **15**, 239–43.

Wang Y., Xu J., Pierson T., O'Malley B.W. & Tsai S.Y. (1997b) Positive and negative regulation of gene expression in eukaryotic cells with an inducible transcriptional regulator. *Gene Ther.* **4**, 432–41.

Wang Z., Engler P., Longacre A. & Storb U. (2001) An efficient method for high-fidelity BAC/PAC retrofitting with a selectable marker for mammalian cell transfection. *Genome Res.* **11**, 137–42.

Wang Z.F., Morris J.C., Drew M.E. & Englund P.T. (2000) Inhibition of *Trypanosoma brucei* gene expression by RNA interference using an integratable vector with opposing T7 promoters. *J. Biol. Chem.* **275**, 40174–9.

Ward A., Etessami P. & Stanley J. (1988) Expression of a bacterial gene in plants mediated by infectious geminivirus DNA. *EMBO J.* **7**, 1583–7.

Ward W.W. & Bokman S.H. (1982) Reversible denaturation of *Aequorea green* fluorescent protein: physical separation and characterization of the renatured protein. *Biochemistry* **21**, 4535–50.

Waterhouse P.M., Graham M.W. & Wang M.B. (1998) Virus resistance and gene silencing in plants can be induced by simultaneous expression of sense and antisense RNA. *Proc. Nat. Acad. Sci. USA* **95**, 13959–64.

Watson B., Currier T.C., Gordon M.P., Chilton M.-D. & Nester E.W. (1975) Plasmid requirement for virulence of *Agrobacterium tumefaciens*. *J. Bacteriol.* **123**, 255–64.

Watson J.D. (1972) Origin of concatameric T7 DNA. *Nature New Biol.* **239**, 197–201.

Watson N. (1988) A new revision of the sequence of plasmid pBR322. *Gene* **70**, 399–403.

Weidle U.H., Lenz H. & Brem G. (1991) Genes encoding a mouse monoclonal antibody are expressed in transgenic mice, rabbits and pigs. *Gene* **98**, 185–91.

Weiler E.W. & Schroder J. (1987) Hormone genes and crown gall disease. *TIBS* **12**, 271–5.

Weiss R., Teich N., Varmus H. & Coffin J. (1985) *RNA Tumour Viruses*, 2nd edn. Cold Spring Harbor Laboratory, Cold Spring Harbor, New York.

Weitzman M.D., Kyostio S.R., Kotin R.M. & Owens R.A. (1994) Adeno-associated virus Rep proteins mediate complex formation between AAV DNA and ins integration site in human DNA. *Proc. Nat. Acad. Sci. USA* **91**, 5808–12.

Welch P.J., Barber J.R. & Wong-Staal F. (1998) Expression of ribozymes in gene transfer systems to modulate target RNA levels. *Curr. Opin. Biotechnol.* **9**, 486–96.

Wells J.A., Vasser M. & Powers D.B. (1985) Cassette mutagenesis: an efficient method for generation of multiple mutations at defined sites. *Gene* **34**, 315–23.

Welsh J., Chada K., Dalal S.S. *et al.* (1992) Arbitrarily primed PCR fingerprinting of RNA. *Nucl. Acids Res.* **20**, 4965–70.

Wen X., Fuhrman S., Michaels G.S. *et al.* (1998) Large scale temporal gene expression mapping of central nervous system development. *Proc. Nat. Acad. Sci. USA* **95**, 334–9.

Wensink P.C., Finnegan D.J., Donelson J.E. & Hogness D.S. (1974) A system for mapping DNA sequences in the chromosomes of *Drosophila melanogaster*. *Cell* **3**, 315–25.

Wettereau J.R. & many others (1998) An MTP inhibitor that normalizes atherogenic lipoprotein levels in WHHL rabbits. *Science* **282**, 751–4.

Wetzel R., Perry L.J., Baase W.A. & Becktel W.J. (1988) Disulphide bonds and thermal stability in T4 lysozyme. *Proc. Nat. Acad. Sci. USA* **85**, 401–5.

Wheeler G.N., Hamilton F.S. & Hoppler S. (2000) Inducible gene expression in transgenic *Xenopus* embryos. *Curr. Biol.* **10**, 849–52.

Wianny F. & Zernicka-Goetz M. (2000) Specific interference with gene function by double stranded RNA in early mouse development. *Nature Cell Biol.* **2**, 70–5.

Wiberg F.C., Sunnerhagen P. & Bjursell G. (1987) Efficient transient and stable expression in mammalian cells of transfected genes using erythrocyte ghost fusion. *Exp. Cell Res.* **173**, 218–31.

Widera G., Gautier F., Lindenmaier W. & Collins J. (1978) The expression of tetracycline resistance after insertion of foreign DNA fragments between the *Eco*RI and *Hind*III sites of the plasmid cloning vector pBR322. *Mol. Gen. Genet.* **163**, 301–5.

Wigler M., Silverstein S., Lee L.S. *et al.* (1977) Transfer of purified herpes virus thymidine kinase gene to cultured mouse cells. *Cell* **11**, 223–32.

Wigler M., Sweet R., Sim G.K. *et al.* (1979) Transformation of mammalian cells with genes from procaryotes and eucaryotes. *Cell* **16**, 777–85.

Wigler M., Perucho M., Kurtz D. *et al.* (1980) Transformation of mammalian cells with an amplifiable dominant acting gene. *Proc. Nat. Acad. Sci. USA* **77**, 3567–70.

Wilcox A.S., Khan A.S., Hopkins J.A. & Sikela J.M. (1991) Use of 3′ untranslated sequences of human cDNAs for rapid chromosome assignment and conversion to STSs: implications for an expression map of the genome. *Nucl. Acids Res.* **19**, 1837–43.

Wiles M.V., Vauti F., Otte J. *et al.* (2000) Establishment of a gene-trap sequence tag library to generate mutant mice from embryonic stem cells. *Nature Genet.* **24**, 13–14.

Wilkins B.M., Chilley P.M., Thomas A.T. & Pocklington M.J. (1996) Distribution of restriction enzyme recognition sequences on broad host range plasmid RP4: molecular and evolutionary implications. *J. Mol. Biol.* **258**, 447–56.

Wilkinson A.J., Fersht A.R., Blow D.M., Carter P. & Winter G. (1984) A large increase in enzyme–substrate affinity by protein engineering. *Nature* **307**, 187–8.

Wilkinson J.Q., Lanahan M.B., Clark D.G. *et al.* (1997) A dominant mutant receptor from *Arabidopsis* confers ethylene insensitivity in heterologous plants. *Nature Biotechnol.* **15**, 444–7.

Wilks A.F. (1989) Two putative protein-tyrosine kinases identified by application of the polymerase chain reaction. *Proc. Nat. Acad. Sci. USA* **86**, 1063–7.

Willadsen S.M. (1989) Cloning of sheep and cow embryos. *Genome* **31**, 956–62.

Williams B.G. & Blattner F.R. (1979) Construction and characterization of the hybrid bacteriophage lambda Charon series for DNA cloning. *J. Virol.* **29**, 555–62.

Williams D.C., Van Frank R.M., Muth W.L. & Burnett J.P. (1982) Cytoplasmic inclusion bodies in *Escherichia coli* producing biosynthetic human insulin proteins. *Science* **215**, 687–8.

Williams T.M., Moolten D., Burlein J. *et al.* (1991) Identification of a zinc finger protein that inhibits IL-2 gene expression. *Science* **254**, 1791–4.

Willmitzer L., Simons G. & Schell J. (1982) The Ti DNA in octopine crown gall tumours codes for seven well-defined polyadenylated transcripts. *EMBO J.* **1**, 139–46.

Willmitzer L., Dhaese P., Schreier P.H. *et al.* (1983) Size, location and polarity of transferred DNA encoded transcripts in nopaline crown gall tumours: common transcripts in octopine and nopaline tumours. *Cell* **32**, 1045–6.

Wilmut I., Schnieke A.E., McWhir J., Kind A.J. & Campbell K.H.S. (1997) Viable offspring derived from fetal and adult mammalian cells. *Nature* **385**, 810–13.

Wilson C., Bellen H.J. & Gehring W.J. (1990) Position effects on eukaryotic gene expression. *Ann. Rev. Cell Biol.* **6**, 679–714.

Wilson L.E., Wilkinson N., Marlow S.A., Possee R.D. & King L.A. (1997) Identification of recombinant baculoviruses using green fluorescent protein as a selectable marker. *Biotechniques* **22**, 674–81.

Wilson R. & 54 others (1994) 2.2Mb of contiguous nucleotide sequence from chromosome III of *C. elegans*. *Nature* **368**, 32–8.

Wilson R.K., Koop B.F., Chen C. *et al.* (1992) Nucleotide sequence analysis of 95 kb near the 3′ end of the murine T-cell receptor α/δ chain locus: strategy and methodology. *Genomics* **13**, 1198–208.

Winans S.C. (1992) Two-way chemical signaling in *Agrobacterium*–plant interactions. *Microbiol. Rev.* **56**, 12–31.

Winter G., Fersht A.R., Wilkinson A.J., Zoller M. & Smith M. (1982) Redesigning enzyme structure by site-directed mutagenesis: tyrosyl tRNA synthetase and ATP binding. *Nature* **299**, 756–8.

Winter G., Griffiths A.D., Hawkins R.E. & Hoogenboom H.R. (1994) Making antibodies by phage display technology. *Ann. Rev. Immunol.* **12**, 433–55.

Winter J.A., Wright R.L. & Gurley W.B. (1984) Map locations of five transcripts homologous to TR-DNA in tobacco and sunflower crown gall tumours. *Nucl. Acids Res.* **12**, 2391–406.

Winzeler E.A. & 25 others (1999) Functional characterization of the *S. cerevisiae* genome by gene deletion and parallel analysis. *Science* **285**, 901–6.

Wong C.-H., Chen S.-T., Hennen W.J. *et al.* (1990) Enzymes in organic synthesis: use of subtilisin and a highly stable mutant derived from multiple site-specific mutation. *J. Am. Chem. Soc.* **112**, 945–53.

Wong T.-K. & Neumann (1982) Electric field mediated gene transfer. *Biochem. Biophys. Res. Commun.* **107**, 584–7.

Woo S.-S., Jiang J., Gill B.S., Paterson A.H. & Wing R.A. (1994) Construction and characterisation of a bacterial artificial chromosome library of *Sorghum bicolor*. *Nucl. Acids Res.* **22**, 4922–31.

Wood K.V. & DeLuca M. (1987) Photographic detection of luminescence in *Escherichia coli* containing the gene for firefly luciferase. *Anal. Biochem.* **161**, 501–7.

Wood W.I., Capon D.J., Simonsen C.C. *et al.* (1984) Expression of active human factor VIII from recombinant DNA clones. *Nature* **312**, 330–7.

Wright G., Carver A., Cottom D. *et al.* (1991) High-level expression of active human alpha-1-antitrypsin in the milk of transgenic sheep. *Biotechnology* **9**, 830–4.

Wu C.Y., Adachi T., Hatano T. *et al.* (1998) Promoters of rice seed storage protein genes direct endosperm-specific gene expression in transgenic rice. *Plant Cell Physiol.* **39**, 885–9.

Wu R. & Taylor E. (1971) Nucleotide sequence analysis of DNA. II. Complete nucleotide sequence of the cohesive ends of bacteriophage λ DNA. *J. Mol. Biol.* **57**, 491–511.

Wu R., Bahl C.P. & Narang S.A. (1978) Chemical synthesis of oligonucleotides. *Prog. Nucl. Acid Res. Mol. Biol.* **21**, 101–38.

Wu S.-Y. & Chiang C.-M. (1996) Establishment of stable cell lines expressing potentially toxic proteins by tetracycline-regulated and epitope tagging methods. *Biotechniques* **21**, 718–25.

Wu X., Holschen J., Kennedy S.C. & Parker-Ponder K. (1996) Retroviral vector sequences may interact with some internal promoters and influence expression. *Hum. Gene Ther.* **7**, 159–71.

Wurm F.M., Gwinn K.A. & Kingston R.E. (1986) Inducible overproduction of the mouse c-Myc protein in mammalian cells. *Proc. Nat. Acad. Sci. USA* **83**, 5414–18.

Wurst W., Rossant J., Prideaux V. *et al.* (1995) A large-scale gene-trap screen for insertional mutations in developmentally regulated genes in mice. *Genetics* **139**, 889–99.

Xiang C.C. & Chen Y. (2000) cDNA microarray technology and its applications. *Biotechnol. Adv.* **18**, 35–46.

Xiang Z.Q., Yang Y., Wilson J.M. & Ertl H.C. (1996) A replication-defective human adenovirus recombinant serves as a highly efficacious vaccine carrier. *Virology* **219**, 220–7.

Xiao J.H., Davidson I., Matthes H., Garnier J.M. & Chambon P. (1991) Cloning, expression, and transcriptional properties of the human enhancer factor TEF-1. *Cell* **65**, 551–68.

Xiong C., Levis R., Shen P. *et al.* (1989) Sindbis virus: an efficient broad host range vector for gene expression in animal cells. *Science* **243**, 1188–91.

Xu C.W., Mendelsohn A. & Brent R. (1997) Cells that register logical relationships among proteins. *Proc. Nat. Acad. Sci. USA* **94**, 12473–8.

Xu L., Sanchez A., Yang Z.-Y. *et al.* (1998) Immunization for Ebola virus infection. *Nature Med.* **4**, 37–42.

Xu X.P. & Li B.J. (1994) Fertile transgenic Indica rice plants obtained by electroporation of the seed embryo cells. *Plant Cell Rep.* **13**, 237–42.

Yadav N.S., Vanderleyden J., Bennet D., Barnes W.M. & Chilton M.-D. (1982) Short direct repeats flank the T-DNA on a nopaline Ti plasmid. *Proc. Nat. Acad. Sci. USA* **79**, 6322–6.

Yager T.D., Dunn J.M. & Stevens J.K. (1997) High-speed DNA sequencing in ultrathin slab gels. *Curr. Opin. Biotechnol.* **8**, 107–13.

Yamao M., Katayama N., Nakazawa H. *et al.* (1999) Gene targeting in the silkworm by use of a baculovirus. *Genes Devel.* **13**, 511–16.

Yang C.C., Xiao X., Zhu X. *et al.* (1997) Cellular recombination pathways and viral terminal repeat hairpin structures are sufficient for adeno-associated virus integration *in vivo* and *in vitro*. *J. Virol.* **71**, 9231–47.

Yang M., Wu Z. & Fields S. (1995) Protein–peptide interactions analyzed with the yeast two-hybrid system. *Nucl. Acids Res.* **23**, 1152–6.

Yang W., Musci T.S. & Mansour S.L. (1997) Trapping genes expressed in the developing mouse inner ear. *Hearing Res.* **114**, 53–61.

Yang X.C., Karschin A., Labarca C. *et al.* (1991) Expression of ion channels and receptors in *Xenopus* oocytes using vaccinia virus. *FASEB J.* **5**, 2209–16.

Yang Y.-G., DeGoma E., Ohdan H. *et al.* (1998) Tolerization of anti-Galα1–3Gal natural antibody-forming B cells by induction of mixed chimerism. *J. Exp. Med.* **187**, 1335–42.

Yansich-Perron C., Vieira J. & Messing J. (1985) Improved M13 phage cloning vectors and host strains: nucleotide sequences of the M13 mp18 and pUC19 vectors. *Gene* **33**, 103–19.

Yansura D.G. & Henner D.J. (1984) Use of the *Escherichia coli lac* repressor and operator to control gene expression in *Bacillus subtilis. Proc. Nat. Acad. Sci. USA* **81**, 439–43.

Yao T.P., Segraves W.A., Oro A.E., McKeown M. & Evans R.M. (1992) *Drosophila* ultraspiracle modulates ecdysone receptor function via heterodimer formation. *Cell* **71**, 63–72.

Yao T.P., Forman B.M., Jiang Z. *et al.* (1993) Functional ecdysone receptor is the product of EcR and ultraspiracle genes. *Nature* **366**, 476–9.

Yates J.L., Warren N., Reisman D. & Sugden B. (1984) A *cis*-acting element from the Epstein–Barr genome that permits stable replication of recombinant plasmids in latently infected cells. *Proc. Nat. Acad. Sci. USA* **81**, 3806–10.

Yates J.L., Warren N. & Sugden B. (1985) Stable replication of plasmids derived from Epstein–Barr virus in various mammalian cells. *Nature* **313**, 812–15.

Ye X.D., Al-Babili S., Kloti A. *et al.* (2000) Engineering the provitamin A (beta-carotene) biosynthetic pathway into (carotenoid-free) rice endosperm. *Science* **287**, 303–5.

Ymer S., Schofield P.R., Draguhn A. *et al.* (1989) GABA receptor beta-subunit heterogeneity: functional expression of cloned cDNAs. *EMBO J.* **6**, 1665–70.

Yoder J., Walsh C. & Bestor T. (1997) Cytosine methylation and the ecology of intragenomic parasites. *Trends Genet.* **13**, 335–9.

Yokoyama T., Copeland N.G., Jenkins N.A. *et al.* (1993) Reversal of left right asymmetry: a *situs inversus* mutation. *Science* **260**, 679–82.

Young B.D., Birnie G.D. & Paul J. (1976) Complexity and specificity of polysomal poly (A)⁺ RNA in mouse tissues. *Biochemistry* **15**, 2823–8.

Young J.M., Cheadle C., Foulke J.S. Jr, Drohan W.N. & Sarver N. (1988) Utilization of an Epstein–Barr virus replicon as a eukaryotic expression vector. *Gene* **62**, 171–85.

Young R.A. & Davis R.W. (1983) Efficient isolation of genes by using antibody probes. *Proc. Nat. Acad. Sci. USA* **80**, 1194–8.

Young R.A., Bloom B.R., Grosskinsky C.M. *et al.* (1985) Dissection of *Mycobacterium tuberculosis* antigens using recombinant DNA. *Proc. Nat. Acad. Sci. USA* **82**, 2583–7.

Young S.M., McCarty D.M., Degtyareva N. & Samulski R.J. (2000) Roles of adeno-associated virus Rep and human

chromosome 19 in site-specific recombination. *J. Virol.* **74**, 3953–66.

Yu D., Ellis H.M., Lee E.-C. *et al.* (2000) An efficient recombination system for chromosome engineering in *Escherichia coli. Proc. Nat. Acad. Sci. USA* **97**, 5978–83.

Yu S.F., Von Ruden T., Kantoff P.W. *et al.* (1986) Self-inactivating retroviral vectors designed for transfer of whole genes into mammalian cells. *Proc. Nat. Acad. Sci. USA* **83**, 3194–8.

Yuan R., Hamilton D.L. & Burckhardt J. (1980) DNA translocation by the restriction enzyme from *E. coli* K. *Cell* **20**, 237–44.

Zabner J., Couture L.A., Graham S.M., Smith A.E. & Welsh M.J. (1993) Adenovirus-mediated gene transfer transiently corrects the chloride transport defect in nasal epithelia of patients with cystic fibrosis. *Cell* **75**, 1–20.

Zaenen I., Van Larbeke N., Teuchy H., Van Montagu M. & Schell J. (1974) Super-coiled circular DNA in crown-gall inducing *Agrobacterium* strains. *J. Mol. Biol.* **86**, 109–27.

Zambrowicz B.P., Friedrich G.A., Buxton E.C. *et al.* (1998) Disruption and sequence identification of 2000 genes in mouse embryonic stem cells. *Nature* **392**, 608–11.

Zambryski P., Depicker A., Kruger H. & Goodman H. (1982) Tumor induction by *Agrobacterium tumefaciens*: analysis of the boundaries of T-DNA. *J. Mol. Appl. Gent.* **1**, 361–70.

Zambryski P., Joos H., Genetello C. *et al.* (1983) Ti plasmid vector for the introduction of DNA into plant cells without alteration of their normal regeneration capacity. *EMBO J.* **2**, 2143–50.

Zehetner G. & Lehrach H. (1994) The reference library system – sharing biological material and experimental data. *Nature* **367**, 489–91.

Zeitlin L., Olmsted S.S., Moench T.R. *et al.* (1998) A humanized monoclonal antibody produced in transgenic plants for immonoprotection of the vagina against genital herpes. *Nature Biotechnol.* **16**, 1361–4.

Zhang J. & Deutscher M.P. (1992) A uridine-rich sequence required for translation of prokaryotic mRNA. *Proc. Nat. Acad. Sci. USA* **89**, 2605–9.

Zhang L., Cui X., Schmitt K. *et al.* (1992) Whole genome amplification from a single cell: implications for genetic analysis. *Proc. Nat. Acad. Sci. USA* **89**, 5847–51.

Zhang L.-J., Cheng L.-M., Xu N. *et al.* (1991) Efficient transformation of tobacco by ultrasonication. *Biotechnol.* **9**, 996–7.

Zhang X. & Studier F.W. (1997) Mechanism of inhibition of bacteriophage T7 RNA polymerase by T7 lysozyme. *J. Mol. Biol.* **269**, 10–27.

Zhao H. & Arnold F.H. (1999) Directed evolution converts subtilisin E into a functional equivalent of thermitase. *Protein Eng.* **12**, 47–53.

Zhao J. & Lemke G. (1998) Selective disruption of neuregulin-1 function in vertebrate embryos using ribozyme-tRNA transgenes. *Development* **125**, 1899–907.

Zhao J.J.G. & Pick L. (1983) Generating loss-of-function phenotypes of the *fushi tarazu* gene with a targeted ribozyme in *Drosophila*. *Nature* **365**, 448–51.

Zhong Z., Liu J.L.-C., Dinterman L.M. *et al.* (1991) Engineering subtilisin for reaction in dimethylformamide. *J. Am. Chem. Soc.* **113**, 683–4.

Zhou G., Weng J., Zheng Y. *et al.* (1983) Introduction of exogenous DNA into cotton embryos. *Methods Enzymol.* **101**, 433–81.

Zhou H.S., O'Neal W., Morral N. & Beaudet A.L. (1996) Development of a complementing cell line and a system for construction of adenovirus vectors with E1 and E2a deleted. *J. Virol.* **70**, 7030–8.

Zhu H. & Dean R.A. (1999) A novel method for increasing the transformation efficiency of *Escherichia coli*–application for bacterial artificial chromosome library construction. *Nucl. Acids Res.* **27**, 910–11.

Zhu Z., Huges K.W., Huang L. *et al.* (1994) Expression of human alpha-interferon cDNA in transgenic rice plants. *Plant Cell Tissue Organ Cult.* **36**, 197–204.

Ziegler A., Cowan G.H., Torrance L., Ross H.A. & Davies H.V. (2000) Facile assessment of cDNA constructs for expression of functional antibodies in plants using the potato virus X vector. *Mol. Breeding* **6**, 327–35.

Zimmer A. & Gruss P. (1989) Production for chimaeric mice containing embryonic stem cell (ES) cells carrying a homeobox *Hox1.1* allele mutated by homologous recombination. *Nature* **338**, 150–3.

Zimmerman S.B. & Pheiffer B. (1983) Macromolecular crowding allows blunt end ligation by DNA ligases from rat liver or *Escherichia coli*. *Proc. Nat. Acad. Sci. USA* **80**, 5852–6.

Zimmerman U. & Vienken J. (1983) Electric field induced cell to cell fusion. *J. Membr. Biol.* **67**, 165–82.

Zimmermann J., Schillberg J., Liao Y.C. & Fischer R. (1998) Intracellular expression of TMV-specific single-chain Fv fragments leads to improved virus resistance in *Nicotiana tabacum*. *Mol. Breeding* **4**, 369–79.

Zoller M.J. & Smith M. (1983) Oligonucleotide-directed mutagenesis of DNA fragments cloned into M13 vectors. *Methods Enzymol.* **100**, 468–500.

Zolotukhin S.M., Potter W.W., Hauswirth J.G. & Muzyczka N. (1996) A humanized green fluorescent protein cDNA adapted for high level expression in mammalian cells. *J. Virol.* **70**, 4646–54.

Zoubenko O.V., Allison L.A., Svab Z. & Maliga P. (1994) Efficient targeting of foreign genes into the tobacco plastid genome. *Nucl. Acids Res.* **22**, 3819–24.

Zubko E., Scutt C. & Meyer P. (2000) Intrachromosomal recombination between *attP* regions as a tool to remove selectable marker genes from tobacco transgenes. *Nature Biotechnol.* **18**, 422–45.

Zupan J., Muth T.R., Draper O. & Zambryski P. (2000) The transfer of DNA from *Agrobacterium tumefaciens* into plants: a feast of fundamental insights. *Plant J.* **23**, 11–28.

Index

Note: Page numbers in *italics* represent figures, tables and boxes. Entries with a numeric or Greek prefix are to be found under the first letter of the main word, e.g. "β-galactosidase" is listed under "g".

Ace locus 107
acetosyringone *227, 235*
Acinetobacter calcoaceticus 139
actin-1 promoter *230*
actinorhodin *308*
acute transforming retroviruses 193
Ada gene *179, 181*
adaptors 38–40, *39*
addition vectors 244
adeno-associated virus vectors 189, *190, 198*
adenosine deaminase 290
adenovirus
    as recombinant vaccine 294
    vectors 188–9, *188, 198*
adh gene 219
ADH1 promoter 165
Aeromonas spp. *275*
*Afl*II site 73
*Afl*III 145
African cassava mosaic virus 242, 314
agarose gel electrophoresis 9–11, *9, 10*
    contour-clamped homogeneous electrical field *10*
    pulsed field 10
    field inversion 10, 89
*Age*I enzyme 32
α-agglutinin 168
Agrobacterium rhizogenes 237
Agrobacterium tumefaciens 142, 147, 156, 220, 224, 225, 269

Agrobacterium-mediated transformation 224–37
    crown-gall disease 224, *225–7*
    disarmed Ti plasmid derivatives as plant vectors *228–34*
    experimental procedure 234, *235*
    high capacity binary vectors 236–7
    monocots 235–6
    and Ri plasmids 237
    T-DNA transfer 225–8, *228*
    tumour-inducing plasmids 224–5, *227*
agroinfection 241, 243
agroinoculation 241
ags gene *228*
ajmalicine *310*
alphavirus
    as recombinant vaccine 294
    vectors 195–7, *196*
*Alu*I enzyme *31, 32*
amino acid sequences 1
aminopterin 177, *179*
amplification 9
amplified genomic libraries 90
amyE gene 143
amylopectin 316
animals
    gene transfer into animal cells 174–201, *198, 199, 200, 201*
        DNA-mediated transformation 174–87
        strategies for 174
        viral transduction 187–99
    genetic manipulation 202–20
        fish 217–18
        invertebrates 218, *219, 220*
        mammals 203–15, *203, 204, 206, 207, 209–11, 214*
        Xenopus laevis 215–17, *216*
antimicrobial agents 296–7

Antirrhinum majus 268
antisense RNA transgenes 260–1
α$_1$-antitrypsin
    animal bioreactors producing *284*
    oxidation resistant *300*
    plant bioreactors producing *286*
AOX1 promoter 166
*Apa*I enzyme *31*
*Ap*R gene 51
Aprt gene *179*
Arabidopsis thaliana 239
    genome *2, 86, 266*
araC gene 76
archaebacteria 139
ars see autonomously replicating sequences
artificial recombinant molecules 8–9, 30
as gene *180, 181*
Aspergillus spp. 158
Aspergillus fumigatus 318
*Asu*I site 61
atrazine *312*
Atropa belladonna 311
Autographa californica multiple nuclear polyhedrosis virus 190
automated sequencing 124–5, *124*
autonomously replicating sequences 158
autoradiography *14, 15*
auxins 221, *222*
*Ava*I site 61
*Ava*II site 63
avermectin *307*
avian myeloblastosis virus 97
    reverse transcriptase 21
*Avr*II enzyme *31*
azaserine *179*

Bac-to-Bac system 191
Bacillus amyloliquefaciens 29
Bacillus circulans 79
Bacillus licheniformis 152
Bacillus megaterium 55
Bacillus stearothermophilus 299
Bacillus subtilis 5, 7, 139
  cloning in 148
  controlled expression 151, 152
  transformation 19, 140
    by plasmid rescue 140–1
    protoplast 141
Bacillus thuringiensis 230, 315
bacmids 191
BACs *see* bacterial artificial
  chromosomes
bacteria
  low-GC
    controlled expression in 151,
    152
    secretion vectors for 152
bacterial artificial chromosomes
  (BACs) 67–9, 68, 69, 89,
  138, 155, 236
bacterioferritin 81
bacteriophage λ 53–60
  essential features 53, 55, 56
  genome size 2
  packaging of DNA *in vitro* 58, 59,
  60
  promoters and control circuits
  53–7, 57
  vector DNA 57
  vectors for cDNA cloning 93, 94
bacteriophage φX174, genome size 2
baculovirus vectors 190–4, 193,
  198
Bam adaptor molecule 38, 39
*Bam*HI enzyme 29, 31, 40, 57, 58,
  68, 153, 158
*Bam*HI site 39, 67, 68
band shift assay 10
*Ban*II 145
BBR1 147
*Bcl*I 153
begomoviruses 242
Bemisia tabaci 242
*Bgl*I site 63
*Bgl*II enzyme 31, 153, 297
bialaphos 313
BIBAC2 vector 236
binary vectors 232–4, 234
  high-capacity 236–7
bioreactors
  plants 285, 286
  transgenic animals 283–4, 284
  transgenic plants 283–4, 284

biotechnology 4, 5
biotin 79
BIOTRX protein 82
birA gene 79
Bithorax gene 107
BK replicon 185–6
ble gene 180
blotting *see* nucleic acid blotting
blue/white screening assay 35, 84,
  193
Bombyx mori multiple nuclear
  polyhedrosis virus 190
Bordatella bronchiseptica 147
border sequences 225
Borrelia burgdorferi 43
Botrytis cinerea 314
boundary elements 210, 211
BPV replicon 185–6, 198
brome mosaic virus 244
bromoxynil 312
bsd gene 180
Bst site 145
*Bst*EII, cleavage sites 145

φC31 155
Cad gene 179, 181
Caenorhabditis elegans, genome size 2
calcium phosphate coprecipitate
  method 174–5
callus culture 221, 222
Campylobacter spp. 275
CaMV 35S 230, 238
cancer, gene therapy for 291–2
Candida albicans, promoters 165
Candida boidnii, promoters 165
capillary blotting 17
CAPture method for cDNA cloning
  99
cassette mutagenesis 132
cat gene 182, 238, 243
Catharanthus roseus 309
cauliflower mosaic virus 241, 242
caulimoviruses 241, 242
*cdc*10 locus 159
cDNA
  cloning 91–101
    CAPture method 99
    full-length 95–100, 99, 100
    PCR as alternative to 100–1
    strategies for 95, 96, 97, 98
  libraries 92–5, 93, 94
  preparation for library
    construction 95, 96–100
  properties 92
  rapid amplification of cDNA ends
    (RACE) 101, 102
  synthesis 95

centromere 158, 159
cer sequence 47
chain termination sequencing 120,
  121, 122
  modifications of 123–4, 123
chalcone 261
chaperons 81
Charon vector series 89
chemically induced dimerization
  251, 252
chi sites 58, 89
chimeras 8, 69, 163
chitinase 314
Chladosporium fulvum 315
Chlamydomonas reinhardtii 240
chloramphenicol acetyltransferase
  74
chloramphenicol transacetylase
  238
chloroplast transformation 240–1
chromosome engineering 258, 259
chromosome jumping 107, 108
chromosome microdissection 91
chromosome walking 107–9, 108,
  267
circular DNA replication 149, 150
circularization 38, 39
  preventing with alkaline
  phosphatase 39
*Cla*I enzyme 31
clo DF13 plasmid 45
cloned gene expression 73, 163–5,
  164
cloned gene products, purification
  80, 82
cloning 1, 2, 4, 9, 85–119, 85,
  139–74
  cDNA 92–101
  cosmids 65, 66, 67
  difference 114–19
  functional 112–14, 113
  in fungi 156–73
  genomic DNA 86–92
    genomic DNA libraries 86–91,
    86, 87, 88, 90
    polymerase chain reaction
    91–2
  Gram-positive bacteria 148–53
  limitations of 95–7
  non-enteric bacteria 144–7
  positional 107
  single-stranded vectors 60–3
  in streptomycetes 153–4
  subtractive 115
cloning vehicles *see* vectors
Clostridium difficile 275
Clostridium welchii 275

Cochliobolus 158
codons 77
cointegrate vectors 232, *233*
Col E1 plasmid 45, 50
combinatorial biosynthesis 4, 306, *307, 308*
comparative sequence analysis *277, 278, 279*
complementary cell lines 188
complex probes 115
concatemerization 190
conditional knockouts 253
consensus sequence 70, *71*
contour-clamped homogeneous electrical field electrophoresis *10*
control circuits 53, 55, 56, *57*
copF gene 151
copy number 45–7, *46, 47*
Corynebacterium spp. 306
cosmids 18, 64–7, *65, 66, 67,* 89
   cloning *65, 66, 67*
cosuppression 261–4, *262, 263*
cotransformation *177, 178*
cre gene 68
cre promoter *317*
Cre recombinase 256, *257, 258*
Cre-*loxP* site-specific recombinase system 212, 256, *257*
Cre-mediated conditional mutants 258–60, *259*
Creutzfeldt-Jakob disease 3
crossover hot-spot instigator sites 58
crown-gall disease 224, *225–7*
cry1Ab gene 315
cry1Ac gene 315
cryptic plasmids 43
crystal proteins 315
CUP1 promoter 171, *172*
cycle sequencing 124
cystic fibrosis 288
cystic fibrosis gene *108*
cystic fibrosis transmembrane conductance regulator 74, 291
cytokinins 221, *222*

Dam methylase 33
Danio rerio 217
λDASH replacement vector 89, *90, 107*
databases 126–7, 130–2, *131*
daunorubicin *307*
dcm gene 33
Dcm methylase 33

debrisoquine metabolism 292
defensins 314
definition of gene manipulation 1
degenerate primers 109
delayed ripening 316
deletions 163
Desulfovibrio spp. 144
dhfr gene *181*
diazobenzyloxymethyl paper 16
dideoxynucleoside triphosphates *121, 122, 123*
diethylamino-ethyl dextran 175
difference cloning 114–19
   by PCR 116, *117, 118, 119*
   with DNA libraries 114–15
difference screening 114–15, *116*
differential fluorescence induction 298
direct DNA transfer 174
directed evolution 301–2
disease modelling 286
DNA
   agarose gel electrophoresis 9
   chimeras 8
   double-stranded 111
   introduction into bacterial cells 139–44, *140, 141, 143, 144*
   introduction into fungi 156
   non-replicating 176–7
   passenger 9
   plasmid 48–9, *48*
   recombinant 3, 4, 30
   satellite 279, *280*
   single-stranded 60–3, 70
   stuffer 189
   target 9
   transferred *see* T-DNA
DNA chips *116, 278*
DNA Databank database 126
DNA fingerprinting 274, *281*
   probes for *280*
DNA fragmentation 26–36
   E. coli Dam and Dcm methylases 33
   enzyme quality 34, *35*
   host-controlled restriction and modification 26, *27, 28*
   nomenclature 29
   number/size of restriction fragments 31
   recognition sequences 30, 31
   restriction and modification systems 27, *29*
DNA libraries, genomic DNA 86–91, *86, 87, 88, 90*

DNA ligase 36, 37, *38*
   joining DNA molecules without 41, *42*
DNA polymerase 123
   Klenow fragment 122
DNA recombination 36–42
   adaptors 38–40, *39*
   DNA ligase 36, 37, *38*
   extra sequences 41
   homopolymer tailing 40–1, *40*
   PCR products 41
   without DNA ligase 41–2
DNA sequencing *see* sequencing
DNA shuffling 302–3, *302*
DNA topoisomerase 41
DNA transfer
   to plants 237–9
   see also gene transfer
DNA vaccines 296
DNA vectors 57
   single-stranded 60–3
DNA-mediated transformation 174–87
   non-replicating DNA 176–84, *178–83*
      cotransformation and selection of stable transformants 177
      non-replicating plasmid vectors for transient transformation 182–3, *182, 183*
      plasmid vectors 181, *182*
      selectable markers 177–9, *178, 179, 180*
      transgene amplification 179–81, *181*
      transient and stable transformation 176–7
   replicon vectors 184–7, *184, 185*
      BK and BPV 185–6
      EBV 186–7
      runaway polyomavirus replicons *184, 185*
   techniques 174–6
      direct transfer 176
      DNA/calcium phosphate coprecipitate method 174–5
      electroporation 176
      phospholipids as gene-delivery vehicles 175–6
dnaJ gene 152
DnaK protein 81
Dolly the sheep 202, *314*
dominant selectable markers 178
dominant-negative mutants 265–6
Down's syndrome 288

downstream activating sequences 164
*Dpn*I enzyme *31*
*Dra*I enzyme *31*
*Dra*II, cleavage sites 145
Drosophila melanogaster 202
  Antennapedia gene 106
  Bithorax gene 107
  ecdysone 251
  enhancer traps 269, *270*
  eye-colour markers 219–20
  fushi tarazu gene 106
  gene transfer to 218–20, *219*
  genome *2*, 86, 266
  heat-shock promoter 247–8
  P-elements 218–20, *219*
    vectors for gene transfer 219, *220*
  rosy gene 107, 219

EBV replicon 186–7, *198*
ecdysone 251
Eco site 145
*Eco*K enzyme 33
*Eco*Ki enzyme 34
*Eco*RI enzyme 30, 31, *37*, 57, 58, 69, 86
  cleavage sites *98*, 145
  star activity 31
*Eco*RII 33
*Eco*RV 32
ecr gene 251
electrophoretic transfer 17
electroporation 18–19, 142, 146, 154, 176
λEMBL cloning vectors 89
EMBL3 cloning vector 58, *88*
EMBL4 cloning vector 58
embryoids 223
embryonic germ cells 213
embryonic stem cells 202
  gene targeting with 205
  transfection of 204–5
*end*A gene 48
endogenous inducible promoters 247–8
endonucleases *see* restriction endonucleases
engrailed-1 gene 212
engrailed-2 gene 212
enhancer trapping 269, *270*, 271
Ent P 307 plasmid 45
Entamoeba histolytica 275
enteric adenoviruses 275
Enterococcus faecalis 142
enterokinase 78
enteroviruses 275

entrapment constructs 269–73
  enhancer traps in *Drosophila* 269, *270*
  function-specific trapping 273
  gene traps 271, *272, 273*
entrapment vectors 271
env gene 194
environomics 3, 299
Epstein-Barr nuclear antigen 1 186
Epstein-Barr virus genome 2
Erwinia spp. 306
erythromycin 307
erythropoietin
  animal bioreactors producing 284
  plant bioreactors producing 286
  recombinant 284
erythrose-4-phosphate 303
ES cells *see* embryonic stem cells
Escherichia coli 5, 275
  bacteriophage λ 53, 55, 56, 57
  cloning in 5
  Dam and Dcm methylases 33
  genome size 86
  as hosts for recombinant molecules 33–4, *34*
  restriction endonuclease 26
  transformation 17–19
ethidium bromide *10*, 23, *43, 44*
European Bioinformatic Institute database 126
explants 221
expressed sequence tags (ESTs) *103*
expression vectors 70, *71*
exteins *81*

F plasmid 45
ferritin 318
field-inversion gel electrophoresis 89
filamentous phages 60–1
  advantages of 61
  development of 61, *62, 63*
firecracker complex 228
fish, gene transfer to 217–18
λFIX replacement vector 89, *90*, 107
FK-506 251, *252*
FKBP12 protein 251
fluorescence-activated cell sorter 298
forensic applications of VNTRs 279–82, *280, 281, 282*
fragile-X syndrome 288
fragment libraries 91–2
frame-shift mutations 163
frs gene *228*
function-specific trapping 273

functional cloning 112–14, *113*
functional complementation 112–14, *113*
functional expression cloning 215, *216*
fungi
  cloning in 156–73
  DNA introduction 156
  see also yeast

gag gene 194
gain of function screening 114
GAL genes 166, *167*
GAL4 protein 166
galactose metabolism 166, *167*
β-galactosidase 35, 254
Galanthus nivalis agglutinin 316
gam gene 58, *88*, 89
GAP promoter 165
gel retardation assay 10
geminiviruses 242–4
gene augmentation therapy 289, 290–2, *291*
gene cloning *see* cloning
gene expression 2
  transient 216
gene inactivation vectors 152, *153*
gene knockout 206
gene manipulation
  animals 202–20
  fish 217–18
  invertebrates 218, *219, 220*
  mammals 203–15, *203, 204, 206, 207, 209–11, 214*
  plants 221–46
  Xenopus laevis 215–17, *216*
gene manipulation techniques 6, 7, 8–25
  agarose gel electrophoresis 9–11, *9, 10*
  nucleic acid blotting 11–17, *11*
    northern blotting 16
    southern blotting 11–16, *12, 13, 14, 15, 16*
    western blotting 16–17
  polymerase chain reaction 19–25, *20, 21, 22, 23, 24*
  transformation of *E. coli* 17–19
    electroporation 18–19
gene marking 290
gene shuffling 302–3, *302*
gene tagging 267–9, *267, 268*
  vectors 269
gene targeting 211–12
  embryonic stem cells 205
  selection strategy 206–7
  vector design 205, *206*

gene therapy 286, 288–92
    cancer therapy 291–2
    gene-augmentation therapy for
        recessive diseases 290, 291
    scope of 288–90, 289
gene transfer 174–201, 198, 199,
    200, 201
    direct 176
    DNA-mediated transformation
        174–87
        non-replicating DNA 176–84,
            178–83
        replicon vectors 184–7, 184,
            185
        techniques 174–6
    strategies for 174
    to fish 217–18
    to invertebrates 218–20
    to plants 221–46
        Agrobacterium-mediated
            transformation 224–8,
            224–35
        chloroplast transformation
            240–1
        direct DNA transfer 237–9
        gene-transfer strategies 223
        in planta transformation
            239–40
        plant callus and cell culture
            221–4
        plant viruses as vectors 241–6
    to Xenopus laevis 215–17
        functional expression cloning
            215, 216
        heterologous expression system
            215
        transgenic Xenopus 216–17
        transient gene expression in
            embryos 216
    viral transduction 187–99
        adeno-associated virus 189,
            190
        adenovirus 188–9, 188
        baculovirus 190–3, 192
        herpesviruses 193
        retrovirus 193–5, 194
        sindbis virus and Semliki Forest
            virus 195–7, 196
        vaccinia and poxvirus 197–9
gene traps 271, 272, 273
Genentech 4, 5
general export pathway 83
genes
    association with cellular activity
        171, 172
    association with function 171–3,
        172, 173

genetic diseases, therapy for
        283–92, 459
    gene therapy 288–92, 289, 291
    genomics 286, 287
    pharming 283–5, 284
    plants as bioreactors 285, 286
    proteins as drugs 283, 284
    transgenic animals as models of
        human disease 286–8
genetic engineering 9
genome sequencing 126
genome size 2, 86
genomic DNA cloning 86–92, 86
    genomic DNA libraries 86–91,
        86, 87, 88, 90
    PCR as alternative to 91–2
genomic DNA libraries 86–91,
        86–8, 90
    λ cloning vectors 86–9, 87, 88
    creation of 88
    high-capacity vectors 89, 90
    λ replacement vectors 89
    subgenomic libraries 91
genomic sequence databases 128
genomics 3, 286, 287
germinal vesicle 215
Gerstmann-Straussler-Scheinker
        syndrome 287
Giardia lamblia 275
gibberellins 221, 222
glnAp2 promoter 308
glucocorticoids, use in plants 251
β-glucuronidase 254
glutathione-S-transferase gene
        171, 172
Glycine max 222
glyphosate 312
gpt gene 180, 181
green fluorescent protein 254, 255,
        298
GroEL protein 81
GroES protein 81
growth hormone see human growth
        hormone
λgt11, immunological screening
        93, 94
guessmers 107
gusA gene 197, 245
gutless vectors 188

HaeIII enzyme 29, 31
Haemophilus aegyptius 29
Haemophilus influenzae 26, 29, 139
    genome 2, 127
handcuffing 47
Hansenula polymorpha, promoters
        165

HAT medium 177
HAT tag elements 172
heat-shock promoters 247–8
helper plasmids 188
helper viruses 188
helper-dependent vectors 188
helper-independent vectors 187
herbicide resistance 311, 312, 313
herpes simplex virus, Tk gene 179,
        181, 206, 207
herpesvirus vectors 193
heterochromatin 210
heterologous expression systems
        215, 216
α-hydroxyacetosyringone 227
HindI 28
HindII enzyme 28, 29
HindIII enzyme 28, 29, 41, 51, 57,
        68
    cleavage sites 145
HinfI restriction sites 280
His3 gene 158
hisD gene 180
histidine patch 82
histopine 226
hit-and-run strategy 207
homologous recombination 155,
        163, 164
homopolymer tailing 40–1, 40
host-controlled restriction and
        modification 26, 27, 28
Hox gene clusters 106
HpaI enzyme 31
    cleavage sites 145
HpaII enzyme 31, 32, 33
HpaII restriction site 40
HphI enzyme 41
Hprt gene 179, 181
hpt gene 180
hsdM gene 29
hsdR gene 29
hsdS gene 29
human genome 2, 86
Human Genome Project 4, 5
human growth hormone 3
    animal bioreactors producing 284
    plant bioreactors producing 286
    recombinant 284
human immunodeficiency virus 91,
        213
hybrid dysgenesis 218
hybridization see nucleic acid
        hybridization
hygromycin 236
hypoxanthine-guanine
        phosphoribosyltransferase
        174

hypoxanthine-guanine phosphoribosyltransferase deficiency 288

imidazolinones *312*
immunity to superinfection 55
immunological screening 109, *110, 111*
Impdh gene *181*
in planta transformation systems 239–40
in vivo expression technology 297–8, *297*
inclusion bodies 79
IncP-group plasmids 145–7, *146, 147*
IncW-group plasmids *147*
indigo, synthesis of 304, *305*
inducible expression systems 247–53
  endogenous inducible promoters 247–8
  recombinant inducible systems 248–53
inducible protein activity 252, *253*
induction ratio 248
infectious diseases, therapy for 293–9
  differential fluorescence induction 298
  environomics 299
  in vivo expression technology 297–8, *297*
  new antimicrobial agents 296–7
  novel routes to vaccines 293–6
    DNA vaccines 296
    plants as edible vaccines 295–6, *295*
    recombinant bacterial vaccines 293–4
    recombinant viruses as vaccines 294, *295*
  signature-tagged mutants 298, *299*
inosine monophosphate 177
insects, resistance to 315–16
insertional mutagenesis 266–7
insertional vectors 57, 269
Institute for Genome Research 126
insulators *210, 211*
inteins 79, *81*
interferons
  animal bioreactors producing *284*
  plant bioreactors producing *286*
intermediate vectors 232
intermolecular association *30*

internal ribosome entry sites *272*
Internet 1, 2
intracellular antibodies *265*
intracytoplasmic sperm injection 213
intramolecular association *30*
intron splicing 2
introns 92, 163
iron deficiency 318
isatic acid 306
isatin 306
isatin hydrolase 306
isopentyl transferase *232*
isopropyl-β-D-thiogalactoride 153, 248
isoschizomers *30*

jumping libraries 108
jumpstarter element 269

kanamycin tolerance *231*
2-ketogulonic acid 306
killer toxin 166
Klenow fragment 122, 123
Kluyveromyces lactis, promoters 165
*Km*R gene 145
*Kpn*I enzyme *31*
KT230 plasmid 145
KT231 plasmid 145

lac promoter 74, *94*
lac repressor system 248, *249*
lacA gene *152*
lacR gene *152*
Lactobacillus lactis 151, 152
lacYZ gene *297*
lacZ gene 35, 54, 69, 234, 270, 308
lea gene *317*
lectins 17, 316
lentiviruses 213
  vector 198
LEU2+ gene 157
leu2 gene 158
leu2 locus 159
libraries
  amplified genomic 90
  cDNA 92–5, *93, 94*
  fragment 91–2
  genomic DNA 86–91, *86, 87, 88, 90*
  jumping 108
  screening expression 109–14
  subgenomic 91
  subtracted cDNA 115
lipofection 19, 176
liposomes 175

LITMUS vectors *72, 73,* 84
Lolium multiflorum 238
long accurate PCR 21–2, *23,* 91
long terminal repeats 194
loxP site 68, *212, 256, 317*
luciferase 254
lycopene 308–9
lys2 gene 158
lysS gene 74
lysogens 55, 56
lysopine *226*

M13 phage *62, 63*
maize streak virus 243
mammals, gene manipulation in 203–15, *203*
  applications of genetically modified mice 208–12
    gene targeting 211–12
    transgenic mice 208, *209*
    YAC transgenic mice 209, *210, 211*
  embryonic stem cells 205–8
    design of targeting vectors 205, *206*
    introduction of subtle mutations 207–8
    selection strategy 206, *207*
  intracytoplasmic sperm injection 213
  nuclear transfer technology 213–15, *214*
  traditional techniques 212–13
  transgenic mice 203–5
    applications of 208, *209*
    pronuclear microinjection 203, *204*
    recombinant retroviruses 204
    transfection of ES cells 204–5
mannopine synthase *230*
mas gene 228
  promoter *230*
α-mating factor 166
matrix attachment regions *210, 211*
*Mba*I enzyme *31*
*Mbo*I 32, 33
mcr system 29, 34
medermycin 308
mederrhodin, synthesis 308
medical benefits of gene manipulation 3–4, *3*
megaprimers 135, *136*
messenger RNA *see* mRNA
metabolic control engineering 308, *309*

metabolic engineering 303–11
  combinatorial biosynthesis 306,
    307, 308
  designed overproduction of
    phenylalanine 303, *304*
  new routes to small molecules
    304, *305, 306*
  plant cells 309–11, *310*
  recombinant pathways 308,
    309
metallothionein promoter 208,
  209, 248
MGH fusion gene 208
mice
  Cre-mediated conditional mutants
    in 258–60
  transgenic *see* transgenic mice
micro-Ti plasmid 232
microbial pathogens, resistance to
  314–15
minerals, improved content of
  318–19
mini-satellite probes 16
mini-Ti plasmid 232
mitochondrial cardiomyopathy
  288
MK gene 208
MMT gene 208
mob gene 145
modification systems
  host-controlled 26–7, *27*
  types of 27, *29*
molecular bar-codes *173*
molecular breeding 302–3, *302*
molecular cloning *see* cloning
Moloney murine leukaemia virus
  reverse transcriptase 21
monocots 235–6
  drug resistance *231*
mouse genome, size of *86*
Mres gene *181*
mRNA 92, *93*
  fingerprint 116
  selection of 5' ends 97
mrr system 29, 34
*Msp*I enzyme *31, 33*
M.*Sss*I 33
Mt-1 gene *181*
multiple cloning sites 53, *54,* 79
mutagenesis *see* site-directed
    mutagenesis
mutagens 120
mutant peptides, selection of
  136–8, *137*
mutants 2, 120
mutations 2
  subtle 207–8, *207*

mutator element 269
*mut*H mutation 134
*mut*L mutation 134
*mut*S mutation 134
*Myf-*5 gene 212
MyoD gene 212

*Nae*I enzyme 31
Narcissus pseudonarcissus 319
*Nar*I enzyme 31
*Nat*I enzyme *31*
National Center for Biotechnology
    Information database 126
*Nci*I enzyme 32
Neisseria gonorrhoeae 139
Neisseria meningitidis 294
neo gene *180*
neoschizomers 30
nested primers 23
nic gene 145
Nicotiana benthamiana 314
Nicotiana tabacum 222
nin deletion 57
nisA gene 152
nisF gene 152
nisin 152
non-enteric bacteria, cloning in
    144–7
non-replicating DNA 176–7
non-replicating plasmid vectors
    198
nopaline synthase 229, 230, 285
nopalines 224, 225, 226
nopalinic acid 226
north-western screening 111–12
northern blotting 16
nos gene 228, 229
  promoter 230
*Not*I enzyme 67, 68, 145
*npt*II gene 236
*Nsi*I, cleavage sites 145
nuclear occlusion bodies 190
nuclear transfer 213–15, *214*
  sheep 202, *314*
nucleic acid blotting 11–17, *11*
  northern blotting 16
  southern blotting 11–16, *12, 13,*
    14, 15, 16
nucleic acid hybridization *12, 13,*
    103–7, *104, 105, 106*
  probe design 105
  screening by 103–7, *104, 105,*
    106
nucleic acid sequences 274–83
  detection at gross level 274–7,
    275, 276
  historical genetics 282–3

single-nucleotide polymorphisms
    277, 278, 279
  variable number tandem repeat
    polymorphisms 279, *280*
  forensic applications 279–82,
    280, 281, 282
nucleic acids 4
  hybridization *12, 13*
nucleotide sequence 1

ocs gene *228, 229*
  promoter *230*
octopine synthase 229, 230
octopines 225, 226
octopinic acid 226
Odc gene *181*
Odontoglossum ringspot virus 244
oil modification 316–18, *317*
oligo-capping 99, *100*
oligonucleotide chips 116
oligonucleotides *13*
  doped *132*
oligotide-directed mutagenesis 133,
    134
omaline 226
oncoretrovirus vectors 198
open reading frames (ORFs) 1, 127,
    128, 159
  finder *128*
  yeast 171, *172*
opines 226
optimum translation 77
ORFs *see* open reading frames
organogenesis 223
ori gene 145
ori region 8, 45, 47, 64, 70
*ori*P gene 186
*ori*S gene 69
*ori*T gene 146
origin of replication *see* ori
Oryzias latipes 217
overdrive sequence 227
overexpression systems 165–6,
    165

P1-derived artificial chromosome
    (PAC) 69, 138, 236
PA170 promoter 152
PAC *see* P1-derived artificial
    chromosome
pac gene *180*
pac site 68
pacase 68
palindromes 30
pAMβ1 plasmid 148–51, *149, 150*
par region 47
particle bombardment 238–9

passenger DNA 9
Pasteurella spp. 144
pathogenesis-related proteins 314
pBAC108L 69
pBAD vector 75, 76
   structure 78
pBBR1 plasmid 147
pBeloBAC11 69
pBR322 plasmid 50–3, 50, 52, 53, 81, 147
   genome 2, 52
   improved vectors derived from 53
   origins of 50
   sequence of 51
   as vector 51, 53
PC194 plasmid 148
pcDNA1.1/Amp vector 185
PCR *see* polymerase chain reaction
pE194 plasmid 148
pEB10 plasmid 149
pECBAC1 69
pectinase 222
pET vectors 74, 75
Petunia hybrida 312
Pfu polymerase 91
pgk locus 159
pGPA14 plasmid 149
pGPB14 plasmid 149
pGreen plasmid 234
pGV3850 plasmid 229
phage display of random peptides 136–8, 137
phage genomes 2
phage M13, life cycle and DNA replication 62
phage P1 67–9, 68, 69
phage promoters 53, 55, 56, 57
   regulated 74
phagemid display vectors 147
phagemids 64, 70, 136
phages
   filamentous 60–1
   restricted 26, 27
pharmacogenomics 3, 4, 292
pharming 283–5, 284
phasmids 64, 94, 136–8, 137
phenotypic lag 18
phenylalanine, designed overproduction of 303, 304
phosphinothricin 312, 313
phosphinothricin acetyltransferase 312
phosphoenol pyruvate 303
phosphoglycerate kinase promoter 164
phospholipids, as gene-delivery vehicles 175–6

pHP3 plasmid 149
pHP3Ff plasmid 149
pHP13 plasmid 149
pHV14 plasmid 149
pHV15 plasmid 149
pHV33 plasmid 149
pHV1431 plasmid 149
pHV1432 plasmid 149
pHV1436 plasmid 149
phytic acid 319
phytoene 319
phytohormones 221, 222
Phytophthora infestans 314
Pichia spp., specialised vectors 168
Pichia methanolica, promoters 165
Pichia pastoris, promoters 165, 166
Pichia stipitis, promoters 165
pIJ101 plasmid 154
PinPoint vector 84
pJV1 plasmid 154
plant breeding 311–19
   improving agronomic traits 311–16
      herbicide resistance 311, 312, 313
      resistance to insects and other pests 315–16
      resistance to microbial pathogens 314–15
      virus resistance 313–14
   modification of production traits 316–19, 317
      delayed ripening 316
      improving vitamin and mineral content 318–19
      starch and oil modification 316–18
plant viruses, as vectors 241–6
   DNA viruses 241–4
      cauliflower mosaic virus 241, 242
      geminiviruses 242–4
   RNA viruses 244–6
      potato virus X 245–6, 245
      tobacco mosaic virus 244
plantibodies 265, 286
plants
   as bioreactors 285, 286
   as edible vaccines 295–6, 295
   gene transfer to 221–46
      Agrobacterium-mediated transformation 224–37
      Agrobacterium rhizogenes and Ri plasmids 237
      crown-gall disease 224, 225–7

      disarmed Ti plasmid derivatives as plant vectors 228–34
      experimental procedure 234, 235
      high-capacity binary vectors 236–7
      monocots 235–6
      T-DNA transfer 225–8, 228
      tumour-inducing plasmids 224–5, 227
   chloroplast transformation 240–1
   direct DNA transfer 237–9
      particle bombardment 238–9
      protoplast transformation 237–8
   in planta transformation 239–40
   plant callus and cell culture 221–4
      callus culture 221, 222
      cell-suspension culture 222
      gene-transfer strategies 223
      protoplasts 222, 223
      regeneration of fertile plants 222–3, 224
   plant viruses as vectors 241–6
      DNA viruses 241–4, 242
      RNA viruses 244–6, 245
   metabolic engineering in 309–11, 310
   resistance to insects in 315–16
   virus resistance in 313–14
plaque lift 104, 105
plasmid DNA 48–9, 48
plasmid pBR322 genome 2
plasmid rescue 140–1, 268
plasmid vectors
   desirable properties of 49–53
   from pBR322 53
plasmids 43–63
   biology 43, 44, 45
   conjugative 43–4
   copy number 45–7, 46, 47
   cryptic 43
   distribution 43
   DNA-mediated gene transfer 181, 182
   helper 188
   host range of 45
   incompatibility 47–8
   non-conjugative 43–4
   partitioning and segregative stability 47
   promiscuous 45, 142

pUC *54*
purification of DNA for *48–9, 48*
relaxed *43–4*
self-transmissible *142*
shuttle *149*
Streptomyces *154*
stringent *43–4*
tumour-inducing *224–5, 227*
yeast centromere *158, 159*
yeast episomal *158*
yeast integrating *157*
yeast replicating *158–9*
see also individual plasmid types
pLB5 plasmid *149*
pMB8 plasmid *50*
pMB9 plasmid *50*
pMUTIN vector *152, 153*
polyadenylation signals *200*
polyhedrin *190, 191, 199*
polyketides *307*
polylinkers *53, 54, 146*
polymerase chain reaction *19–25, 20–4*
  as alternative to cDNA cloning *100–1*
  as alternative to genomic DNA cloning *91–2*
  difference cloning *116, 117–19*
  hot-start protocol *23*
  inverse *267*
  joining of products *41*
  key factors affecting *22–3*
  long accurate *21–2, 23*
  multiplex *276*
  nested primers *23*
  real-time quantitative *23–5, 24*
  reverse transcription PCR *21, 22*
  screening by *109–10*
  site-directed mutagenesis *135, 136*
position effects *210, 211*
positional cloning *107*
post-transcriptional gene silencing *262, 264*
potato virus X *245–6, 245*
poxvirus vectors *197–9*
pPICZ vector *168*
Pribnow box *51*
primers *42*
  degenerate *109*
  nested *23*
probes *12*
prodrug activation therapy *292*
promiscuous plasmids *45, 142*
promoters
  endogenous inducible *247–8*
  fungal *165*

phage *53, 55, 56, 57*
phage λ *53, 55, 56, 57*
phosphoglycerate *164*
plant *230*
  see also individual promoters
pronuclear microinjection *203, 204*
protein engineering *299–303*
  directed evolution *301–2*
  gene shuffling *302–3, 302*
  rational approach to *301*
  single amino acid changes *299, 300*
  subtilisin *300, 301*
protein purification vectors *76, 78, 79*
protein sequence databases *129*
protein splicing *81*
protein three-dimensional structure database *129*
protein-peptide interactions *170*
protein-protein interactions *169, 170, 171*
proteome *266*
proteomics *3*
proto-oncogenes *193*
protoplast fusion *222*
protoplast transformation *141, 154, 237–8*
protoplasts *222, 223*
protoxins *315*
provitamin A *319*
pRSV-neo *182*
pSAM2 plasmid *154*
pSC101 plasmid *50*
Pseudomonas spp. *5*
Pseudomonas putida *139, 145, 304*
Pseudomonas syringae *314*
pseudotyping *195*
pSG5 plasmid *154*
Pspac promoter *153*
Pst site *145*
*Pst*I enzyme *30, 40*
  quality control *34*
pSV2-dhfr *182*
pTB19 plasmid *149–50*
pUB110 plasmid *148*
pUC plasmid *54*
pulsed-field gel electrophoresis *10, 89*
purA gene *297–8*
*Pvu*I enzyme *31, 163, 164*
*Pvu*II enzyme *31, 163, 164*
*Pvu*II site *63*
pYeLeu10 *157*

quality control *34, 35, 125, 126*
quelling *262*

R6K plasmid *45*
Rana pipiens *213*
random fragments *87*
rapid amplification of cDNA ends (RACE) *101, 102*
*rec*BC gene *17*
recognition sequences *30, 31*
recombinant bacterial vaccines *293–4*
recombinant DNA *3, 4*
  integration of *143, 144*
  maintenance in new hosts *142*
recombinant DNA technology *274–375*
  infectious diseases *293–9*
    differential fluorescence induction *298*
    environomics *299*
    in vivo expression technology *297–8, 297*
    novel routes to vaccines *293–6, 295*
    signature-tagged mutagenesis *298, 299*
    targets for new antimicrobial agents *296–7*
  metabolic engineering *303–11*
    combinatorial biosynthesis *306, 307, 308*
    designed overproduction of phenylalanine *303, 304*
    new routes to small molecules *304, 305, 306*
    plant cells *309–11, 310*
    recombinant pathways *308, 309*
  nucleic acid sequences as diagnostic tools *274–83*
    detection of sequences at gross level *274–7, 275, 276*
    forensic applications *279–82, 280, 281, 282*
    historical genetics *282–3*
    single-nucleotide polymorphisms *277, 278, 279*
    variable number tandem repeat polymorphisms *279, 280*
  plant breeding *311–19*
    improving agronomic traits *311–16, 312, 313*
    modification of production traits *316–19, 317*
    protein engineering *299–303*
      directed evolution *301–2*
      gene shuffling *302–3, 302*
      rational approach to *301*

recombinant DNA technology
  protein engineering (*cont.*)
    single amino acid changes
      299, *300*
    subtilisin 300, *301*
  therapies for genetic diseases
    283–92
    gene therapy 288–92, *289,*
      291
    genomics 286, *287*
    importance of SNPs 292
    pharming 283–5, *284*
    plants as bioreactors 285, *286*
    proteins as drugs 283, *284*
    transgenic animals as models of
      human disease 286–8
recombinant inducible systems
  248–53
  chemically induced dimerization
    251, *252*
  heterologous use of steroids 251
  inducible protein activity 252,
    253
  lac and *tet* repressor systems 248,
    249
  tet activator and reverse activator
    systems 249–51, *250*
recombinant molecules 33–4, *34*
recombinant proteins as drugs 283,
  284
recombinant retroviruses 204
recombinant viruses as vaccines
  294, *295*
recombinase activation of gene
  expression (RAGE) 257
red gene 58, *88, 89*
relaxed plasmids 43–4
reorientation angle 10
*rep*A gene 47
*rep*E gene *69*, 151
repeat sequences 1
repeat-induced gene silencing *262,*
  263
replacement vectors 57
replication-competent vectors 187
replication-defective vectors 187
replicons 8, 43, 151, *154*, 184–7,
  184, 185
  BK and BPV 185–6
  EBV 186–7
  runaway polyomavirus *184,*
    185
  viral *198*
  see also plasmids
reporter genes 171, *182, 183*
  visible markers *254, 255*
  see also individual reporter genes

representational difference analysis
  118, *119*
reptation 9
resolvase 298
response elements *182*
restriction endonucleases 26, *27*
  average fragmentation size *31*
  preferential cleavage 31
  quality of *34, 35*
  recognition sites 1
  types of *27*
  see also individual enzymes
restriction fragment length
    polymorphisms *278, 279*
restriction fragments
  blunt ends 30
  cohesive ends *30*
    preformed 38
  number and size *31*
restriction sites *31*
  generation of *32*
  mapping of *16*
restriction systems
  host-controlled 26–7, *27*
  types of *27, 29*
restriction-enzyme-mediated
    integration 158, 217
retrofitting 69
retrovirus vectors 193–5, *194*
retrovirus-like vectors 159–63,
  160, 161, 162
retroviruses
  in production of transgenic
    chickens 212–13
  recombinant 204
reverse dot blot *275, 276*
reverse *tet* transactivator system
  250
reverse transcriptase PCR 21, *22*
reverse two-hybrid systems *170*
Rhizoctonia solani 314
Ri plasmids 45, 237
ribozyme constructs 261
RNA I 46
RNA II 46
RNA interference 264–5
RNA polymerase 70, 71
RNA probes 71, *72, 73*
RNA-mediated virus resistance 314
rol genes 237
rolling-circle replication 149, *150,*
  151
root-inducing plasmids *see* Ri
  plasmids
Rop protein 46
rosy gene 107, 219
rotational symmetry 30

rotaviruses *275*
RP4 plasmid 146
RSF1010 plasmid 45, *145*

Sa plasmid *147*
Saccharomyces cerevisiae
  cloning in 156–73
    choice of vector 161–3, *162*
  galactose metabolism 166, *167*
  genome *2, 7, 86,* 266
  promoters 165
  specialised vectors *168*
  VMA1 gene 79
*Sac*I enzyme, cleavage sites 145
*Sac*II enzyme 31, *67,* 145
*Sal*I enzyme 180
*Sal*I site 98
Salmonella spp. *275*
Salmonella typhimurium 293–4,
  298
salvage pathway 177, *178*
satellite DNA *279, 280*
*Sau*3AI enzyme 31, *33, 88, 89*
Sbf micro I enzyme *31*
Schizosaccharomyces pombe 156
Schwanniomyces occidentalis,
  promoters 165
SCP2* plasmid *154*
screenable marker genes 171, *182*
screening 86, 101, 103–14
  expression cloning 109–14
    functional cloning 112–14,
      113
    immunological screening 109,
      110, 111
    screening with alternative
      ligands 112
    south-western and north-
      western screening 111–12
  sequence-dependent 103–9
    chromosome walking 107,
      108
    screening by hybridization
      103–7, *104, 105, 106*
    screening by PCR 108–9
screening expression libraries
  109–14
  immunological screening 109,
    110, 111
*Scr*FI enzyme 32
sec gene products 83
secologanin *310*
selectable markers 177–9, *179,*
  180, 231, *232*
  drug resistance markers 231
  herbicide resistance markers
    231

and transgene amplification
179–80, *181*
self-inactivating vectors 195
self-transmissible plasmids 142
Semliki Forest virus
as recombinant vaccine 294–5
vectors 195–7, *196*, *198*
Sequenase 123
sequence alignment *130*
sequence alignment databases *130*
sequence analysis 1–2
comparative *277*, *278*, *279*
sequence identification databases
129
sequence translation *129*
sequence translation databases *129*
sequence-dependent screening
103–9
by hybridization 103–7, *104*,
105, *106*
by PCR 108–9
chromosome walking 107, *108*
sequencing 120–32
accuracy of 125, *126*
automated 124–5, *124*
chain terminator 120, *121*, *122*
modifications of 123–4, *123*
cycle 124
data analysis
database searches 127,
129–32, *131*
DNA sequence databases
126–7
sequence data 127, *128–30*
shotgun 70
whole-genome 126
serial analysis of gene expression
(SAGE) 103
Serratia marcescans 29
*Sfi*I 67, 68, 145
shaker-2 gene *113*
sheep, produced by nuclear transfer
202, 314
Shigella spp. 275
Shine-Dalgarno sequence 77, 151
shotgun sequencing 70
shuttle plasmids 149
shuttle vectors 142
signal peptides *201*
signature-tagged mutagenesis 173,
298, *299*
sindbis virus
as recombinant vaccine 295
vectors 195–7, *196*, *198*
single-nucleotide polymorphisms
(SNPs) *277*, *278*, *279*,
292

single-primer method 132, *133*,
134
deficiencies of 134
site-directed mutagenesis 120,
132–8
cassette mutagenesis *132*
PCR methods *135*, *136*
selection of mutant peptides
136–8, *137*
single-primer method 132, *133*,
134
site-specific integration *211*
site-specific recombination
253–60, *254*, *255*
chromosome engineering 258,
259
Cre-mediated conditional mutants
in mice 258–60
deletion of transgene sequences
256–8, *257*
site-specific transgene integration
258
SLP1 plasmid *154*
*Sma*I enzyme 29, 30, 67
*Sma*I restriction site 40
*Sm*R gene 145
SNPs *see* single nucleotide
polymorphisms
sodA gene *152*
sodium dodecyl sulphate 48
somaclonal variation 222
somatic embryogenesis 222–3
south-western screening 111–12
southern blotting 11–16, *12–16*
Spac system 151
specialist vectors 166–8, *167*, *168*
*Spe*I enzyme *31*
*Spe*I site 73
Spiroplasma 33
*Spl*I 145
splicing 2, 92, 163
Spodoptera frugiperda 190–1
*Sst*I, cleavage sites 145
stable transformants, cotransformation and selection
177, *178*
stable transformation 177
Staphylococcus aureus 17, 45, 142
plasmids *148*
star activity 31
starch modification 316–18, *317*
steroids, heterologous use of 251
streptavidin 17, 79
Streptococcus faecalis 149
Streptococcus gordonii 294
Streptomyces spp. 43
Streptomyces coelicolor 153

Streptomyces hygroscopicus 312
streptomycetes
cloning in 153–4
vectors for 154–5, *154*
stricosidine *310*
stringent plasmids 43–4
stuffer DNA 58, 189
subcloning 85
subgenomic libraries 91
substance K 215
subtilisin 132, 300, *301*
subtle mutations 207–8, *207*
subtracted cDNA library 115, *117*
subtractive cloning 115
suicide genes 292
sulphonylurea *312*
superbinary vectors 236
SuperScript II enzyme 97
supervirulent bacterial strains 236
SV40 replicon *184*, *198*
SWISS-PROT database 127

T-DNA gene function *228*, *229*
T-DNA transfer 225–8
genes required for 227, *228*
sequences required for 225,
227
tac promoter 74
tacrine 292
TACs *see* transformation-competent
bacterial artificial
chromosomes
tag proteins 76, *80*
tag-and-exchange strategy 206,
207
*Tai*I enzyme 31
Taq DNA polymerase 20, 91, 123,
136
target DNA 9
TATA box 164, 269
*Tc*R gene 51
technology protection 317
telomeres 2, 159, *160*
terminal deoxynucleotidyltransferase 40
terminator technology 317
terminators 200
tet activator/reverse activator
systems 249–51, *250*
tet repressor system 248, *249*
tetracycline 249
β-thalassaemia 288
Thermus spp. 144
Thermus aquaticus 20
DNA polymerase 123
thioredoxin fusion proteins 81
three-component systems *170*

thyA gene 144, 297–8
thymidine kinase 208, 292
thymidine monophosphate 177
Ti plasmids 224–5, *227*
  gene maps *227*
  structure *229*
Ti-vectors, disarmed *229–32*
tissue culture
  cell-suspension 222
  plant callus *221, 222*
Tk gene *179, 181*, 206, 207
tml gene *228*
tmr gene *228*
tms gene *228*
tnpA gene 146
tnpR locus 298
tobacco genome, size of *86*
tobacco mosaic virus 238, 244–5,
  313
tobacco ringspot virus 314
tomato golden mosaic virus 242
Torpedo marmorata 215
transcription 151
transcription terminators *71*
transcriptional gene silencing
  262
transcriptome 266
transduction 174
transfection 176–7
  of embryonic stem cells 204–5
  transient 177
transfer RNA 100
transferred DNA *see* T-DNA
transformants
  bacteria 154, *155*
  yeasts *157*
transformation 174
  Agrobacterium-mediated 224–37
    crown-gall disease 224,
    225–7
    disarmed Ti plasmid derivatives
    as plant vectors *228–34*
    experimental procedure 234,
    235
    high capacity binary vectors
    236–7
    monocots 245–6
    and Ri plasmids 237
    T-DNA transfer 225–8
    tumour-inducing plasmids
    224–5, *227*
  Bacillus subtilis 19, *140*
  by plasmid rescue *140, 141*
  chloroplasts 240–1
  DNA-mediated 174–87
    non-replicating DNA 176–84,
    178–83

replicon vectors 184–7, *184,*
  185
  techniques 174–6
Escherichia coli 17–19
in planta 239–40
protoplasts *141*, 154, 237–8
stable 177
transient 177, 182–4, *182, 183*
transformation-competent bacterial
  artificial chromosomes
  (TACs) 236
transgene activation and switching
  257–8, *257*
transgene amplification 179–81,
  181
transgene integration 258
transgene rescue *211*
transgene sequence deletion
  256–8, *256, 257*
transgene silencing 262, *263*
  homology-dependent 262, *263*
  position-dependent 262
transgenes 174
  dominantly acting *211*
  large genomic *211*
transgenic animals
  as bioreactors 283–5, *284*
  as human disease models 286–8
  see also individual animal species
transgenic birds 212
transgenic chickens 212–13
transgenic fish 217–18
transgenic flies 218–20, *219, 220*
transgenic mice 203–5
  applications of 208, *209*
  pronuclear microinjection 203,
  204
  recombinant retroviruses 204
  transfection of ES cells 204–5
  YAC 209, *210, 211*
transgenic plants 221–46
  Agrobacterium-mediated
    transformation 224–37
    crown-gall disease 224,
    225–7
    disarmed Ti plasmid derivatives
    as plant vectors *228–34*
    experimental procedure 234,
    235
    high capacity binary vectors
    236–7
    monocots 235–6
    and Ri plasmids 237
    T-DNA transfer 225–8, *228*
    tumour-inducing plasmids
    224–5, *227*
  as bioreactors 283–5, *284*

chloroplast transformation
  240–1
direct DNA transfer 237–9
  particle bombardment 238–9
  protoplast transformation
  237–8
as edible vaccines 295–6, *295*
in planta transformation 239–40
plant callus and cell culture
  221–4
  callus culture 221, *222*
  gene-transfer strategies 223
  protoplasts 222, *223*
  regeneration of fertile plants
  222–3, *224*
plant viruses as vectors 241–6
  DNA viruses 241–4, *242*
  RNA viruses 244–6, *245*
transgenic sheep 215
transgenic technology 247–73
  functional genomics 266–73
    entrapment constructs
    269–73, *270, 272, 273*
    gene tagging 267–9, *267, 268*
    insertional mutagenesis
    266–7
  gene inhibition at protein level
    265–6, *265*
    dominant-negative mutants
    265–6
    intracellular antibodies *265*
  gene inhibition at RNA level
    260–5
    antisense RNA transgenes
    260–1
    cosuppression 261–4, *262,*
    263
    ribozyme constructs 261
    RNA interference 264–5
  inducible expression systems
    247–53
    endogenous inducible
    promoters 247–8
    recombinant inducible systems
    248–53, *249, 250, 252*
  site-specific recombination
    253–60, *254, 255*
    chromosome engineering 258,
    259
    Cre-mediated conditional
    mutants in mice 258–60
    deletion of transgene sequences
    256–8, *256, 257*
    site-specific transgene
    integration 258
transgenic *Xenopus laevis* 216–17
transient gene expression 216

transient transfection 177
transient transformation 177,
182–4, *182, 183*
translation *77*, 151
transplastomic tobacco plants 240
transposable elements 266
transposons 41, 143, 171
trc promoter 74
triparental matings 232, *233*
triplets *77*
Tro1 gene 158
trp promoter 74
trpB gene *180*
trpE gene *152*
tryptamine *310*
L-tryptophan *310*
tumour necrosis factor 290
tumour-inducing plasmids *see* Ti
plasmids
two-hybrid system 169, *170, 171*
bait and hook *170*
dual-bait *171*
post-translational modifications
170
protein–peptide interactions *170*
reverse *170*
three-component systems *170*
Ty elements 159, *160, 161, 162*
tylosin *307*

ubiquitin-1 promoter *230*
Umps gene *181*
upstream activating sequences 164
upstream repressing sequences 164
Ura3 gene 158
usp gene 251

vaccines 3–4
DNA 296
edible, plants as 295–6, *295*
novel routes to 293–6
recombinant bacterial 293–4
recombinant viruses as 294, *295*
vaccinia virus
as recombinant vaccine 294, *295*
vectors 197–9, *198*
vacuum blotting 17
variable number tandem repeat
polymorphisms 279, *280*
forensic applications of 279–82,
280, 281, 282
vector DNA 57
vectors 6, *8*
addition 244
bacterial 154–5, *154*
binary 232–4, *234*
high-capacity 236–7

choice of 69–70, *69*
cointegrate 232, *233*
DNA-mediated gene transfer
181, *182*
expression 70, *71*
from IncP-group plasmids 145,
146, 147
gene inactivation 152, *153*
gene tagging 269
gutless 188
helper-dependent 188
helper-independent 187
insertional 57, 269
intermediate 232
maximization of protein synthesis
73, *74, 75, 76*
phage λ *see* bacteriophage λ
PinPoint 84
plant viruses 241–6
promotion of protein export 82–4
protein purification 76, *78, 79*
pseudotyping 195
replacement 57
replication-competent 187
replication-defective 187
retrovirus 193–5, *194*
retrovirus-like 159–63, *160,
161, 162*
for RNA probes 71, *72, 73*
self-inactivating 195
shuttle 142
for single-stranded DNA 70
solubilization of expressed
proteins 79–82
for streptomycetes 154–5
superbinary 236
viral 187–99
see also cosmids; phages; plasmids;
and individual vectors
Vibrio spp. 275
Vibrio cholera 294
vindoline *310*
virA gene 227
viral oncogenes 193
viral replicons 198
viral transduction vectors 187–99,
198
adeno-associated virus 189, *190*
adenovirus 188–9, *188*
baculovirus 190–3, *192*
herpesvirus 193
retrovirus 193–5, *194*
sinbis virus and Semliki Forest
virus 195–7, *196*
vaccinia and poxviruses 197–9
virE1 gene 236
virG gene 227, 236

virus resistance 313–14
virus-like particles 159–60
viruses, recombinant, as vaccines
294, *295*
visible marker genes 254, *255*
vitamin A deficiency 318
vitamin C synthesis *306*
vitamins, improved content of
318–19
VNTRs *see* variable number tandem
repeat polymorphisms

western blotting 16–17
wheat dwarf virus 243
wheat genome, size of *86*
wings-clipped element 269

Xenopus laevis 202
gene transfer to 215–17
functional expression cloning
215, *216*
heterologous expression system
215
transgenic *Xenopus* 216–17
transient gene expression in
embryos 216
genome, size of *86*
nuclear transfer 213
Xenopus tropicalis 217
xenotransplantation 286, *287*
*Xho*I enzyme 89
*Xho*RI, cleavage sites 145
*Xma*I enzyme 30
cleavage sites 145

YACs *see* yeast artificial
chromosomes
Yarrowia lipolytica, promoters *165*
YCp *see* yeast centromere plasmids
yeast artificial chromosomes (YACs)
68, 69, 89, 138, 158, 159,
160, 162
transgenic mice 209, *210, 211*
yeast centromere plasmids 158,
159, *162*
yeast chromosome III genome *2*
yeast episomal plasmids 158, *162*
yeast integrating plasmids 157,
162
yeast replicating plasmids 158–9,
162
yeast surface display 168–71, *169,
170, 171*
yeast transformants 157
yeasts
choice of vectors for cloning
161–3, *162*

yeasts (*cont.*)
  open reading frames 171, *172*
  specialist vectors 166–8, *167*,
    168
  see also Saccharomyces cerevisiae

YEp *162*
Yersinia enterocolitica 275
YES vectors *168*
Yip *see* yeast integrating plasmids
YRp *see* yeast replicating plasmids

λZAP vector 93, 94
  immunochemical screening
    111
Zea mays genome, size of 86
ZEBRA 187